# Environmental Challenges

Edited by

SHIMSHON BELKIN

*Laboratory for Environmental Microbiology*
*Division of Environmental Sciences*
*The Fredy and Nadine Herrmann Graduate School of Applied Science*
*The Hebrew University of Jerusalem*
*Israel*

(Administrative editor: Shoshana Gabbay)

Reprinted from *Water, Air, and Soil Pollution* 123: 1–4, 2000

SPRINGER-SCIENCE+BUSINESS MEDIA, B.V.

A C.I.P. Catalogue record for this book is available from the Library of Congress.

ISBN 978-94-010-5878-0 ISBN 978-94-011-4369-1 (eBook)
DOI 10.1007/978-94-011-4369-1

*Printed on acid-free paper*

# TABLE OF CONTENTS

In memory of Holger W. Jannasch,
who taught me the value of transdisciplinarity without using the word even once

With many thanks to members of the board of the Israel Society for Ecology and Environmental Quality Sciences, as well as to the organizing committee of its 7th International Conference: H. Almog, A. Brenner, E. Doron, N. Galil, S. Gelber, A. Goren, N. Gressel, Y. Hadar, M. Haran, J. Koch, R. Laster, R. Ortal, M. Peleg, A. Rosen, M. Sandigurski, H. Shuval and M. Waldman.

*Water, Air, and Soil Pollution* **123:** vii, 2000.

# PREFACE

Can a safe and healthy environment be maintained in a rapidly changing world? In countries where a positive response may be ventured, what are the prospects for remedying past injuries, caused by ignorance or neglect? In other parts of the world, where environmental problems are far out-weighed by more immediate, often life-threatening concerns, is it at all possible to give environmental issues the attention they deserve? Does the attractive concept of sustainable development have a chance to be implemented on anything but a local scale?

Answers to many of these questions will not be forthcoming for many years to come; by the time that some of the others will be even seriously considered, irreparable harm may already be caused. There is thus a growing urgency to find at least partial answers to some of the burning issues, with the degree of urgency often inversely proportional to the level of national development.

In a modest effort to respond to these needs, a group of world-renowned scientists met in Jerusalem, Israel, in the summer of 1999, in an attempt to integrate state-of-the-art science with some of the more timely issues in international environmental cooperation. Nearly 800 participants from 39 countries took part in this intense multidisciplinary effort. Approximately 50 of the invited presentations are included in this volume. They address major issues in vital areas of concern including water and wastewater quality and treatment, air pollution and control, conservation and risk assessment, legal and economic aspects, environmental education and international cooperation.

This unique combination of the exact, legal, social and political sciences can only coexist under an environmental umbrella that is able to harmonize different languages and make them accessible and understood to practitioners in other disciplines. It is our fervent hope that the studies presented in this volume will further this understanding and will provide us with better tools for handling the environmental challenges that face us at the dawn of the third millennium.

*Shimshon Belkin*
*Editor*

*Water, Air, and Soil Pollution* **123:** ix, 2000.

# STATUS OF THE DRINKING WATER STANDARDS PROGRAM IN THE UNITED STATES

H. J. BRASS

*United States Environmental Protection Agency, Office of Ground Water and Drinking Water, Technical Support Center, 26 West Martin Luther King Drive, Cincinnati, Ohio, 45268 , USA*

**Abstract.** The 1996 Amendments to the Safe Drinking Water Act permit standards to be set on a risk management basis. They are driven by: sound peer reviewed science; availability of adequate data and information; prioritized rule making based on risk; increased stakeholder involvement and regulatory negotiations; cost-benefit analysis; better clarity; timely guidance as to provisions of the standards; and implementation assistance though training and guidance materials. The priority rulemaking activities include: standards for harmful microbiological contaminants, disinfectants and disinfectant byproducts; a ground water rule to protect ground water systems from microbiological pathogens; and standards for radon, radionuclides, and arsenic. Additionally, a contaminant candidate list (CCL) must be published every five years to identify potential substances for subsequent regulation. From the CCL, at least five candidates must be identified to consider for regulation within a five year period. A maximum of thirty contaminants for monitoring by water systems must be published by 1999 and every five years thereafter. Monitoring information derived serves as a basis for future standard setting activities. Drinking water standards are set based on health effects and occurrence information through a two step process. First USEPA establishes a non-enforceable maximum contaminant level goal (MCLG) which is the maximum permissible level of a contaminant where no adverse health effects occur. Once the MCLG is established, a maximum contaminant level (MCL) is promulgated as close to the MCLG as feasible.

**Keywords**: maximum contaminant level, regulations, Safe Drinking Water Act, standards, treatment

## 1. Introduction

In 1974, Congress passed the Safe Drinking Water Act (SDWA) and assigned responsibility for its administration to the Environmental Protection Agency (USEPA) (United States Public Law, 1974). The purpose of the SDWA is to protect public health by ensuring that tap water in the United States is safe for drinking and bathing. SDWA requires USEPA to set drinking water standards that must be met by the Public Water Systems (PWS) that deliver drinking water to the tap. The standards are set to ensure the water is fit to drink, with an adequate margin of safety.

Within USEPA, the Office of Ground Water and Drinking Water (OGWDW) writes regulations and works with its regional offices to oversee SDWA implementation by individual states within the United States. In developing drinking water regulations, OGWDW uses risk-based priority setting, sound science, quality data from reliable databases, suitable treatment

*Water, Air, and Soil Pollution* **123:** 1–9, 2000.

technologies, and accurate analytical methods. OGWDW uses finished water standards, treatment requirements and prevention of contamination, as effective approaches to ensuring safe drinking water.

The purpose of this paper is to present these processes in further detail.

## 2. The Safe Drinking Water Act's 1996 Amendments

The Safe Drinking Water Act was enacted by Congress on December 16, 1974 and amended in 1986 and 1996 (United States Public Law, 1986, 1996). The Act authorizes USEPA to set drinking water standards called National Primary Drinking Water Regulations (NPDWRs). NPDWRs protect people against experiencing adverse effects from consuming drinking water containing contaminants.

Prior to the 1996 Amendments, the pace of regulation development was unrealistic in terms of meeting required deadlines and the needs of good science. The earlier amendments posed a significant challenge, particularly for small water systems, in keeping pace with the range of new contaminants being regulated, in terms of cost and treatment.

On August 6, 1996 (United States Public Law, 1996), Congress made dramatic changes and improvements in the structure of the Act. The new amendments reflect greater recognition of prevention, as opposed to treatment, as an effective approach to ensuring safe drinking water. Most important among these changes are:

- A Drinking Water State Revolving Fund;
- Source Water Assessment and Protection;
- Sound, Peer Reviewed Science;
- Improved Standard Setting;
- Capacity Development;
- Public Information.

The 1996 Amendments require USEPA to use the best available, peer-reviewed science to:

- Identify those most vulnerable to adverse health effects from contaminant exposure, such as infants, children, pregnant women, the elderly and those with serious illness. Considering risk to these "sensitive subgroups" when setting standards is required.
- Establish a national occurrence database of chemical, microbial, radiological and other contaminants which are known or likely to occur in public water systems.
- List 30 (maximum) unregulated contaminants for monitoring by public water systems by 1999, and every five years thereafter. Monitoring data will be made publicly accessible in the database. USEPA promulgated the first Unregulated Contaminant Monitoring Rule in

September, 1999 (United States Environmental Protection Agency, 1999).

Regarding Standard Setting, the 1996 Amendments authorize USEPA to:

- Prioritize rulemaking based on risk. Contaminants are considered first for regulation which present the greatest public health risk. Contaminants are selected based on health criteria and occurrence, and whether a regulation would be effective in reducing risk.
- Establish a Contaminant Candidate List (CCL), to be published every five years. This system replaces the cycle of regulating 25 contaminants every three years. From the CCL, USEPA must identify at least five contaminants to consider for regulation within a five-year schedule. USEPA published the first CCL in March 1998 (United States Environmental Protection Agency, 1998a).
- Increase stakeholder involvement in a range of areas (for example, cost-benefit analysis, peer-reviewed science, and public comment on rules).
- Consider whether the health benefits of a regulation would justify the cost.

## 3. Priority Rules of Interest

Consistent with the new Congressional deadlines, USEPA is working with stakeholders to develop the following regulations first:

- Ground Water Rule (see below);
- Microbials and Disinfection Byproducts (see below);
- Radon - NPDWR due August 2000;
- Radionuclides – NPDWR for radium, uranium, and alpha, beta, and photon emitters, due November 2000. (The radionuclides requirement is due to a court settlement and not specifically addressed in the 1996 amendments);
- Arsenic – NPDWR due January 2001.

## 4. Ground Water Rule

USEPA has the responsibility to develop a ground water rule which not only specifies the appropriate use of disinfection but, just as importantly, addresses other components of ground water systems to assure public health protection. This general provision is supplemented with an additional requirement that USEPA develop standards specifying the use of disinfectants for ground water systems as necessary. To meet these requirements, USEPA is working with stakeholders to develop a Ground Water Rule (GWR) proposal in the spring of calendar 2000, and a final rule by the winter of calendar year 2000.

## 5. Microbial and Disinfection Byproduct Standards

To control disinfectants and disinfection byproducts and to strengthen control of microbial pathogens in drinking water, USEPA is developing a group of interrelated regulations, as required by the SDWA. These standards, referred to collectively as the microbial-disinfection byproduct (M-DBP) rules, are intended to address risk trade-offs between the two different types of contaminants.

USEPA finalized the first two of these M-DBP rules, the Stage 1 Disinfectants and Disinfection Byproducts Rule (DBPR) and the Interim Enhanced Surface Water Treatment Rule (IESWTR), in December 1998 (United States Environmental Protection Agency, 1998b, c). These M-DBP rules are: the first standards under the 1996 Amendments to the Safe drinking Water Act; the first in the United States to address *Cryptosporidium*; the first to address non-chlorine byproducts; the first that use treatment techniques for control of DBPs; and the first to limit the levels of disinfectant residuals.

The Stage 1 DBP rule has the following features.

- The standard for total trihalomethanes (TTHMs) is tightened to 80 ug/L from 100 ug/L;
- A standard of 60 ug/L was promulgated for the sum of five haloacetic acids (HAA5);
- A standard of 10 ug/L was promulgated for bromate;
- Limits were set on the maximum concentrations of disinfectant residuals – 4 mg/L for chlorine, 4 mg/L for chloramines, and 0.8 mg/L for chlorine dioxide;
- A treatment technique was established – enhanced coagulation/enhanced softening;
- Existing pre-disinfection credits were retained as allowed under the original Surface Water Treatment Rule (United States Environmental Protection Agency, 1989).

The Stage 1 IESWTR rule has the following features:

- A *Cryptosporidium* Maximum Contaminant Level Goal of zero is established and a 2-log removal of *Cryptosporidium* by filtration is required;
- The turbidity standard is strengthened (0.3 NTU 95$^{th}$ percentile; 1 NTU maximum);
- Continuous monitoring of individual filters is required;
- Disinfectant profiling/benchmarking is required as are sanitary surveys for all surface water systems;
- *Cryptosporidium* is included in the definition of ground water under the direct influence of surface water and in the watershed control requirements for unfiltered public water systems.

The Agency also finalized and has implemented a third rule, the Information

Collection Rule (United States Environmental Protection Agency, 1996), that will provide data to support development of subsequent M-DBP regulations. These subsequent rules include a Stage 2 DBPR and a companion "Long-Term 2" Enhanced Surface Water Treatment Rule (LT2ESWTR). The Agency must also develop a "Long-Term 1" Enhanced Surface Water Treatment Rule (LT1ESWTR), which will focus primarily on small surface water systems (serving fewer than 10,000 people). In addition, USEPA must develop a regulation to address issues related to recycling filter backwash. The deadlines for these rules are as follows: LT1ESWTR – November 2000; Filter Backwash Rule – August 2000; Stage 2 DPBR – May 2002; and LT2ESWTR – May 2002.

## 6. How USEPA Sets Drinking Water Standards

When setting a drinking water standard for a particular contaminant, it is first necessary to understand the effects of exposure. Contaminants cause adverse health effects in a variety of ways. Exposure to microbial pathogens can cause illness in some people after exposure to only one cyst, virus or bacterium. Effects of chemical contaminants depend on the potency, or strength, of the dose and the length of exposure to the contaminant. The two basic categories of health effects are acute and chronic.

Acute health effects can occur from a single exposure to a toxic substance. This single exposure may result in severe biological harm or death. Single exposures are usually characterized as lasting no longer than a day. All public water systems must monitor for contaminants that can pose an immediate health threat upon ingestion. These contaminants include nitrate, nitrite and microbiological contaminants, such as total coliforms. Compliance monitoring for these contaminants is required for all public water systems.

Chronic health effects occur from long-term, low-level exposure to a toxic contaminant. Standards for contaminants that can pose a chronic health threat (ingested over a lifetime) apply to all public water systems.

Primary drinking water standards are set through a two-step process. First, USEPA determines a maximum contaminant level goal (MCLG), the maximum level of a contaminant in drinking water at which no known or anticipated adverse effect on the health of persons would occur, and which allows an adequate margin of safety (United States Environmental Protection Agency, 1991a). MCLGs are non-enforceable health goals.

Once the MCLG is determined, USEPA sets a maximum contaminant level (MCL) as close as feasible to the MCLG. The MCL is the maximum permissible level of a contaminant in water which is delivered to any user of a public water system. MCLs are enforceable standards.

SDWA defines "feasible" as the level that may be achieved with the use of the best technology, treatment techniques, and other means which USEPA finds (after examination for efficiency under field conditions and not solely under laboratory conditions) are available, taking cost into consideration. Determining the feasible level is the "heart" of the decision process for any regulation.

## 7. Setting MCLGs

For chemicals that can cause adverse non-cancer health affects, the MCLG is based on the reference dose (RfD). The RfD is an estimate of the amount of a chemical that a person can be exposed to on a daily basis that is not anticipated to cause adverse health effects over a person's lifetime. In RfD calculations, sensitive subgroups are included, and uncertainty may span an order of magnitude.

For carcinogens, if there is evidence that a chemical is carcinogenic, and there is no dose below which the chemical is safe, the MCLG is set at zero. If a chemical is carcinogenic and a safe dose can be determined, the MCLG is set at a level above zero that is safe. For existing rules, MCLGs for carcinogens are set at zero.

For *Cryptosporidium***,** a microbiological contaminant, an MCLG was set at zero, based on the genus level, due to potential infectivity causing death by ingestion of a single viable oocyst (United States Environmental Protection Agency, 1998c).

## 8. Treatment Techniques

If USEPA determines that it is not economically or technologically feasible to ascertain the level of a contaminant, it may require the use of a treatment technique rather than an MCL, to prevent adverse effects on human health. This occurs when there is no reliable and economic method to measure the contaminant at particularly low concentrations, or when other factors obtain which make establishing an MCL undesirable.

A treatment technique is an enforceable procedure or level of technological performance which water systems must follow to ensure that a contaminant is controlled in their drinking water supplies.

The Surface Water Treatment Rule (SWTR) (United States Environmental Protection Agency, 1989) and the Lead and Copper Rule (LCR) (United States Environmental Protection Agency, 1991b) are two examples of the use of a treatment technique. For the SWTR, which focused on microbial controls for *Giardia* and viruses, the rule requires that systems properly filter water, unless they can meet certain strict criteria. The rule also requires that systems disinfect the water. Because *Giardia* and viruses are very difficult to reliably measure at low concentrations, the SWTR set filtration and disinfection performance standards. Day-to-day water plant compliance with these treatment techniques could be reliably measured, and thus, that assured the removal or inactivation of these microbes.

The LCR required treatment techniques rather than MCLs because water corrosivity is the key factor in controlling lead and copper in drinking water. An action level serves as a screen for whether systems have adequately reduced corrosivity. The rule required systems that do not meet the action levels at the tap to optimize corrosion control treatment.

## 9. Cost-Benefit Analysis

For all future drinking water standards, USEPA is required to conduct a cost-benefit analysis and provide comprehensive, informative, and understandable information to the public (United States Public Law,1996). USEPA must determine whether the benefits of a standard justify the costs, and is required to use the "best available, peer reviewed science and supporting studies" to support the development of standards.

The standard setting procedure established in the 1996 Amendments has new flexibility compared to the previous law. After determining an MCL or treatment technique based on affordable technology, USEPA must assess whether the benefits of that standard justify the costs. If not, USEPA may adjust the MCL for a particular class or group of water systems to a level that "maximizes health risk reduction benefits at a cost that is justified by the benefits." USEPA may not adjust the MCL if the benefits justify the costs to large systems, and if small systems are likely to receive variances. The rationale is that expense to those small systems should not change a national standard if those systems can receive a variance from a national standard anyway, based on affordability. The Small Business Regulatory Enforcement Fairness Act (SBREFA) is one mechanism for USEPA to discuss and accurately determine cost-benefit impacts to small water systems with owners and operators of such systems.

## 10. Implementation

Once USEPA has proceeded through the standard setting process and established the MCL for a specific contaminant, the Agency must determine a process to implement the regulation. The first step in regulatory compliance for an MCL is contaminant monitoring.

Monitoring requirements for most inorganic and organic contaminants follow a Standardized Monitoring Framework (United States Environmental Protection Agency, 1991a,c). The framework is a synchronized schedule that is intended to simplify monitoring for existing and upcoming rules.

Monitoring frequency for contaminants regulated under the framework generally depends upon the source water, surface or ground water, and upon water quality. Certain rules allow for increased or decreased monitoring. USEPA sets "detection limits" for inorganic and organic contaminants. These detection limits are generally contaminant-specific. "Detection" of a contaminant generally suggests that a system must conduct increased monitoring until the state decides that the system is "reliably and consistently" below the MCL.

Some systems may conduct decreased monitoring if the state grants a waiver. The state may grant a waiver based on previous sampling results or it may grant a waiver based on the system's vulnerability to a specific contaminant. The two types of vulnerability waivers include "use waivers" and "susceptibility waivers." A "use waiver" can be granted if the system can show that the contaminant has not been used, manufactured and/or stored within a certain area of the system's water source. A "susceptibility waiver" may be granted based on source protection, wellhead protection, previous sample results, and environmental fate/transport analysis.

## 11. Determining Compliance with MCLs and with Treatment Techniques

Water systems are in compliance with organic and inorganic MCLs if the water sample does not exceed the MCL. There is a standardized compliance monitoring strategy for each MCL. For instance systems must monitor for trihalomethanes at a particular point in the distribution system. Most chemical contaminants, however, are sampled at the entry point to the distribution system, and compliance is based on annual averages.

Treatment technique rules differ from MCL rules since the contaminant usually cannot be directly monitored. In these rules, the burden is upon the water system to show the state that the treatment technique is not needed. This is not so, however, for the Surface Water Treatment Rule, for which a treatment technique is always needed (United States Environmental Protection Agency, 1989). This rulemaking applies to all public water systems using surface water or ground water under the direct influence of surface water. This rule requires all systems using those sources of water to use disinfection and filtration as part of the treatment process unless they can meet specific criteria to avoid filtration.

## For More Information

To learn more about the Safe Drinking Water Act, call the Safe Drinking Water Hotline in the US at 202-260-5543 (or 1-800-426-4791), or visit USEPA's OGWDW Home Page at http://www.USEPA.gov/OGWDW.

## Acknowledgements

Sherri Umanski and Corry Westbrook are thanked for providing information used in this paper.


## References


United States Environmental Protection Agency: 1999, *Federal Register*, **64,** 50556-50620, Final Rule.

United States Environmental Protection Agency: 1998a, *Federal Register*, **63**, 10274-10287, Final Rule.

United States Environmental Protection Agency: 1998b, *Federal Register*, **63**, 69390-69476, Final Rule.

United States Environmental Protection Agency: 1998c, *Federal Register*, **63**, 69478-69521, Final Rule.

United States Environmental Protection Agency: 1996, *Federal Register*, **61,** 24354-24388, Final Rule.

United States Environmental Protection Agency: 1991a, *Federal Register*, **56,** 3526-3597, Final Rule.

United States Environmental Protection Agency: 1991b, *Federal Register*, **56,** 26460-26564, Final Rule.

United States Environmental Protection Agency: 1991c, *Federal Register*, **52,** 25690-25717.

United States Environmental Protection Agency: 1989, *Federal Register*, **54,** 27486-27541, Final Rule.

United States Public Law: 93-523, December 16, 1974.

United States Public Law: 99-339, June 19, 1986, as amended.

United States Public Law: 93-523, August 6, 1996, as amended.

# RECENT DEVELOPMENTS IN MICROBIOTESTING AND EARLY MILLENNIUM PROSPECTS

C. BLAISE, F. GAGNÉ and M. BOMBARDIER
*Aquatic Toxicology, Centre Saint-Laurent, Environment Canada, Quebec Region, 105 McGill Street, Montreal, QUE., Canada, H2Y 2E7*

**Abstract.** Small-scale toxicity testing with microbiotests is a rapidly-expanding component of the field of aquatic toxicology which contributes diverse contamination assessment tools and approaches for a variety of environmental (liquid and solid) media. In this short review on microbiotesting, some of the recent developments conducted under the second St.Lawrence River Action Plan (1993-1998) at the St. Lawrence Centre (Environment Canada, Quebec Region, Montreal) are recalled. These include 1) employing the SOS Chromotest to determine the genotoxic status of major industrial effluents discharging to the St. Lawrence River and their potential impact on downstream biota, 2) developing an algal solid phase assay to predict the toxic potential of freshwater sediments, 3) developing a microplate-based cnidarian assay to screen for toxicity of chemicals and environmental samples, 4) developing an alternative assay to whole fish acute (sub)lethal toxicity testing with the help of rainbow trout primary hepatocytes, 5) developing a microplate-based phagocytosis assay to check for immunocompetence of feral bivalve shellfish and 6) conducting a major investigation to develop a cost-effective multitrophic bioanalytical battery to assess the (geno)toxicity of freshwater sediments. In addition, integrative tools with specific microbiotests were respectively constructed to determine the toxic potential of industrial effluents (PEEP : Potential Ecotoxic Effects Probe) and that of sediments (SED-TOX). Such examples illustrate the diversity of on-going endeavors in the field of small-scale toxicity testing internationally, as further corroborated by recent books entirely dedicated to the subject. It is undeniable that many important challenges still lie ahead for this field early into the third millennium and likely well beyond.

**Keywords**: Algal solid phase assay, effluents, genotoxicity, *Hydra* assay, immunocompetence assay, microbiotests, PEEP Index, sediments, SED-TOX Index, SOS Chromotest, toxicity, Trout hepatocyte assay.

## 1. Introduction

Bioassays are at the forefront of the daily battles we wage against ecotoxicity. In this respect, their utility to circumscribe the toxic potential of a myriad of anthropogenic chemicals, as well as that of samples taken from a variety of environmental media (liquid and solid), is undeniable (Blaise *et al.*, 1988). With time, growing pressure to diagnose an ever-increasing number of environmental samples to detect their (sub)lethal toxic potential also enhanced the need for cost-effective testing, which simple microbiotests could provide (Blaise, 1998; Blaise *et al*, 1998). As a result, microbiotesting is now contributing to more cost-

*Water, Air, and Soil Pollution* **123:** 11–23, 2000.

efficient delivery of environmental programs and it is now clear that the era of small scale aquatic toxicology is well established (Wells *et al.*, 1998).

At Environment Canada's Montreal-based St. Lawrence Centre (SLC), multidisciplinary environmental research is undertaken, on its own or through partnership, under specific St. Lawrence River Action Plans (SLAP) to generate relevant (biological, chemical, physical) data which provide updated information on the health status of the St. Lawrence River ecosystem, the *raison d'être* of our Centre. SLAP goals are to protect, conserve and restore this economically-significant and biologically-rich fluvial system which, over the years, has suffered ecosystemic setbacks owing to environmental negligence. Industrial pollution, stemming from several important sectors (pulp and paper, metallurgy, chemical production, mining, oil refinery, metal finishing, and textiles), constitutes a major source of toxic wastes to the St. Lawrence River. Reduction of this toxic input, therefore, comprises a major objective of these Action Plans. With the termination of SLAP I in March of 1993, marked reduction in toxic input of industrial origin to the St. Lawrence River had been achieved. Under the second five-year Action Plan, known as the St. Lawrence Vision 2000 (SLV 2000) Action Plan, which ran from April 1993 to March of 1998, reduction of industrial contaminants discharging into the River continued to be a top priority. In the context of the Action Plans, the SLC employed small-scale aquatic toxicology as one important approach to investigate chemical contamination to the St. Lawrence River.

In this short review, some of the major developments and applications undertaken at the SLC during the past five years in the area of microbiotesting are reported (Table I). Future challenges for microbiotesting in the upcoming 21st century are also discussed.

## 2. Recent Developments and Applications in Micro-Scale Ecotoxicity Assessment

### 2.1. SOS - INDUCING EFFLUENTS

A major study was undertaken under the first and second SLAPs (1988-98) in which organic extracts of 50 industrial effluent samples stemming from 42 priority industries were screened with the SOS Chromotest for their genotoxic potential (White el al., 1996). Developed by Pasteur Institute researchers (Quillardet *et al*, 1982 ; Quillardet and Hofnung, 1985), this test takes advantage of *Escherichia coli*'s capacity to respond to DNA-damaging events via the SOS response pathway (Walker, 1984). Thanks to a specific *sulA :: lac z* gene fusion,

TABLE I

Recent studies/activities in microbiotesting undertaken at the St. Lawrence Centre of Environment Canada, under the second St. Lawrence River Action Plan (1993-98)

| Microbiotesting level | Type of bioassay or activity | Major objective | References |
|---|---|---|---|
| Bacterium | *Escherichia coli* PQ37 (SOS Chromotest) | Assessment of the genotoxic status of major industrial effluents discharging to the St. Lawrence River and their potential impact on downstream biota | White *et al.*, 1996 and 1998 |
| Micro-alga | *Selenastrum capricornutum* (Algal solid phase assay) | Development of an algal solid phase assay to predict the toxic potential of freshwater sediments | Blaise and Ménard, 1998 |
| Micro-invertebrate | *Hydra attenuata* toxicity assay | Development of a microplate-based cnidarian assay to screen for toxicity of chemicals and complex environmental samples | Blaise and Kusui, 1997 ; Trottier *et al.*, 1997 |
| Fish cells | Primary hepatocytes of *Oncorhynchus mykiss* | Development of an alternative to whole fish acute (sub)lethal toxicity testing with the aid of Rainbow trout primary hepatocytes | Gagné and Blaise, 1995 ; Gagné *et al.*, 1995 ; Gagné and Blaise, 1996a,b,c ; Gagné and Blaise, 1997 ; Gagné and Blaise, 1998 |
| Multitrophic assessment | Several (micro)organisms at various levels of biological organization | Development of a cost-effective multitrophic bioanalytical battery to assess the (geno)toxicity of freshwater sediments | Côté *et al.*, 1998a and 1998b |
| Integrated battery : effluent assessment | PEEP index | Development of a bioassay-based index, as an aid to decision-making, to determine the toxic potential of industrial effluents | Costan *et al.*, 1993 ; Bermingham *et al.*, 1996 ; Blaise, 1996 ; Kusui and Blaise, 1999 |
| Integrated battery : sediment assessment | SED-TOX index | Development of a bioassay-based index, as an aid to decision-making, to determine the toxic potential of sediments | Bombardier and Bermingham, 1999 |

the production of functional ß-galactosidase is placed under the sole control of the SOS repair pathway (*sulA*), itself induceable by DNA-damaging agents (Walker, 1984). Industrial sectors investigated included pulp and paper, chemical manufacturing, metal refining, and metal surface treatment; municipal wastewater treatment was also considered. Major findings revealed that a marked proportion of suspended solid extracts (70%) and of acid (62%) and base (43%) partitioning aqueous extracts proved to be genotoxic. While genotoxicity was related to sample/industry type, effluent particulates (suspended solids) were generally four orders of magnitude more potent than aqueous filtrates, when potency values were expressed in equivalent units of original sample (White *et al.*, 1996).

A parallel study with the SOS Chromotest on tissue extracts of downstream biota demonstrated exposure to genotoxicants (White *et al.*, 1998), with lipid normalized values indicating that genotoxin concentrations in invertebrate tissues were significantly higher than those in fish (Figure 1). This result may reflect enhanced detoxification processes of lipophilic contaminants in fish, as compared to invertebrates. While the ultimate hazard of genotoxicants of industrial origin is presently unknown, the long-term risk to exposed biota also remains to be addressed. Meaningful in the context of this paper is the wealth of relevant information obtained on the genotoxic activity of industrial discharges and the related transfer to biota based on the application of a single microbiotest.

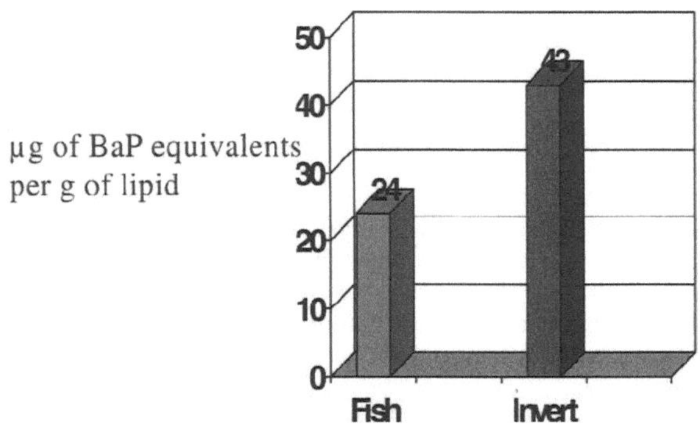


[1] Based on 152 samples (120 fish: walleye, perch, pike; 32 invertebrates: gastropods, clams/mussels, amphipods, crayfish) collected from the St. Lawrence River and Saguenay Fjord ecosystems. No SOS positive responses from biota (fish and invertebrates) at reference sites.

Fig. 1. Exposure of downstream biota located in areas prone to genotoxic point source pollution (adapted from White *et al.*, 1998).[1]

## 2.2. A NEW ALGAL SOLID PHASE ASSAY (ASPA) TO ASSESS THE TOXIC POTENTIAL OF FRESHWATER SEDIMENTS

Phytoplankton plays a crucial role in stabilizing aquatic ecosystems but it can be adversely affected by toxicity resulting from sediment resuspension owing to both natural (e.g., flood scouring) and man-made activities (e.g., dredging, openwater deposition). Hence, we felt there was justification in developing a direct contact microassay which could react to both readily hydrosoluble and ad(ab)sorbed contaminants. While there are numerous microbiotests to assess liquid media (Day *et al.*, 1995; Côté *et al.*, 1998a), few are presently adapted to properly measure the toxic potential of solid media and there is clearly a gap to fill in this respect in terms of micro-scale bioassay development. With the exception of this new algal solid phase assay (ASPA), other available direct contact microtests are exclusively conducted with bacteria (Brouwer *et al.*, 1990 ; Kwan and Dutka, 1992 ; Microbics, 1992 ; Corbisier *et al.*, 1996 ; Bitton *et al.*, 1996).

Conducted in rotating tubes containing serial dilutions of the test sediment and cells of the green alga *Selenastrum capricornutum*, the assay runs for a 24 h exposure period (Blaise and Ménard, 1998). The capacity of cell esterases to cleave the (non polar) fluorescein diacetate stain and liberate fluorescein, a polar and fluorescent molecule, then becomes the criterion which allows us to determine the extent to which algae have been intoxicated by the sediment. Individual cell fluorescence (from each of 2000 cellular events) from each test tube can then be precisely quantified with the help of a flow cytometer. The resulting toxicity endpoint (IC50 or NOEC and LOEC) relates to both esterase inhibition and cell membrane integrity. In applying the ASPA on a series of contaminated sediments sampled from various locations in the Quebec portion of the St. Lawrence River, we found that it was able to correctly discriminate sediments on the basis of their contamination level (Blaise and Ménard, 1998). Within the same study, sediments taken from a highly industrialized area of concern generated ASPA toxicity responses that matched those generated with standardized macrobenthic bioassays (the amphipod *Hyalella azteca* and the midge *Chironomus riparius*). Further validation studies aimed at generating data with additional sediment samples are on-going. We are also attempting to modify our present testing procedure, which requires the use of flow cytometry, so as to replace it with a microfluorimetric endpoint. This more cost-effective instrumentation alternative would enable ASPA testing by interested laboratories which do not possess a flow cytometer.

## 2.3. TOXICITY TESTING WITH A FRESHWATER CNIDARIAN

Following its initial use as a teratogenicity screening tool (Johnson and Gabel, 1982), the freshwater cnidarian *Hydra attenuata* was only recently exploited to assess the acute lethal toxicity of wastewaters (Fu *et al.*, 1994). Advantages of using a *Hydra* animal model for bioassessment include its wide distribution in freshwater environments, thereby making it a representative animal for conducting environmental hazard assessment, as well as the particular robustness of *Hydra attenuata* which makes it easy to manipulate, rear and maintain in the laboratory. When exposed to bioavailable toxicants, this organism undergoes marked morphological changes which are manifested by sublethal and, eventually, lethal effects (Johnson and Gabel, 1982). From their normal appearance, the animals progressively exhibit bulbed (clubbed) tentacles as an initial sign of toxicity, followed by shortened tentacles and body. After these sublethal manifestations, and if toxicity continues to prevail, *Hydra* reach the so-called « tulip phase » when death becomes an irreversible event. The *post-mortem* stage is finally indicated by the disintegration of the organism. Noting *Hydra* morphology during exposure allows for simple recording of (sub)lethal toxicity effects. We have recently developed and described a 96-h exposure microplate-based procedure which was first employed to assess the toxic potential of Japanese industrial effluents (Blaise and Kusui, 1997). A detailed technical protocol was also published as an aid for those wishing to apply this procedure in their laboratory (Trottier *et al.*, 1997). Two recent comparative studies with batteries of microbiotests clearly demonstrated that the *Hydra* assay displayed good sensitivity in detecting effluent (Kusui and Blaise, 1998) and freshwater sediment interstitial water (Côté *et al.*, 1998b) toxicity. While additional investigations are on-going to confirm its scope of usefulness to appraise the toxic potential of other complex environmental matrices (urban snow, organic extracts of particulates from water and air), present information suggests that this simple microassay has its place as an effective toxicity detection system in the field of aquatic toxicology.

## 2.4. AN ALTERNATIVE TO WHOLE FISH TESTING

Toxicity characteristics of industrial wastewaters in Canada are presently assessed with a standardized Rainbow trout (*Oncorhynchus mykiss*) acute lethality assay (Environment Canada, 1990). This regulatory bioassay, employed for compliance monitoring of specific types of industrial effluents, requires sacrificing more than 100 fingerling trout whenever an effluent sample is analyzed. For ethical reasons, as well as those linked to cost- and time-effectiveness, labour-intensiveness, analytical output and effluent sample volume requirements, there is unquestionable value in searching for alternative procedures which would eliminate (or at least reduce) the drawbacks associated with whole animal testing. In this respect, we have recently turned to the application and validation of an *in vitro* cell system which could greatly decrease the need for the *in vivo* trout model. We chose Rainbow trout primary hepatocytes instead of an immortal cell system because, unlike the latter which can lose tissue-specific characteristics when they are grown over long periods of time, the former do not. As well as more closely matching the *in vivo* properties of hepatic tissue, primary hepatocytes play a key role in metabolic transformation and are often the target of chemical aggression (Baksi and Frazier, 1990). In our laboratories, we have shown that the hepatocyte cell model can correctly predict (sub)lethal toxicity to fingerling Rainbow trout when both are tested with effluent samples representing various industrial sectors (Gagné and Blaise, 1997). Promotion of the primary hepatocyte cell assay as a viable adjunct (or alternative) to the whole fish assay for effluent toxicity assessment is presently on-going in Canada. In addition to effluent appraisal, we have used this cell system succesfully to evaluate the (geno)toxic potential of sediment organic extracts (Gagné and Blaise, 1995; Gagné *et al.*, 1995). Because primary hepatocytes produce vitellogenin when stimulated hormonally, their potential as a screening tool to detect xeno-estrogens was also examined; such an assay has recently been developed in our laboratory (Gagné and Blaise, 1998; Gagné *et al.*, 1999). Overall, this primary cell line, relying on miniaturized protocols employing 96-well microplates to measure various (sub)lethal toxicity parameters after periods of exposure varying from 24 to 48 h at 15$^0$C, offers versatile possibilities for microbiotesting, as confirmed by a variety of studies conducted in recent years (Figure 2).

## 2.5. ADEQUACY OF SMALL-SCALE ASSAYS IN PREDICTING FRESHWATER SEDIMENT TOXICITY

Displacement of sediments (e.g., dredging, deposition, *ex-situ* treatment) is a major source of concern for aquatic ecosystems because it may cause the mobilization/resuspension of contaminants with potential consequences on water quality and aquatic biota. The effects of sediment-associated contaminants need to be determined if remediation procedures are to be designed and implemented. A battery of several microbiotests with several test species representing different trophic levels may provide a means for making decisions in respect to selecting displacement options for sediments. The rationale is that a single test species can never adequately predict contaminant effects for all biota. For this purpose, we have recently evaluated several microbiotests to assess their adequacy in being able to appraise the (geno)toxicity of freshwater sediments. The major objective of the study was to develop and recommend a cost-effective multitrophic bioanalytical battery. An assemblage of 20 bioassays, comprising 18 microbiotests and 2 whole sediment bioassays conducted with macroinvertebrates (*Chironomus riparius* and *Hyalella azteca*), was applied to assess test and endpoint adequacy in detecting the toxic potential of 15 freshwater sediment samples collected in the St. Lawrence River/Great Lakes systems. The performance of the microbiotests was evaluated by comparing their toxicity responses with those of benthic organism bioassays and benthic community structure indices, as well as with sediment contaminant characteristics. Details on this major investigation, including the approach employed to rank the suitability of the microbiotests to predict the toxic potential of freshwater sediments, have recently been reported (Côté *et al.*, 1998a). Based on selected scientific and practical criteria, the two benthic assays and seven micro-scale assays successfully passed the rigorous assessment process. From this work, we were able to propose two battery approaches, utilizing the aforementioned tests coupled with physico-chemical characterization, for cost-effective appraisal of freshwater sediment toxicity (Côté *et al.*, 1998a).

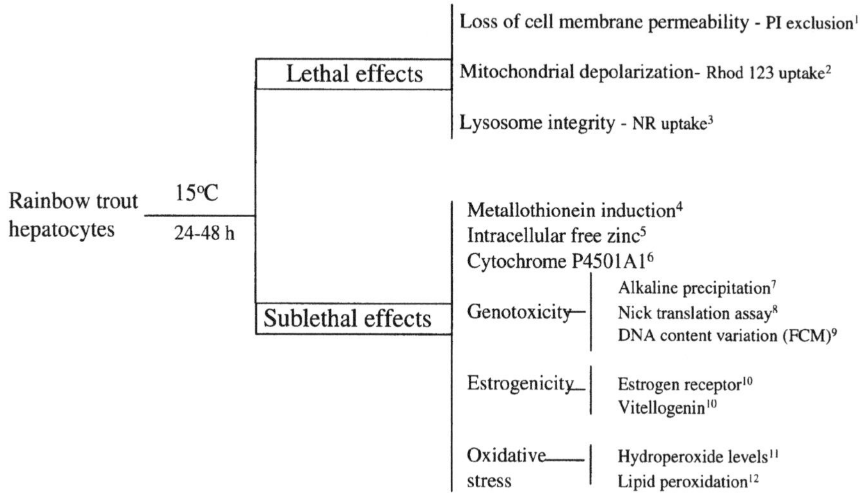


[1] Zucker *et al.*, 1988; [2] Wu *et al.*, 1990; [3] Zhang *et al.*, 1990; [4] Gagné *et al.*, 1990, [5] Gagné and Blaise, 1996a ; [6] Lee and Dasmahapatra, 1993; [7] Olive, 1988; [8] Gagné and Blaise, 1995; [9] Prosperi *et al.*, 1993; [10] Anderson *et al.*, 1993; [11] Lebel *et al.*, 1992; [12] Latour *et al.*, 1992.

Fig. 2. Multi-effect screening with rainbow trout hepatocytes.

## 2.6. EFFLUENT (PEEP INDEX) AND SEDIMENT (SED-TOX INDEX) ASSESSMENTS WITH INTEGRATED BATTERIES OF BIOLOGICAL TESTS

At the SLC, the PEEP (Potential Ecotoxic Effects Probe) Index allowing the assessment and comparison of the toxic potential of industrial effluents was developed to serve the purposes of environmental management (Costan *et al.*, 1993). To be effective, this effluent assessment index is dependent on the use of an appropriate suite of multitrophic bioassays (decomposers, primary producers and consumers) enabling the measurement of various types (acute, chronic) and levels (lethal, sublethal) of toxicity. At the time of its conception, this index integrated the results of a selection of practical small-scale screening bioassays (*Vibrio fischeri* Microtox$^R$ test, *Selenastrum capricornutum* growth inhibition microtest, *Ceriodaphnia dubia* lethality and reproduction inhibition tests, *Escherichia coli* genotoxicity in the SOS Chromotest), and took into account the persistence of toxicity (i.e., biotests were performed on an effluent before and after a five-day biodegradability procedure), (multi)specificity of toxic impact (number of bioindicators affected by an effluent), as well as toxic loading (effluent flow in $m^3/h$). The resulting PEEP Index number is reflected by a $\log_{10}$ value that may vary from 0 to 10. The structure of the mathematical formula generating PEEP values is simple and "user-friendly" in that it can accommodate any number and type of bioassays to fit particular needs (Costan *et al.*, 1993). The approach is novel in that it combines information on 1) the biodegradability/persistence of effluent toxicity (indicative of its possible fate in receiving waters), 2) the trophic levels targeted by effluent toxicity (indicative of the ecological scope of impact) and 3) the flow characteristics of the effluent (indicative of toxic loading released to the environment). The integration of these concepts into a PEEP scale or Index is clearly an unparalleled attempt to bring together factors of relevant ecotoxicological importance into a simple, practical and useful management tool to literally "peep" into the hazardous potential of industrial effluents *via* an initial bioanalytical screening strategy. Its effectiveness in predicting the overall hazard potential of wastewaters was revealed for 77 priority effluents investigated under the first two SLAPs, as well as for 20 Japanese effluents located in the Toyama Prefecture (Kusui and Blaise, 1998). In quantifying the toxicity of industrial discharges, the PEEP Index unambiguously points a finger at the most problematic ones requiring priority attention in terms of clean-up action, such that environmental protection effectiveness can be achieved. It is evident by observing the range of PEEP values determined for the industrial effluents studied under SLAPs that « all effluents are not created equal » and that some are irrefutably more noxious than others (Figure 3).

It was recently felt that the principle of the PEEP Index could realistically be extended to evaluate solid media. Hence, a sediment toxicity (SED-TOX) Index was developed which integrates toxicity data generated with a test battery reflecting different trophic levels, various acute and chronic (sub)lethal endpoints, as well as routes of exposure (sediment and interstitial water). Based on a formula which summarizes and then reduces toxicity responses into a single value (SED-TOX score), where the latter is subsequently assigned to one of four classes of toxic intensity (no hazard potential, marginal hazard potential, moderate hazard potential, high hazard potential), relative hazard for individual sediment samples taken from a group of geographically-connected sites can be ascertained (Bombardier and Bermingham, 1999). In one investigation conducted since its recent development, the SED-TOX index has shown that it is able to discriminate sediment noxiousness based on its degree of (in)organic contamination (Figure 4).

SED-TOX validation studies in St. Lawrence River sediment sites of interest are presently on-going.

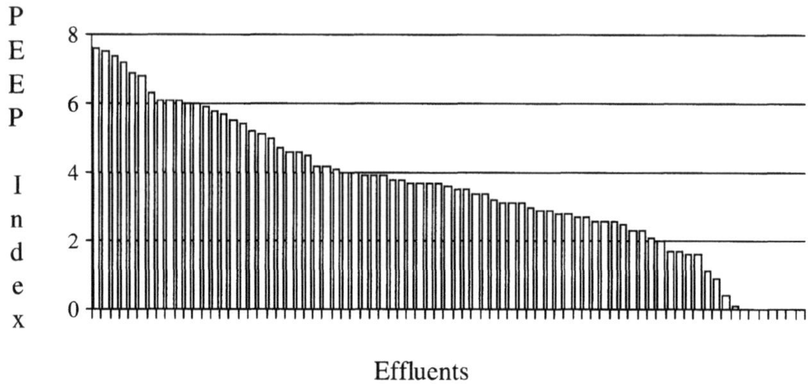


Fig. 3. PEEP values for 77 Canadian-based effluents investigated under two Saint-Lawrence River Action Plans (1988-1998).

## 2.7. BOOKS ON MICROBIOTESTS

Contemporary endeavors in the field of microbiotesting, exemplified to some extent by those given herein, are planetary in scope, as confirmed by authoritative texts now appearing which are entirely dedicated to this topic. A recent publication, resulting from a collaboration struck between Environment Canada, Fisheries and Oceans Canada and CRC/Lewis Press, comprises over 40 chapters provided by international experts (Wells *et al.*, 1998). It is organized in a fashion to provide information on novel techniques and their applications to research scientists, environmental managers, academics, and the private sector (industry, consultants). This book significantly promotes the use of small-scale testing of chemicals and environmental samples and should stimulate even more developments and applications in this area for the future. An upcoming second book , catalyzed by the holding of the recent *International Symposium on New Microbiotests* (Brno, Czech Republic, June 1998), is scheduled for release later this year (Persoone *et al.*, 1999). With over 60 chapters, it documents diverse creative and imaginative work of a practical nature which is being conducted in different parts of the world with small-scale assays to evaluate toxicity of pure chemicals, biological toxins, as well as toxic wastes released to aquatic and terrestrial environments. It confirms once again that development and application of micro-scale toxicity testing are undeniably broad-based in scope.

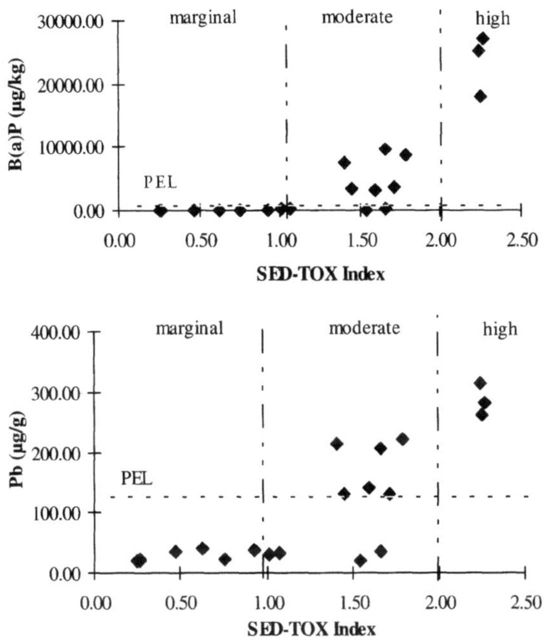


Fig. 4. SED-TOX scores for a series of marine sediment samples (collected from a harbor site in Eastern Canada) showing a gradient of (in)organic contamination with B(a)P and Pb.

## 3. Future Challenges and Conclusions

Future investigations involving microbiotests can be expected to assess issues linked to genotoxicity, endocrine toxicity and immunotoxicity, which are clearly understudied effects at this time. Optimization of microbiotest batteries for hazard/risk assessment of specific (liquid and solid) media is also an area requiring further attention to determine and validate adequate numbers of (non redundant) tests and endpoints to generate relevant ecotoxicological information. Within this concern, the suitability of microbiotests to appraise several types of yet poorly investigated environmental matrices (e.g., groundwaters, surface waters, soils/sediments, suspended solids, ice, snow, air) will also be of particular interest. Development and use of micro-techniques that could relate to both ecosystem and human health, while seemingly an unattainable goal to achieve in the short term, may unfold as interrelationships between these two spheres become better documented. So-called *biomarker-bioassays* may represent an important first step in this direction (Lee *et al.*, 1998). Promotion of currently validated microbiotests as alternatives to more costly tests, while demanding in terms of time and less glamorous *per se* as a scientific challenge, nevertheless constitutes a worthwhile initiative for the future, as it would contribute to more cost-effective testing both for initial characterization purposes of varied sources of contaminants, as well as in the regulatory arena. Lastly, transfer of microbiotesting techniques internationally should be strongly encouraged to ensure a basic ecotoxicity screening capacity for key laboratories engaged in ecosystem protection and conservation activities. The International Development Research Centre (Ottawa, Canada), through its recently-implemented WaterTox program, appears to have taken the lead in this area by disseminating cost-effective technology transfer of several microbiotests to scientists of developing countries. In conclusion, it is clear that the still fledgling field of microbiotesting already offers practical and versatile environmental tools for ecotoxicology. Important avenues of research remain to enhance our knowledge of the potential toxic hazards linked to environmental contaminants. In this light, we anticipate that this area of expertise, a genuine *offshoot* of the field of aquatic toxicology, can be expected to grow significantly well into the third millennium.


## Acknowledgements

This short review on microbiotesting development and applications conducted at the SLC, essentially under the second St. Lawrence Action Plan (1993-1998), results from the first author's invited presentation given during the holding of the $7^{th}$ *International Conference of the Israeli Society for Ecology and Environmental Quality Sciences : Environmental Challenges for the Next Millennium* (Jerusalem, Israel, June 13-18, 1999). The first author is grateful to Dr. Shimshon Belkin (Professor at the Hebrew University of Jerusalem and Confrence Chairman) for his invited talk. All authors are also thankful to management of the SLC for supporting this scientific initiative.


## References


Anderson, M. J., Olsen, H., Matsumara, F. and Hinton, D. E.: 1996, *Toxicol. Appl. Pharm.* **137**, 210-218.

Baksi, S. M. and Frazier, J. M.: 1990, *Aquat. Toxicol.* **16**, 229-259.

Bitton, G., Garland, E., Kong, I. C., Morel, J. L. and Koopman, B.: 1996, *J. Soil Contam.* **5**, 385-394.

Blaise, C., Sergy, G., Wells, P., Bermigham, N., van Coillie, R.: 1988, *Toxicol. Assess.* **3**, 385-406.

Blaise, C. and Kusui, T.: 1997, *Environ. Toxicol. Water. Qual.* **12**, 53-60.

Blaise, C.: 1998, *Ecotox. Environ. Safety* **40**, 115-119.

Blaise, C. and Ménard, L.: 1998, *Water Qual. Res. J. Canada* **33**, 133-151.

Blaise, C., Wells, P. and Lee, K.: 1998, *Microscale testing in Aquatic Toxicology Advances, Techniques and Practice*, in P. Wells, K. Lee and C. Blaise (eds), CRC Lewis Publishers, Boca Raton, Florida, pp. 1-9.

Bombardier, M. and Bermingham, N.: 1999, *Environ. Toxicol. Chem.* **18**, 685-698.

Brouwer, H., Murphy, T. and McArdle, L.: 1990, *Environ. Toxicol. Chem.* **9**, 1353-1358.

Corbisier, P., Thiry, E. and Diels, L.: 1996, *Environ. Toxicol. Water Qual.* **11**, 171-177.

Costan, G., Bermingham, N., Blaise, C. and Férard, J. F.: 1993, *Environ. Toxicol. Water Qual.* **8**, 115-140.

Côté, C., Blaise, C., Schroeder, J., Douville M. and Michaud J.-R.: 1998a, *Water Qual. Res. J. Canada* **33**, 253-277.

Côté, C., Blaise, C., Michaud, J.-R., Ménard, L., Trottier, S., Gagné, F., Riebel P. and Lifschitz, R.: 1998b, *Environ. Toxicol. Water Qual.* **13**, 93-110.

Day, K. E., Dutka, B. J., Kwan, K. K., Batista, N., Reynoldson T. B. and Metcalfe-Smith, J. L. : 1995, *J. Great Lakes Res.* **21**, 192-206.

Environment Canada: 1990, Biological Test Method: Acute Lethality Test Using Rainbow Trout, Conservation and Protection, Environment Canada, Ottawa, Ontario, pp. 53.

Fu, L. J., Staples, R. E. and Stahl, Jr., R. G.: 1994, *Environ. Toxicol. Chem.*, **13**, 563-569.

Gagné, F., Marion, M. and Denizeau, F.: 1990, *Fund. Appl. Toxicol.* **14**, 429-437.

Gagné, F. and Blaise, C.: 1995, *Environ.Toxic. Water Qual.* **10**, 217-229.

Gagné, F., Trottier, S., Blaise, C., Sproull J., and Ernst B.: 1995, *Toxicol. Letters* **78**, 175-182.

Gagné, F. and Blaise, C.: 1996a, *Environ. Toxic. Water Qual.* **11**, 319-325.

Gagné, F. and Blaise, C.: 1996b, *Environ. Toxicol. Water Qual.* **11**, 53-63.

Gagné, F. and Blaise, C.: 1996c, *Toxicol. Letters* **87**, 85-92.

Gagné, F. and Blaise, C.: 1997, *Environ. Toxicol. Water Qual.* **12**, 305-314.

Gagné, F. and Blaise, C.: 1998, *Aquatic Toxicol.* **44**, 83-91.

Gagné, F., Pardos, M. and Blaise, C.: 1999, *Bull. Environ. Contam. Toxicol.* **62**, 723-730.

Johnson, E. M. and Gabel, B. E. G.: 1982, *J. Am Coll. Toxicol.* **1**, 57-71.

Kwan, K. K. and Dutka, B. J.: 1992, *Bull. Environ. Contam. Toxicol.* 49, 656-662.

Kusui, T. and Blaise, C.: 1998, in R. Salem (ed), *Impact Assessment of Hazardous Aquatic Contaminants : Concepts and Approaches*, Ann Arbor Press, Michigan, USA, pp. 161-181.

Latour, I., Pregaldien, J.-L. and Calderon, P. B.: 1992, *Arch. Toxicol.* **66**, 743-749.

Lebel, C. P., Ischiropoulos, H. and Bondy, S. C.: 1992, *Chem. Res. Toxicol.* **5**, 227-231.

Lee, P. C. and Dasmahapatra, A.: 1993, *Comp. Biochem. Physiol. C.* 106, 649-653.

Lee, K., Wells, P. and Blaise, C.: 1998, in P. Wells, K. Lee and C. Blaise (eds), *Microscale testing in Aquatic Toxicology Advances, Techniques and Practice*, CRC Lewis Publishers, Boca Raton, Florida, pp. 647-649.

Microbics: 1992, *Microtox[R] update manual,* Microbics Corporation, Carlsbad, CA, U.S.A.

Olive, P. L.: 1988, *Envir. Molec. Mutagen.* **11**, 487-495.

Persoone, G., Janssen, C. and De Coen W.: 1999, *New Microbiotests for routine toxicity screening and biomonitoring*, Kluwer Academic/Plenum Publishers (in press).

Prosperi, E., Supino, R. and Bottiroli, G.: 1993, *Cytometry* **14**, 53-58.

Quillardet, P., Huisman, O., D'Ari, R. and Hofnung, M.: 1982, *Proc. Natl Acad. Sci. USA* **79**, 5971-5975.

Quillardet, P. and Hofnung, M.: 1985, *Mutat. Res.* **147**, 65-78.

Trottier, S., Blaise, C., Kusui, T. and Johnson, E. M.: 1997, *Environ. Toxicol. Water Qual.* **12**, 265-271.

Walker, G. C.: 1984, *Microbiol. Rev.* **48**, 60-93.

Wells, P., Lee, K. and Blaise, C.: 1998, *Microscale testing in Aquatic Toxicology Advances, Techniques and Practice.* CRC Lewis Publishers, Boca Raton, Florida.

White, P., Rasmussen, J. and Blaise, C.: 1996, *Environ. Molec. Mutagen.* **27**, 116-139.

White, P., Rasmussen, J. and Blaise, C.: 1998, *Environ. Toxicol. Chem.* **17**, 304-316.

Wu, E. Y., Smith, M. T., Bellomo, G. and Di Monte, D.: 1990, *Arch. Biochem. Biophys.* **282**, 358-362.

# BIOLOGICAL STABILITY: A MULTIDIMENSIONAL QUALITY ASPECT OF TREATED WATER

D. VAN DER KOOIJ
*Kiwa N.V., Groningenhaven 7, POB 1072, 3430 BB Nieuwegein, theNetherlands*

**Abstract.** Regrowth processes in drinking water distribution systems may lead to hygienic, aesthetic and technical problems. These complex processes depend on interactions between micro-organisms and (i), compounds serving as energy sources; (ii), environmental conditions (temperature, hydraulics) and (iii), physico-chemical processes (sedimentation, corrosion, disinfection), respectively. The concentration of growth-promoting compounds is considered as the main driving force for regrowth and a large variety of tests has been developed to assess the growth-promoting properties of treated water. These methods range from determining the decrease of the concentration of dissolved organic carbon in a batch test to the assessment of the Biofilm Formation Rate (BFR) in a flow-through test. Biostability assessment of treated water in the Netherlands includes the AOC test in combination with the BFR test. The growth-promoting properties of synthetic materials in contact with treated water are determined with the Biofilm Formation Potential (BFP) test. A complete understanding of regrowth processes enabling to define appropriate control measures requires further research including: (i), the effect of reactive surfaces on the availability of slowly degradable compounds, and (ii), improvement of mathematical models describing regrowth processes.

**Keywords:** biofilm, biological stability, drinking water, easily assimilable organic carbon (AOC), materials, regrowth

## 1.Introduction

Drinking water supply companies make many efforts to achieve a good quality of drinking water from water sources containing a variety of chemical and biological contaminants. Multiple treatment barriers, including (combinations of) physical, chemical and biological processes, are applied to reduce the concentrations of these contaminants below the Maximum Acceptable Concentration (MAC) values as defined in national legislation. Maintaining water quality during distribution is another challenge, because many factors may impair the quality of treated water in the distribution system. Factors of major importance are: (re)contamination from outside the distribution system, effects of materials in contact with treated water, and biological and/or physico-chemical processes. Good engineering practices, e.g., selection of proper materials and appropriate working procedures, are essential to protect drinking water from contamination. Water quality monitoring is used to detect the efficacy of the preventive measures and to intervene when needed to maintain water quality. Biological processes in the distribution system may cause hygienic (growth of opportunistic pathogens), aesthetic (taste and odour)

*Water, Air, and Soil Pollution* **123:** 25–34, 2000.

or technical problems (corrosion). The main approaches in controlling these processes are: (i), maintaining a disinfectant residual and (ii), distributing biologically-stable drinking water in a system consisting of biostable materials in contact with water. Biostability is defined as the inability of water or a material in contact with water to support microbial growth in the absence of a disinfectant (Rittmann and Snoeyink, 1984). Maintaining a disinfectant residual can affect taste and odour (Bryan *et al.,* 1973) and/or results in the formation of disinfection by products with toxic properties (Bull and Kopfler, 1991). On the other hand, achieving biological stability, which implies a far reaching removal of biodegradable compounds, makes high demands on water treatment and on the quality of distribution system materials coming into contact with treated water, respectively. Methods and criteria to define biostability are required to either design or optimise water treatment and distribution systems for this purpose. Furthermore, a fundamental understanding of regrowth processes is needed for the development and interpretation of tests to assess the growth-promoting properties of water and materials in contact with water. The need to collect data about effects of measures in treatment and distribution has resulted in the development of a multitude of methods for measuring the growth-potential of drinking water. To a lesser extent also methods for determining the biostability of materials in contact with water have become available. This paper summarizes the complexity of regrowth processes and lists a number of tests developed to assess biostability of treated water. Also a brief description is given of the methods used in the Netherlands for biostability assessment and further research needs are identified.

## 2. Regrowth Processes and Methods for Biostability Assessment

### 2.1. REGROWTH PROCESSES

Problems caused by regrowth of micro-organisms range from colour caused by iron-precipitating bacteria (Clark *et al.,* 1967) to disease caused by opportunistic pathogens e.g., *Legionella* (Wadowsky *et al.,* 1982). Key processes in regrowth are: formation of biofilms, detachment of (micro)organisms from the pipe wall and accumulation of sediments, respectively. The presence of biodegradable compounds is a major driving force in these regrowth processes but also other environmental conditions (e.g., temperature, system hydraulics) and physico-chemical processes (adsorption, oxidation/reduction, sedimentation) have a significant effect. Elucidation of microbial growth processes in drinking water distribution systems is hampered by a number of reasons. These reasons include: (i), the large variety of suspended and attached micro-organisms involved; (ii), undefined compounds serving as energy sources at unknown concentrations; (iii), effects of different materials in contact with water; (iv), the hydraulic complexity of distribution systems; (v), inaccessibility of sampling locations. Properties of both the predominant and the problem organisms are insufficiently known to predict their behaviour (growth, survival) in the described environment. Examples of problem organisms include *Legionella* spp. growing in hot water systems and coliforms multiplying in distribution systems (LeChevallier, 1990). Effects of only a few factors (mainly water temperature and disinfectants) have been studied for these organisms. For certain types of bacteria, e.g., *Pseudomonas* spp., *Spirillum* sp., *Flavobacterium* spp., *Aeromonas* spp. and also coliforms, some information has been collected about their ability to multiply at low substrate concentrations (Van der Kooij *et al.*, 1982; Van der Kooij and Hijnen, 1984, 1985, 1988; Camper *et al.*, 1991). However, substrate utilisation kinetics and conditions favouring growth or survival remain largely unknown for most bacteria multiplying in distribution systems. Based on the assumption that the concentration of biodegradable organic compounds is the main factor limiting regrowth, microbiologists have developed methods to assess the growth potential of treated water.

## 2.2. METHODS AND PARAMETERS TO ASSESS THE GROWTH-PROMOTING PROPERTIES OF TREATED WATER

Studies of bacterial multiplication in treated water aiming at determining the growth potential began at the end of the 19$^{th}$ century (Frankland and Frankland, 1894). It was observed that the number of culturable bacteria increased much more in stored samples of high quality deep well water with a very low concentration of organic compounds than in river water samples. The assumption was made that well water contained a relatively high concentration of biodegradable compounds, because of the absence of bacteria. However, already at that time it was known that well-closed bottles should be used for growth measurements instead of bottles with cotton plugs. (Frankland and Frankland, 1894). In the Netherlands, Beijerinck (1891) suggested to study the growth-promoting properties of water by inoculating selected pure cultures in sterilised samples of the water to be tested. Heymann (1928) described a method based on determining the effect of a series of passages through sand columns on the concentration of organic compounds in the water, measured as permanganate value. However, this method was not widely applied. Interest in regrowth phenomena strongly increased after 1970 when ozonation and granular activated carbon filtration were introduced in water treatment for the removal of persistent organic pollutants. In 1982, a method for determining the heterotrophic growth potential of treated water was described, which was based on determining the maximum level of growth of a selected pure culture in a water sample collected and contained in thoroughly cleaned, glass-stoppered Erlenmeyer flasks (Van der Kooij *et al.*, 1982). Within one decade many alternative methods for assessing the growth potential of drinking water were developed in Europe and in the USA (Huck, 1990). Additional methods have been developed since 1990.

The methods can be classified on the basis of the test parameters and test conditions including:

- chemical (e.g., Dissolved Organic Carbon) or biological (biomass) parameter;
- suspended or attached (biofilm) growth;
- batch test (static) or plug-flow (dynamic) system;
- short /medium/long test period;
- on location (in situ) or in the laboratory;
- nature of biomass (pure cultures or mixed population);
- nature of biomass parameter (heterotrophic plate count, total direct count, adenosinetriphosphate, turbidity).

Table I lists the characteristics of the most commonly used tests.

TABLE I
Characteristics of methods for assessing the growth potential of treated water

| Method | Characteristics | Reference |
|---|---|---|
| AOC | B, PC/HPC; SG/ST; L/MD/LD | Van der Kooij *et al.*, 1982,1984 |
| | B, MP/ATP; SG/ST; L/MD | Stanfield and Jago, 1986 |
| | B, PC/ATP; SG/ST; L/MD | LeChevallier *et al.*, 1993 |
| | B, MP/TU; SG/ST; L/MD | Werner, P. 1985 |
| BGP | B, MP/TDC; SG/ST, L/LD | Servais, *et al.*, 1987 |
| BDOC | C (DOC), MP; AG/ST; L/MD | Joret and Levy, 1986 |
| | C (DOC), MP; AG/DT; IS/SD | Lucena *et al.*, 1990 |
| BFR | B, MP/ATP; AG/DT; IS/LD | Van der Kooij *et al.*, 1995 |

AG, attached growth (biofilm); AOC, assimilable organic carbon; ATP, adenosinetriphosphate; B, biomass-based parameter; BDOC, biodegradable dissolved organic carbon; BFR, biofilm formation rate; BGP, bacterial growth potential; C, chemical parameter (organic carbon); DT, dynamic (flow-through) test; DOC, dissolved organic carbon; HPC, heterotrophic plate count; IS, in situ; L, laboratory; MP, mixed population (indigenous flora); PC, pure culture(s); SG, suspended growth; SD, MD, LD, short (hours), medium (days), long (weeks) duration of test; ST, static (batch); SU, suspended growth; TDC, total direct count; TU, turbidity.

All these tests have their advantages and limitations, which will not be evaluated in this paper. Only a few comments will be made here:

- Most tests are batch tests conducted in the laboratory under defined conditions. These tests give information about the growth potential;
- The growth potential of treated water may also be affected by the presence of inorganic compounds (e.g., ammonia and sulfides) and methane. These compounds are not included in AOC and BDOC tests;

- Biofilm formation as occurring in distribution systems can be simulated with a variety of techniques and devices under conditions enabling quantitative measurements of biofilm parameters, e.g., Rototorque system (Van der Wende *et al.,* 1989), Robbins device, coupon test and a biofilm monitor (Van der Kooij *et al.,* 1995);
- A combination of growth tests, determining the (effects of) concentrations of rapidly and more slowly available compounds, as well as chemical analysis (e.g., for ammonia and methane), may be needed to assess the biostability of treated water. In the Netherlands biostability assessment of water is conducted using the AOC test and the BFR test;
- Assessment of biological stability of treated water may also require the inclusion of other parameters e.g., the concentration of biomass in the water entering the distribution system;
- The growth-promoting properties of materials in contact with drinking water should also be assessed.

## 3. Biostability Assessment in the Netherlands

### 3.1. AOC TEST

Assessment of the concentration of easily assimilable organic ca0rbon (AOC) is based on growth measurements of two selected pure cultures in a sample of pasteurised water contained in a thoroughly cleaned glass stoppered Erlenmeyer flask. The strains used in the test are: *Pseudomonas fluorescens* strain P17, which is capable of utilising a wide range of low molecular weight compounds at very low concentrations ( Van der Kooij *et al.,* 1982) and a *Spirillum* sp. strain NOX, which only utilises carboxylic acids (Van der Kooij and Hijnen, 1984). These compounds appear to dominate the easily biodegradable low molecular weight compounds; carbohydrates were observed in much lower concentrations (Van der Kooij and Hijnen, 1985). In most types of drinking water the AOC test requires an incubation period of 1 to 2 weeks before the test strains reach their maximum level of growth. This period exceeds the residence time of drinking water in the distribution system. However, observations on water sampled from a number of distribution systems have shown that AOC concentrations decline rapidly in the distribution system, indicating that biofilm processes play an important role in AOC uptake. AOC reduction was limited at concentrations below 10 μg of C/l and at these low levels the heterotrophic plate counts (HPC) remained low. Based on these observations it was concluded that AOC values below 10 μg of C/l are indicative for drinking water with a limited regrowth potential for bacteria contributing to HPC values (van der Kooij, 1992).

The availability of AOC is also demonstrated in an AOC-reduction test. In such a test additional water samples are collected and incubated at 15°C without pasteurisation. After a defined incubation period (e.g., 7 days) these samples are also pasteurised and the AOC concentration is determined. Typical results are presented in Figure 1, which shows the decrease in AOC-concentration in ozonated water to a level of about 10 μg of C/l. Even in treated water with AOC concentrations below 10 μg of C/l some further reduction was observed.

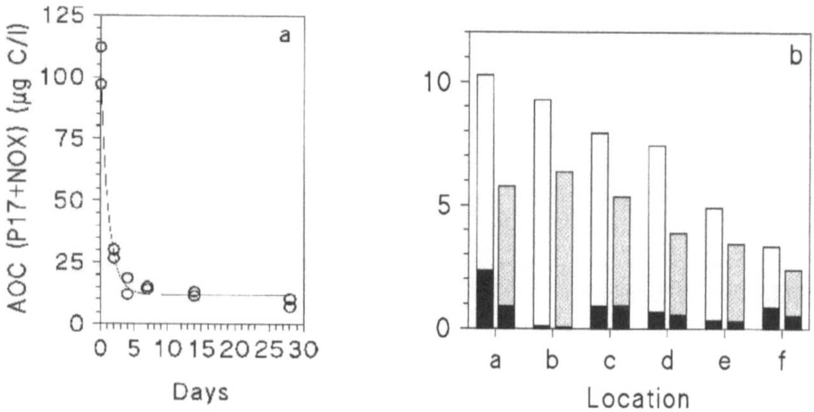


Fig. 1. a: AOC decrease in unpasteurised samples of ozonated water stored at 15 °C; b: AOC decrease in samples of 6 types of treated water stored at 15°C for 7 days. Black proportion of the bars is the fraction of AOC utilised by strain P17; the dotted bars are AOC values in the stored samples.

### 3.2. BIOFILM FORMATION

The Biofilm Formation Rate (BFR) value of (treated) water is assessed with a biofilm monitor, consisting of a vertically-placed glass column (L = 60 cm; Ø 2.5 cm) containing glass cylinders on top of each other. Water flows downward through this column with a flow rate of 0.2 m/s. Cylinders are collected from the column at regular intervals, sonicated in sterile water and the suspended biomass concentration is assessed with ATP-analysis (Van der Kooij *et al.*, 1995). Subsequently the BFR value is calculated from the linear increase of the biomass concentration on the surface of glass cylinders as a function of time. BFR values in treated water usually range from below 1 pg ATP/cm$^2$.d (slow sand filtrate) to about 75 pg ATP/cm$^2$.d. The highest BFR values in treated water have been observed in ground water supplies using anaerobic ground water as the raw water source. This water is treated by intensive aeration(s) and filtration(s), respectively. The AOC concentration in treated water in most cases is clearly below 10 μg C/l. Dosage experiments revealed that a concentration of 10 μg acetate-C/l causes a BFR value of 365 pg ATP/cm$^2$.d (Van der Kooij *et al.*, 1995b). Hence, concentrations of a few μg of easily available compounds can cause a significant biofilm formation. Such compounds may also include methane, ammonia and sulfides. Therefore, combining the AOC test with an assessment of the BFR value gives information about the growth potential for heterotrophic bacteria and the rate at which the water can promote biofilm formation. In unchlorinated ground water supplies a clear relationship has been found between BFR values and the regrowth of aeromonads in the distribution system. From this relationship it was calculated that the risk of exceeding the guideline value for aeromonads in the distribution system (200 CFU/100 ml as 90-percentile over a one-year period) was less than 20% when the BFR value was less than 10 pg ATP/cm$^2$.d (Van der Kooij *et al.*, 1999).

## 3.3. EFFECTS OF SYNTHETIC MATERIALS

Biofilm formation on the wall of the pipes is caused by compounds present in the water and by compounds originating from the pipe material. In the Netherlands PVC is the main pipe material. The Biofilm Formation Potential (BFP) test has been developed to assess the growth promoting properties of synthetic materials in contact with water (Van der Kooij *et al.*, 1994). In this test the ATP concentration on pieces of material incubated in slow sand filtrate at 25°C is assessed as a function of time. Typical results are presented in Figure 2.

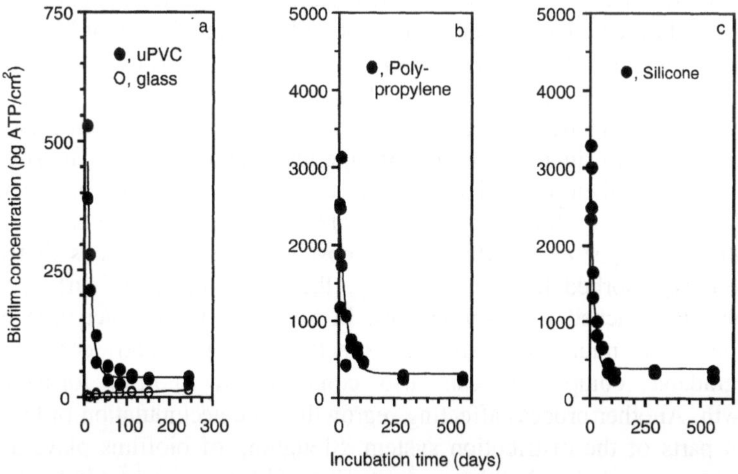


Fig. 2. The biofilm concentration (pg ATP/cm$^2$) on glass, unplasticized PVC (a), polypropylene (b), silicone tubing (c), as a function of time in the BFP test conducted at 25 °C.

The ATP-concentration on the material declines as a function of time. The plateau level, which for many materials is reached after about 8 to 12 weeks is defined as the BFP-value. Unplasticized PVC has a very low BVP value (< 100 pg ATP/cm$^2$). BFP values of PE materials range from about 350 to 2700 pg ATP/cm$^2$.

### 3.4. UNIFIED BIOFILM APPROACH

Assessment of the concentration of biofilm on the pipe wall gives direct information about the biological stability of the involved distribution system. Biofilm concentrations ranging from about 40 pg ATP/cm$^2$ to 5800 pg ATP/cm$^2$ (median value: 650 pg ATP/cm$^2$) have been observed on segments of PVC pipes of about 20 distribution systems, mainly ground water supplies. Biofilm concentrations on pipe walls, BFR values of the water and the BFP-value of materials provide data based on one and the same parameter (biomass activity). This suit of cohering parameters has been defined as the Unified Biofilm Approach for assessing the biostability of water and materials in contact with water (Van der Kooij *et al.*, 1999).

## 4. Evaluation

In the Netherlands water supply companies aim at limiting regrowth by distributing biologically stable water in distribution systems using biostable materials in contact with water. Biostability assessment of water includes the AOC test and the BFR test. The biostability of materials is tested with the BFP test. From the data obtained in the investigations, values for the test parameters have been derived which either are indicative for a high degree of biostability or for situations where biofilm formation causes problems. The use of one simple biomass parameter (ATP) in the biofilm-formation based tests and in observations in practice makes it easier to draw conclusions from the obtained data, especially in situations with inert pipe material, e.g., PVC. The situation is more complex when a reactive material, e.g., cast iron comes in contact with water. It has been shown that such material considerably enhances biofilm formation (Camper *et al.*, 1996). The most likely explanation is that organic compounds absorbed from the water by the corroding metal surface become available to attached micro-organisms. In this way, the contact time between organics and micro-organisms is greatly enlarged and certain slowly biodegradable compounds can also contribute to biofilm formation and regrowth. Another process affecting regrowth is the accumulation of biomass in certain parts of the distribution system. Sloughing of biofilms plays a role in sediment accumulation, but the concentration of biomass present in treated water may also contribute to this process. Sediment accumulation is also affected by hydraulic conditions (e.g., fluctuations in flow rate).

The described regrowth processes and test procedures clearly show that biostability is a multidimensional quality characteristic of treated water. One single test is not sufficient to define biostability. Further investigations needed to elucidate regrowth processes and biostability assessment include:

- the effect of surface reactivity on biofilm formation and biostability;

- the relationship between AOC uptake and biofilm formation in distribution systems.

Mathematical modelling of biofilm formation and regrowth processes in distribution system is another approach in elucidating regrowth phenomena (Dukan *et al.*, 1996; Servais *et al.*, 1995). However, a number of processes as described above are not included in these models, despite their complexity. Furthermore, in these models only BDOC is used as the critical parameter for regrowth potential. Improvement of mathematical modelling of regrowth processes is needed to enable improvement of distribution system design and/or the definition of effective preventive and curative measures for regrowth limitation.

**Acknowledgements**

The methods for determining the biostability of drinking water and materials in the Netherlands have been developed in the framework of the Joint Research Program of the Water Supply Companies in the Netherlands. Many people, including Harm Veenendaal, Hans Vrouwenvelder and Wim Hijnen contributed to the investigations.

## References

Beijerinck, M. W.:1891, "Qualitative und quatitative microbiologische Analyse," *Centralblatt für Bakteriologie und parasitenkunde* **X,** 723-727.

Bryan, P. E., Kuzminski, L. N., Sawyer, F. M. and Feng, T. H.: 1973, "Taste thresholds of halogens in water", *Jour. Am. Water Works Assoc.* **65**, 363-368.

Bull, R. J. and Kopfler, F. C.:1991, *Health effects of disinfectants and disinfection by-products,* American Water Works Research Foundation, Denver, USA.

Camper, A. K., McFeters, G. A., Characlis, W. G. and Jones, W. L.:1991, "Growth kinetics of coliform bacteria under conditions relevant to drinking water distribution systems", *Appl. Environ. Microbiol.* **57,** 2233-2239.

Camper, A. K., Jones, W. L. and Hayes, J. T.: 1996, "Effect of growth conditions and substratum composition on the persistence of coliforms in mixed-population biofilms," *Appl. Environ. Microbiol.* **62,** 4014-4018.

Clark, F. M., Scott, R. M. and Bone, E.: 1967, "Heterotrophic iron precipitating bacteria.," *Jour. Am. Water Works Assoc.* **59**, 1036-1042.

Dukan, S., Levy, Y., Piriou, P., Guyon, F. and Villon, P.: 1996, "Dynamic modelling of bacterial growth in drinking water networks," *Wat. Res.* **30**, 1991-2002.

Frankland, P. and Frankland, P.: 1894, *Micro-organisms in water; their significance, identification and removal,* Longmans, Green and Co. London, New York.

Heijmans, J. A.: 1928, "De organische stof in het waterleidingbedrijf", *Water en Gas* **12,** 61-65/69-72.

Huck, P. M.: 1990, "Measurement of biodegradable organic matter and bacterial growth potential in drinking water", *Jour. Am. Water Works Assoc.* **82**(7), 78-86.

Joret, J. C. and Levy, Y.: 1986, "Méthode rapide d'évaluation du carbone éliminable des eaux par voie biologique", *Trib. Cebedeau* 510, **39**, 3-9.

LeChevallier, M. W.: 1990, "Coliform regrowth in drinking water: a review", *Jour. Am. Water Works Assoc.* **82** (11), 74-86.

LeChevallier, M. W., Shaw, N. E., Kaplan, L. E. and Bott, T.: 1993, "Development of a rapid assimilable organic carbon method for water", *Appl. Environ. Microbiol.* **59**, 1526-1531.

Lucena, F., Frais, J. and Ribas, F.: 1990, "A new dynamic approach to the determination of biodegradable dissolved organic carbon in water", *Environmental Technol.* **12**, 343-347.

Rittmann, B. E. and Snoeyink, V. L.:.1984, "Achieving biologically stable drinking water", *Jour. Am. Water Works Assoc.* **76:10**, 106-114.

Servais, P., Billen, G. and Hascoët, M. C.: 1987, "Determination of the biodegradable fraction of dissolved organic matter in waters", *Wat. Res.* **21**, 4:445-450.

Servais, P., Laurent, P., Billen, G. and Gatel, D.: 1995 "Development of a model of BDOC and bacterial biomass fluctuations in distribution systems", *Revue des Sciences de l'Eau.* **8**, 427-462.

Stanfield, G. and Jago, P. H.: 1987, *The development and use of a method for measuring the concentration of assimilable organic carbon in water*, WRc-report PRU 1628, WRc Medmenham.

Van der Kooij, D., Visser, A and Oranje, J. P.: 1982, "Multiplication of fluorescent pseudomonads at low substrate concentrations in tap water", *Antonie van Leeuwenhoek* **48**, 229-243.

Van der Kooij, D., Visser, A., and Hijnen, W. A. M.: 1982, "Determining the concentration of easily assimilable organic carbon in drinking water", *Jour. Am. Water Works Assoc.* **74**, 540-545.

Van der Kooij, D. and Hijnen, W. A. M.: 1984, "Substrate utilisation by an oxalate-consuming Spirillum species in relation to its growth in ozonated water", *Appl. Environ. Microbiol.* **47**, 551-559.

Van der Kooij, D. and Hijnen, W. A. M.: 1985, "Determination of the concentration of maltose- and starch-like compounds in drinking water by growth measurements with a well-defined strain of a Flavobacterium species", *Appl. Environ. Microbiol.* **49**, 765-771.

Van der Kooij, D. and Hijnen, W. A. M.: 1988, "Nutritional versatility and growth kinetics of an Aeromonas hydrophila strain isolated from drinking water", *Appl. Environ. Microbiol.* **54**, 2842-2851.

Van der Kooij, D.: 1992, "Assimilable organic carbon as an indicator of bacterial regrowth", *Jour. Am.* Water Works Assoc. **84**(2), 57-65.

Van der Kooij, D., and Veenendaal, H. R.: 1994, "Assessment of the biofilm formation potential of synthetic materials in contact with drinking water during distribution", Proc. Am. Water Works Assoc. Water Qual. Technol. Conf. Miami Florida 1993, pp. 1395-1407.

Van der Kooij, D., Veenendaal, H. R., Baars-Lorist, C., van der Klift, H. W. and Drost, Y. C.: 1995, "Biofilm formation on surfaces exposed to treated water", *Wat.Res.* **29**, 1655-1662.

Van der Kooij, D., Vrouwenvelder, H. S. and Veenendaal, H. R.: 1995b, "Kinetic aspects of biofilm formation on surfaces exposed to drinking water", *Wat. Sci. Tech.* **32**, 61-65.

Van der Kooij, D., van Lieverloo, J. H. M., Schellart, J. A. and Hiemstra, P.: 1999, "Maintaining quality without a disinfectant residual", *Jour. Am. Water Works Assoc.* **91**(1), 55-64.

Van der Wende, E., Characklis, W. G. and Smith, D. B.: 1989, "Biofilms and bacterial drinking water quality", *Wat. Res.* **23**, 1313-1322.

Wadowsky, R. M., Yee, R. B., Mezmar, L., Wing, E. J., and Dowling, J. N.: 1982, "Hot water systems as sources of Legionella pneumophila in hospital and non-hospital plumbing fixtures", *Appl. Environ. Microbiol.* **43**, 1104-1110.

Werner, P.: 1985, "Eine Methode zur Bestimmumg der Verkeimungsneigung von Trinkwasser," *Vom Wasser,* **65**, 257-270.

# FROM MEDIA TO MOLECULES: NEW APPROACHES TO THE DETECTION OF MICRO-ORGANISMS IN WATER

C.R. FRICKER
*Thames Water Utilities Ltd., Spencer House Laboratory, Manor Farm Rd., Reading, Berkshire RG2 OJN, UK*
*Tel:0118 923 6211; Fax:0118 923 6311; e-mail:CFricker@compuserve.com*

**Abstract.** Present methods for the detection of micro-organisms in the environment are slow, inefficient and often unreliable. Alternative approaches which are reliable and rapid, enabling results to be obtained within one working day, are required. The development of molecular techniques, in particular *in situ* hybridisation offers the potential for rapid and specific assays. This paper describes the use of oligonucleotide probes targeted against the 16S and 18S ribosomal RNA molecules of *Escherichia coli* and *Cryptosporidium parvum* for rapid and specific detection.

*In situ* hybridisations with biotinylated peptide nucleic acid (PNA) probes in combination with the tyramide signal amplification (TSA) system enabled the specific detection of *Escherichia coli* and *Cryptosporidium parvum* within 3 hours. The *C. parvum* assay provided a species-specific alternative to the currently available fluorescent antibody approach.

**Keywords**: probes, *C.parvum*, *E.coli*, *in situ* hybridisation, rapid methods, water

## 1. Introduction

Identification and detection of coliforms and *E. coli* are the most commonly performed tests within the water industry. At present microbiological water quality is determined using one of a number of methods including membrane filtration, Colilert or most probable number methods (Anon, 1994; Cowburn *et al.*, 1994; Walter *et al.*, 1994). Other methods have been investigated such as the polymerase chain reaction (Fricker and Fricker, 1994) and impedance monitoring by using enzyme specific substrates (Timms *et al.*, 1996), but these methods have produced unreliable results. Colilert is quick and easy to perform, produces favourable results when compared to membrane filtration and can be used by any routine laboratory. However, it requires 18 hours for a result to be obtained. There is a need for rapid tests which will detect and correctly identify *E. coli* so that water providers can respond quickly to bacteriological failures of water quality and thereby help protect public health.

Routine monitoring of water samples also involves screening for the presence of the waterborne parasite *Cryptosporidium parvum.* In the United Kingdom, regulations are in force which compel water providers to test treated water for the presence of these organisms on a daily basis at treatment plants

*Water, Air, and Soil Pollution* **123:** 35–41, 2000.

which are considered to be at risk. Current methods for the detection of *C. parvum* rely heavily on the use of immunological techniques which employ fluorescently-labelled monoclonal antibodies and epifluorescence microscopy or flow cytometry (Vesey *et al.*, 1993; Hofman *et al.*, 1997; Clancy *et al.*, 1999). These methods are labour intensive and may give misleading results as the antibodies are genus specific and thus other members of the genus *Cryptosporidium* may be detected.

The development of molecular techniques based on ribosomal RNA has enabled the specific detection micro-organisms in natural environments, and methods have improved considerably in the last few years (Giovannoni *et al.*, 1988; De Long *et al.*, 1989).The use of 16S and 18S ribosomal RNA sequences for phylogenetic studies (Woese 1987) has provided a large database from which species, genus and strain specific regions can be identified. There are many advantages to the use of rRNA as a target for species identification. Ribosomal RNA is present in all organisms, there is no evidence to date of lateral gene transfer and there is a large amount of sequence data available. The ribosomal RNA molecule contains regions which are conserved, semi-conserved and highly variable thus allowing probes of varying specificity to be designed..

The development of fluorescence *in situ* hybridisation techniques (FISH) for bacterial detection has enabled some organisms to be identified in their natural habitat without the need for enrichment techniques (Amann *et al.*, 1990; Manz *et al.*, 1993). The method involves the use of nucleic acid probes labelled with fluorochromes to target rRNA within morphologically intact cells. Individual cells with hybridised probes can then be identified by the use of epifluorescence microscopy and flow cytometry enabling not only species-specific detection but quantitative detection.

The application of *in situ* hybridisation techniques for the detection of specific bacteria in environmental samples has been limited by the low ribosome content of some cells found in water which has resulted in very low levels of probe-conferred fluorescence. As a result, more sensitive and reliable methods are necessary.

Previous studies have shown the potential use of biotinylated Peptide Nucleic Acid (PNA) oligonucleotide probes directed against the 16S rRNA molecule for the *in situ* detection of *E. coli* in water (Prescott and Fricker, 1999). As PNA oligonucleotide probes have been shown to bind to complementary RNA with increased efficiency, the use of biotinylated PNA oligonucleotides in combination with the tyramide signal amplification (TSA) signal amplification system might provide a more rapid and sensitive method for the detection of *E. coli in situ*. The specific detection of *C. parvum* with oligonucleotide probes labelled with fluorescein and Texas Red , directed against the 18S molecule has also been examined (Vesey *et al.*, 1998).

This paper describes the use of PNA probes for the specific detection of *C.parvum* and *E.coli* in water.

## 2. Materials and Methods

### 2.1. DETECTION OF *E. coli* IN TAP WATER

Routine samples of tap water (500ml) were taken from a variety of sites within the Thames water distribution system into bottles containing sodium thiosulphate to neutralise the effects of chlorine. Samples were then spiked with approximately $10^3$ *E.coli* cells/ml, and 10ml was then filtered through 'Costar' metallic membrane filters and hybridisation performed with the biotinylated PNA specific probe (Table I) Serial dilutions were plated onto $R_2A$ agar plates which were incubated for 3 days at 37°C to enable accurate counts of viable *E.coli* to be obtained. In addition to the experiments with pure cultures of *E.coli* experiments were conducted in which other organisms such as *Klebsiella*, *Enterobacter* and other *Escherichia* species were added to the tap water along with *E.coli*.

### 2.2.HYBRIDISATION WITH *E.coli* SPECIFIC PROBE

Membrane filters were placed on a membrane filter pad (Whatman, Maidstone, UK) which had been pre-soaked with 6% paraformaldehyde and the cells allowed to "fix" at room temperature for 20 minutes. Membranes were then rinsed three times with sterile distilled water and the cells lysed by overlaying the filter with 300 µl of lysozyme solution containing 50µg/ml of lysozyme and incubating at room temperature for 10 minutes. After a further wash in sterile distilled water, the membranes were then hybridised at $50^0$C in 10mM sodium phosphate buffer pH 7.2 and 10µmoles of PNA specific probe for 30 minutes. Following hybridisation, the membranes were washed at $50^0$C for 15 minutes in 500µl of 20mM Tris-HCl, 0.01%SDS, 180mM NaCl, 5mM EDTA. The membranes were given three additional rinses in sterile distilled water.

### 2.3. DETECTION OF *IN SITU* HYBRIDISED CELLS

The PNA biotinylated probe was detected by use of the TSA signal amplification kit (NEN Life Sciences). A modification of the method provided with the TSA Kit was used. Briefly membranes were blocked by incubation in 500µl of 0.1% blocking agent, 0.1% Tris-HCl pH 7.5, 0.15 M NaCl (TNB buffer) at RT for 15 minutes. A 1:100 dilution of streptavidin-HRP in TNB buffer was then added and incubation continued for a further 30 minutes. The membranes were washed three times for 5 minutes each in 0.1M Tris-HCL pH 7.5, 0.15M NaCl, 0.05% Tween 20 (TNT) buffer. Fluorescein tyramide (300 µl of a 1:50 dilution) in 1X amplification diluent (provided in the TSA Kit) was then added and the membranes incubated in the dark at room temperature for 10 minutes. After three 5 minute washes at room temperature in TNT, buffer membrane filters were mounted on glass microscope slides in a low fluorescence mounting medium, a coverslip placed on top and the membranes viewed by epifluorescence microscopy. The images were captured by use of a cooled charge-coupled device camera (CCD) and analysed with image analysis software (Digital Pixel, Brighton, UK). The images were then saved in TIFF format for importing into Photodeluxe version 2.0 (Adobe, Mountain View, California).

## 2.4. 18S rRNA TARGETED FISH FOR *C.parvum*

Aliquots (10µl ) of a fresh (<4 weeks old) stock suspension of $1x10^8$ *C. parvum* oocysts (Harley Moon strain, University of Arizona) were suspended in 100µl of DEPC treated phosphate buffered saline (PBS, pH7.2). The cell wall was permeabilised by the addition of 50% ethanol followed by heating at $80^0$C for 10minutes.The oocysts were then washed twice in 1ml of ice cold DEPC treated PBS containing RNASEin and resuspended in hybridisation buffer containing 0.9M NaCl, 20mM Tris-HCl pH 7.2, 0.5% SDS, 30U RNASEin/ml and 1pmol/µl of species specific probe IA/IIA as shown in Table I. Hybridisation was performed at $48^0$C for two hours with probes labelled with Texas Red or FITC. Oocysts hybridised with probes labelled with Texas Red were then washed and resuspended in 1ml of PBS and stained with 10ul of FITC-labelled anti-*Cryptosporidium* monoclonal antibody (Cellabs.) on ice for 10minutes, filtered through a 2µm polycarbonate filter membrane and washed with 500µl of DEPC treated PBS. Oocysts labelled with FITC-labelled probe were washed and concentrated on membranes. The membranes were placed on glass slides, mounted in low fluorescence mounting medium and examined by epifluorescence microscopy.

## 2.5. USE OF THE CHEM*SCAN* SYSTEM TO DETECT FLUORESCENTLY LABELLED *CRYPTOSPORIDIUM*

Chem*Scan* (Chemunex, France) is a laser scanning device used to detect fluorescent events on a membrane surface. A laser is directed across the surface of a membrane by oscillating mirrors which ensure that the entire membrane surface is scanned without the possibility that any area of the membrane is missed. Fluorescent events are detected and depending on the size of the particle, the amplitude of the signal and the ratio of red to green fluorescence are rejected or marked as potential *Cryptosporidium* oocysts. A membrane map is created by the instrument which accurately marks the exact co-ordinates of any particles having characteristics similar to *Cryptosporidium* oocysts. Membranes are then transferred to an epifluorescence microscope with a motorised stage and the Chem*Scan* moves in turn to each of the co-ordinates which are present on the membrane map. Particles are then viewed and confirmed (or not) as oocysts. Each particle confirmed as an oocyst is then viewed with appropriate filters to determine if hybridisation with the Texas Red-labelled probe had occurred.

TABLE I

Species specific primers targeted against the 16S and 18S rRNA molecules used in the hybridisation assays

| Primer Sequence | Organism |
|---|---|
| 5'GCAAAGCAGCAAGCTC | *Escherichia coli* NCTC 9001 |
| 5'GACAATTATTAAATATATTTTATAAAACTAC | *Cryptosporidium parvum* IA |
| 5'ACAACATAT TGGACATTATAATGA TT | *Cryptosporidium parvum* IIA |

## 3. Results and Discussion

*In situ* hybridisation experiments with the biotinylated PNA probe in combination with TSA were performed on tap water samples that were spiked with *E. coli*. Brightly labelled cells were detected in under 3 hours. The 30 minute hybridisation time needed for annealing of the PNA oligonucleotide is considerably shorter than that used for biotinylated DNA probes resulting in a significant reduction in the detection time. Some initial experiments on the probe specificity were also carried out *in situ*. When mixed populations of cells, containing *E. coli*, *Klebsiella*, *Enterobacter* and other *Escherichia* species were examined, some cells fluoresced whilst others did not. The number of cells detected by fluorescence was in agreement with those detected with plate counts on $R_2A$ agar medium and whilst these specificity experiments need to be continued, the fact that other *Escherichia* species were not detected suggests that the probe may be completely species specific. More work is required however to determine if the probe reacts with all strains of *E.coli*.

The use of PNA probes targeted against the 16S rRNA molecule for the specific detection of *E. coli* offers a fast, efficient alternative to conventional

DNA approaches, and results can be available in 3 hours. However, the procedure is labour intensive and the method would be improved dramatically if the number of steps can be reduced.The possibility of utilising probes labelled with fluorochromes and quenchers might allow this. In these systems only a probe which has hybridised will fluoresce and this removes the need for a number of wash steps. Despite these shortcomings, the ability to perform this assay quickly will enable response times to be improved, enabling better protection of public health.

The 18S rRNA sequence of *C. parvum* is known to be highly variable and it is now thought that at least two genotypes exist. In the present study, two oligonucleotide probes were assessed for their ability to hybridise to the *C.parvum* isolates examined. When the probe designated IIA was used, no hybridisation signal was obtained. However when probe IA was used oocysts were brightly labelled. It is likely therefore that these probes are able to distinguish between the two genotypes effectively. This warrants further investigation. Oocysts that were hybridised with oligonucleotide probes labelled with Texas Red were significantly brighter than those oocysts that were labelled with fluorescein. A universal probe, used as a positive control produced fluorescent signals which were significantly brighter than the species specific probes. It is thought that this is a result of the binding efficiency of these probes and its target within the rRNA molecule.

The use of FISH may enable rapid detection of micro-organisms but the use of epifluorescence microscopy for end point detection is too time consuming for routine use. The use of the Chem*Scan* system enables detection of *Cryptosporidium* oocysts based on their reaction with FITC-labelled monoclonal antibody and the 18S rRNA conferred species-specific fluorescence. This system is promising for the routine detection of *Cryptosporidium* in water.

## 4. Conclusion

This study has demonstrated that the use of fluorescently labelled PNA probes can be used for the rapid detection of microbes in water. Probes were used effectively to demonstrate the presence of both *E.coli* and *C.parvum* which are two of the most commonly sought organisms in water. The technology is relatively simple and could be applied to the detection of other organisms. Use of the Chem*Scan* system enables automated detection of fluorescently labelled organisms.

## References


Amann, R.I., Binder, B.J., Olsen, R. J., Chisolm, S. W., Devereux, R. and Stahl, D. A.: 1990, *Appl. Envir. Microbiol.* **56**,1919-1925.

Anon,: 1994, The Microbiology of Water 1994, Part 1 - Drinking Water, HMSO, London.

Clancy, J. L., Bukhari, Z., McCuin, R. M., Matheson, Z. and Fricker, C. R.: 1999, *JAWWA*. **91**, 60-68.

Cowburn, J. K., Goodall, T., Fricker, E., Walter, K. S. and Fricker, C. R.: 1994, *Lett. Appl. Microbiol.* **19**, 50-52.

De Long, E. F., Wickham, G. S. and Pace N. R.: 1989, *J. Bact.* **170**,720-726.

Fricker, E. J. and Fricker, C. R.: 1994, *Lett Appl. Microbiol.*, **19**, 44-46.

Giovannoni, S. J., DeLong, E. F., Olsen, G. N. and Pace, N. R.: 1988, *J. Bact.* **170**,720-726.

Hoffman, R. M., Standridge, J. H., Prieve, A. F., Cucunato, J. C. and Bernhardt, M.: 1997, *JAWWA* **89**, 104-111.

Lindquist, H. D. A.: 1997, *Water Research* **31**,2668-2671.

Manz, W., Szewzik, U., Ericsson, P., Amann, R. I., Schleifer, K. H. and Stenstrom T. A.: 1993, *Appl. Environ. Microbiol.* **59**,2293-2298.

Prescott, A. M. and Fricker, C. R.: 1999, *Mol. Cell. Probes* **13**, 261-268.

Timms, S., Colquhoun, K. O. & Fricker, C. R.: 1996, *J. Microbiol. Meth.* **26**, 125-132.

Vesey, G., Slade, J. S., Byrne, M., Shepherd, K., Dennis, P. J. and Fricker, C. R.: 1993, *J. Appl. Bact.* **75**, 87-90.

Walter, K. S., Fricker, E. J. and Fricker, C. R.: 1994, *Lett. Appl. Microbiol.* **19**, 47-49.

Woese, C. R.: 1987, Bacterial Evolution. *Microbiol. Rev.* **51**, 221-271.

# ENZYMATIC DETECTION OF COLIFORMS AND *Escherichia coli* WITHIN 4 HOURS

H. NELIS and S. VAN POUCKE

*University of Ghent, Laboratory for Pharmaceutical Microbiology, Harelbekestraat 72, B - 9000 Gent, Belgium*

**Abstract.** Commercial enzymatic membrane filtration (MF) tests for total coliforms (TC) and *Escherichia coli* in drinking water, based on ß-galactosidase (TC) or ß-glucuronidase (*E. coli*) detection take 18-24 h to complete. They either use chromogenic substrates (e.g., X-gal, X-gluc) or fluorogenic ones, particularly 4-methylumbelliferyl-glycosides. The latter afford improved detectability and, under particular conditions reduce the detection time to approximately 8 h. As part of our efforts to further enhance the speed of enzymatic MF tests, we examined two more advanced, two-step approaches. The first one involves microcolony formation in a short growth phase, followed by a separate enzyme assay using 4-trifluoromethylumbelliferyl-glycosides plus a membrane permeabilizer, and reading under a UV-lamp. For *E. coli* this type of test takes 5.5 h but field studies revealed a high false-negative rate. Switching from visual microcolony detection to instrumental single cell detection with the aid of a laser scanner (ChemScan) resulted in tests for *E. coli* and TC of 3.5 h. Both procedures involve enzyme induction (3 h) in the target cells, followed by cell labelling (30 min) using fluorescein-di-ß-D-glycosides and laser scanning (3 min) of the membrane filter. Application of the ChemScan *E. coli* test to naturally contaminated well and surface, as well as uncontaminated well and tap (n=149) water samples has indicated >90 % agreement, and equivalence with reference methods. The false-negative rate was 3.0-6.0% versus Chromocult Coliform and m FC agar, respectively. In 10 out of 149 samples, low numbers (1-3) of the target organisms were detected by laser scanning and not by conventional plate methods. These one-working day tests for TC and *E. coli* would be particularly useful for emergency monitoring of drinking water.

**Keywords:** ChemScan, coliforms, drinking water,*E. coli*, enzymatic tests, solid phase cytometry

## 1. Introduction

The routine monitoring of drinking water for the absence of pathogenic microbes is practically not feasible. An accepted alternative is to look for indicator bacteria that always occur in association with pathogens of fecal origin (Greenberg *et al.*, 1992). Among the proposed indicator bacteria, the total coliforms (TC) do not necessarily indicate a health risk as their presence may result from regrowth in the distribution system or from environmental sources (Allen and Edberg, 1997; Geldreich, 1997). The group of thermotolerant or "fecal" coliforms (FC) also represents a poor indicator of fecal contamination because it includes species of the genera *Klebsiella* and *Enterobacter* that lack significance for human health (Allen and Edberg, 1997). Only *E. coli*, the main representative of the thermotolerant coliforms, always originates from the intestine of humans and warm-blooded animals and is generally considered to be the preferred indicator of the fecal contamination of drinking water (Hofstra and Huis in 't Veld, 1988). However, TC continue to be maintained in the legislation of most countries as a parameter of general water quality.

Water utilities test for TC and *E. coli* as part of their monitoring of raw and partially purified waters, to show compliance of their final product with legal standards and to guarantee its safety in particular emergency situations. The latter include accidental pollution of the distribution system, construction works, shutdowns of pipes or major breakdowns in the supply. Traditionally, the detection of coliforms has relied on lactose fermentation in a selective broth or agar medium, at 35-37°C (TC) or 44.5°C (FC) (Greenberg *et al.*, 1992). Three test formats can be used, i.e. most probable number (MPN), presence-absence (P/A) and membrane filtration (MF). As these procedures are presumptive; subsequent confirmation of the results is required to preclude false positives. In recent years, enzymatic tests for the detection of TC and *E. coli*, based on the demonstration of ß-galactosidase and ß-glucuronidase, respectively, have become increasingly popular (Edberg and Edberg, 1988; Brenner *et al.*, 1993; Jermini *et al.*, 1994). Compared to the fermentative procedures, they are more rapid (18-24h versus 48-96h) as they do not require subsequent confirmation and they allow a simultaneous detection of TC and *E. coli*. To demonstrate the marker enzymes, chromogenic (usually indoxyl derivatives) or fluorogenic (4-methylumbelliferyl derivatives) substrates are incorporated in a selective broth or agar medium and cleaved in the course of bacterial growth. In the widely used Colilert® and Colisure® tests, which exist both in MPN and in P/A format, additional selectivity for the target bacteria is ensured by the defined substrate technology (Edberg and Edberg, 1988). Examples of commercial enzymatic MF tests include CHROMagar ECC®, Chromocult® and ColiScan® (Manafi, 1998).

However, while the enzymatic methods offer convenience and sufficient speed for compliance testing, more rapid techniques may be desirable to accommodate the above mentioned emergency situations. In these cases, testing should be feasible within a working day or preferably less than 5 hours (Geldreich, 1997), to allow sufficient time for sample collection and transport. One of the earliest successful developments in the area of rapid methods was the seven hour fecal coliform test (Reasoner *et al.*, 1979). This method was still based on lactose fermentation but involved a more sensitive detection of the acid formed. Recent and more sophisticated experimental approaches for TC, *E. coli* or *Enterobacteriaceae* in general include polymerase chain reaction (PCR) based on the demonstration of the *lacZ* (TC), the *lamB* (*E. coli*) or *uidA* (*E. coli*) genes, an enzyme linked immunosorbent assay (ELISA) and impedance measurement (Bej *et al.*, 1990; Bej *et al.*, 1991a,b; Obst *et al.*, 1992; Colquhoun *et al.*, 1995; Iqbal *et al.*, 1997). However, all these methods take more than a working day. According to Fricker and Fricker (1994), PCR is poorly suitable for the direct application to water, but its potential can be conveniently exploited for the rapid confirmation of presumptive *E. coli* colonies in MF tests.

In view of the available extensive evidence regarding their excellent performance (Edberg *et al.*, 1988; Sartory and Howard, 1992; Brenner *et al.*, 1993,1996; Grant, 1997), the enzymatic methods seem to be the best candidates for further speed enhancement. Several investigators have tried to improve the P/A tests by using colorimetric and particularly fluorimetric instead of visual end point detection, often as part of automated systems (Berg and Fiksdal, 1988; Apte and Batley, 1994; Park *et al.*, 1995; Peterson *et al.*, 1995). The fluorimetric tests monitor the cleavage by ß-galactosidase of 4-methylumbelliferyl-ß-D-galactopyranoside (MU-gal) and reportedly take 6 (FC) to 8 (TC) hours. Some of them have been commercialized, e.g., the Colifast-6®.

As part of our efforts to develop even more rapid enzymatic procedures, we identified and optimized some speed-limiting parameters (Van Poucke and Nelis, 1995, 1997b; Nelis, 1999). In particular, we found the nature and the rate of cellular uptake of the substrate to be crucial to the detection of the enzyme acitivities in the low numbers of cells generated after the very short incubation times. Thus, to enhance the intrinsic detection sensitivity for ß-galactosidase (TC), we replaced the fluorogenic MU-gal by the chemiluminogenic Galacton®, a fenylgalactose substituted 1,2-dioxetane derivative (Van Poucke and Nelis, 1995, 1997b). Likewise, Glucuron® was used instead of 4-methylumbelliferyl-ß-D-glucuronide (MUG) in an experimental chemiluminometric P/A test for *E. coli* based on ß-glucuronidase detection (Van Poucke and Nelis, 1997a). In addition, a two-step concept was devised in which an initial growth/target enzyme induction ("preincubation") phase was conducted separately from the actual demonstration of the enzyme activity ("assay") (Nelis, 1999). Advantages of this dissociation include the lack of competition between the enzyme substrate and the inducer (e.g., isopropyl-ß-D-thiogalactopyranoside [IPTG] for ß-galactosidase) and, most important, the possibility to add a bactericidal membrane permeabilizer in the assay to promote the uptake of the substrate. This optimization resulted in a prototype P/A test that had the potential to detect 1 CFU of *E. coli* per 100 ml of water in 6 hours (Van Poucke and Nelis, 1995). However, high numbers of other ß-galactosidase or ß-glucuronidase containing waterborne bacteria, mostly with low optimal growth temperature, also reacted in the sensitive enzyme assay, even though they did not multiply under the conditions used (Van Poucke and Nelis, 1997a). Defined substrate methods, in which the enzyme substrate is actively metabolized exclusively by coliforms suffer much less from such "passive" interference of non-target bacteria (Covert *et al.*, 1989; Cowburn *et al.*, 1994; Edberg *et al.*, 1989). We postulated that this phenomenon would also be absent in MF tests where the non-target are physically separated from the target bacteria and only the latter form (micro)colonies under selective conditions of medium composition and incubation temperature. Several investigators have used the microcolony approach and fluorescence labelling as part of one-step enzymatic procedures, i.e. with the enzyme substrate included in the growth medium. Berg and Fiksdal (1988) reported a prototype MF test, subsequently commercialized as

Envirofast® (Eccles *et al.*, 1998), for the visual detection of fluorescent FC in 6-8 h, using MU-gal as a substrate for ß-galactosidase. Sarhan and Foster (1991) added MUG to a selective medium and achieved a visual detection of fluorescent *E. coli* microcolonies in 7.5 hours. This substrate, incorporated in lauryl sulphate broth was also used by Sartory *et al.* (1999) who, however, visualized the microcolonies by epifluorescence microscopy.

We have developed a two-step MF approach, consisting of a multiplication/enzyme induction phase followed by fluorescence labelling of the microcolonies on the basis of their ß-galactosidase (TC) or ß-glucuronidase (*E. coli*) activity, in conjunction with visual or instrumental detection of fluorescent microcolonies. The latter involved the use of solid phase cytometry (SPC), a laser scanning technique that is capable of detecting single fluorescent cells on a membrane filter. The present paper reports on the application in the field of the visual and the SPC tests for *E. coli*.

## 2. Materials and Methods

### 2.1. PROCEDURE WITH VISUAL DETECTION

Water samples were filtered over 47-mm 0.4-µm Nylaflo™ (Gelman) membrane filters and incubated for 5 - 8 h on ILM (Nelis, 1999) in a water bath at 41.5°C (*E. coli*) or 37°C (TC). The membrane filter was subsequently transferred to a cellulose pad impregnated with 1.8 ml of a solution containing (per ml of 0.05 M sodium phosphate buffer, pH 7.3) 0.1 mg 4-trifluoromethylumbelliferyl-ß-D-glucuronide (TFMUG; *E.coli* detection) or 4-trifluoromethylumbelliferyl-ß-D-galactopyranoside (TFMU-gal; TC detection) (Molecular Probes), 1 µmol of $MgCl_2.6H_2O$ and 1 mg of polymyxin B sulfate (Sigma) and reincubated at 44.5°C. Yellow-green fluorescent microcolonies were counted manually under a 366-nm UV-lamp.

### 2.2. SPC PROCEDURES

Water samples were filtered over 25-mm 0.4-µm Cycloblack™-coated polyester membrane filters (Chemunex, Ivry-sur-Seine, France). The membrane filter was subsequently placed on a 25-mm cellulose pad (Millipore), impregnated with 650 µl of a mixture of Colicult Base Medium, Counterstain E (CSE) and Enzyme Inducer Cocktail (89:10:1,vol:vol:vol) (all from Chemunex), and both were incubated at 37°C for 3h. After incubation, the filter was transferred to a second cellulose pad impregnated with 600 µl Colicult Base Medium containing 200 µM fluorescein-ß-D-di-glucuronide (FDG, Molecular Probes), 10% CSE and 1mM $MgCl_2.6H_2O$ (*E.coli* detection) or 400 µM fluorescein-ß-D-di-galactopyranoside (FD-gal, Molecular Probes), 10% CSE, 0.01% IPTG (Sigma) and 1mM $MgCl_2.6H_2O$ (TC detection). The filter-pad combination was left for

30 min on a glass plate cooled to 0°C with the aid of a circulating water:ethylene glycol mixture (70:30, vol:vol) kept at -7°C. Finally, the membrane filter was removed from the pad, mounted on top of a 25-mm black nitrocellulose filter (Millipore) in a metal holder (Chemunex) and analyzed in the solid phase cytometer. The instrument was a Chem*Scan* C or a *Scan* RDI (Chemunex), equipped with a 488-nm argon laser and photomultipliers sensitive to the green and amber regions of the light spectrum. A computer visualized all detected fluorescent events in a primary and, following software discrimination, a secondary scan map. The results displayed in the latter were visually confirmed using an epifluorescence microscope with a moving stage, connected to the laser scanner, to differentiate bacteria from other fluorescent events that possibly escaped the software discrimination process. Highlighting of each spot on the computer screen directed the motorized stage to the corresponding position on the membrane filter for inspection of the appearance of each detected event.

## 2.3. REFERENCE PROCEDURES

The numbers of *E. coli* were determined on m FC agar at 44.5°C (24 h), with confirmation using the indole test at 44.5°C, and on Chromocult Coliform agar (Merck), supplemented with 10 µg of cefsulodin per ml (CC-agar). On the latter medium, *E. coli* appeared as purple colonies after 24 h of incubation at 37°C. The total number of ß-glucuronidase containing bacteria that were able to grow at 22°C was determined by membrane filtration and incubation for 7 days on R2A agar containing 0.1 mg/ml of 5-bromo-4-chloro-3-indolyl-ß-D-glucuronide (X-gluc; Biosynth).

## 2.4. INTERPRETATION OF THE RESULTS

Due to the inability to carry out a biochemical confirmation of the results in both rapid tests, false-positives and -negatives were defined as follows. If the target bacterium was detected in both the experimental and the reference methods, the response was considered true positive (TP). If, however, no *E. coli* were present on the reference medium, unlike on the membrane in the experimental test, the response was considered false-positive (FP). Likewise, a response was considered true-negative (TN) when typical (micro-)colonies/cells were absent in both the experimental test and the reference method, whereas a false-negative (FN) response was defined as a case where only the reference medium yielded typical colonies. Correlation and equivalence were calculated according to Bland and Altman (1986), who proposed an approach to compare data from two methods without knowledge of the true value. To this end, bias-plots are constructed by plotting the difference between the counts in the two methods versus the respective average counts. Only data pairs yielding an average count of below or equal to 100/100 ml were taken into consideration. The mean difference (D), being an estimation of bias, and the width of its 95% confidence interval (D ± 2 SD; SD = standard deviation) are a measure of the correlation between the two tests. In this study, *E. coli* counts in the experimental tests were subtracted from those obtained on the reference media. To check for equivalence, the bias from zero was calculated. Results of two tests were considered significantly different if "zero" was not included in the interval D ± 2 SE (SE = standard error), with the sign of D indicating the nature of the discrepancy. A negative value of D indicated overestimation, while a positive one corresponded with an underestimation.

## 3. Results and Discussion

### 3.1. CONCEPTS AND SPEED

In enzymatic MF tests for *E. coli* and TC taking less than 8 hours, the target bacteria form fluorescent microcolonies following the cleavage, by ß-glucuronidase or ß-galactosidase, of the fluorogenic substrates MUG or MU-gal, respectively, in the course of bacterial propagation (Berg and Fiksdal, 1988; Sarhan and Foster, 1991; Sartory *et al.*, 1999). The speed of such tests is primarily restricted by the capability of the naked eye or an instrument to capture the weak light signals emitted by the microcolony. The light output in turn is dependent on the size of the microcolony, its level of enzymatic activity and the quantum efficiency of the fluorogenic substrate at the given wavelength of excitation. The number of cells in the microcolony cannot be significantly enhanced, except possibly by the addition of growth stimulants, due to the incubation time restriction inherent in a rapid test. The level of cellular enzyme activity can be boosted by efficient induction of the target enzyme, whereas the result of the enzyme assay in terms of fluorescence signal depends on the nature and the rate of uptake of the substrate. In accordance with the philosophy of our experimental P/A test (Van Poucke and Nelis, 1995, 1997b), we addressed each parameter and optimized its contribution to the fluorescence yield. We found that the incorporation of certain proteins in the growth medium intensified the fluorescence of the microcolonies, which may be explained by a higher number of cells and, hence, a reduction of the generation time. A substantial increase in fluorescence resulted from the addition of a proprietary cocktail of enzyme inducers and amplifiers to the growth medium. The enzyme assay itself was optimized by using a superior substrate, i.e. TFMUG for ß-glucuronidase or TFMU-gal for ß galactosidase. TFMU-substrates afford yellow-green fluorescence and, unlike their MU-counterparts, do not require the inconvenient spraying of the plate with alkali to intensify fluorescence. The uptake of the substrate was greatly enhanced in the presence of the membrane permeabilizer polymyxin B. However, due to its bactericidal action, polymyxin B could not be added to the growth medium. Consequently, the test had to be broken down in two steps, consisting of a preincubation phase, ensuring cell multiplication and enzyme induction, and an enzyme assay phase affording the fluorescence labelling. An additional advantage of the two-step approach is that it minimizes the competition between the inducers and the substrate for the target enzyme. This supposedly further contributes to the higher speed of our test with reference to one-step procedures. The intracellular retention of fluorescein in the SPC test was improved by conducting the labelling at low temperature (Nolan *et al.*, 1988). Following the combined optimization of all parameters, fluorescent microcolonies of waterborne *E. coli* could be visually detected on the basis of their ß-glucuronidase activity in 5.5 hours (5 hours of preincubation and 0.5 hours of enzyme assay). When the ß-galactosidase of *E. coli* was assayed, 6.5-7.5 hours were needed. For other TC possessing lower ß-galactosidase activity such as *Enterobacter* and *Citrobacter* sp., the detection time increased to 8-9 hours. Below the indicated incubation times the microcolonies were too small and/or their light output too weak to be detected under a UV-lamp. The use of a fluorescence imager (CCD camera) did not enhance the detectability and led to comparable analysis times. For lack of further perspectives to optimize the fluorescence output of the microcolonies, only the use of a more sophisticated detection device than the naked eye and a CCD camera could increase the speed of the test. While a CCD camera still requires microcolonies, the Chem*Scan* (Chemunex) is capable of detecting fluorescence at the single cell level (Brailsford, 1996; Mignon-Godefroy *et al.*, 1997). This technique, recently entitled solid phase cytometry (SPC), involves laser scanning of a membrane filter in 3 minutes and has a detection limit of 1 fluorescent cell per filtered volume of water. So far, it has mainly been used for the determination of the total viable count of waters and, occasionally for the detection of specific microorganisms, i.e. *Cryptosporidium* and *E. coli* O157:H7 (Brailsford, 1996; Guyomard, 1997; Reynolds *et al.*, 1997; Wallner *et al.*, 1997; Pyle *et al.*, 1999; Zanelli *et al.*, 1999). A SPC prototype test for *E. coli* based on the labelling by fluorescence *in situ* hybridization has been described by Prescott *et al.* (1998). Although SPC has been designed to detect single cells, in our test some multiplication of *E. coli*/TC could not be avoided in the course of the 3 hours that were required to induce a sufficient level of ß-glucuronidase or ß-galactosidase. As a result, mixtures of single cells and small microcolonies were obtained. In 149 water samples tested, approximately 81% of *E. coli* grew out to microcolonies, whereas 19%, probably representing a less vital fraction of the population, remained as single cells. Both microcolonies and single cells are, however, seen by the Chem*Scan* as single spots and, hence equally contribute to the count. The total analysis time of the SPC test was approximately 3.5 hours, which is broken down in the steps of membrane filtration (taking less than 1 min), enzyme induction (3 hours), labelling (30 minutes), laser scanning (3 minutes) and microscopic confirmation of the result (a few seconds to several minutes). The latter allowed to distinguish the target cells with their typical rod morphology from fluorescent particles or spherical protozoan-like structures. Autofluorescent particles were commonly found when large volumes (e.g., 100 ml) of waters, including some coliform-free ones, were filtered. The protozoan-like organisms have, so far, only been observed in association with *E. coli*.

## 3.2. EVALUATION

The 5.5-h test with visual detection and the 3.5-h SPC test for *E. coli* were applied in two field studies to 89 and 149, respectively, surface, partially purified, well and tap waters. Table I presents the comparative evaluation of the results. The visual test tended towards an underestimation of the counts and, particularly, yielded an extremely high false-negative rate (20.0% against the results on CC-agar, 37.5% against the results on m FC agar). In contrast, the bias of the SPC test was not significantly different from zero, showing equivalence with each of the reference methods, and the false-negative rate was low (3.0% against results on CC-agar, 6.0% against results on m FC agar). The confidence interval ($\alpha = 0.05$) on the bias for counts up to 100/100 ml was much narrower with reference to those obtained for the visual *E. coli* test. The figures for the false-positive rate should be considered with caution. Given the inability to confirm the results in SPC, their true or false positivity could not be ascertained. In addition, >70% of the samples yielding a presumed false-positive result represented borderline cases where SPC detected only 1-2 *E. coli*'s per 100 ml.

Evidence for the specificity of the SPC *E. coli* test came from the analysis of *E. coli*-free samples but from which high numbers ($10$-$10^4$ per 100 ml) of ß-glucuronidase containing non-target bacteria were recovered on R2A agar supplemented with X-gluc. Although these potential interferences do not grow on selective reference media (Van Poucke and Nelis, 1997a; Tryland and Fiksdal, 1998) they could potentially have been detected in SPC after concentration on the membrane filter. Their escape from detection is probably due to their lower enzyme activities under the conditions of the test.

TABLE I

Evaluation of the visual 5.5-h test and the SPC 3.5-h test for *E. coli*, for average counts up to 100/100 ml.

| | N[a] | Correlation analysis (counts) | | | Bias analysis (counts) | | Agreement (%) | |
|---|---|---|---|---|---|---|---|---|
| | | Bias | D-2SD[b] | D+2SD | D-2SE[c] | D+2SE | FN[d] | FP[e] |
| CC-agar as reference medium | | | | | | | | |
| Visual test | 82 | +5.6 | -18.4 | +29.6 | +3.0 | +8.2 | 20.0 | 4.1 |
| SPC test | 134 | -0.1 | -12.7 | +12.4 | -1.2 | +1.0 | 3.0 | 12.2 |
| m FC-agar as reference medium | | | | | | | | |
| Visual test | 85 | +7.9 | -35.5 | +51.1 | +3.2 | +12.6 | 37.5 | 2.0 |
| SPC test | 142 | +0.2 | -26.3 | +26.6 | -2.1 | +2.4 | 6.0 | 12.2 |

[a] Number of samples analyzed; [b] Standard deviation; [c] Standard error; [d] False-negative rate; [e] False-positive rate

## 4. Conclusion

Our results demonstrate the feasibility to enumerate *E. coli* and TC within 6 hours as fluorescent microcolonies using a two-step procedure for enzymatic labelling followed by visual detection. However, in order to be visible to the naked eye under a UV-lamp, a microcolony should contain a minimal number of fluorescent cells, which depends on the *lag*-phase and the generation time of the bacteria in question. Stressed and injured cells will obviously fail to multiply sufficiently in such short time and, hence, will escape detection. This accounts for the positive bias and the high false-negative rate of the visual *E. coli* test. SPC does not suffer from this limitation because it detects both microcolonies and single cells that fail to multiply. The combination of this powerful technique with a two-step enzymatic procedure for the fluorescence labelling of *E. coli* and TC has resulted in tests of 3.5 h. A field study for the *E. coli* test has shown good performance and agreement with reference methods. To the best of our knowledge, these are currently the most rapid existing tests for *E. coli* and TC in drinking water.

## Acknowledgements

We thank Dr. L. Foissac, Dr. J.-L. Drocourt and Dr. P. Cornet of Chemunex, and The Flemish Center for Water Research (SVW, Antwerp, Belgium) for financial support. Mrs. L. Mels (SVW) is acknowledged for excellent technical support.

## References

Allen, M. A. and Edberg, S. C: 1997, in *Coliforms and E. coli. Problem or Solution?* The Royal Society of Chemistry, Cambridge,176-181.

Apte, S. C. and Batley, G. E.: 1994, *Sci. Total Environ.* **141,** 175-180.

Bej, A. K., DiCesare, J. L., Haff, L. and Atlas, R. M.: 1991a, *Appl. Environ. Microbiol.* **57,** 1013-1017.

Bej, A. K., Mahbubani, M. H., DiCesare, J. L. and Atlas, R. M.: 1991b, *Appl. Environ. Microbiol.* **57,** 3529-3534.

Bej, A. K., Steffan, R. J., DiCesare, J., Haff, L. and Atlas, R. M.: 1990, *Appl. Environ. Microbiol.* **56,** 307-314.

Berg, J. D. and Fiksdal, L.: 1988, *Appl. Environ. Microbiol.* **54,** 2118-2122.

Bland, J. M. and Altman, D. G.: 1986, *The Lancet* February 8, 307-310.

Brailsford, M.: 1996, *Microbiol. Europe* **4,** 18-20.

Brenner, K. P., Rankin, C. C., Roybal, Y. R., Stelma, Jr., G. N., Scarpino, P. V. and Dufour, A. P.: 1993, *Appl. Environ. Microbiol.* **59,** 3534-3544.

Brenner, K. P., Rankin, C. C., Sivaganesan, M. and Scarpino, P. V.: 1996, *Appl. Environ. Microbiol.* **62,** 203-208.

Colquhoun, K. O., Timms, S. and Fricker, C. R.: 1995, *J. Appl. Bacteriol.* **79,** 635-639.

Covert, T. C., Shadix, L. C., Rice, E. W., Haines, J. R. and Freyberg, R. W.: 1989, *Appl. Environ. Microbiol.* **55,** 2443-2447.

Cowburn, J. K., Goodall, T., Fricker, E. J., Walter, K .S. and Fricker, C. R.: 1994, *Lett. Appl. Microbiol.* **19,** 50-52.

Eccles, J. P., Fricker, E. J. and Fricker, C. R.: 1998, (On CD-ROM) *Proceedings of the Water Quality Technology Conference*, American Water Works Association, San Diego, Calif., USA.

Edberg, S. C., Allen, M. J., Smith, D. B., and The National Collaborative Study: 1988, *Appl. Environ. Microbiol.* **54,** 1595-1601.

Edberg, S. C., Allen, M. J., Smith, D. B., and The National Collaborative Study: 1989, *Appl. Environ. Microbiol.* **55,** 1003-1008.

Edberg, S. C. and Edberg, M. M.: 1988, *Yale J. Biol. Med.* **61,** 389-399.

Fricker, E. J. and Fricker, C. R.: 1994, *Lett. Appl. Microbiol.* **19,** 44-46.

Geldreich, E. E.: 1997, In *Coliforms and E. coli. Problem or Solution?* The Royal Society of Chemistry, Cambridge, 218-234.

Grant, M. A.: 1997, *Appl. Environ. Microbiol.* **63,** 3526-3530.

Greenberg, A. E., Clesceri, L. S., Eaton, A. D. (eds): 1992. *Standard methods for the examination of water and wastewater,* 18th ed., American Public Health Association, Washington, D.C.

Guyomard, S.: 1997, *Pharm. Technol. Europe* **9,** 50-52.

Hofstra, H. and Huis in 't Veld, J. H. J.: 1988, *J. Appl. Bacteriol. Symp. Suppl.*, 197S-212S.

Iqbal, S., Robinson, J., Deere, D., Saunders, J. R., Edwards, C. and Porter, J.: 1997, *Lett. Appl. Microbiol.* **24,** 498-502.

Jermini, M., Domeniconi, F. and Jäggli, M.: 1994, *Lett. Appl. Microbiol.* **19,** 332-335.

Manafi, M.: 1998, *De Ware(n)-Chemicus* **28,** 12-17.

Mignon-Godefroy, K., Guillet, J.-G. and Butor, C.: 1997, *Cytometry* **27,** 336-344.

Nelis, H.: 1999 January, U.S. Patent 5, 861,270.

Nolan, G. P., Fiering, S., Nicolas, J.-F. and Herzenberg, L. A.: 1988, *Proc. Natl. Acad. Sci. USA* **85,** 2603-2607.

Obst, U., Hübner, I., Wecker, M. and Bitter-Suermann, D.: 1989, *Aqua* **38,** 136-142.

Park, S.J., Lee, E., Lee, D., Lee, S. and Kim, S.: 1995, *Appl. Environ. Microbiol.* **61,** 2027-2029.

Peterson, S.M., Apte, S.C., Davies, C.M., Johns, I., Fitch, P., Swan, G., Prus-Wisniowski, T. and Skyring, G.: 1995, In *Proceedings of the Australian Water and Wastewater Association 16th Federal Convention*, Australian Water and Wastewater Association, Sydney.

Prescott, A. M., Reynolds, D. and Fricker, C. R.: 1998, (On CD-ROM) *Proceedings of the Water Quality Technology Conference*, American Water Works Association, San Diego, Calif., USA.

Pyle, B. H., Broadaway, S. C. and McFeters, G. A.: 1999, *Appl. Environ. Microbiol.* **65,** 1966-1972.

Reasoner, D. J., Blannon, J. C. and Geldreich, E. E.: 1979, *Appl. Environ. Microbiol.* **38,** 229-236.

Reynolds, D. T., Fricker, E. J., Purdy, D. and Fricker, C. R.: 1997, *Wat. Sci. Tech.* **35,** 433-436.

Sarhan, H. R. and Foster, H. A.: 1991, *J. Appl. Bacteriol.* **70,** 394-400.

Sartory, D. P., Parton, A. and Rackstraw, C.: 1999, Presentation at the International Conference on "Rapid Detection Assays for Food and Water", Royal Society of Chemistry, York, UK, March 15-17 1999.

Sartory, D. P. and Howard, L.: 1992, *Lett. Appl. Microbiol.* **15,** 273-276.

Tryland, I. and Fiksdal, L.: 1998, *Appl. Environ. Microbiol.* **64,** 1018-1023.

Van Poucke, S. O. and Nelis, H. J.: 1995, *Appl. Environ. Microbiol.* **61,** 4505-4509.

Van Poucke, S. O. and Nelis, H. J.: 1997a, *Appl. Environ. Microbiol.* **63,** 771-774.

Van Poucke, S. O. and Nelis, H. J.: 1997b, *J. Biolumin. Chemilumin.* **12,**165-176.

Wallner, G., Tillmann, D., Haberer, K., Cornet, P. and Drocourt, J.-L.: 1997, Eur. J. Parent. Sci. **2,** 123-126.

Zanelli, F., Compagnon, B., Joret, J. C. and de Roubin, M. R.: 1999, Poster presented at the international conference on "Minimising the Risk from *Cryptosporidium* and other Waterborne Particles," International Association on Water Quality, Paris, France.

# DETECTION OF INFECTIOUS *Cryptosporidium Parvum* OOCYSTS IN ENVIRONMENTAL WATER SAMPLES USING AN INTEGRATED CELL CULTURE-PCR (CC-PCR) SYSTEM

M. W. LECHEVALLIER[1], M. ABBASZADEGAN[2], and G. D. DI GIOVANNI[2]
*[1]American Water Works Service Company, Inc., 1025 Laurel Oak Rd. P.O. Box 1770, Voorhees, New Jersey, 08043 USA. [2]American Water Works Service Company, Inc., Quality Control and Research Laboratory, 1115 S. Illinois St., Belleville, Illinois, 62220 USA*

**Abstract.** Examination of naturally occurring *C. parvum* oocysts from environmental water samples has previously been hampered by the inability to determine the public health significance of detected organisms. As a result the safety of drinking water supplies was in question. These limitations have been resolved through the development and application of a method that incorporates immunomagnetic separation (IMS) and an infectivity determination using an integrated cell culture, polymerase chain reaction assay (CC-PCR). Briefly, the method concentrates water samples by filtration or centrifugation and isolates oocysts by IMS. An acidified Hank's balanced salt solution (HBSS) containing 1% trypsin was used for the dissociation of captured oocysts from the IMS beads. *In vitro* HCT-8 cell culture of purified oocysts was performed in 96-well microtiter plates and infected cells were detected using PCR primers specific for *C. parvum*. A total of 242 raw source waters or filter backwash water samples from twenty five sites in the U.S. were analyzed to validate the procedure. Oocyst seeded in raw and filter backwash water samples were used to evaluate recovery efficiencies and performance of the CC-PCR protocol with different water quality matrices. The CC-PCR detected infectious *Cryptosporidium parvum* in 6 of 121 (5.0%) raw and 9 of 122 (7.4%) filter backwash water samples. All CC-PCR positive samples were confirmed by cloning and DNA sequence analysis of the PCR products. Isolates were shown to originate from human and animal sources. Grouping of genotypes permitted evaluation of strain diversification and variation. Current studies are using this technique to examine oocysts in various watersheds and in the finished drinking water of over 80 surface water treatment plants.

**Keywords:** cell culture, *Cryptosporidium parvum*, filter backwash water, genotype, immunomagnetic separation, polymerase chain reaction, raw water

## 1. Introduction

The uncertainty regarding the adequacy of conventional water treatment practices (coagulation, filtration, and chlorine disinfection) for controlling *Cryptosporidium parvum* in water is related to the fact that methods used for the U. S. Environmental Protection Agency Information Collection Rule (ICR) (USEPA, 1996) or the proposed method 1622 (USEPA, 1999) do not specifically detect the human pathogen *Cryptosporidium parvum* or determine viability or infectivity of recovered oocysts. Recently several PCR-based methods for the detection of *Cryptosporidium parvum* have been described (Johnson *et al.*, 1995; Wagner-Wiening and Kimmig, 1995; Rochelle *et al.*, 1997; Deng *et al.*, 1997; Kaucner and Stinear, 1998; Gibbons

*Water, Air, and Soil Pollution* **123:** 53–65, 2000.

*et al.*, 1998), including an infectivity assay based on *in vitro* cell culture of the parasite using Caco-2 cells and detection of *C. parvum* infected cells by targeting *C. parvum hsp70* mRNA using reverse transcription polymerase chain reaction (RT-PCR) (Rochelle *et al.*, 1997). We have developed another strategy for a *Cryptosporidium parvum* infectivity assay which integrates immunomagnetic separation (IMS) to recover *Cryptosporidium* oocysts from water samples followed by *in vitro* cell culture using HCT-8 cells and detection of *C. parvum* infected cells by targeting *C. parvum hsp70* DNA using standard PCR (DiGiovanni *et al*, 1997). We previously reported a detection limit of less than 5 purified infectious oocysts for the CC-PCR *C. parvum* infectivity assay used and an approximate 10-fold growth of the parasite in cell culture. The CC-PCR strategy permits the assay of raw, filter backwash and finished water samples; utilizes *in vitro* cell culture in a 96-well format for high sample throughput; and standard PCR detection of genomic *C. parvum hsp70* DNA, which is less time consuming and more suitable than RT-PCR for attempts to quantitate the number of infectious oocysts present in a sample.

This communication reports that the CC-PCR *C. parvum* infectivity assay was field tested using raw source water and filter backwash water grab samples from twenty five sites in the U.S. Oocyst seeded raw and filter backwash water samples were used to evaluate recovery efficiencies and performance of the CC-PCR protocol with different water quality matrices. Confirmation of CC-PCR positive samples by DNA sequence analysis of the *hsp70* PCR products showed that the isolates were similar to known *C. parvum* sequences and permitted analysis of the heterogeneity of infectious waterborne isolates. Now the method is ready for application to finished drinking water samples to determine the safety of treated water.

## 2. Materials and Methods

### 2.1. OOCYSTS AND MICROSCOPY

Purified live *C. parvum* oocysts (bovine isolate LA-1) were obtained from Waterborne, Inc. (New Orleans, LA). Oocyst stocks and Percoll-sucrose flotation purified oocysts were enumerated by immunofluorescence assay (IFA) microscopy as described in the *ICR Microbial Laboratory Manual* (USEPA, 1996). Oocysts purified by immunomagnetic separation (IMS) were enumerated by fluorescence microscopy as follows. IMS purified samples (40 µl sample plus 10 µl deionized water microfuge tube wash) or purified oocyst stocks (50 µl) were placed into individual wells of treated three-well microscope slides (Meridian Diagnostics, Inc., Cincinnati, OH) and dried at 42°C. Samples were fixed with one drop of room temperature methanol and air dried, then 75 µl of FITC labeled anti-*Cryptosporidium* monoclonal antibody (FITC-mAb; Waterborne, Inc., New Orleans, LA) added to each well and slides incubated at 37°C for 30 minutes in a humid chamber. Excess FITC-mAb was removed by aspiration with a micropipettor followed by a single wash with one drop of deionized water and aspiration. Slides were placed in the dark until dry and then 10 μl of mounting medium (10% glycerol, 80% PBS, 5% 5M NaCl, 5% formalin, 2.5% DABCO, pH 8.6) added to each well; a coverslip applied, and slides examined by epifluorescence microscopy at x 200 magnification. Presumptive oocysts were confirmed using x 400 magnification epifluorescence microscopy and x 1000 magnification Nomarski DIC microscopy to determine internal morphology.

## 2.2. RECOVERY OF OOCYSTS

Approximately 10 L grab samples of raw and filter backwash water were concentrated by centrifugation at 1800 x *g*, 4°C , for 10 min. For finished drinking water samples, 100 L were filtered through the Envirochek filter (Pall Gleman Laboratory, Ann Arbor, MI) according to published procedures (USEPA, 1999). Seeded samples were spiked with 1080 to 3612 purified live *C. parvum* oocysts prior to concentration. *Cryptosporidium* oocysts were recovered from up to three 0.5 mL packed pellet portions of each water sample concentrate by IMS (Dynabeads anti-*Cryptosporidium*; Dynal A.S., Oslo, Norway). IMS was performed according to the manufacturer's suggestions with the exception of the dissociation step. The manufacturer's acidified Hank's balanced salt solution (AHBSS, pH 2.75) dissociation protocol for viability testing was originally used but was found to have a variable endpoint pH (ca. 3.0 -6.0) due to residual IMS SL buffer A. To remove residual IMS buffer a second wash in 1 x PBS (pH 7.2) was performed after the samples had been separated using a microfuge tube magnetic particle concentrator (MPC-M). To improve recovery, a combined acid and enzymatic dissociation method which may digest the antibody-oocyst complexes directly off of the IMS beads and serve as an excystation trigger for subsequent cell culture was used. For each IMS reaction 200 μl of AHBSS/1% trypsin (type II-S porcine pancreas, Sigma Chemical Co., St. Louis, MO) was added, samples incubated at 37°C for 1 h with 10 sec of vortexing every 15 min. Following separation of the samples in a MPC-M concentrator, the supernatants containing the dissociated oocysts were transferred to a microfuge tube. To ensure recovery of oocysts a second wash of the IMS beads with 100 μl of AHBSS/1 % trypsin was performed. Dissociated samples were neutralized with 0.5 N NaOH (4.0 μl); replicate IMS reactions pooled and centrifuged at maximum speed in a microcentrifuge for 2 min with no brake; and aspirated down to 20 μl for CC-PCR or 40 μl for microscopy.

Percoll-sucrose flotation for the recovery of *Cryptosporidium* oocysts was performed on up to 2.0 mL packed pellet of each water sample concentrate as described in the *ICR Microbial Laboratory Manual* (USEPA, 1996), with the exception that a 1.15 specific gravity Percoll-sucrose solution was used.

## 2.3. *IN VITRO* CELL CULTURE OF *C. parvum*

Human ileocecal adenocarcinoma (HCT-8; ATCC CCL-244) cells were cultivated as previously described (Woods *et al.*, 1995). Cell culture maintenance medium consisted of RPMI 1640 with L-glutamine and 5% FBS, pH 7.2. Growth medium used for the *in vitro* development of *Cryptosporidium parvum* contained 10 % FBS, 15 mM HEPES (N-[2-hydroxyethyl]piperazine-N'-[2-ethanesulfonic acid]), 50 mM glucose, 35 μg $mL^{-1}$ ascorbic acid, 1.0 μg $mL^{-1}$ folic acid, 4.0 μg $mL^{-1}$ 4-aminobenzoic acid, and 2.0 μg $mL^{-1}$ calcium pantothenate. Twenty four hours prior to inoculation, 96-well cell culture microplates were seeded with 5 x $10^4$ HCT-8 cells per well. Plates were then incubated at 37°C in a 5% $CO_2$ humidified incubator to allow for the development of 100% confluent monolayers. Just prior to inoculation of monolayers 50 μl of maintenance medium was removed. Samples for CC-PCR were resuspended in 180 μl of prewarmed growth medium immediately following IMS dissociation and used to inoculate two HCT-8 cell monolayers (100 μl inoculum each). AHBSS/1 % trypsin treated *C. parvum* oocysts were used as positive controls and single cycle freeze/thaw killed oocysts were used as negative controls. The inoculated cell monolayers were incubated at 37°C in a 5% $CO_2$ humidified incubator for up to 72 hours. After incubation the cell monolayers were washed five times with 200 μl of PBS to remove unexcysted oocysts. Cell monolayers were harvested by the addition of 200 μl 1 x Tris-EDTA (TE, pH 8.0) buffer and resuspended cells transferred to microfuge tubes. Cell harvests were centrifuged at maximum speed in a microcentrifuge for 2 min, aspirated down to a 5-10 μl volume, and frozen at -20°C until PCR analysis.

## 2.4. DETECTION OF INFECTIOUS OOCYSTS

HCT-8 cell harvests and *C. parvum* oocyst PCR positive controls were lysed by eight cycles of freezing in liquid nitrogen and thawing in a 98°C heated block. Aliquots of lysed samples were used directly for PCR without further purification. PCR primers specific for the *C. parvum hsp70* gene were used resulting in a 361 bp product (20). PCR was performed using a Perkin-Elmer model 9600 thermal cycler (PE Applied Biosystems, Foster City, CA). Each 50 μl PCR mixture contained 5.0 μl of 10 x amplification buffer with Mg (1.5 mM final; Boehringer Mannheim, Indianapolis, IN); 200 μM of each dATP, dTTP, dCTP, and dGTP (Boehringer Mannheim, Indianapolis, IN); 200 nM of each of forward and reverse CPHSP2 primer; 2.5 μl of 30 mg $mL^{-1}$ bovine serum albumin (BSA; Sigma Chemical Co., St. Louis, MO); and varying amounts of *C. parvum* template DNA. Amplification conditions were as follows: initial denaturation at 95°C for 5 minutes; samples held at 80°C while 2.0 U *Taq* DNA polymerase (Boehringer Mannheim, Indianapolis, IN) was added (hot start); 40 cycles of denaturation at 94°C for 30 seconds; annealing at 59°C for 1 minute; extension at 72°C for 30 seconds; followed by a single final extension at 72°C for 10 minutes and a 4°C soak. Amplification products were separated by horizontal gel electrophoresis on a 2.0 % agarose gel (Amresco, Solon, OH) containing 0.5 µg $mL^{-1}$ ethidium bromide (Sigma Chemical Co.,St. Louis, MO) and visualized under UV light. Gel images were captured using a gel documentation system (UVP, Inc., Upland, CA).

### 2.5. CLONING AND DNA SEQUENCE ANALYSIS

CC-PCR products were cloned and sequenced for confirmation of homology to the *C. parvum hsp70* gene. Products were cloned using a TOPO TA cloning kit (Invitrogen, Carlsbad, CA) according to the manufacturer's instructions. Cloned products were sequenced commercially (ACGT, Northbrook, IL and DNA Sequencing Service, The University of Arizona, Tucson, AZ) and sequence homology to the *C. parvum* KSU-1 *hsp70* gene (Khramtsov *et al.*, 1995) confirmed using Gene Runner version 3.0 (Hastings Software, Inc., Hastings, NY). Duplicate clones of each PCR product were sequenced and analyzed to identify sequencing errors. Clones obtained from independent CC-PCR of the same sample were analyzed to exclude random nucleotide misincorporation by *Taq* DNA polymerase. The *C. parvum* KSU-1 *hsp70* reference sequence used in this study has GenBank accession number U11761.

## 3. Results

### 3.1. OOCYST RECOVERIES USING SEEDED WATER SAMPLES

Flotation and IMS recoveries of oocysts from raw and filter backwash water samples seeded with live *Cryptosporidium parvum* LA-1 are summarized in Table I. Results of CC-PCR analysis of IMS purified seeded samples are also included in Table I. While IMS had higher mean recoveries than flotation for both raw and filter backwash waters, differences were not significant ($P > 0.05$). Mean IMS and flotation oocyst recoveries were lower for filter backwash water than for raw water. This was most likely due to interference by the high amounts of debris in the backwash samples.

CC-PCR results agreed with all IFA oocyst counts of IMS purified seeded samples. This included a positive CC-PCR assay for the seeded filter backwash sample C, which had an IFA count of only 3 oocysts. IMS purified filter backwash samples B and F had no detectable oocysts by IFA and were negative by CC-PCR. Deionized water negative controls ($n = 8$) were all found to be negative by IFA and CC-PCR.

Recovery of *C. parvum* from 20 spiked 100 L finished water samples averaged 37% (range 15 to 41%) (Figure 1). The turbidity of the filtered water was typically <0.1 NTU.

TABLE I
Oocyst recoveries from seeded water samples and CC-PCR results

| Sample | % Oocyst recovery | | CC-PCR |
|---|---|---|---|
| | Flotation IFA | IMS IFA | |
| Raw water A | 18.3 | 56.6 | Positive |
| Raw water B | 14.9 | 33.7 | Positive |
| Raw water C | 14.2 | 10.4 | Positive |
| Raw water D | 22.3 | 6.9 | NA[a] |
| Raw water E | 13.6 | 27.8 | Positive |
| Raw water F | 16.1 | 21.0 | Positive |
| Backwash water A | 9.4 | 17.2 | Positive |
| Backwash water B | <7.7 | <1.7 | Negative |
| Backwash water C | <3.9 | 6.2 | Positive |
| Backwash water D | 2.0 | 7.2 | NA |
| Backwash water E | <1.2 | 19.8 | Positive |
| Backwash water F | <10.9 | <2.9 | Negative |
| Deionized water negative control ($n = 8$) | ND[b] | ND | Negative |
| Mean % recovery | Raw 16.6 ± 3.0 Backwash 5.8 ± 3.7 | Raw 26.1 ± 16.5 Backwash 9.1 ± 6.9 | 100% Agreement with IFA |

[a]NA, data not available, invalid CC-PCR assays due to power failure during PCR and an incubator malfunction, respectively [b]ND, none detected

## 3.2. DETECTION OF *CRYPTOSPORIDIUM* IN ENVIRONMENTAL SAMPLES

Flotation-IFA detected oocysts in 13.2% (16 of 121) of the raw and 5.7% (7 of122) of the filter backwash water samples, while CC-PCR detected infectious *C. parvum* in 5.0% (6 of 121) of the raw and 7.4% (9 of 122) of the filter backwash water samples. Raw water mean equivalent volumes assayed were 1.57 L by the Percoll-sucrose flotation and 2.68 L by IMS. Filter backwash water mean equivalent

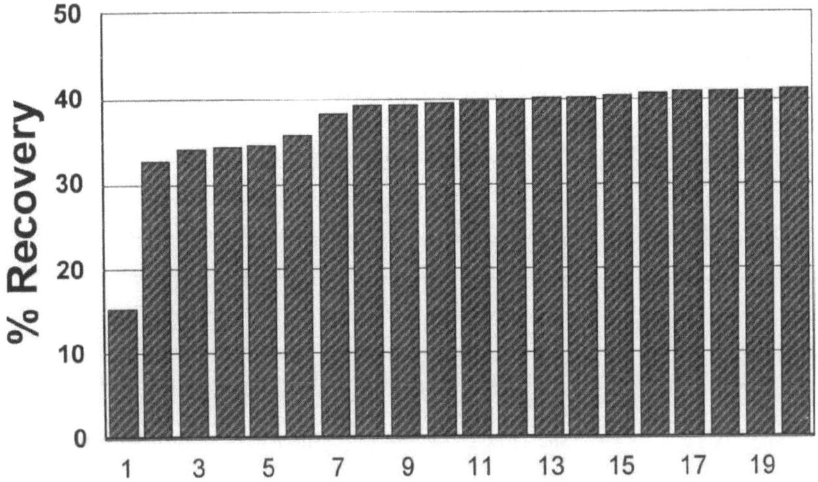


Fig. 1. Recovery efficiency of *C. parvum* from 20 spiked finished drinking water samples. Average recovery was 37%.

assayed were 0.75 L by flotation and 1.01 L by IMS. Sites with positive samples for *Cryptosporidium* oocysts by flotation IFA and infectious *C. parvum* by CC-PCR are listed in Table II. Several sites had multiple positive samples and sites 11 and 25 had samples which were positive by both flotation IFA and CC-PCR. Using both methods *Cryptosporidium* was detected at 19 of the 25 sites.

### 3.3. DNA SEQUENCE ANALYSIS OF CC-PCR PRODUCTS

All CC-PCR products were confirmed by cloning and DNA sequence analysis to be >98 % homologous to the *C. parvum* KSU-1 *hsp70* gene. Five of the CC-PCR products were found to be 100% homologous to the *C. parvum* KSU-1 *hsp70* reference sequence while the other products contained nucleotide substitutions which represented six different *hsp70* genotypes (Figure 2). *C. parvum* LA-1, which was used as our laboratory quality control strain, was also found to differ from the *C. parvum* KSU-1 *hsp70* reference sequence at three positions and a single environmental *C. parvum* strain also had this *hsp70* genotype (Figure 2, AWS-11).

### 3.4. FINISHED WATER TESTING

Beginning April 1999, utility subsidiaries of the American Water Works Company began testing the filtered drinking water from conventional surface water treatment plants for the presence of infectious *C. parvum*. Eventually, all 80 treatment plants will be tested monthly for 24 months. When completed, this data base of 1,920 samples will provide an accurate assessment of the risk of cryptosporidiosis from filtered surface water (typically <0.1 NTU). Initial results from the analysis of 50 samples from 26 sites has not detected infectious *C. parvum* in drinking water samples.

TABLE II
Sites positive for *Cryptosporidium* oocysts by flotation IFA and infectious *C. parvum* by CC-PCR

| Flotation IFA | | CC-PCR | |
|---|---|---|---|
| Raw | Backwash | Raw | Backwash |
| Site 2 (x 2)[a] | Site 11 (x 2) | Site 7 | Site 3 |
| Site 3 | Site 13 | Site 9 | Site 8 |
| Site 4 | Site 17 | Site 11* | Site 9 |
| Site 5 | Site 25 (x 3)* | Site 12 | Site 10 |
| Site 8 | | Site 13 | Site 15 |
| Site 9 | | Site 14 | Site 17 (x 2) |
| Site 11* | | | Site 21 |
| Site 14 | | | Site 25* |
| Site 16 | | | |
| Site 18 | | | |
| Site 20 | | | |
| Site 21 | | | |
| Site 25 | | | |

[a]Number of positive sample occurrence in parentheses if more than one
*Sites which had a sample positive by both flotation IFA and CC-PCR

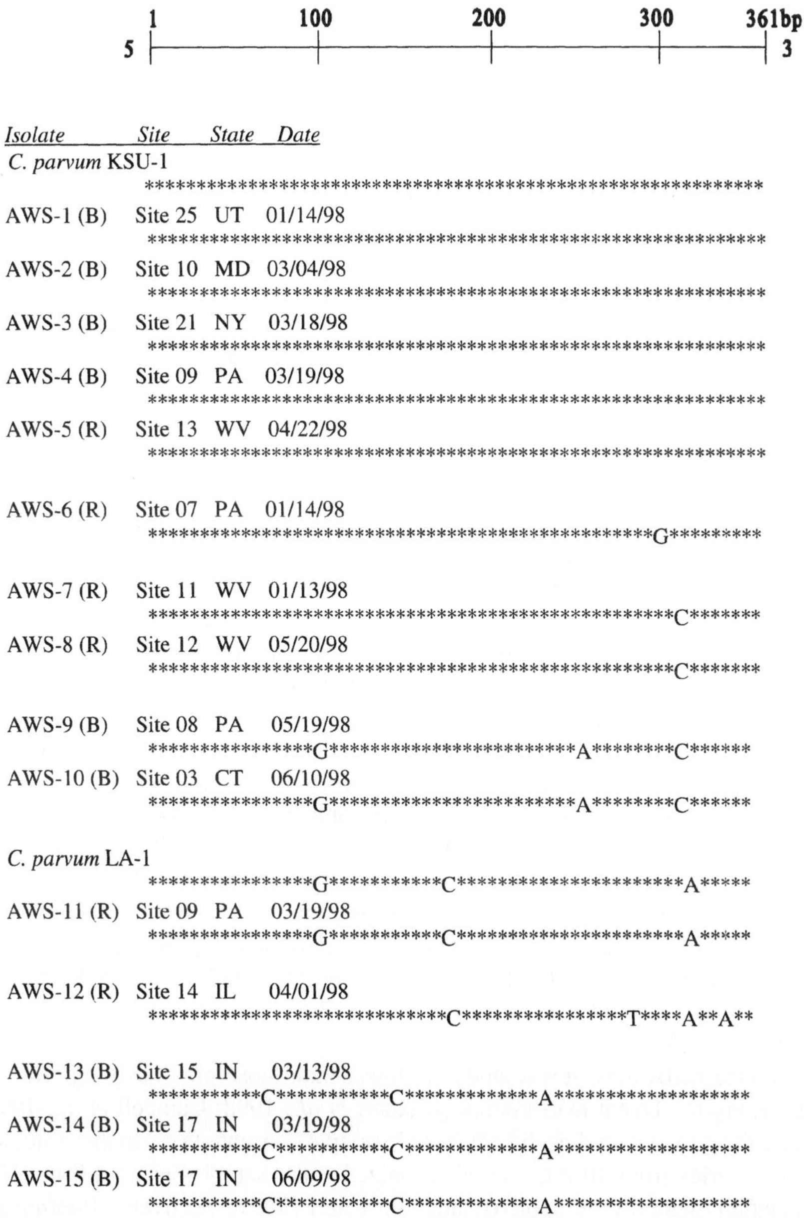


Fig. 2. DNA sequence alignment of *hsp70* CC-PCR products of infectious *C. parvum* recovered from raw (R) and filter backwash (B) water. *C. parvum hsp70* PCR products were 361bp in length and the *C. parvum* KSU-1 *hsp70* sequence was the reference sequence (GenBank accession U11761, nucleotides 2423-2783). *C. parvum* LA-1 served as the laboratory quality control strain. Nucleotide substitutions from the 5′ end were as follows: nt 2507, T→C; nt 2528, A→G; nt 2563, T→C; nt 2591, T→C; nt 2646, C→A; nt 2682, G→A; nt 2714, C→T; nt 2731, A→G; nt 2740, T→C; nt 2747, G→A; nt 2762, G→A.

## 4. Discussion

To make accurate human health risk assessments for *Cryptosporidium*, it is critical to have a method to determine viable and infective oocysts in drinking water. Such assessments are necessary for policy makers to develop appropriate regulations to ensure the safety of potable water supplies, and for engineers and scientists to evaluate the adequacy of treatment barriers. In this study we present the results of the first field testing of an integrated cell culture-PCR *C. parvum* assay that provides measurement of oocyst viability and infectivity.

Comparison of oocyst recoveries from seeded samples showed that the mean IMS oocyst recoveries were higher than Percoll-sucrose flotation recoveries for both raw and filter backwash samples, although the differences were not statistically significant (Table I). Despite the similar recoveries of flotation and IMS with seeded samples revealed in this study, there are several advantages of IMS over flotation which are critical for the infectivity assay strategy employed in this study. First, IMS is superior to flotation in removing debris and allows greater equivalent volumes of water concentrate to be assayed. This was particularly true for the raw water samples assayed in this study. This difference was not as evident for the filter backwash samples assayed in this study since the experimental design limited IMS to the purification of 1.5 mL of packed pellet and flotation to 2.0 mL packed pellet; and flotation was more effective at removing the large debris particles from filter backwash water than the fine particles from the raw water samples. Another advantage of IMS over flotation is that the final IMS purified sample volume is typically 50 μl compared to 5 mL for flotation purified samples. This is critical for the CC-PCR assay since large volume purified samples are not amenable to analysis. CC-PCR results for IMS purified seeded samples agreed with all IFA oocyst counts. This included a positive CC-PCR assay for a seeded filter backwash sample which had an IFA count of only 3 oocysts. In addition, no cytotoxic effects to cell monolayers of trace water debris present in IMS purified samples was observed for any of the water samples. Data for recovery of oocysts from finished drinking water samples (Figure 1) yielded an average 37% recovery efficiency with a coefficient of variation of 15.3%.

The IMS recoveries in this study are lower than those previously reported by others using the Dynal IMS system (Bukhari *et al.*, 1998; Campell *et al.*, 1997) because the previous studies did not include sample concentration. In those studies, IMS recoveries from 10 mL seeded deionized water samples using a 0.1 N HCl dissociation method were approximately 93% and 77% respectively. Preliminary trials in our laboratory using 10 mL deionized water samples seeded directly with oocysts revealed a mean recovery of 92.6% ($n = 3$) using the 0.1 N HCl dissociation method and a mean recovery of 61.2% ($n = 6$) using the AHBSS/trypsin dissociation method. Despite the lower dissociation efficiency, the AHBSS/trypsin dissociation method was used in this study instead of the 0.1 N HCl method because of its compatibility with the CC-PCR infectivity assay. The AHBSS/trypsin dissociation serves as an excystation trigger and pretreatment for cell culture and has been shown to have a less adverse impact than 0.1 N HCl on the infectivity of environmentally stressed oocysts.

Comparison of detection of *Cryptosporidium* oocysts by flotation IFA and infectious *C. parvum* by CC-PCR (Table 2) showed that of the 15 CC-PCR positive samples only 2 were IFA positive for oocysts. It was anticipated that there would be a greater number of IFA positive samples than CC-PCR positive samples since IFA detects all oocysts, dead or infectious. Indeed, IFA detected oocysts in 13.2% of the raw water samples compared to the detection of infectious *C. parvum* by CC-PCR in 5.0% of the raw water samples. Unexpectedly, there were more filter backwash water samples positive for infectious *C. parvum* by CC-PCR (7.4%) than total oocysts detected by IFA (5.7%). These results are likely a reflection of the splitting of samples containing very low numbers of oocysts between the IFA and CC-PCR assays; the similar equivalent volumes of filter backwash water examined by IFA and IMS (0.75 L and 1.01 L respectively).

Previous studies have shown *Cryptosporidium* oocysts present in treatment plant filter backwash water (Richardson *et al.*, 1991; Cornell and Lee, 1993; Karanis *et al.*, 1996). The degree of concentration of oocysts by sand filtration was reported to range from less than one (Karanis *et al.*, 1996) up to three orders of magnitude (Richardson *et al.*, 1991; Cornell and Lee, 1993). In this study, IFA positive raw water samples had a mean concentration of 192 oocysts $100L^{-1}$ (range 37-1,463), while filter backwash water samples had a mean concentration of 224 oocysts $100L^{-1}$ (range 37 -556). If the IFA data is adjusted using oocyst recovery efficiencies determined from seeded samples, then positive filter backwash samples contained about 3.3 times as many oocysts than positive raw water samples. This level is reasonable given the removal of oocysts by sedimentation and the volume of backwash water.

Perhaps more important than the levels of organisms, is the determination that 7.4% of the filter backwash samples contained live, infectious oocysts. Previous studies revealed that oocysts may be present in treatment plant effluents (LeChevallier *et al.*, 1991 and 1993), including two sites from this study which had samples positive for infectious *C. parvum*. These results demonstrate that viable *C. parvum* are present in source waters and are able to penetrate treatment barriers to the point of filtration. It is not inconceivable that oocysts might breakthrough and enter finished drinking water supplies if the treatment process is not carefully controlled. On-going studies are evaluating this possibility, however, viable oocysts have not yet been detected in the limited number of samples processed.

DNA sequence analysis of the CC-PCR products revealed a total of six different *C. parvum hsp*70 genotypes (Figure 2). Our laboratory quality control strain, *C. parvum* LA-1, differed from the *C. parvum* KSU-1 *hsp*70 reference sequence at three nucleotide positions. These differences were reproducible with different lots of oocysts obtained from the supplier which were harvested from different experimentally infected immunosuppressed gerbils. These results suggested that *hsp*70 sequences may be useful for differentiating strains of *C. parvum*. Five of the CC-PCR products (AWS-1-5) were identical to the *C. parvum* KSU-1 *hsp*70 reference sequence and came from samples from five different states, suggesting widespread occurrence of this *C. parvum hsp*70 genotype. Similarly, AWS-9 and AWS-10, which represented a novel *C. parvum hsp*70 genotype, came from samples collected in two different states. In contrast, other *C. parvum hsp*70 genotypes appeared to have limited geographic distribution. For example, the *C. parvum hsp*70 genotype represented by AWS-13, AWS-14, and AWS-15, came only from samples collected in Indiana. There was also evidence of the occurrence of mixed *C. parvum hsp*70 genotypes within watersheds. Two samples from site 9 (AWS-4 and AWS-11) were collected at the same time, yet contained two different *C. parvum hsp*70 genotypes. In contrast, two samples from site 17 (AWS-14 and AWS-15) were collected at different times and contained identical *C. parvum hsp*70 genotypes. These results exemplify the complex ecology of *C. parvum* and analysis of additional environmental strains is necessary to clarify the significance of our findings.

Using the PCR primers of Johnson *et al.* (1995) we attempted to amplify *Cryptosporidium* 18S rRNA genes from the genomic DNA present in the completed *hsp*70 CC-PCR samples to gain further information about the infectious *C. parvum* detected in this study. Sequence polymorphism within this amplified region differentiates *C. parvum* from *C. baileyi*, *C. muris* and *C. wrairi* (Johnson *et al.*,1995); as well as differentiates human and animal *C. parvum* strains (Morgan *et al.*, 1997; Xiao *et al.*, 1998). Thus far, we have been able to successfully amplify and sequence only one sample (AWS-12) and the 18S rRNA gene fragment was identical to a novel human *C. parvum* strain characterized by colleagues at the Centers for Disease Control and Prevention (unpublished results). The AWS-12 human *C. parvum* strain represented a novel *hsp*70 genotype (Figure 2).

Results of this study support the utility of IMS as a sample purification method and the high sensitivity and specificity of the CC-PCR assay using water samples of diverse water qualities. Future research is aimed at optimization of oocyst recovery and application of the CC-PCR infectivity assay to finished water samples.

## Acknowledgements

This research was supported by grants from the American Water Works Research Foundation and the American Water Works Service Company, Inc. We gratefully acknowledge the efforts of Ramon Aboytes-Torres, and Dale Young for sample processing.

## References


Bukhari, Z., McCuin, R. M., Fricker, C. R. and Clancy, J. L.: 1998, *Appl Environ Microbiol* **64,** 4495-9.

Campbell, A. T., Gron, B. and Johnsen, S. E.: 1997, in C.R. Fricker, J.L. Clancy, P.A. Rochelle (eds), American Water Works Association; Newport Beach, CA, 91-96.

Cornwell, D. and Lee, R.: 1993, *Recycle stream effects on water treatment,* American Water Works Association Research Foundation, Denver, CO.

Deng, M. Q., Cliver, D. O. and Mariam, T. W.: 1997, *Appl Environ Microbiol* **63,** 3134-8.

Di Giovanni, G. D., Le Chevallier, M., Battigelli, D., Campbell, A. and Abbaszadegan, M.: 1997, in *Proceedings of the Water Quality Technology Conference,* American Water Works Association, Denver, CO.

Gibbons, C., Rigi, F. and Awadelkariem, F.: 1998, *Protist* **149,** 127-134.

Johnson, D. W., Pieniazek, N. J., Griffin, D. W., Misener, L. and Rose, J. B.: 1995, *Appl Environ Microbiol* **61,** 3849-55.

Karanis, P., Schoenen, D. and Seitz, H. M.: 1996, *Zentralbl Bakteriol* **284,** 107-14.

Kaucner, C. and Stinear, T.: 1998, *Appl Environ Microbiol* **64,** 1743-9.

Khramtsov, N., Tilley, M., Blunt, D., Monteleone B. and Upton, S.: 1995, *J Euk Microbiol* **42,** 416-422.

LeChevallier, M. W. and Norton, W. D.: 1993, *Monitoring of Giardia and Cryptosporidium in the American Water System,* American Water Works Service Co., Inc., Voorhees, NJ.

LeChevallier, M. W., Norton, W. D. and Lee, R. G.: 1991, *Appl Environ Microbiol* **57**, 2617-21.

Morgan, U. M., Constantine, C. C., Forbes, D. A. and Thompson, R. C.: 1997, *J Parasitol* **83,** 825-30.

Richardson, A,. Frankenberg, R., Buck, A., Selkon, J., Colbourne, J. and Parsons, J.: 1991, *Epidemiol Infect* **107,** 485-495.

Rochelle, P. A., Ferguson, D. M., Handojo, T. J., De Leon, R., Stewart, M. H. and Wolfe, R. L.: 1997, *Appl Environ Microbiol* **63,** 2029-37.

USEPA: 1996, *ICR Microbial Laboratory Manual,* Office of Research and Development, Government Printing Office, Washington, DC.

USEPA: 1999, *Method 1622: Cryptosporidium in Water by Filtration/IMS/FA*, Office of Research and Development, Government Printing Office, Washington, DC.

Wagner-Wiening, C. and Kimmig, P.: 1995, *Appl Environ Microbiol* **61,** 4514-6.

Woods, K. M., Nesterenko, M. V. and Upton, S. J.: 1995, *FEMS Microbiol. Lett.* **128,** 89-94.

Xiao, L., Sulaiman, I, Fayer, R. and Lal, A. A.: 1998, *Mem. Inst. Oswaldo Cruz.* **93,** 687-692.

# GENOTOXICITY IN GERMAN SURFACE WATERS - RESULTS OF A COLLABORATIVE STUDY

G. REIFFERSCHEID[1] and T. GRUMMT[2]
[1]*AMMUG, University of Mainz, Germany,* [2]*Federal Environmental Agency, Bad Elster Branch, Germany*

**Abstract.** As surface waters are widely used for the preparation of drinking water, appropriate test systems are required for the monitoring of possible genotoxic contaminations. In the course of a BMBF-funded collaborative project several methods (Ames test, *umu* test, alkaline elution, DNA unwinding assay, Comet assay and unscheduled DNA-synthesis test) have been examined for their ability to measure genotoxicity particularly in natural surface waters. The project was subdivided into two parts: In the first part appropriate test versions were developed (sensitized Ames test, luminometric *umu* test, alkaline elution using clams, Comet assay with fish cells or aquatic plants), adapted to the test subject and validated regarding their sensitivity towards standard genotoxins. All test results were statistically evaluated. In the course of the second part both natural and concentrated samples of the rivers Rhine, Elbe, Mulde, Wupper and one drinking water resource (Wahnbachbarrage) were tested. In parallel all samples were chemically analyzed. Among the unconcentrated samples several statistically positive test results were obtained both for the rivers Elbe and Rhine especially with the Ames test and the Comet assay. Only one river (Wupper) showed significant genotoxicity in the *umu* test. In this case chemical analysis revealed concentrations of about 41 and 47 µg/l of fluoroquinolonic acid, a bacterial gyrase inhibitor which may be responsible for this effect. No genotoxicity could be found in the drinking water resource even after concentration. The water extracts clearly showed different background genotoxicity in the *umu* test corresponding to increased pollution along the Rhine (Karlsruhe < Köln < Düsseldorf). The eucaryotic in vitro tests revealed comparable results with nonconcentrated water samples. As a conclusion of the study we propose a graduated testing battery consisting of a bacterial (*umu* or Ames test) and an eucaryotic test like the Comet assay or the alkaline elution assay followed by an additional eucaryotic test (UDS test or micronucleus test) in a decisive function. There is a need for further evaluation by effect-orientated chemical analysis which should be done in the case of positive genotoxicity results in at least two tests or persistent positive results in only one.

**Keywords:** alkaline elution, Ames test, Comet assay, DNA unwinding assay, genotoxicity, indicator tests, surface water, *umu* test

## 1. Introduction

Genotoxins are widely distributed in the human environment. Their occurrance can be of natural origin but anthropogenic input into the air, the soil and water via combustion or industrial and municipal waste waters has to be considered.

Because of their ability to damage the primary information carrier of living organisms, the DNA, genotoxins are potentially hazardous. The most important reaction of the affected cells or organisms is repair of DNA damage. But repair

*Water, Air, and Soil Pollution* **123:** 67–79, 2000.

competence and capacity differs from organ to organ and varies in different organisms.

Consequences of continuous genotoxic stress can be initiation of mutations, accelerated aging of cells, ineffective adaption to changing environmental conditions or as worst case, carcinogenesis.

In recent years several studies have tried to answer the question about the responsibility of genotoxins in drinking waters as well as of natural or waste waters for enhanced cancer incidences. Genotoxic and carcinogenic effects of arsenic were discussed by several authors (Smith *et al.*, 1993, Gebel 1997, Moore *et al.*, 1997). Anderson and Wild (1994) found that genotoxins can influence the stability of ecosystems. Although the genotoxic potency of industrial wastes can vary over 10 orders of magnitude (Claxton *et al.*, 1998) municipal wastewaters can achieve loading values that are several orders of magnitude greater than those of most industries. In this context White and Rasmussen (1998) calculated that the volumetric proportion of the daily discharge that is of industrial origin rarely exceeds 30%.

Nevertheless it is very difficult to find correlations between data from acute toxicity tests and genotoxicity assays. Moreover, in most cases where genotoxic potentials were found, no correlations could be observed between genotoxicity results and chemical analysis. Complex mixtures like waste waters and surface waters are composed of a multitude of chemical substances. In order to evaluate the potential genotoxic risk of such mixtures test systems using living biological cells or organisms are needed. Chemical analysis can evaluate neither adverse effects of chemicals nor possible additive, synergistic or antagonistic events.

The purpose of the current study was to adapt and modify already known methods for the detection of genotoxicity to biological systems which belong to different biological levels including aquatic organisms or cells of those organisms. Furthermore the systems had to be modified for testing of natural surface water. The different systems were compared and evaluated in order to find an appropriate testing battery. The present contribution tries to give a survey of the results of the collaborative project. Details will be published by the participants (in preparation).

## 2. Materials and Methods

The project was subdivided into two parts: In the first part appropriate test versions were developed. The participants of the cooperative project and their test systems are listed in Table I. The bacterial tests were: conventional and sensitized high-dose Ames test and DIN-standardized and luminometric or fluorometric *umu* tests (DIN 38415-3). The eucaryotic assays were: alkaline elution with clams with fish cell lines (RTG-2) or aquatic plants (green algae),

TABLE I
Test systems and participants of the collaborative project (abbreviations see Table III)

| Test systems | Organisms | Participants |
|---|---|---|
| ***Bacterial test systems*** | | |
| *Umu* test (DIN/ISO, modif.) | *Salmon. typhim.* | Univ. Mainz, G. Reifferscheid<br>LfU Karlsruhe, J. Zipperle |
| Ames test (DIN/ISO, modif.) | *Salmon. typhim* | Univ. Heidelberg, L. Erdinger<br>FEA Bad Elster, T. Grummt |
| ***Eucaryotic test systems*** | | |
| DNA unwinding assay | cl, fla, fli | TU Berlin, P.-D. Hansen<br>Stadtw. Düsseld., J. Schubert |
| Alkaline elution | cl, alg | Univ. Mainz, P. Waldmann<br>GEW Köln, I. Hübner |
| Comet assay | prot., alg, cl, fli, fhe | WFM Mainz, U. Obst<br>Univ. Heidelberg, T. Brauneck<br>UFZ Leipzig, H. Segner<br>LfU Karlsruhe, K. Deventer<br>CCR Roßdorf, H.-G.Miltenburger |
| UDS test | fhe, cho | FEA Bad Elster, T. Grummt<br>CCR Roßdorf, H.-G.Miltenburger |
| ***Chemical analysis*** | | TZW Karlsruhe, B. Hambsch |
| ***Statistical analysis and Project coordination*** | | MHH Hannover, B. Schneider<br>FEA Bad Elster, T. Grummt |

DNA unwinding assays with fish cell lines, clams and fish larvae and unscheduled DNA-synthesis assay (UDS) with fish hepatocytes (a short description of organisms and cells is given in Table II). All test systems used were indicator tests which detect primary DNA-damage excepting the Ames assay. Whereas the Ames test measures reversions of frameshift or base pair mutations in the histidine operon, the *umu*-assay is based on the measurement of the genotoxin dependent induction of the mutator gene *umuC* in bacteria (Shinagawa *et al.* 1983; Oda *et al.*, 1985). The UmuC gene is - in connection with other genes capable of inducing DNA damage - responsible for the generation of genotoxin induced mutations in response to genotoxic stress. The alkaline Comet assay (single cell gel electrophoresis) as well as the alkaline elution technique (Kohn *et al.*, 1991) detect DNA strand breaks as the most frequent DNA-damage, alkaline-labile sites and incomplete excision repair sites. The alkaline DNA unwinding assay uses a hydroxylapatit elution technique to separate DNA double strands from single strands. The unscheduled DNA-Synthesis test (UDS) observes DNA repair synthesis after excision of damaged sites.

All test systems were adapted to test natural water and validated regarding their sensitivity towards standard genotoxins. Appropriate methods for assessment of cytotoxicity (viable cell count, measurement of cell growth, measurement of cell viability by vital staining) were applied. The genotoxic model chemicals were 4-nitroquinoline-N-oxide (4-NQO), nitrofurantoin (NF), benzo(a)pyrene, 2-acetylaminofluorene (2-AAF) and dimethylnitrosamine (NDMA) (Table II). All test results were statistically evaluated. Using the Dunnett-test lowest significant concentrations were determined for each assay (p=0.95). A water sample was considered genotoxic when the results exceeded the 95 percentage of the significance factors determined in a one year period.

TABLE II
Genotoxic substances used for evaluation and comparison of the test systems

| Substance | Metabolic activation (± S9) | Genotoxic profile and metabolic activation | Genotoxicity | Mutagenicity | Carcinogenicity* |
|---|---|---|---|---|---|
| 4-Nitroquinoline-N-oxide | - | formation of DNA-adducts; reduction by flavoproteins; esterization by seryl- and prolyl-tRNA-synthetase forms electrophilic molecules; alkylation of guanine and adenine | + | + | + |
| Nitrofurantoin | - | formation of DNA-adducts (covalent binding to tissue molecules); oxidative stress by formation of oxygen radicals | +** | +** | - |
| Benzo(a)-pyrene | + | formation of DNA-adducts; epoxydation by cytochrom P450, formation of electrophilic carbeniumion | + | + | + |
| Dimethylnitrosamine | + | alkylation; activation by microsomal cytochrom P450-enzymes leads to the formation of unstabile products, disintegrating to reactive, alkyating species | + | + | + |
| 2-Acetylaminofluoren | + | formation of DNA-adducts; activation by deacetylation and N-hydroxylation by cytochrom P450-enzymes | + | + | + |

* rodent carcinogenesis studies ** genotoxicity and mutagenicity particularly in bacterial test systems; valuation according to CHEM-BANK™ 1998

In the course of the second part samples from the rivers Rhine, Elbe, Mulde, Wupper and one drinking water resource (Wahnbachbarrage) were analysed with the biotests. In addition XAD-extracts (maximum concentration: 750 times) were tested with the *umu* test. Statistically positive results were verified by repeated testing. In parallel all samples were chemically analyzed.

Figure 1 shows the sampling sites of the surface waters. Three sites were chosen for the most important German river, the Rhine, and three sites were located at the Elbe and Mulde in the eastern part of Germany. The samples were collected over a one year period every 2 months, immediately refrigerated and shipped to the testing laboratories within 24 hours in order to preserve the stability of the samples as much as possible. Additional samples were taken from the river Wupper.

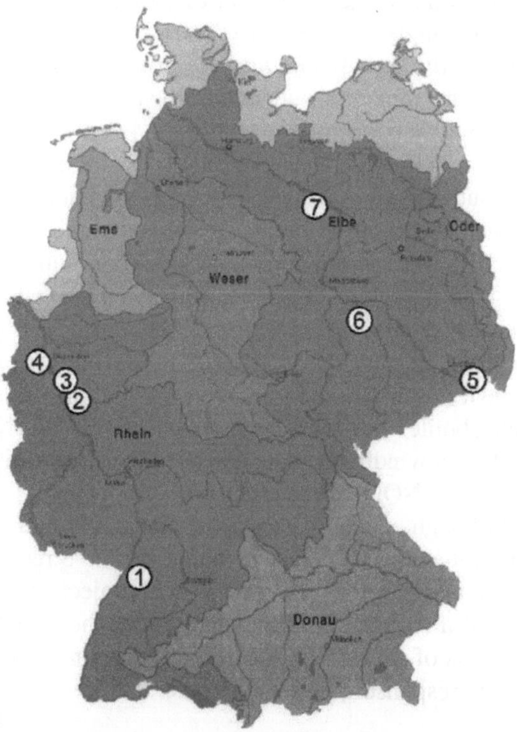


Fig. 1. Sampling sites of natural surface water.
1Karlsruhe 2Wahnbachdam 3Köln 4Düsseldorf 5Schmilka 6Dessau 7Schnackenburg

## 3. Results and Discussion

### 3.1. EFFECTS OF GENOTOXINS

The bacterial assays detected genotoxicity of 4 substances out of 5 (Table III). The high dose Ames test as well as luminometric and fluorometric *umu* tests which use low amounts of inoculated bacteria significantly showed more sensitve test results compared to the conventional tests. Especially genotoxicity of the nitrocompounds 4-NQO and NF was measured at very low concentrations. All test systems of the project succeeded in detecting the genotoxicity of the carcinogen NQO in the low microgram range. As expected, NF which acts as antibiotic was genotoxic in the bacterial tests at between 1.6 and 100 µg/l. Surprisingly the Comet assays with lower eucaryotes *Chlamydomonas reinhardtii* and *Acanthamoeba castellanii* were as sensitive as the bacterial cells. In the remaining eucaryotic test systems NF either didn't show genotoxicity or only at concentrations above 2.4 mg/l.

Most Comet assays, independent of the cell types, were unable to measure genotoxicity of the substances AAF and B(a)p up to concentrations of mg/l. In contrast primary hepatocytes (+S9) and V79 cells (as well as Ames and *umu* bacteria + S9) clearly showed genotoxicity of AAF at concentrations of between 2 and 22 mg/l. The clam *Corbicula fluminea* was able to metabolize AAF into its reactive species and showed effects at 23 µg/l. The lowest effect concentration was observed with the *umu* test strain NM2009 expressing high levels of *o*-acetyltransferase which is responsible for the transformation of AAF into its genotoxic metabolite.

Generally the DNA unwinding assay showed rather insensitive test results. With the exception of 4-NQO all substances were genotoxic only at high concentrations. Although not very practicable and subject to methodological difficulties the UDS test revealed sensitive results for 4-NQO, AAF and B(a)p. Summarizing the first part of the project, it was possible to sensitize both bacterial methods, while the adaption of the alkaline elution and Comet assays using cells of aquatic organisms in vivo or in vitro was successful and revealed graduated responses and sensitivities.

### 3.2. GENOTOXICITY OF SURFACE WATERS

The results of the testing of natural surface waters were not necessarily expected. Whereas the samples did not show any cytotoxicity, genotoxic potentials were measured at several places. An overview of positive test results is provided in Table IV. Altogether 408 assays were performed with

TABLE III

LOEC concentrations [µg/l] of the genotoxins used in the collaborative project (data taken from abstracts of the BMBF-Statusseminar Forschungszentrum Karlsruhe April 27, 1999)

| Test system | LOEC [µg/l] | | | | |
|---|---|---|---|---|---|
| | AAF | B(a)p | DMNA | NF | NQO |
| ***umu* test** | | | | | |
| DIN/ISO | 1,400; 22* | 80 | $9.3 \times 10^6$ | 20 | 10 |
| modified | n.d.;10* | 26 | $8 \times 10^6$ | 7 | 1 |
| **Ames test** | | | | | |
| DIN/ISO | 2,000 | 400 | - | 100 | 100 |
| modified | 600 | 60 | n.d. | 5 | 1 |
| **DNA unwinding** | | | | | |
| fli (RTG-2) | 7,800 | 5,000 | $125 \times 10^3$ | 7,800 | 78 |
| cl (gills) | - | 25 | - | - | n.d. |
| fla | 250 | 100 | n.d. | 1,250 | 78 |
| **Alkaline elution** | | | | | |
| cl | 23 | n.d. | 70 | 2,400 | 5 |
| alg | - | n.d. | n.d. | n.n. | 38 |
| **Comet assay** | | | | | |
| alg | - | - | 1,000 | 5 | 1 |
| prot | - | - | 10 | 1 | 10 |
| cl (gills) | - | - | $2.6 \times 10^4$ | $3.3 \times 10^4$ | 18.6 |
| fli (RTG-2) | - | 605.5 | - | $3.1 \times 10^4$ | 5 |
| fli (RTG-2) | - | 151.4 | $55.6 \times 10^3$** $1.16 \times 10^3$*** | $2.38 \times 10^3$ | 2.4 |
| fli (RTL-W1) | n.d. | 471 | - | n.d. | 2,5 - 20 |
| fhe (in vivo) | - | - | 7,400 | - | 9,5 |
| fi (gills, in vivo) | - | - | - | - | 9,5 |
| fhe | - | - | - | 2,380 | 1.9 |
| fhe (+S9) | 2,230 | 252 | 74 | n.d. | n.d. |
| V79 | $2.23 \times 10^5$ | 75.7 | $370 \times 10^3$** $1.16 \times 10^3$*** | $238.2 \times 10^3$ | 76.1 |
| **UDS test** | | | | | |
| fhe | 5.6 | 3.7 | 1,000 | - | 20 |

*umu* DIN/ISO: test performed according to DIN 38415-3 and ISO/FDIS 13829
*umu* modified: luminometric or fluorometric detection of *umuC* gene expression
Ames DIN/SIO: test performed according to DIN 38415-4
Ames modified: high dose plate incorporation Ames test using up to 4 ml sample per plate
alg: algae (*Chlamydomonas reinhardtii*)
prot: protozoa (*Acanthamöba castellanii*)
fli: fish cell line (RTG-2 or RTL-W1 rainbow trout)
cl: clams (gill cells; *Corbicula fluminea* or *Dreissena polymorpha*)
fla: fish larvae (*Brachydanio rerio*)
fhe: fish primary hepatocyte (*Brachydanio rerio or Oncorhynchus mykiss*)
V79: chinese hamster lung fibroblast cell line
cho: chinese hamster ovary cell line
*: *umu* strain NM2009 expessing high *o*-acetyltransferase activity
**: S9 from rat livers
***: S9 from hamster livers
-: no effective concentration
n.d.: not determined

water of the Rhine and 321 with samples of Elbe and Mulde. Contrary to all expectations, the highest percentage of positive test results was found at the sampling site Düsseldorf (Rhine). There, nearly 25% of all assays revealed statistically significant genotoxic potentials. The reasons for the increase of positive test results between the sampling sites Köln and Düsseldorf, which are close to each other, remain unknown. Surprisingly at the drinking water dam Wahnbachbarragapproximately 7% of all assays showed positive results, although chemical analysis did not indicate any chemical pollution. Table V summarizes all positive results of the routine testing programme (without additional samples) in a test system specific way. Considering the low sensitivities of the DNA unwinding assay in the testing of chemical substances and the differing test results of the Comet assay with fish cell lines (Table V) this table should not be regarded as a ranking list.

Because it is impossible to characterize complex mixtures of chemical substances in surface waters in their entirety, 41 industrial relevant and genotoxic aromatic nitrocompounds were selected for analysis. In samples of the Rhine (Köln, Düsseldorf) the genotoxins *o*-toluidine, *o*-anisidine, *m*-cresidine and 2-amino-4-nitrotoluene were found at concentrations below 30 ng/l. At the river Elbe, Schmilka was the most polluted site concerning nitroaromatic genotoxins. In some cases concentrations of more than 100 ng/l of *o*-toluidine, *o*-anisidine, *m*-cresidine and 2-amino-4-nitrotoluene were measured (data not shown). 2,4-dinitrotoluene was detected at concentrations of up to 500 ng/l.

However, correlations between genotoxicity and chemical data were not observed in the routine testing programme. The only example of a possible connection between genotoxic events and chemical analysis was found for the river Wupper which had been tested twice by an additional sample. The bacterial *umu* test showed statistically significant genotoxicity results at both sampling times (Figure 2). The induction rates were approximately 1.5 as high

TABLE IV
Overview of 'positive' test results and percentage of 'positive' test results of the natural river samples

| Rhine | | | | Elbe and Mulde | | | |
|---|---|---|---|---|---|---|---|
| sampling site | n | 'positive' test results | % 'positive' test results | sampling site | n | 'positive' test results | % 'positive' test results |
| Wahnbach barrage | 102 | 7 | 6.7 | Schmilka | 136 | 21 | 15.4 |
| Karlsruhe | 102 | 14 | 13.7 | Schnackenburg | 102 | 14 | 13.7 |
| Köln | 102 | 16 | 15.7 | Dessau | 85 | 14 | 16.4 |
| Düsseldorf | 102 | 25 | 24.5 | | | | |

TABLE V
Test specific comparison of 'positive' test results of the routine testing programme (additional samples excluded; abbreviations see table 3) (Rhine).

| Test system | Organism | Number of 'positive' test results Rhine | Elbe | Total |
|---|---|---|---|---|
| Comet assay | Fli | 11 | 5 | 16 |
| Ames (modified) | Bacteria | 8 | 7 | 15 |
| Comet assay | Prot | 10 | 3 | 13 |
| Comet assay | Alg | 7 | 3 | 10 |
| DNA unwinding | Fla | 4 | 6 | 10 |
| Comet assay | Sl | 5 | 5 | 10 |
| UDS test | Fhe | 3 | 4 | 7 |
| Comet assay | Fhe | 1 | 3 | 4 |
| Comet assay | Fli | 2 | 1 | 3 |
| DNA unwinding | Cl | 2 | 1 | 3 |
| alkaline elution | Cl | 0 | 2 | 2 |
| DNA unwinding | Fli | 0 | 1 | 1 |
| alkaline elution | Cl | | | 0 |
| Comet assay | Cl | | | 0 |
| Ames (according to DIN) | Bacteria | | | 0 |
| *umu* test (according to DIN; modified) | Bacteria | | | 0 |

as those of the negative control. The growth factor of 0.8 (20 % growth inhibition) suggested only slight cytotoxicity. The genotoxic potential was found to be attributable to the compound 1-cyclopropyl-1,4-dihydro-4-oxo-6-fluoro-7- chloro-3-quinolonic acid (FCS; between 41 and 48 μg/l). FCS is a so-called bacterial gyrase (topoisomerase) inhibitor and a potent inducer of the *umu*-genes (Figure 3). The biological function of these enzymes is resolution of topological problems in the course of DNA replication. Gyrases insert single- and doublestrand breaks into the DNA and splice them again in a controlled manner. Inhibitions of the enzyme function lead to the generation of DNA strand breaks and mutations. Normally FCS is a typical bacterial toxin but genotoxicity can obviously be modulated by UV irradiation leading to the production of DNA active oxygen species. Thus eucaryotic cells can be affected by fluoroquinolones as previously shown with the chromosomal aberration test (V79 cells) and the Comet assay using mouse lymphoma cells (Chetelat *et al.*, 1996). With the exception of the lower eucaryotic organisms *Acanthamoeba castellanii* (protozoon) and *Chlamydomonas reinhardtii* (green alga) which showed Comet formation after incubation with Wupper water (Weßler *et al.*, 1999) no eucaryotic test system revealed genotoxicity. These findings indicate that lower eucaryotic organisms could be more sensitive to fluoroquinolones than higher eucaryotic cells, which may have ecotoxicological consequences. This supposition has to be verified by additional experiments.

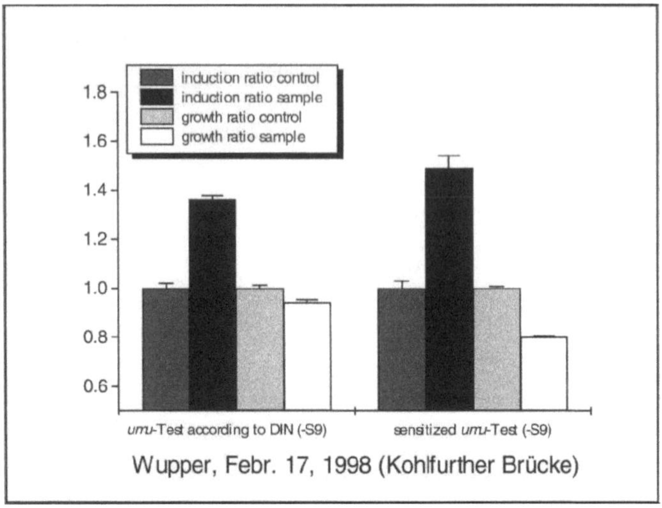


Fig. 2. Genotoxicity of original water of the Wupper measured with the *umu* test

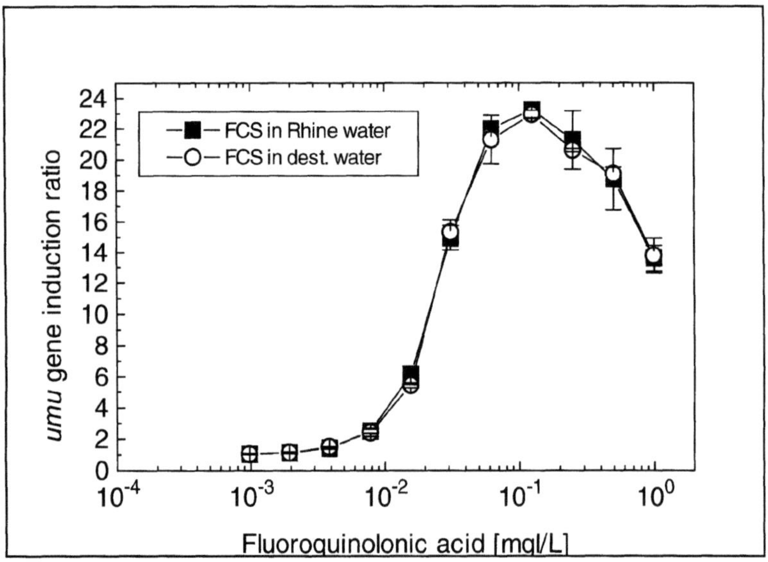


Fig. 3. Genotoxicity of fluoroquinolonic acid (FCS) in the *umu* assay (according to DIN 38415-3)

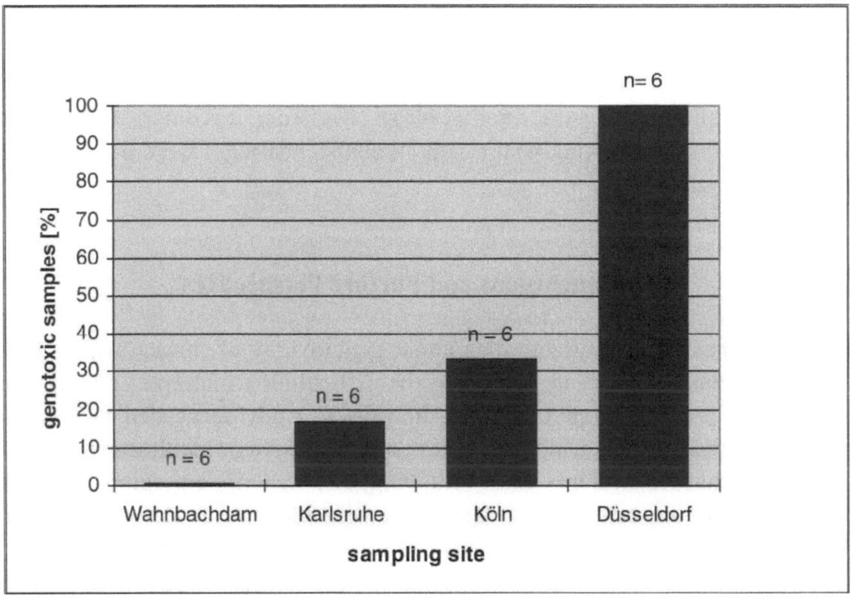


Fig. 4a. Percentage of genotoxic extracts of the Rhine samples in the *umu* assay

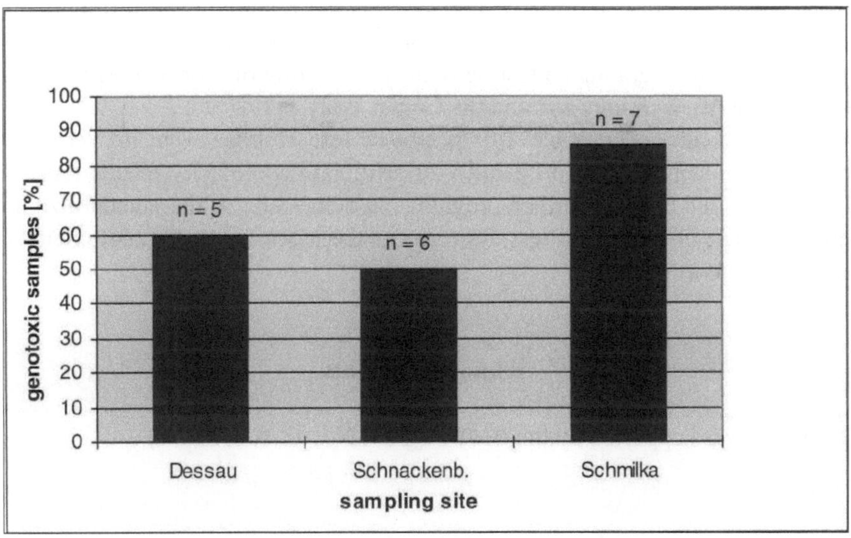


Fig. 4b. Percentage of genotoxic extracts of the Elbe and Mulde samples in the *umu* assay

Although extraction experiments are regarded as problematic because of the possible generation of misleading results in biotests by loss of components of the mixture or modulating effects due to matrix effects, they clearly showed background genotoxicity in the Rhine and Elbe measured with the *umu* assay (Figures 4a and b). Especially the genotoxicity of Rhine extracts reflected an increasing pollution over the river's course (Karlsruhe < Köln < Düsseldorf). The extracts of the drinking water dam (Wahnbachbarrage) were not genotoxic in the *umu* assay.

## 4. Conclusions and Further Perspectives

In the past nearly all available data about genotoxicity of surface waters came from the testing of extracts. Realizing the difficulties to interpret results from assays with preconcentrated samples the present study succeeded in applying appropriate test systems which are sufficiently sensitive to monitor genotoxicity of natural surface waters. The following test criteria emerging from the complexity of the test subject 'natural water' had to be considered in the project: practicability, reproducibility, sensitivity, taxonomical and ecological diversity, possibility of statistical evaluation, cost-effectiveness, transferability of results, protection of animals and compatibility of the test systems. For the screening of complex mixtures with unknown genotoxic potentials sensitive, reproducible, accurate and robust test systems are required. The results of the project indicate that the basis for a reliable detection of genotoxicity of surface waters should be a graduated testing battery consisting of 2 or 3 short term test systems like Ames or *umu* test and the Comet assay as first step.

Regarding the occurrence of 'positive' test results even in very low contaminated samples, the probability of artificial genotoxicity results must be clarified. In this context further progress must be made in the standardization and validation of biological test systems for the assessment of genotoxicity in the aquatic environment.

## Acknowledgments

This study was supported by the BMBF (grant No. 02-WU9562/0).

## References

Anderson, S. L and Wild, G. C.: 1994, *Environ. Health Perspect.* **102**, 9-12.

Claxton, L. D., Houk, V. S. and Hughes, T. J.: 1998, *Mutation Res.* **410**, 237-243.

DIN 38415-3/12.96, *Deutsche Einheitsverfahren zur Wasser-, Abwasser und Schlammuntersuchung - Suborganismische Testverfahren (Gruppe T) - Teil 3: Bestimmung des erbgutverändernden Potentials von Wasser- und Abwasserinhaltsstoffen mit dem umu-Test.*

Chetelat, A., Albertini, S. and Gocke, E.: 1996, *Mutagenesis* **11**, 497-504

Gebel, T.: 1997, *Chem. Biol. Interact.* **107**, 131-144.

Kohn, K. W.: 1991, Pharmac. Ther. **49**, 55-77.

Moore, L. E., Smith, A. H., Hopenhayn-Rich, C., Biggs, M. L., Kalman, D. A. and Smith, M. T.: 1997a, *Cancer Epidemiol. Biomarkers Prev.* **6**, 31-6.

Moore, L. E., Smith, A. H., Hopenhayn-Rich, C., Biggs, M. L., Kalman, D. A. and Smith, M. T.: 1997b, *Cancer Epidemiol. Biomarkers Prev.* **6**, 1051-1056.

Oda, Y., Nakamura, S-I., Oki, I, Kato, T. and Shinagawa, H. 1985, *Mutation Res.* **147**, 219-229.

Shinagawa, H., Kato, T., Ise, T., Makino, K. and Nakata, A.: 1983, *Gene* **23**, 167-174.

Smith, A. H., Hopenhayn,-Rich, C., Warner, M., Biggs, M. L., Moore, L. and Smith, M. T.: 1993, *J. Toxicol. Environ. Health* **40**, 223-234.

Weßler, H. and Obst, U.: 1999, Abschlußbericht BMBF-Projekt 02-WU9553/1

White, P. A. and Rasmussen, J. B.: 1998, *Mutation Res.* **410**, 223-236.

# BLAZING TOWARDS THE NEXT MILLENNIUM: LUCIFERASE FUSIONS TO IDENTIFY GENES RESPONSIVE TO ENVIRONMENTAL STRESS

D. C. ALEXANDER[1], M. A. COSTANZO[2], J. GUZZO[2], J. CAI[3], N. CHAROENSRI[1], C. DIORIO[1], and M. S. DUBOW[1*]

[1] *Department of Microbiology and Immunology, McGill University, 3775 University Street, Montreal, Quebec, Canada, H3A 2B4,* [2] *Current address: Department of Medical Genetics and Microbiology, University of Toronto, King's College Circle, Toronto, Ontario, Canada, M5S 1A8,* [3] *Current address: Department of Biochemistry, McGill University, 3655 Drummond Street, Montreal, Quebec, Canada, H3G 1Y6*

[*] *To whom correspondence should be addressed: Phone: 514 398 3926; Fax: 514 398 7052; e-mail: msdubow@microimm.mcgill.ca*

**Abstract.** Contamination of the environment by toxic compounds is a problem of global concern and in addressing this problem, it is necessary to identify the mechanisms by which specific agents exert their toxic effects, and develop effective, inexpensive strategies for detecting compounds in their biologically active and available form. We have used reporter gene fusion technology to identify genes, in the genetically well-characterized bacterium *Escherichia coli*, whose expression is affected by specific environmental toxins. As an added benefit of our approach, we have elaborated methods to use these gene fusion clones as biosensors to detect specific toxic agents, such as arsenic oxyanions. Arsenic is an abundant and useful element which is also an environmental toxin that can pose severe risks to health. Arsenic toxicity varies with oxidation state, organometallic form, and bioavailability. The *Escherichia coli ars*B fusion strains respond specifically to arsenic in its toxic, oxyanionic form, and can detect bioavailable amounts of these oxyanions in contaminated water samples. Combinations of different biosensor clones and assay automation will augment the use of luminescent biosensors for the detection of specific toxic agents in the environment.

**Keywords:** arsenic, biomonitoring, biosensors, environmental genetics, *Escherichia coli* luciferase

## 1. Introduction

Environmental contamination of water, air, and soil is a problem of global concern. Every year, vast quantities of natural and synthetic compounds are introduced into the environment. Contamination from both natural and anthropogenic sources is associated with extensive environmental damage and can affect the health of all living organisms. Both acute insults and long term chronic exposure can have detrimental effects, but the mechanisms by which many compounds exert their deleterious effects is not yet clear. However, in response to a toxic insult, an organism may reorganize its physiology, involving changes in gene expression.

*Water, Air, and Soil Pollution* **123:** 81–94, 2000.

The proteins encoded by most genes do not catalyze visible reactions, or produce distinctive phenotypes and changes in gene expression are difficult to measure. However, some genes, known as reporter genes, encode protein products which catalyze easily monitored reactions. Production of β-galactosidase from the *lac*Z gene of *Escherichia coli* can be measured with a variety of chromogenic substrates (Miller, 1972). The *lux*AB genes from the marine bacteria, *Vibrio harveyi*, encode luciferase, a heterodimeric protein which catalyzes a reaction that generates visible blue-green light (Meighan, 1993). With reporter gene fusion technology a promoterless reporter gene that lacks regulatory signals of its own, is coupled to the regulatory signals of a gene of interest. Expression of the gene of interest results in expression of the reporter gene, which can be readily measured by assaying for the reporter gene product. We have used reporter gene fusion strategies to identify genes whose expression is altered upon exposure to specific toxic agents (Diorio *et al.*,1995; Guzzo *et al.*, 1992; Guzzo and DuBow, 1994; Guzzo *et al.*, 1991; Briscoe *et al.*, 1996). By identifying and characterizing these genes, it will be possible to better understand the effects of pollutants and the mechanisms of toxicity.

Different chemical forms of a compound, or a single form in different environments, can have markedly different effects. In addition to being reliable, sensitive and inexpensive, methods to detect contamination should be able to assess if a compound is biologically available and present in an active, toxic form. Toxin-inducible gene fusion clones may be adapted as biosensors for the detection of specific chemical compounds in the environment. Because these clones are biological organisms, their gene expression is affected only when a toxic compound is present in a biologically relevant form and concentration. In addition, these clones may be sensitive to antagonistic and synergistic effects that occur when toxic agents interact with one another and in different environments. It is also advantageous that the reporter gene assays that have been developed for laboratory use are reliable, sensitive and inexpensive

In this paper, we demonstrate that toxin-inducible reporter gene fusion clones can be used as biosensors to detect pollutants in the environment. Our model system uses *Escherichia coli* gene fusion biosensor clones that are induced by arsenic. These biosensor clones were generated by constructing *lux*AB and *lac*Z reporter gene fusions to the *ars*B gene in the chromosomal *ars* operon of the bacterium, *Escherichia coli* (Cai and DuBow, 1996; Diorio *et al.*, 1995). Arsenic is a ubiquitous compound with numerous chemical forms. Arsenobetaine is not dangerous, but arsenic oxyanions are toxic and this toxicity varies with the pH of the environment (Malachowski, 1990). Transcription of the *ars* operon, as well as the *ars*B::*lux*AB and *ars*B::*lac*Z gene fusions is augmented specifically in the presence of arsenite and antimonite oxyanions (Costanzo, 1997; Diorio *et al.*, 1995). We have shown that it is possible to correlate light production by an *E. coli ars*B::*lux*AB reporter gene fusion strain in the presence of toxic oxyanions that are available to bacterial cells (Costanzo,

1997; Diorio *et al.*, 1995). This work demonstrates the potential for using luminescent biosensors to detect specific toxic agents in the environment.

## 2. Arsenic in the Environment

### 2.1. ARSENIC - SOURCES AND DISTRIBUTION

Arsenic is an abundant metalloid. Worldwide, an estimated 155,000 metric tonnes of arsenic are released into the environment each year. Approximately 30% of this is derived from natural sources, such as geothermal vents, volcanic activity, and the weathering of rocks (Ferguson and Gavis, 1972). Mining is the major anthropogenic source of arsenic. Arsenic is released during the smelting of copper, gold and other ores, as well as from the combustion of fossil fuels and through leaching of mining waste (Vahter, 1986; Chivers and Peterson, 1987). In freshwater and marine environments, arsenic concentrations in the range of 1 to 10 μg/L (part per billion) are considered average (Sanders, 1980). Arsenic concentrations in soil tend to be higher and are usually in the range of 2 to 20 mg/kg (part per million) (Huang, 1994). However, the levels of arsenic in soil and water vary dramatically with local geology and proximity to industrial sources. In some areas, arsenic levels of several hundred parts per billion have been found in drinking water (Chen *et al.*, 1995; Vahter *et al.*, 1995). Recently, in Sydney, N.S., Canada, arsenic-contamination from a coke ash waste site prompted the relocation of local residents. Soil arsenic concentrations of 52 to 435 mg/kg were reported (CanTox, 1998).

### 2.2. USES AND TOXICITY

The toxicity of arsenic is infamous. In addition to serving as the poison often used by mystery writers (Kesselring, 1940), arsenic is the active component of Lewisite, a chemical warfare agent. Medical applications of arsenic include Salvarson 606, the first synthetic antimicrobial agent, used to treat syphilis, and melarson oxide, which is employed in the treatment of protozoan infections (e.g., African sleeping sickness and Kala azar) (Doak and Freedman, 1970; Fairlamb *et al.*, 1989). Arsenicals have been used by the agricultural industry as pesticides, and in sheep and cattle dipping solutions. Chromated copper arsenate is widely employed as a wood preservative (Weis and Weis, 1995).

Arsenic exposure is associated with morbidity and mortality. Ingested doses of 50-300 mg of arsenic oxide can be fatal to humans (Chen and Lin, 1994). Acute effects include gastrointestinal damage, muscular cramps, and cardiac abnormalities (Done and Pert, 1971; Hine *et al.*, 1977). Chronic exposure to arsenic is associated with a plethora of health problems including damage to the skin, liver, lungs, cardiovascular system, and nervous system as well as an increased incidence of cancer (Tseng *et al.*, 1968; Hameda and Horiguchi, 1977; Cebrian *et al.*, 1983; Chen and Lin, 1994).

The toxicity of arsenic varies with its oxidation state and organometalloid form. Inorganic arsenite (AsIII) and arsenate (AsV) compounds are the most hazardous. Arsenite interferes with sulfur metabolism and can inactivate important sulfhydryl-containing enzymes (Squibb and Fowler, 1983). Arsenate competes with phosphate for uptake and substitution of arsenate for phosphate results in the uncoupling of oxidative phosphorylation (Malachowski, 1990). Methylation can reduce the toxicity of inorganic arsenicals. Organic arsenicals, such as arsenobetaine, are not considered toxic (Malachowski, 1990; Tamaki and Frankenberger, 1992).

## 3. Monitoring Arsenic Contamination

### 3.1. ANALYTICAL METHODS

Many analytic methods are available for the detection of toxic agents in the environment. The best methods have high specificity for the compound of interest and low, but biologically relevant, detection limits. For compounds such as arsenic, where toxicity varies with oxidation state, it is important to measure the contribution of individual species. Hydride generation atomic absorption spectroscopy (HG-AAS) is an efficient means to determine speciation of inorganic arsenic in water and has a detection limit of 0.05 nM (Masscheleyn *et al.*, 1991). A method coupling high performance liquid chromatography (HPLC) to HG-ASS has been used to measure both inorganic and organic arsenicals in urine (Lopezgonzalvez *et al.,* 1996). Although analytical methods allow accurate measurement of total arsenic concentration, they tend to neglect factors affecting bioavailability. For example, arsenic toxicity is greatly reduced by arsenic-sequestering ligands and elements, particularly phosphate, that mediate arsenic uptake (Cox *et al.*, 1996). However, current analytical methods may not consider these variables.

### 3.2. BIOMONITORING METHODS

Assay systems that employ living organisms have become increasingly popular. Biological monitoring methods are advantageous because they can measure toxins at concentrations that are biologically relevant and effective under *in situ* conditions. The most basic biomonitoring methods correlate the morbidity or mortality of test organisms (e.g., fish, invertebrates, algae, or bacteria) with the presence of toxic pollutants (Peltier and Weber, 1985). For example, the Microtox system correlates water pollution with inhibition of light emission from a bioluminescent bacteria (Bulich and Isenberg, 1981). Although such methods allow a measure of environmental health, they may not provide information about the nature of the toxic insult, the type of compounds present, or possible antagonistic or synergistic effects. Some tests measure specific types of toxin mediated damage. For example, the *Salmonella*/microsome assay (the Ames test) and the λ Inductest are used to monitor genotoxicity (Maron and Ames, 1983; Moreau *et al.,* 1976). Tests for monitoring specific compounds also exist. In the MetPad and MetPlate assays, inhibition of an *Escherichia coli* enzyme is correlated with the presence of toxic heavy metals (Bitton and Koopman, 1992; Bitton *et al.*, 1994). Zaman *et al.* (1995) developed an insect model for monitoring oxidative stress caused by arsenic. This system suggested that arsenite, but not arsenate, induced oxidative stress. Most importantly, the results varied when different insect species were used. This indicates that tests of this type need to be validated and compared with the results from a variety of species. The most useful biomonitoring strategies should allow simple, rapid, cost-effective identification of specific toxic compounds in a manner that is representative for a wide variety of organisms.

### 3.3. THE LUX Tn*5* TRANSPOSON SYSTEM

We have elaborated a gene fusion system for identifying bacterial genes whose expression is affected by environmental changes (Guzzo and DuBow, 1991). Briefly, a library of *E. coli* clones, each containing a single, random chromosomal *lux*AB transcriptional gene fusion was created via integration of a modified Tn*5* transposon (Figure 1a). This transposon has an intact IS*50*R end, which encodes the transposase function required for random transposition into the *E. coli* chomosome. The IS*50*L end is truncated, but retains the *cis*-acting sequences required for transposition. Transcription, from an exogenous, upstream promoter, can proceed through the IS*50*L and the promoterless *lux*AB genes. The *lux*AB genes encode a mixed function oxidase which, in the presence of oxygen and aldehyde, catalyzes a light-generating reaction. By exposing the library to media containing toxic compounds, and monitoring light emission (Figure 1b), we have identified genes whose expression is augmented upon exposure of *E. coli* cells to arsenic, aluminum, nickel, selenium, and tributyl tin (Diorio *et al.*, 1995; Guzzo *et al.*, 1992; Guzzo and DuBow, 1994; Guzzo *et al.*,1991; Biscoe *et al.,* 1996). Although 'gene chip' and 'gene microarray' methods for identifying changes in gene transcription are in development, such methods remain very expensive and time consuming. The *lux* Tn*5* *tet* transposon method is sensitive, specific, and simple to perform, and uses inexpensive components. Perhaps most importantly, light emission is rapid and can be measured in real time.

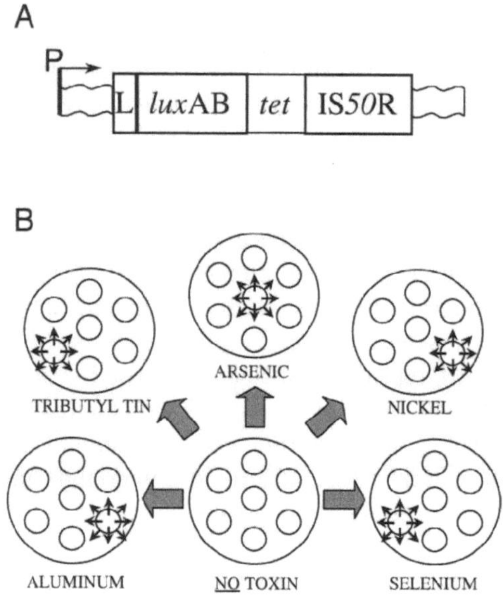


Fig.1.a. The *lux*AB Tn*5* *tet* transposon. This modified Tn*5* transposon has an intact IS*50*R end (wavy boxes), and a IS*50*L end (L) truncated such that it still contains the *cis*-acting sequences required for transposition, but does not interfere with transcription from an exogenous promoter (P). The tetracycline resistance (*tet*) gene aids selection of cells in which transposition has occurred. b. Changes in light emission can be monitored with photographic film or a photon counting luminometer.

## 4. Biomonitoring with *ars*B Gene Fusions

### 4.1. THE *ars* OPERON AND CONSTRUCTION OF *E. coli* *ars*B FUSION STRAINS

The first arsenic-tolerant bacteria were isolated from a sheep dipping solution by H.H. Green in 1918. Fifty years later, plasmid-mediated arsenic and antimony resistance was described for both Gram-positive and Gram-negative bacteria (Novick and Roth, 1968; Hedges and Baumberg, 1973). Chromosomal resistance loci have also been found (Diorio *et al.*, 1995; Carlin *et al.*, 1995; Cai *et al.*, 1998). This resistance is mediated by the highly conserved *ars* operon, which encodes an arsenite- and antimonite-inducible efflux pump. This operon contains from 3 to 5 genes. ArsR is an arsenite- and antimonite-dependent trans-acting repressor that controls basal expression of the operon (Wu and Rosen, 1991). ArsB is an arsenite oxyanion-specific efflux pump. ArsC mediates reduction of arsenate to arsenite and so is necessary for arsenate-resistance (Oden *et al.,* 1994). ArsA is an arsenite- and antimonite-inducible ATPase which, when present, enhances efflux efficiency (Rosen *et al.*, 1988). ArsD is a second trans-acting repressor. It responds to higher concentrations of inducer than ArsR, and serves to establish the upper expression limit for the *ars* operon (Wu and Rosen, 1993).

We identified the chromosomal *E. coli ars*RBC operon and generated an *E. coli ars*B::*lac*Z fusion strain (Diorio *et al.*, 1995) through insertion of a transposable bacteriophage Mu*d*I (*lac* $Ap^R$) *lac*Z fusion vector into the *E. coli* chromosome (Casadaban and Cohen, 1979). Subsequently, an *E. coli ars*B::*lux*AB fusion strain was constructed (Cai and DuBow, 1996). The *ars*B locus is highly conserved, with homology at the DNA level found among several bacteria and homology at the protein level widespread among procaryotes and eucaryotes (Table I). The eucaryotic homologs mediate transport across the cell membrane, but they do not appear to be specific for arsenic. However, it is conceivable that the phosphate transport pumps from *Saccharomyces cerevisiae* and *Caenorhabditis elegans* could be involved with arsenic uptake. The *ars*B gene product is integral to arsenic resistance. The *E. coli ars*B fusions strains, in which *ars*B is disrupted, were shown to be 10 to 100 fold more sensitive to sodium arsenite and sodium arsenate than wild type strains. Similarly, expression of the *ars* operon from a multicopy plasmid was shown to enhance arsenic resistance 3 to 10 fold (Diorio *et al.,* 1995).

## 4.2. ARSENIC DETECTION WITH *E. coli ars*B FUSION STRAINS

The *E. coli ars*B fusion strains respond to arsenic in a reproducible manner that reflects speciation, toxicity and bioavailability. Diorio *et al.* (1995) demonstrated that the *E. coli ars*B::*lac*Z fusion responds to sodium arsenite at concentrations of 5 μg As/L. Sodium arsenate, which is less toxic, is detected at concentrations of 100 μg As/L or more. However, cacodylic acid, a relatively non-toxic organic arsenical, does not induce expression, even at concentrations of 1000 μg As/L (Diorio *et al.,* 1995). The *E. coli ars*B::*lux*AB fusion strain was used to detect chromated copper arsenate (CCA) (Cai and DuBow, 1997). CCA, a toxic wood preservative that confers a green tint to treated wood and is readily leached into the environment, was detected at concentrations of 10 μg As/L. This is well within the part per billion range that is relevant to environmental monitoring. It should be noted that this response is arsenic specific. The *E. coli ars*B::*lux*AB strain did not respond to a solution that contained only chromated copper (Cai and DuBow, 1997). Costanzo (1997) from our lab, demonstrated that the *E. coli ars*B fusion strains are also capable of detecting arsenic in

TABLE I
Conservation of the ArsB Protein

Homologs of the *E. coli* chromosomal ArsB protein were identified by searching the GenBank database (Benson *et al.*, 1999) using the BLASTn algorithm (Atschul *et al.*, 1997). Homology refers to the combined amino acid similarity and identity within the homologous region. Identity refers to the amino acid identity within the homologus region.

| STRAIN | LOCATION | % HOMOLOGY | % IDENTITY |
|---|---|---|---|
| *Escherichia coli K-12 MG1655* | Chromosome | 100 | 100 |
| *Escherichia coli* | Plasmid R773 | 84 | 80 |
| *Escherichia coli* | Plasmid R46 | 87 | 84 |
| *Salmonella typhi* | Chromosome | 39 | 21 |
| *Acidiphilum multivorum* | Plasmid pKW301 | 86 | 84 |
| *Pseudomonas aeruginosa* | Chromosome | 74 | 62 |
| *Yersinia enterocolitica* | Transposon Tn2502 | 83 | 80 |
| *Yersinia pestis* | Chromosome | 82 | 75 |
| *Mycobacterium tuberculosis* | Chromosome | 44 | 26 |
| *Mycobacterium tuberculosis* | Chromosome | 40 | 22 |
| *Mycobacterium bovis* | Chromosome | 44 | 26 |
| *Mycobacterium avium* | Chromosome | 45 | 28 |
| *Mycobacterium leprae* | Chromosome | 39 | 20 |
| *Bacillus subtilis* | Chromosome | 67 | 51 |
| *Staphylococcus aureus* | Plasmid pl258 | 72 | 52 |
| *Staphylococcus aureus* | Chromosome | 71 | 53 |
| *Staphylococcus epidermidis* | Chromosome | 69 | 49 |
| *Staphylococcus xylosus* | Plasmid pSX267 | 68 | 52 |
| *Staphylococcus pneumoniae* | Chromosome | 47 | 25 |
| *Thiobacillus ferrooxidans* | Chromosome | 66 | 52 |
| *Campylobacter jejuni* | Chromosome | 51 | 59 |
| *Thermotoga maritima* | Chromosome | 42 | 22 |
| *Synechocystis PCC6803* | Chromosome | 42 | 22 |
| *Archaeoglobus fulgidus* | Chromosome | 42 | 22 |
| *Pyrococcus horikoshii OT3* | Chromosome | 44 | 21 |
| *Methanococcus jannaschii* | Chromosome | 48 | 23 |
| *Shewanella putrefaciens* | Chromosome | 66 | 52 |
| *Rhodobacter capsulatus* | Chromosome | 36 | 22 |
| *Saccharomyces cerevisiae* | Chromosome | 43 | 19 |
| *Arabidopsis thaliana* | Chromosome | 10 | 43 |
| *Caenorhabditis elegans* | Chromosome | 53 | 27 |
| *Mus musculus* | Chromosome | 45 | 22 |
| *Homo sapiens* | Chromosome | 46 | 24 |

contaminated environmental samples (Figures 2 and 3). Atomic absorption spectroscopic analysis of the Goodfellow Effluent by Environment Canada indicated that the samples contained a total arsenic concentration of 280 μg/L (Brian Walker, Personal Communication). The *E. coli ars*B fusion strains indicated that arsenic was present, but comparison with arsenite and arsenate standards suggested that the bioavailable levels were much less than 280 μg/L (Figures 2 and 3). Moreover, the *E. coli ars*B::*lux*AB and *E. coli ars*B::*lac*Z strains provided comparable results. There was a concern that undissolved matter in the effluent might interfere with the biological analysis, so both filtered and unfiltered samples were tested. Undissolved matter does not interfere with analysis, and the filtered solution was equally able to induce expression of the fusions (Costanzo, 1997). The arsenic operon is an attractive candidate for biosensor use because its transcription is mediated by specific compounds. Moreover, its widespread conservation across species boundaries suggests that results obtained with the *E. coli ars*B fusion strains will be relevant to other organisms.

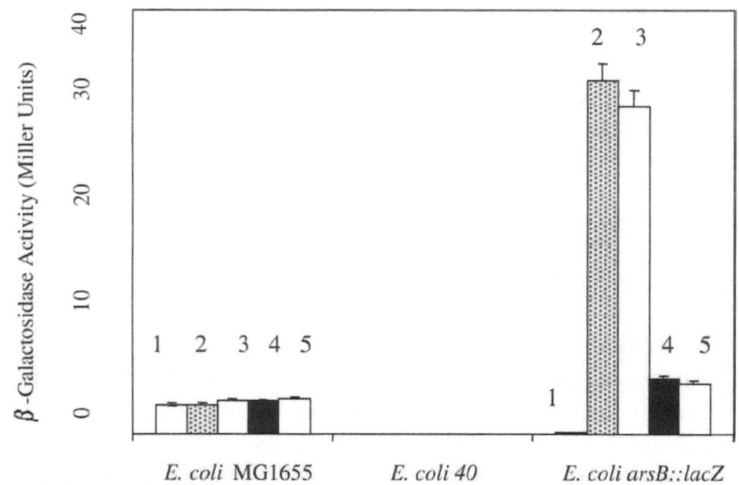


Fig. 2. Detection of arsenic oxyanions with *E. coli* MG1655 (wild type), *E. coli* 40 (Δ*lac*) or *E. coli ars*B::*lac*Z gene fusion biosensor exposed to equal volumes of distilled deionized water (1), distilled deionized water containing arsenite (100 μg/L) (2) or arsenate (1000 μg/L) (3), and filtered (4) or unfiltered (5) samples of an arsenic-contaminated environmental isolate. β-galactosidase activity, in Miller units (Miller, 1972) was measured 90 minutes after exposure (Costanzo, 1997). Error bars indicate standard deviation.

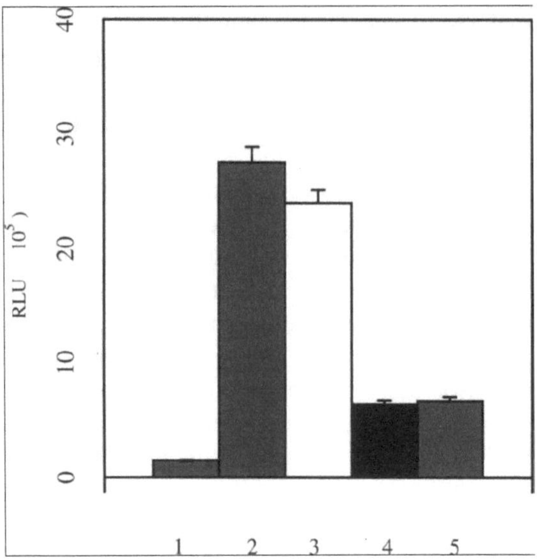


Fig. 3. Detection of arsenic oxyanions with *E. coli ars*B::*lux*AB gene fusion biosensor exposed to equal volumes of distilled deionized water (1), distilled deionized water containing arsenite (100μg/L) (2) or arsenate (1000μg/L) (3), and filtered (4) or unfiltered (5) samples of an arsenic-contaminated environmental isolate. Light emission was measured 90 minutes after exposure (Costanzo, 1997). One relative light unit (RLU) is equivalent to one photon per second. Error bars indicate standard deviations.

## 5. The Future of Biomonitoring

As the new millennium approaches, it is time to pause, and consider the condition of our world. Human activities over the past hundred years have resulted in a dramatic increase in pollution. Contamination of water, air and soil has had detrimental effects on the environment and health. Although the mechanisms by which this damage is inflicted is often unclear, monitoring for changes in gene expression in response to toxic agents can facilitate our understanding of this damage. Indeed, the term 'toxicogenomics' has already been coined to describe this genetic approach (Nuwaysir *et al.*, 1999). The success of the gene fusion techniques that we, and others, have employed demonstrates the feasibility of this strategy. In the near future, the combination of gene chip technology with data from genome sequencing projects and bioinformatics analysis, will accelerate the identification of genes whose expression is modified by toxic agents. A ToxChip, to facilitate the study of the effects of toxins on mice genes, has already been developed (Nuwaysir *et al.*,

1999; Medlin, 1999). With this technology, it is possible to monitor changes in thousands of genes simultaneously. By correlating changes in expression in response to a variety of toxins, it may be possible to identify sets of genes that are co-expressed and elucidate the pathways involved with damage or protection from specific pollutants. Some of these genes may be adapted to serve as the basis for new and specific biosensors.

It has been demonstrated that *E. coli lux* fusion biosensor clones have the potential to be used for the rapid, reliable detection of toxic agents. Other groups have developed a system that uses a set of different luminescent bacteria for detection of water pollutants (Belkin, 1998). The ideal strategy might use a panel of different luminescent biosensor clones that indicate the presence and approximate bioavailable concentration of a wide range of toxins, including heavy metals and metalloids, pesticides and halogenated phenolic compounds

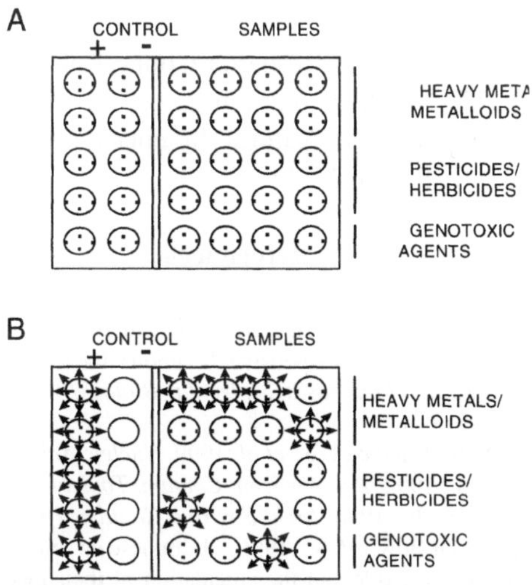


Fig. 4. a. A biomonitoring strategy that uses a matrix of luminescent organisms to detect specific environmental toxins. The 'control' region of the matrix is used to ensure that the system is working correctly. Specific concentrations of toxins may be added to the positive control region (+) to calibrate light emission. The control region could include a constitutively luminescent organism for the detection of toxins that might compromise the viability of, and therefore the light emission by, the biosensor organisms. The aqueous sample is added to organisms in the 'sample' region of the matrix. Sets of organisms may be used to determine the presence, and bioavailable concentration, of specific toxins including heavy metals and metalloids, herbicides and pesticides, and agents with genotoxic effects. b) Possible test results, as measured with a light monitoring device, such as a microplate reader. Light emission in the positive control region (+) and the absence of light emission in the negative control region (-) indicate that the test is working properly. Light emission in the 'Sample' region represents the responses of organisms to specific compounds (e.g., arsenic and mercury, atrazine, and a genotoxic agent). Absence of light emission indicates that other compounds (e.g., lead, cadmium, DDT...) are not present in a bioavailable form, or at significant concentrations.

(Figure 4). Luminescent methods show great promise, because light emission is rapid and is easily measured. Indeed, portable luminometers for field use have been developed.

Although it is essential to reduce our toxic emissions, it is also important to better understand the impact of contaminants on the environment and health. Genetic methods, such as the gene fusion strategy we have described, facilitate the identification of pathways that are involved with responses to specific toxic agents. Moreover, luminescent biosensors provide a rapid, sensitive, inexpensive method for the detection of compounds that are present in a biologically active and available form. Together, these approaches allow us to identify specific environmental pollutants and understand their effects.

## Acknowledgements

DCA is supported by a Medical Research Council of Canada Studentship (#St 46660 AP007480). The work in our laboratory is supported by an operating grant (OGP0003222) from the Natural Sciences and Engineering Research Council of Canada (NSERC) to MSD.

## References

Altschul, S.F., Madden, T.L., Schaffer, A.A. *et al.*: 1997, *Nucleic Acids Res.* **25,** 3389-3402.

Belkin, S.: 1998, 'A panel of stress-responsive luminous bacteria for monitoring wastewater toxicity' in R.A.LaRossa (ed) *Methods in Molecular Biology/Bioluminescence.* Humana Press, Inc., Totowa, N.J., pp. 247-58.

Benson, D.A., Boguski, M.S., Lipman, D.J., *et al.*:1999, *Nucleic Acids Res.* **27,** 12-17.

Bitton, G., and Koopman, B.: 1992, *Rev. Environ. Contam. Toxicol.* **125,** 1-24.

Bitton, G. Jung, K., Koopman, B.: 1994, *Arch. Environ. Contam. Toxicol.* **27,** 25-28.

Briscoe, S.F., Diorio, C., DuBow, M.S.: 1996, 'Luminescent biosensors for the detection of tributyltin and dimethyl sulfoxide and the elucidation of their mechanisms of toxicity' in M. Moo-Young, W.A. Anderson, and A.M. Chakrabarty (eds) *Environmental Biotechnology: Principles and Applications.* Kluwer Academic Publishers, The Netherlands, pp. 645-655.

Buchet, J.P. and Lauwerys, R.R.: 1985, *Arch. Toxicol.* **57,** 125-129.

Bulich, A.A. and Isenberg, D.L.: 1981, *ISA Trans.* **20,** 29-33.

Byrne, A.R., Slejkovec, Z., Stijve, T., *et al.*: 1995, *Appl. Organomet. Chem.* **9,** 305-313.

Cai, J. and DuBow, M.S.: 1996, *Can. J. Microbiol.* **42,** 662-671.

Cai, J. and DuBow, M.S.: 1997, *Biodegradation,* **8,** 105-111.

Cai, J., Salmon, K.S., DuBow, M.S.: 1998, *Microbiol.* **144,** 2705-2713.

CanTox: 1998, 'Human Health Risk Assessment of the Frederick Street Area' Report commissioned by the Nova Scotia Department of Health and Health Canada, August 1989.

Carlin, A., Shi, W., Dey, S., and Rosen, B.P. 1995. *J. Bacteriol.* **177,** 981-986.

Casadaban, M.J. and Cohen, S.N.: 1979, *Proc. Natl. Acad. Sci. USA* **76,** 4530-4533.

Cebrian, M.E., Albores, A., Aquilar, M., Blakely, E.: 1983, *Human Toxicol.* **2,** 121-133.

Chen, C.J. and Lin, L.J.: 1994, 'Human carcinogenicity and atherogenicity induced by chronic exposure to arsenic', in J.O. Nriagu (ed) *Arsenic in the Environment, Part II, Human Health and Ecosystem Effects.* John Wiley and Sons Inc. New York. NY. USA. pp 109-131.

Chen, S.-L. Yeh, S.J., Yang, M.-H. Lin, T.H.: 1995. *Biol. Trace Element Res.* **48,** 263-74.

Chilvers, D.C. and Peterson, P.J.: 1987, 'Global cycling of arsenic', in T.C. Hutchinson an K.M. Meema (eds), *Lead, Mercury, Cadmium and Arsenic in the Environment.* John Wiley and Sons Inc. New York. NY. USA. pp 279-301.

Costanzo, M.A.: 1997, M.Sc. Thesis, Department of Microbiology and Immunology, McGill University. pp. 112.

Cox, M.S., Bell, P.F., Kovar, J.L.: 1996. *J. Plant. Nutr.* **19,** 1599-1610.

Cullen, W.R. and Reimer, K.J.: 1989, *Chem. Rev.* **89,** 713-764.

Diorio, C., Cai, J., Marmor, J., Shinder, R., DuBow, M.S.: 1995. *J. Bacteriol.* **177,** 2050-2056.

Done, A.K. and Pert, A.J.: 1971. *Clin. Toxicol.* **4,** 343-355.

Doak, G.O. and Freedman, L.E.: 1970, 'Arsenicals, antimonials, and bismuthials', in A. Berger (ed), *Medicinal Chemistry*, 3rd ed. Wiley-Interscience, New York, NY, USA. pp 610-626.

Fairlamb, A.H., Henderson, G.B., Cerami, A.: 1989, *Proc. Natl. Acad. Sci. USA* **86**, 2607-2611.

Ferguson,J.F. and Gavis, J.: 1972, *Water Res.* **6,** 1259-1274.

Green, H.H.: 1918, *Rep. Dir. Vet. Res. S. Afr.* **5-6,** 593-599.

Guzzo, A., Diorio, C., DuBow, M.S.: 1991, *Appl. Environ. Microbiol.* **57,** 2255-2259.

Guzzo, A. and DuBow, M.S.: 1994, *Mol. Gen. Genet.* **242,** 455-460.

Guzzo, A. and DuBow, M.S.: 1991, *Arch. Microbiol.* **57,** 2255-2259.

Guzzo, J., Guzzo A., DuBow, M.S.: 1992, *Toxicol Let.* **64/65,** 687-693.

Hameda, T. and Horiguchi, S.: 1977, *Jpn. J. Ind. Health* **18,** 103-115.

Hedges, R.W. and Baumberg, S.: 1973, *J. Bacteriol.* **115,** 459-460.

Hine, C.H., Pinto, S.S., Nelson, K.W.: 1977, *J. Occup. Med.* **19,** 391-396.

Huang, Y.C.; 1994, 'Arsenic distribution in soils', in J.O. Nriagu (ed.) *Arsenic in the Environment, Part I, Cycling and Characterization.* John Wiley and Sons Inc. New York. NY. USA. pp 17-50.

Kesselring, J.: 1940, *Arsenic and Old Lace,* Dramatists Play Service, Inc., New York, NY. USA.

Lopezgonzalvez, M., Gomez, M.M., Palacios, M.A., Camara, C.: 1996, *Chromatographica* **43,** 507-512.

Malachowski, M.: 1990, *Clin. Lab. Med.* **10,** 459-472.

Maron, D.M. and Ames, B.N.: 1983. *Mutation Res.* **113,** 173-215.

Masscheleyn, P.H., Delaune, R.D., Patrick, W.H. Jr.: 1991. *J. Environ. Qual.* **20,** 96-100.

Medlin, J.F. 1999. *Environ. Health Perspect.***107,** A256-A258

Meighen, E.A. 1993. *FASEB J.* **7,** 1016-1022.

Miller, J.H. 1972. 'Assay of β-galactosidase' in *Experiments in Molecular Genetics.* Cold Spring Harbor Laboratories, Cold Spring Harbor, N.Y. USA pp 352-355.

Moreau, P., Bailone, A., Devoret, R.: 1976, *Proc. Natl. Acad. Sci. USA.* **73,** 3700-3704.

Novick, R.P. and Roth, C.: 1968, *J. Bacteriol.* **95,** 1335-1342.

Nuwaysir, E.F., Bittner, M., Trent, J., *et al.* 1999. *Mol. Carcinog.* **24,**153-159.

Oden, K.L., Gladysheva, T.B., Rosen, B.P.: 1994, *Mol. Microbiol.* **12,** 301-306.

Peltier, W.H. and Weber, C.I.: 1985 *Methods for Measuring the Acute Toxicity of Effluents to Freshwater and Marine Organisms*, 3rd ed. United States Environmental Protection Agency, Cincinati, OH, EPA Report-600/4-85/013.

Rosen, B.P., Weigel, U., Karkaria, C, Gangola, P.: 1988, *J. Biol. Chem.* **263,** 3067-3070.

Sanders, J.G.; 1980, *Mar. Environ. Res.* **3,** 257-266.

Shibata, Y., Sekiguchi, M., Otsuki, A., Morita, M.: 1996, *Appl. Organomet. Chem.* **10,**713-719..

Squibb, K.S. and Fowler, B.A.: 1983, 'The toxicity of arsenic and its compounds' in B.A. Fowler (ed) *Biological and Environmental Effects of Arsenic.* Elsevier, Amsterdam, pp. 233-263.

Tamaki, S. and Frankenberger, W.T. Jr.: 1992, *Rev. Environ. Contam. Toxicol.* **124,** 79-110.

Tseng, W.P., Chu, H.M., How, S.W., *et al.*: 1968. *J. Natl, Cancer Inst.* **40,** 453-463.

Vahter, M.; 1986, 'Arsenic', in T.W. Clarkson, L. Friberg, G. Nordberg, and P.R. Sager (eds), *Biological Monitoring of Toxic Chemicals.* Plenum Press. New York, NY, USA. pp 303-321.

Vahter, M., Concha, G., Nermall, B., *et al.*: 1995, *Eur, J. Pharmacol.* **293,** 455-462.

Weis, J.S. and Weis, P.: 1994. *Arch. Environ. Contam. Toxicol.* **26,** 103-109.
Wu, J. and Rosen, B.P.: 1993, *Mol. Microbiol.* **8,** 615-623.
Wu, J. and Rosen, B.P.: 1991, *Mol. Microbiol.* **5,** 1331-1336.
Zaman, K., Macgill, R.S., Johnson, J.E., *et al.*: 1995. *Arch. Insect Biochem. Physiol.* **29,** 199-209.

# IDENTIFICATION OF NEW DRINKING WATER DISINFECTION BY-PRODUCTS FROM OZONE, CHLORINE DIOXIDE, CHLORAMINE, AND CHLORINE

S. D. RICHARDSON[1], A. D. THRUSTON, JR. [1], T. V. CAUGHRAN[1], P. H. CHEN[1], T. W. COLLETTE[1], K. M. SCHENCK[2], B. W. LYKINS, JR. [2], C. RAV-ACHA[3], and V. GLEZER[3]

[1]*National Exposure Research Laboratory, U.S. Environmental Protection Agency, 960 College Station Rd., Athens, GA 30605, USA,* [2]*National Risk Management Research Laboratory, U.S. Environmental Protection Agency, 26 W. Martin Luther King Dr., Cincinnati, OH 45268, USA,* [3]*Research Laboratory of Water Quality, Israel Ministry of Health, 69 Ben-Zvi St., Tel-Aviv 61082, Israel*

**Abstract**. Many drinking water treatment plants are currently using alternative disinfectants to treat drinking water, with ozone, chlorine dioxide, and chloramine being the most popular. However, compared to chlorine, which has been much more widely studied, there is little information about the disinfection by-products (DBPs) that these alternative disinfectants produce. Thus, it is not known if the DBPs from alternative disinfectants are safer or more hazardous than those formed by chlorine. To answer this question, we have set out to comprehensively identify DBPs formed by these alternative disinfectants, as well as by chlorine. The results presented here represent a compilation of the last 8 years of our research in identifying new DBPs from ozone, chlorine dioxide, chloramine, and chlorine. We also include results from recent studies of Israel drinking water disinfected with both chlorine dioxide and chloramine. Over 200 DBPs were identified, many of which have never been reported. In comparing by-products formed by the different disinfectants, ozone, chlorine dioxide, and chloramine formed fewer halogenated DBPs than chlorine.

**Keywords:** chloramine, chlorine, chlorine dioxide, disinfection by-products, drinking water, ozone

## 1. Introduction

Although chlorine has been used to disinfect drinking water for approximately 100 years, there have been concerns raised over its use due to the formation of some potentially hazardous by-products. Of the more than 300 disinfection by-products (DBPs) that have been reported formed from chlorine treatment (Richardson, 1998), at least four of them--chloroform, bromodichloromethane, bromoform, and MX (3-chloro-4-(dichloromethyl)-4-oxobutenoic acid)--have been shown to cause cancer in laboratory animals (Bull and F.C. Kopfler, 1991; Komulainen *et al.*, 1997). Potential adverse health effects for many of the other reported chlorine DBPs are not known, as appropriate health effects studies have not been carried out. Because of these concerns, alternative disinfectants are being explored. Ozone, chlorine dioxide, and chloramine are popular alternatives, as they are effective against microorganisms, and do not form appreciable levels of the chlorine-containing

*Water, Air, and Soil Pollution* **123:** 95–102, 2000.

by-products that are of concern. However, compared to chlorine, which has been much more widely studied, there is little information about the disinfection by-products (DBPs) that these alternative disinfectants produce. Thus, it is not known if the DBPs from alternative disinfectants are safer or more hazardous than those formed by chlorine. To answer this question, we have set out to comprehensively identify DBPs formed by these alternative disinfectants, as well as by chlorine. Most of the drinking water samples were taken from locations in the United States; however, results from recent studies of Israel drinking water, which is disinfected with both chlorine dioxide and chloramine, are also included. Because Israel's source water (Lake Kinneret - the Sea of Galilee) contains natural levels of bromide that are among the highest in the world, it offered a unique opportunity to study the effect of elevated bromide levels on the formation of chlorine dioxide DBPs. DBP studies typically target a few expected chemicals in the drinking water. Due to advanced instrumentation available in our laboratory, we were able to carry out more comprehensive DBP studies, with the ability to identify new DBPs that are not present in mass spectral library databases. In this paper, we report the successful identification of many DBPs from the alternative disinfectants and from chlorine, many of which have not been previously reported.

## 2. Experimental

Ozonated water was collected from three sources: 1) a full-scale ozone treatment plant in Valdosta, GA, 2) a pilot ozonation plant in Jefferson Parish, LA, and 3) laboratory-scale ozonations carried out on Suwannee River humic and fulvic acid-fortified distilled water (5 mg/L), with an ozone dose of 2:1 ozone:dissolved organic carbon (DOC). Samples were concentrated using XAD resins. Additional experimental details are similar to those in a previous study (Richardson *et al.*, 1994). Secondary chlorine and chloramine were applied to some ozone samples collected from the Jefferson Parish pilot plant, in amounts to achieve residuals of 2 to 3 mg/L. The effect of bromide was studied by adding 1.0 ppm sodium bromide to the raw water prior to disinfection.

Chlorine dioxide-treated water was collected from a pilot drinking water treatment plant in Evansville, Indiana, and from two full-scale treatment plants in Israel. For the samples taken in Evansville, IN, secondary chlorine was applied to some of the chlorine dioxide treated samples, in amounts to obtain a free chlorine residual of 2 to 3 mg/L; additional experimental details for these samples can be found elsewhere (Richardson *et al.*, 1994). For water collected from the two full-scale plants in Israel, the following treatment conditions applied. At the first plant: 0.75 ppm chlorine dioxide was applied, along with 0.6 ppm of chloramine, which was generated with 0.8 ppm chlorine and excess ammonia. Residual disinfectant levels at the time of the sampling were 0.0 ppm free chlorine, 0.0 ppm chlorine dioxide, and 0.6 ppm chloramine. At the second plant, 0.85 ppm chlorine dioxide was applied (without additional chloramine treatment). The residual disinfectant level of chlorine dioxide at the time of the sampling was 0.35 ppm. Samples were concentrated using XAD resins.

Low and high resolution gas chromatography-electron ionization mass spectrometry (GC/EI-MS) analyses were performed on a VG 70-SEQ high-resolution hybrid mass spectrometer, equipped with a Hewlett Packard model 5890A gas chromatograph and a DB-5 column. Low resolution experiments were carried out at 1000 resolution; high resolution experiments at 10,000. Chemical ionization mass spectrometry (CI-MS) experiments were carried out on a Finnigan TSQ 7000 triple quadrupole mass spectrometer using methane gas. LC/MS analyses were accomplished using a Hewlett Packard 1050 LC coupled to a Fisons Platform quadrupole mass spectrometer. Electrospray ionization was used with a flow rate of 0.3 mL/min and a source temperature of 160°C. Samples (50 μL) were injected onto a Supelco Supelcosil C18 LC column (5μ m particle size, 150 x 2.1 mm i.d.), which was eluted using a gradient of 50:50 acetonitrile/water to 98:2 acetonitrile/water over 1 hour. Gas chromatography/infrared spectroscopy (GC/IR) analyses were performed on a Hewlett Packard Model 5890 Series II GC (with Restek Rtx-5 column) interfaced to a Hewlett Packard Model 5965B infrared detector (IRD).

## 3. Results and Discussion

### 3.1. OZONE DBPS

To determine the identity of ozone DBPs, we first applied a combination of GC/MS and GC/IR techniques to the XAD resin extracts of ozonated water samples. The criteria we used for listing an identified compound as a DBP was its presence in the treated samples in quantities at least 2 to 3 times greater than in the untreated, raw water (as judged from comparing chromatographic peak areas). Overall, many DBPs were identified, several of which have never been reported previously. Table I lists these by-products. The majority of the ozone by-products identified contained oxygen as part of their structure, with aldehydes, ketones, carboxylic acids, aldo-acids, keto-acids, keto-aldehydes, hydroxy-ketones, and cyano-aldehydes observed. No halogenated by-products were found. Many of the compounds were not present in any spectral library (NIST or Wiley). For compounds that did not contain molecular ions (which provide molecular weight information), chemical ionization mass spectrometry (CI-MS) was used to obtain this information. High resolution electron ionization mass spectrometry (EI-MS) provided the necessary empirical formula information for the molecular ion and fragments, and helped to limit the number of possible structures for each unknown DBP. GC/IR was useful for determining the functional group. When available, standards were purchased to confirm difficult identifications and to determine a particular isomer precisely, when spectra were not conclusive. Because it is believed that many of the unidentified DBPs from ozone are located in the polar fraction, which cannot be easily extracted from water, we also applied two derivatization procedures coupled with either GC/MS or liquid chromatography/mass spectrometry (LC/MS). The first derivatization procedure involved the use of pentafluorobenzylhydroxylamine (PFBHA), which reacts with small, polar aldehydes and ketones to form nonpolar oximes that can be easily extracted into an organic solvent and analyzed by GC/MS (Glaze and Weinberg, 1993). The second derivatization procedure involved the use of 2,4-dinitrophenylhydrazine (DNPH) (Grosjean and Grosjean, 1995), followed by LC/MS analysis, and has not been applied previously for the identification of DBPs. DNPH also reacts with the carbonyl group of aldehydes and ketones, but, unlike PFBHA, it permits the analysis of highly polar compounds (with multiple polar substituents). Using these approaches, we were able to successfully identify some polar aldehydes and ketones that were DBPs of ozone.

TABLE I
Ozone DBPs Identified

| **DBPs** |
|---|
| **Aldehydes:** formaldehyde, acetaldehyde, propanal, butanal, 2-methyl propenal, pentanal, 3-methyl butanal, hexanal, heptanal, octanal, nonanal, decanal, undecanal, dodecanal, tridecanal, benzaldehyde, 2-methoxy-4-hydroxy-benzaldehyde, cyanoformaldehyde |
| **Ketones:** acetone, 2-butanone, 3-methyl-2-butanone, 2-pentanone, 3-hexanone, 2-hexanone, 3-methyl cyclopentanone, C7-ketone (2 isomers), 6-methyl-5-hepten-2-one, 6-hydroxy-2-hexanone |
| **Di-carbonyls:** glyoxal, 2-ketopropanal (methylglyoxal), 2,3-butanedione (dimethylglyoxal), isomer of 2,3-butanedione, C5-dicarbonyl, 5-keto-hexanal |
| **Carboxylic acids**: 2-methyl propanoic acid, butanoic acid, 3-methyl butanoic acid, pentanoic acid, hexanoic acid, heptanoic acid, octanoic acid, nonanoic acid, decanoic acid, undecanoic acid, dodecanoic acid, tridecanoic acid, tetradecanoic acid, pentadecanoic acid, hexadecanoic acid, heptadecanoic acid, octadecanoic acid, phenylacetic acid, benzoic acid, ethanedioic acid, propanedioic acid, butanedioic acid, 2-ethyl-3-methyl maleic acid, *tert*-butyl maleic acid, pentanedioic acid, hexanedioic acid, heptanedioic acid, octanedioic acid, nonanedioic acid, decanedioic acid, undecanedioic acid, tridecanedioic acid, phthalic acid, isophthalic acid, terephthalic acid, 1,2,4-benzenetricarboxylic acid, 1,3,5-benzenetricarboxylic acid, 1,2,4,5-benzenetetracarboxylic acid, 1,2,3,4-benzenetetracarboxylic acid, 1,2,3,5-benzenetetracarboxylic acid |
| **Keto-acids:** 3-keto-butanoic acid, 3-methyl-2-keto-butanoic acid, 9-keto-nonanoic acid |
| **Nitriles:** benzeneacetonitrile, heptanenitrile |

NOTE: Underlined DBPs were confirmed by the analysis of authentic standards; other DBPs listed are tentative identifications.

## 3.2. OZONE-CHLORINE AND OZONE-CHLORAMINE DBPS

When chlorine or chloramine was used as a secondary disinfectant (after treatment with ozone), a greater variety of chlorinated and brominated DBPs was found (Table II). All of these compounds were also found in samples treated with chlorine or chloramine only. A few DBPs, however, were formed at higher levels in the ozone-chlorine and ozone-chloramine samples, indicating that the combination of ozone and chlorine or chloramine is important in their formation. In comparing by-products formed by secondary treatment of chlorine or chloramine, chloramine appeared to form the same types of halogenated DBPs as chlorine, but they were generally fewer in number and lower in concentration.

TABLE II
Halogenated Ozone-Chlorine and Ozone-Chloramine DBPs Identified

| **Halogenated DBPs** |
|---|
| **Halo-alkanes/alkenes:** chloroform, bromoform, bromodichloromethane, dibromochloromethane, dichloroiodomethane, bromochloroiodomethane, dibromoiodomethane, tribromochloromethane, 1,2-dibromo-1-chloroethane, 2,3-dichlorobutane, hexachlorocyclopentadiene |
| **Haloaldehydes:** chloroacetaldehyde, trichloroacetaldehyde (chloral hydrate), dichloroacetaldehyde, bromodichloroacetaldehyde, 2-chloro-2-methyl propanal, 2-bromo-2-methyl propanal |
| **Haloketones:** 1,1-dichloropropanone, 1,3-dichloropropanone, 1-bromo-1-chloropropanone, 1,1-dibromopropanone, 1,1,1-trichloropropanone, 1,1,3-trichloropropanone, 1-bromo-1,1-dichloropropanone, 1,1,3-tribromopropanone, 1,1,3,3-tetrachloropropanone, 1,1,1,3-tetrachloropropanone, 1,1-dibromo-3,3-dichloropropanone, 1,3-dibromo-1,3-dichloropropanone, 1,1,3-tribromo-3-chloropropanone, 1,1,3,3-tetrabromopropanone, 1,1,1,3,3-pentachloropropanone, 1,1,1-trichloro-2-butanone |
| **Halo-dicarbonyls:** 2,2,4-trichloro-1,3-cyclopentenedione |
| **Haloacids:** chloroacetic acid, bromoacetic acid, dichloroacetic acid, trichloroacetic acid, bromochloroacetic acid, dibromoacetic acid, 2-chloropropanoic acid, 2,2-dichloropropanoic acid, 2-chloro-3-methyl maleic acid |
| **Haloacetonitriles:** dichloroacetonitrile, bromochloroacetonitrile, dibromoacetonitrile, bromodichloroacetonitrile, dibromochloroacetonitrile |
| **Haloalcohols:** 2-bromoethanol, 3-chloro-2-butanol |
| **Halo-nitro-methanes:** dibromonitromethane, trichloronitromethane (chloropicrin), tribromonitromethane (bromopicrin) |
| **Other halogenated DBPs:** chloromethyl benzene, bromoxylene, 2-chlorobenzothiazole, 2-bromobenzothiazole, 2,2,2-trichloroacetamide, methane sulfonyl chloride, dichloromethyl methyl sulfone |

NOTE: Underlined DBPs were confirmed by the analysis of authentic standards; other DBPs listed are tentative identifications. Haloaldehydes are likely present in hydrated form (as for chloral hydrate).

## 3.3. DBPS FORMED WITH ELEVATED BROMIDE LEVELS

When elevated levels of bromide (1.0 ppm) were added to the water prior to ozonation, only one brominated by-product--dibromoacetonitrile--was found. Many more by-products were identified when secondary chlorine or chloramine was applied after ozonation. When comparing low-bromide water to water with elevated bromide, a tremendous shift in speciation was observed for samples treated with secondary chlorine or chloramine. In the absence of elevated bromide levels, chlorinated species dominate (e.g., chloroform, trichloroacetaldehyde, tetrachloropropanone, dichloroacetonitrile, trichloronitromethane); with elevated bromide levels, these shift to brominated species (e.g., bromoform, tribromoacetaldehyde, tetrabromopropanone, dibromoacetonitrile, tribromonitromethane). Additional details will be available elsewhere (Richardson *et al.*, in press).

## 3.4. CHLORINE DIOXIDE DBPS

*Evansville, Indiana, drinking water.* Chlorine dioxide treatment, by itself, produced several DBPs that have not been reported previously (Richardson *et al.*, 1994). Table III lists these by-products. Most of the DBPs contained oxygen in their structures: ketones, carboxylic acids, and maleic acids. Only two of the by-products were chlorinated, and no halomethanes were observed. When chlorine was used as the secondary disinfectant (after treatment with chlorine dioxide), a greater number and variety of chlorinated and brominated DBPs were found.

TABLE III
Chlorine Dioxide Disinfection By-products Identified in Evansville, Indiana Drinking Water

| DBPs |
|---|
| **Chlorine-containing compounds:** 1,1,3,3-tetrachloropropanone, (1-chloroethyl)dimethylbenzene |
| **Carboxylic acids:** butanoic acid, pentanoic acid, hexanoic acid, heptanoic acid, 2-ethylhexanoic acid, octanoic acid, nonanoic acid, decanoic acid, undecanoic acid, tridecanoic acid, tetradecanoic acid, hexadecanoic acid, 2-*tert*-butylmaleic acid, 2-ethyl-3-methylmaleic acid, benzoic acid |
| **Ketones:** 2,3,4-trimethylcyclopent-2-en-1-one, 2,6,6-trimethyl-2-cyclohexene-1,4-dione |
| **Aromatic compounds:** 3-ethyl styrene, 2-ethyl styrene, ethylbenzaldehyde, naphthalene, 2-methylnaphthalene, 1-methylnaphthalene |
| **Esters:** hexanedioic acid, dioctyl ester |

### *3.4.1. Israel drinking water*

Figure 1 shows the chlorine dioxide and chlorine dioxide-chloramine DBPs that were identified in the high-bromide Israel drinking waters. Evident from these structures is the high degree of bromine incorporation into the DBP structures. Precise isomers of the brominated aromatic DBPs are not known at this time, and in most cases, multiple isomers were evident. As most of these isomers would be expected to give similar, if not identical mass spectra, standards will be required to give precise isomer assignments. However, for these brominated aromatic compounds, it is known from high resolution EI-MS and CI-MS information that the empirical formula assignments for these compounds are correct. An interesting finding from this study is the presence of 1,1,3,3-tetrabromopropanone, which is the brominated analog of 1,1,3,3-tetrachloropropanone--a DBP found in the Evansville, IN, chlorine dioxide study.

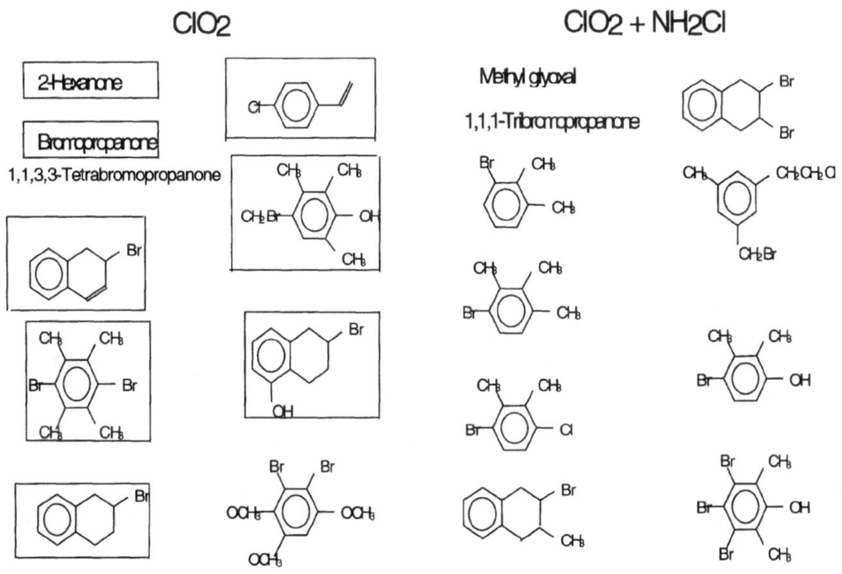


Fig. 1. Chlorine Dioxide and Chlorine Dioxide-Chloramine Disinfection By-products Identified in Israel Drinking Water

NOTE: Structures in boxes were observed in both samples treated with chlorine dioxide only and for samples treated with chlorine dioxide and chloramine. Precise isomers are not known for the brominated aromatic structures (an example structure is given in each case).

## 4. Conclusions

In comparing the different disinfectants studied, chlorine formed the largest number of halogenated by-products. Chloramine produced the same types of halogenated by-products as chlorine, but they were fewer in number and lower in concentration than for chlorine. Ozone produced no halogenated by-products, except for dibromoacetonitrile, which was formed in the presence of elevated bromide levels. In the absence of elevated bromide levels, chlorine dioxide produced relatively few DBPs, with only two halogenated DBPs observed. When elevated bromide levels were present, as in the case of the Israel drinking water, a host of brominated compounds was observed, including 1,1,3,3-tetrabromopropanone, which is the brominated analog of 1,1,3,3-tetrachloropropanone--a DBP found with chlorine dioxide treatment in the absence of elevated bromide. Non-halogenated DBPs were quite similar for all of the different disinfectants studied (e.g., carboxylic acids, aldehydes, etc.).

## References

Bull, R. J. and Kopfler, F. C.: 1991, *Health Effects of Disinfectants and Disinfection By-products*, AWWA Research Foundation, Denver, CO.

Glaze, W. H. and Weinberg, H. S.: 1993, *Identification and Occurrence of Ozonation By-Products in Drinking Water*, American Water Works Association Research Foundation, Denver, CO.

Grosjean, E. and Grosjean, D.: 1995, Liquid Chromatographic Analysis of $C_1$-$C_{10}$ Carbonyls. *Intern. J.Environ. Anal. Chem.* **61**, 47-64.

Komulainen, H., Kosma, V. -M., Vaittinen, S. -L., Vartianinen, T., Kaliste-Korhonen, E., Lotjonen, S., Tuominen, R. K. and Tuomisto, J.; 1997, Carcinogenicity of the Drinking Water Mutagen 3-Chloro-4-(dichloromethyl)-5-hydroxy-2(5H)-furanon, *J. Nat. Cancer Inst.* **89**, 848-856.

Richardson, S. D., Thruston, Jr . A. D., Collette, T. W., Patterson, K. S., Lykins, Jr. B. W., Majetich, G. and Zhang, Y.: 1994, Multispectral Identification of Chlorine Dioxide Disinfection By-products in Drinking Water, *Environ. Sci. Technol.* **28**, 592-599.

Richardson, S. D.: 1998, Drinking Water Disinfection By-products, in R.A. Meyers (ed), *The Encyclopedia of Environmental Analysis & Remediation*, vol. 3, John Wiley & Sons, New York, 1898-1421.

Richardson, S. D., Thruston, Jr. A. D., Caughran, T. V., Chen, P. H., Collette, T. W., Floyd, T. L., Patterson, K. S., Lykins, Jr. B. W., Sun , G. and Majetich, G. (in press), Identification of New Drinking Water Disinfection By-products Formed in the Presence of Bromide, *Environ. Sci. Technol.*

# ECOTOXICOLOGICAL INDICATORS OF WATER QUALITY: USING MULTI-RESPONSE INDICATORS TO ASSESS THE HEALTH OF AQUATIC ECOSYSTEMS

S. M. ADAMS and M. S. GREELEY
*Environmental Sciences Division, Oak Ridge National Laboratory*
*Oak Ridge, Tennessee, 37831, U.S.A.*

**Abstract.** As sensitive and ecologically relevant measures of environmental conditions, bioindicators can be used to assess the health of aquatic ecosystems which may be compromised by a variety of environmental stressors such as contaminants, sediments, nutrients, and varying temperature, salinity, and hydrologic regimes. The bioindicators approach is a proven bioassessment method that uses responses of key (sentinel) aquatic organisms both as integrators of stress effects and as sensitive response (early-warning) indicators of environmental health. This integrated approach involves measuring a suite of selected biological and ecological responses at several levels of biological organization from the biomolecular and biochemical to the community levels. When properly designed and applied in field situations, bioindicator studies can help identify causal mechanisms between environmental stressors and population and community-level effects, and serve as a basis for which the effectiveness of remedial actions on the health of aquatic organisms can be evaluated. Rapidly-responding sensitive biomarkers, such as biomolecular and biochemical responses, and slower- response ecologically relevant bioindicators, such as population and community responses, can be included in field bioassessment programs to provide measurement endpoints for use in environmental compliance, regulatory decision-making, and ecological risk assessments. This bioindicators approach should be particularly relevant in helping to identify and diagnose sources of stressors in environments impacted by multiple stressors. To demonstrate use of bioindicators in addressing water quality issues, spatial and temporal patters in various biological responses are related to spatial and temporal patterns of contaminants in two aquatic systems compromised by different stressors.

**Keywords**: bioindicators, ecosystem health, multiple stressors, water quality

## 1. Introduction

Aquatic ecosystems including streams, rivers, lakes, and estuaries have been subjected to increasing anthropogenic stress over the past several decades. Aquatic organisms can experience a number and variety of natural and man-induced stressors which vary both spatially and temporally. These organisms are typically subjected to variations in physicochemical factors (varying hydraulic, temperature, and salinity regimes), changes in food and habitat availability, exposure to contaminants, and increases in nutrient (eutrophication) inputs. For example, agricultural practices can have adverse impacts on the health of aquatic systems through contaminant and nutrient increases and through modifications in vegetative cover which can result in altered hydrodynamic regimes and sediment loading to these systems. Contaminants and nutrients pose a significant concern to water quality and to the health of aquatic

*Water, Air, and Soil Pollution* **123:** 103–115, 2000.

systems because of not only the varied types of pollutants that impact these systems, but also because of the many ways pollutants can effect the health of aquatic systems. In reality, most aquatic systems experience some level of chronic and/or sublethal exposure from environmental pollutants which impact these systems through point source or non-point source pathways (Schlenk, 1996). While impacts on aquatic systems from point-source pathways are relatively straightforward to assess using standard toxicological and ecotoxicological approaches, effects from non-point sources are much more difficult to assess and quantify primarily due to multiple stressor effects. In recent years eutrophication of aquatic systems from non-point sources such as agricultural runoff and atmospheric deposition, in particular, has become an increasing concern (Pelley, 1998; Charles, 1991).

Classical toxicology has been one of the primary approaches by which the effects of environmental pollutants on aquatic organisms have been assessed. A variety of standardized laboratory toxicology studies have been conducted where contaminant dose, for example, has been related to responses of individual organisms particularly at the biochemical or physiological levels. These types of controlled lab studies, however, provide limited information about how organisms, short of mortality or other survival endpoints, respond to environmental stressors in their natural environment. The experimental conditions of laboratory studies seldom accurately reflect natural conditions (Cairns, 1981) and they lack ecological realism (NRCC, 1985; Lagadic *et al.,* 1994). In addition to these laboratory toxicology studies, efforts to assess and evaluate the effects of environmental stressors on organism health have involved examination of the relationships between contaminant concentrations and specific biological responses, usually at only one or two levels of organization (e.g., Huggett *et al.*, 1992; McCarthy and Shugart, 1990). Ecotoxicological studies have also utilized correlation statistics or even semi-quantitative procedures to try and establish relationships between specific types or groups of contaminants and various biological responses (Schlenk *et al.*, 1996; Adams *et al.*, 1999). Although statistical correlations may suggest cause and effect, they cannot be used, however, to establish definitive causality between a stressor (e.g., contaminant) and specific biological responses.

Because of the problems and limitations of laboratory toxicology studies and most other methods which have been used in attempts to evaluate and quantify the effects of environmental stressors on the health of aquatic systems, better field bioassessment approaches are needed which provide an integrated framework for addressing cumulative and/or synergistic environmental impacts on the health of aquatic systems. One such method, the bioindicators approach, has proven successful in assessing and evaluating the effects of contaminants and other types of environmental stressors on the health and integrity of key components of aquatic systems. The bioindicator approach is based on using sentinel or ecologically important organisms and specific responses of these organisms as integrators and sensitive response indicators of past and existing environmental conditions (Adams, 1990; Adams *et al.* 1999). Both rapidly-responding exposure biomarkers such as bimolecular and biochemical responses (Huggett *et al.*, 1992), and slower-response but ecologically relevant bioindicators such as population and community responses are included in this bioindicator assessment methodology. Thus, this approach allows for biomarkers of exposure to various stressors to be linked or related to the more ecologically significant bioindicator effects which improves our ability to establish possible causal relationships between environmental stressors (e.g., contaminants) and ecologically relevant endpoints. These ecologically relevant endpoints have proven to be useful for addressing environmental compliance and regulatory issues and for use in the ecological risk assessment process (McCarthy and Power, in press; Power and Adams, 1997). The primary advantages of this integrated bioindicators approach, therefore, is that it can provide (1) early warning signals of environmental damage, (2) assessment of the integrated effects of a variety of environmental stressors on the health of organisms, populations, and communities, (3) sentinels of potential hazards to human health based on the responses of fish, shellfish, and other wildlife to environmental stressors, (4) evaluation of the effectiveness of remediation efforts in cleanup of contaminated areas, and (5) scientifically sound information for addressing ecological and possibly human health risk issues at contaminated sites. Within this context, the primary objective of this presentation is to demonstrate, using two brief examples, how bioindicators can be utilized in field situations to address environmental issues related to water quality. This objective will be accomplished by relating spatial and temporal patterns in various biological responses at different levels of biological organization to spatial and temporal patterns in contaminant levels in two aquatic systems impacted by different stressors.

## 2. Approach

### 2.1. CONCEPTUAL BASIS OF BIOINDICATORS

Bioindicators can be defined as anthropogenically-induced variations in biochemical or physiological components or functions (i.e., biomarkers) that have been either statistically correlated or causally-linked, in at least a semiquantitative manner, to biological effects at one or more of the organism, population, community, or ecosystem levels of biological organization (McCarty and Munkittrick, 1996). Adams (1990) more simply defined bioindicators to be multiple measures of organism, population, or community-level health which include several levels of biological organization and time scales of response. A field-based bioindicators approach basically involves measuring a selected suite of responses at each of several levels of biological organization from the molecular to community levels for key or sentinel organisms in the environment. Biological responses measured with this approach fall along two gradients, one of response time and the other of ecological relevance (Figure 1). Environmental stressors first affect sub-organismal components such as cells and tissues, and, if a stressor continues to be of sufficient duration and magnitude, effects at these lower levels will eventually be manifested at increasingly higher levels of biological organization. Responses shown at the left of Figure 1 (e.g., biomolecular/biochemical) are essentially quick-responding and sensitive indicators to environmental stressors but they are generally characterized by low ecological significance. Responses at the right side of Figure 1 (e.g., population-level), however, display very little response sensitivity but possess high ecological relevance. Ecologically-relevant endpoints are important in the ecological risk assessment process and also in environmental compliance and regulatory assessment. The bioindicators approach, therefore, includes the best two features of an environmental health assessment program, a combination of sensitive and rapidly-responding biomarkers and slower-responding bioindicators which have ecological relevance. By using this approach, therefore, we can increase our ability to link or correlate environmental stressors such as contaminants to observed biological effects.

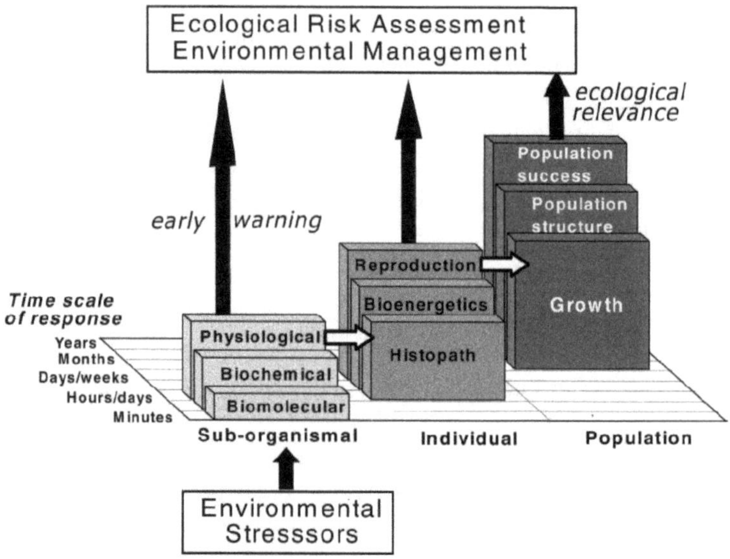


Fig.1. Hierarchial response of organisms to environmental stressors illustrating the more sensitive early warning indicators at the lower levels of biological organization and the slower-responding but more ecologically relevant indicators at the higher levels of organization (modified from Munkittrick and McCarthy, 1995).

## 2.2. BIOINDICATORS OF ENVIRONMENTAL STRESS

Corresponding to the various levels of biological organization in Figure 1, a suite of bioindicators can be measured at each of these levels as shown in Table I. Each of the six main levels of responses in Table I represent increasing levels of biological organization, decreasing levels of response sensitivity to environmental stressors, and increasing ecological relevance.

TABLE I
Representative bioindicators measured at six major levels of biological organization.

| Biochemical | Physiological | Histopath | Individual | Population | Community |
|---|---|---|---|---|---|
| MFO enzymes | Creatinine | Necrosis | Growth | Abundance | Richness |
| Bite metabolites | Transaminase enzymes | Macrophage aggregates | Total body lipid | Size & age distribution | Index biotic integrity |
| DNA integrity | Cortisol | Parasitic lesions | Organo-indices | Sex ratio | Intolerant species |
| Stress proteins | Triglycerides | Functional parenchyma | Condition factor | Bioenergetic parameters | Feeding types |
| Antioxidant enzymes | Steroid hormones | Carcinomas | Gross anomalies | Reproductive Integrity | |

At the biomolecular/biochemical level, mixed function oxidase enzymes (MFOs) have been used as indicators of exposure to a variety of contaminants (Stegeman *et al.*, 1992), bile metabolites have been used to reflect exposure to PAH types of compounds (Spies *et al.* 1996), measures of DNA integrity have proven to be reliable measures of exposure to genotoxic agents (Shugart *et al.*, 1992), stress proteins have been used widely to indicate non-specific response from a variety of environmental stressors (Pyza *et al.*, 1997), and antioxidant enzymes are just coming into wide use as indicators of oxidative stress (particularly from heavy metals) at the cell level (DiGiulio, 1992). Moving up to the next level of organization, a variety of physiological responses have proven valuable for assessing the health of organisms collected from the field including organ dysfunction indicators (creatinine for kidney damage and transaminase enzymes as indicators of liver impairment) (Adams *et al.*, 1999), cortisol as a non-specific alarm reaction stress indicator (Schreck, 1990), triglycerides (neutral lipids) as simple bioenergetic indicators of stress at the physiological level (Adams *et al.*, 1992), and steroid hormones such as female estradiol which reflects reproductive function (Donaldson, 1990). At the next level of organization (Table I and Figure 1), structural changes in tissues and organs are measured as various histopathological indicators such as cell necrosis, macrophage aggregates, and incidences of carcinomas and tumors (Teh *et al.*,1997). At the individual organism level, measures of somatic growth, total body lipid, various organo-somatic indices (liver-somatic index, etc.), and identification of several gross anomalies such as gross body lesions and disease and parasites have proven to be good representative indices of overall organism health (Goede and Barton, 1990). Population-level indicators typically measured are relative abundance, age, and size frequency (indicators of reproductive fitness and recruitment) (Shuter 1990), and various measures of reproductive integrity such as fecundity, egg condition, and age at maturity (Donaldson, 1990). Lastly, several ecologically relevant responses at the community level are used in our bioassessment programs such as species richness, the Index of Biotic Integrity (this index is composed of 10 individual community-related metrics), and distribution of trophic feeding types in the community (Fausch *et al.*, 1990). This list of bioindicators in Table I is not intended to be inclusive for all possible biological effects that can be measured in bioassessment programs, but we have found that these particular parameters have worked best for us in a variety of aquatic systems (streams, rivers, lakes, estuaries) under a variety of environmental stress situations (Adams *et al.*, 1992; Adams *et al.*,1996; Greeley *et al.*,1994).

## 3. Results and Discussion

Two examples are presented in this section to demonstrate how many of the bioindicators listed in Table I have been used in the field to assess the effects of stressors on the health of aquatic systems, to help establish possible causal relationships between environmental stressors and biological effects, and to evaluate the effectiveness of environmental remediation (cleanup). In both examples, the use of bioindicators is demonstrated by applying them within the spatial gradient of a stream or river to investigate the relationship between contaminant loading and biological effects, and also by applying them along a temporal gradient to evaluate how biological responses change as contaminant levels in the environment also change over time.

### 3.1. STREAM IMPACTED BY MIXED CONTAMINANTS

A third-order stream in East Tennessee (USA) receives point-source discharges of various contaminants such as heavy metals (mercury), PCBs, and PAHs from an industrial complex near its headwaters. This stream has a distinct gradient in contaminant loading as evidenced by the decreasing downstream concentrations of both PCBs in sunfish and sediment and also mercury in water (Peterson *et al.*, 1994). Spatial patterns in fish responses to contaminant loading and exposure were investigated at four sites in this stream which were sampled at increasing distances from the industrial outfall. Fish were also collected from three uncontaminated streams in separate watersheds to serve as reference sites. A suite of bioindicators from each of the six major categories in Table 1 were measured for individual fish and for population and community-level analysis over a 10-year time frame in the contaminated stream during a period when remedial (cleanup) activities were occurring. Therefore, as the levels of contaminants decreased in the stream, we would also expect to see concomitant improvements over time in the health of sentinel or indicator organisms in the stream at various levels of biological organization.

Relationships among fish responses at four levels of biological organization were evaluated to determine how well responses at lower levels reflected or tracked responses at higher levels of organization along the spatial gradient of the stream. The patterns of biological responses along the spatial contaminant gradient of this stream (Figure 2) were similar to the patterns in the downstream gradient of contaminant concentrations in both sediment and water which suggests possible causal relationships between contaminant concentrations and the relative intensity of individual biological responses along this spatial gradient. These results demonstrate that bioindicators at several levels of biological organization display similar downstream patterns in their response to contaminant concentrations in this system.

Elevated levels of MFO enzymes (Figure 2) indicate exposure to contaminants, and increased serum creatinine levels in fish reflect kidney impairment, which is possibly related to the downstream gradient in contaminants (mercury exposure). The similar spatial patterns in contaminant loading and in reproductive impairment as demonstrated by increasing (improved) steroid hormone concentrations downstream also suggest that contaminants may have a direct effect on reproductive performance of fish in this stream (Figure 2). At the individual level, impaired lipid function may have resulted from both direct effects of contaminants on physiological disturbances within the organism (altered metabolic performance) and also indirect effects on foodchain dynamics. These downstream patterns in biochemical, physiological, and individual responses are also reflected in the downstream patterns of population-level responses such as growth and, in particular, mean size of females. The downstream trends in community-level responses are also similar to the patterns observed at the other levels of biological organization and to the spatial trends in contaminant concentrations within the stream.

Temporal relationships at four different levels of biological organization and function are also evident for fish health in this contaminated stream (Figure 3). At the physiological level, improvement of organ (kidney) function as shown by decreased creatinine levels (Figure 3A) preceded observed changes at the individual level (growth) (Figure 3B), changes at the fish community level (as shown by species diversity) (Figure 3C), and improvements at the benthic macroinvertebrate community level (species richness) (Figure 3D).

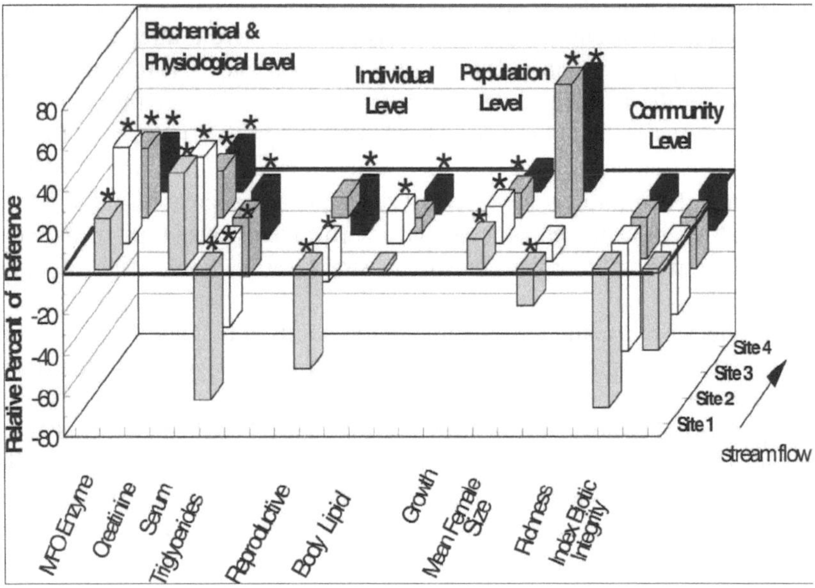


Fig. 2. Spatial gradients in bioindicator responses at four major levels of biological organization which reflect downstream gradients in contaminant loading in a stream with a point-source contaminant discharge. Stars above histograms represent significant differences in reference values ($P<0.05$).

Increases in growth of fish were also observed to coincide with the appearance of the more sensitive (pollution intolerant) fish species in the upper sections of this stream. Similarly, increases in benthic community health, as demonstrated by species diversity (Figure 3D), increased gradually after 1992 concomitant with initiation of dechlorination practices at the industrial facility and improvement in fish health at the individual and community levels. Thus, not only do bioindicators function as sensitive early-warning indicators of improvement in the health of sentinel species in impacted systems (Figure 3A), but they can also track and reflect changes at higher levels of biological organization and function (Figures 3C and 3D), suggesting causal relationships between these levels (Adams *et al.*, in press). Based on the results of this study we were able to (1) document that the health of the stream improved as cleanup activities proceeded within this system, (2) identify possible causal relationships between sources of the stress (in this case mercury and chlorine) and biological effects at several different levels of organization and, (3) provide recommendations to environmental regulators as to future biomonitoring and assessment activities appropriate for this system.

To investigate the relationships between downstream loading of contaminants and responses in the fish community in the Pigeon River at various levels of biological organization, Adams *et al.* (1996) reported that spatial gradients in contaminant loading were related to elevated detoxification enzyme (MFO) activity, several measures of condition and bioenergetic status of fish, growth parameters, and several fish population-and community-level parameters. The patterns observed among many of the response variables along the downstream gradient of this river suggest existence of causal relationships between contaminants, indicators of contaminant exposure, individual-level responses, and population-and-community-level effects. When distinct downstream gradients in biological response are observed at multiple levels of organization, establishing causality between contaminant sources and responses in aquatic organisms becomes relatively more straightforward (Adams *et al.*, 1996).

Temporal patterns in bioindicator response can also provide evidence relating environmental stressors such as contaminants to biological effects. Due primarily to a reduction in paper mill process activities, the levels of contaminants such as dioxin declined rather rapidly in the Pigeon River from 1988-1996. This trend is evident in the temporal decline of dioxin in three species of fish sampled below the paper mill discharge (Figure 4A). Consistent with this decline in body burdens of dioxin (and probably other associated contaminants in the river), the level of the mixed function oxidase enzyme, which is an indicator of contaminant exposure, also exhibited a steady reduction over this same time period (Figure 4B). At the individual and population level, the sex ratio (ratio of males vs females) increased from a low of about 5% females in 1988 to 25% in 1996 (Figure 4C) (note that the normal ratio is about 50/50). This highly skewed sex ratio is significant because dioxin and other polychlorinated compounds are known to be powerful endocrine disrupters (Colburn *et al.*, 1993) which can impair the reproductive function of individuals and therefore the structure and dynamics of wildlife populations. Consistent with the observed decreases in dioxin levels in fish, reductions in the levels of MFO enzyme activity, and improvement in sex ratio, was also a steady improvement in the integrity of the fish community in the river as reflected by the Index of Biotic Integrity (IBI) (Figure 4D). This index, which is composed of 10 individual metrics reflecting fish community status, is an integrative measure of the overall fitness and health of the fish community. Thus, just as relationships between water quality (contaminants) and spatial differences (site differences along a downstream gradient) in bioindicator responses were observed in this river, temporal relationships can also be seen between levels of contaminants in the river over time and a variety of biological responses.

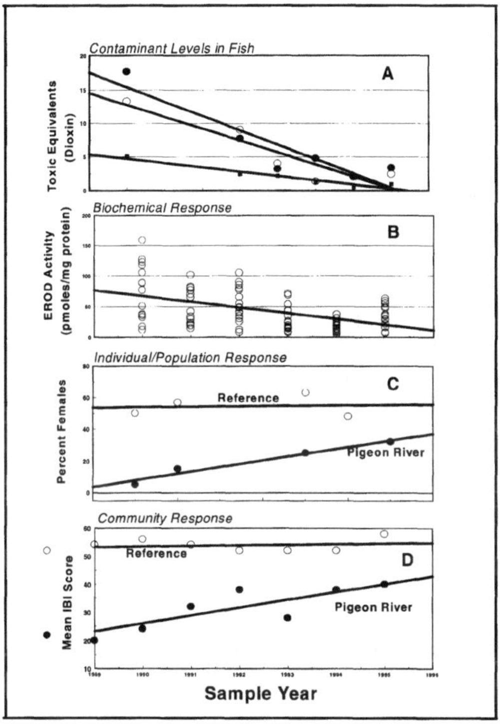


Fig. 4. Temporal response relationships between changing levels of contaminants in the Pigeon River and biological responses at three levels of biological organization.

## 4. Conclusions

In both examples presented here, spatial and temporal patterns in water quality, as reflected by levels of contaminants in the environment, were shown to be related to spatial and temporal patterns for a variety of biological responses at different levels of organization. In both the stream and river example, bioindicators at several levels of biological organization displayed not only similar downstream patterns in their response to contaminant loading, but also reflected the changes in contaminant levels over time. By identifying and establishing relationships between spatial and temporal patterns in water quality and biological effects, we should be better able to understand the mechanisms of stress responses in ecological systems caused by anthropogenic stressors such as environmental pollutants. In turn, a better understanding of the relationships between environmental stressors and effects on biological systems should improve our ability to make more informed environmental management and regulatory decisions particularly in the ecological risk assessment process. In addition, this approach should be useful in helping to identify and differentiate sources of anthropogenic stressors in environments impacted by multiple stressors.

## References


Adams, S. M., Greeley, M. S. and Ryon, M. G. (in press), *Human and Ecological Risk Assessment*.

Adams, S. M., Bevelhimer, M. S., Greeley, M. S., Levine, D. A. and Teh, S. J.: 1999, *Environ. Toxicol. Chem.* **18**, 628-540.

Adams, S. M., Greeley, M. S., Ham, K. D., LeHew, R. F. and Saylor, C. F.: 1996, *Can. J. Fish. Aquat. Sci.* **53**, 2177-2187.

Adams, S. M., Crumby, W. D., Greeley, M. S., Ryon, M. G. and Schilling, E. M.: 1992, *Environ. Toxicol. Chem.* **11**, 1549-1557.

Adams, S. M.: 1990, *American Fisheries Society Symposium* **8**,1-8.

Cairns, J.: 1981, *Water Res.* **15**, 941-952.

Charles, D. F.: 1991, *Acidic Deposition and Aquatic Ecosystems: Regional Case Studies,* Springer-Verlag, New York.

Colborn, T. and Clement, C. (eds).: 1992, *Chemically-induced alterations in sexual and functional development: the wildlife/human connection. Advances Modern Environmental Toxicology, vol 21*, Princeton Sci. Publ., Princeton, NJ.

DiGiulio, R. T.: 1992, in M. A. Mayes, and M. G. Barron (eds), Indices of oxidative stress as biomarkers for environmental contamination, *Aquatic toxicology and risk assessment,* American Society for Testing and Materials, ASTM 1124, vol. 14, Philadelphia, 15-31.

Donaldson, E. M.: 1990, *American Fisheries Society Symposium* **8**,109-122.

Fausch, K. D., Lyons, J., Karr, J. R. and Angermeier, P. L.: 1990, *American Fisheries Society Symposium* **8**, 123-144.

Goede, R. W. and Barton, B. A.: 1990, *American Fisheries Society Symposium* **8**, 93-108.

Greeley, M. S. Jr. and eleven coauthors: 1994, Bioindicator assessment of fish health and reproductive success in Lake Hartwell and Twelve Mile Creek, *Biological investigation for the remedial investigation/feasibility study, Sangamo Weston, Inc./Twelve Mile Creek/Lake Hartwell superfund site.* U.S. Corps of Engineers, Technical Memo No. 8, Savannah, GA.

Hugget, R. J., Kimerle, R. A., Mehrle, P. M. and Bergman, H. L. (eds): 1992, *Biomarkers: Biochemical, physiological, and histological markers of anthropogenic stress.* Lewis Pubs., Boca Raton, FL

Lagadic, L., Caquet, T. and Ramade, F.: 1994, *Ecotoxicology* **3**, 193-208.

McCarthy, J. F. and Shugart, L. R. (eds): 1990, *Biomarkers of environmental contamination,* Lewis Pubs. Boca Raton, FL.

McCarty, L. S. and Munkittrick, K. R.: 1996, *Human and Ecol. Risk Assess.* **2**, 268-274.

McCarty, L. S. and Power, M.: (in press), *Human and Ecol. Risk Assess.*

Munkittrick, K. R. and McCarty, L. S.: 1995, *J. Aquat. Ecosystem Health* **4**, 77-90.

NRCC (National Research Council of Canada).: 1985, *Publ. No. NRCC 24371. National Research Council of Canada*, Ottawa.

Pelley, J.: 1998, *Environ. Sci. and Technol.*, Oct. 1, 1998.

Peterson, M. J., Southworth, G. R. and Ham, K. D.: 1994,*Water, Air, and Soil Poll.* **73**, 169-178.

Power, M. and Adams, S. M.: 1997, *Environ. Manage.* **21**, 803-830.

Pyza, E., Mak, P., Kramarz, P. and Laskowski, R.: 1997, *Ecotoxicol. Environ. Saf.* **38**, 244-251.

Schlenk, D., Perkins, E. J., Hamilton, G., Zhang, Y. S. and Layher, W.: 1996, *Can. J. Fish. Aquat. Sci.* **53**, 2299-2309.

Schreck, C. B.: 1990, *American Fisheries Society Symposium* **8**, 29-37.

Shugart, L. R., Bickham, J., Jackim, G., McMahon, G., Ridley, W., Stein, J. and Steinert, S.: 1992, DNA alterations, in: R. J. Huggett, R. A. Kimerle, P. M. Mehrle, and H. L. Bergman (eds), *Biomarkers*, Lewis Pubs., Boca Raton, FL., 125-153.

Shuter, B. J.: 1990, *American Fisheries Society Symposium* **8**,145-166.

Spies, R. B., Stegeman, J. J., Hinton, D. E., Woodin, B., Smolowitz, R., Okihiro, M. and Shea, D.: 1996, *Aquat. Toxicol.* **34**, 195-219.

Stegeman, J. J., Brouwer, M., DiGiulio, R. T., Forlin, L., Fowler, B. A., Sanders, B. M. and Van Veld, P. A.: 1992, Molecular responses to environmental contamination: Enzyme and protein synthesis as indicators of contaminant exposure and effect, in: R.J. Huggett, R. A. Kimerle, P.M. Mehrle, H. L. Bergman (eds), *Biomarkers*, Lewis Pubs., Boca Raton, FL., 125-153.

Teh, S. J., Adams, S. M. and Hinton, D. E.: 1997, *Aquat. Toxicol.* **37**, 51-70.

# ECOLOGICAL ENGINEERING OF BIOREACTORS FOR WASTEWATER TREATMENT

C. P. L. GRADY JR. and C. D. M. FILIPE

*Department of Environmental Engineering and Science, L. G. Rich Environmental Research Laboratory, Clemson Research Park, Clemson University, Clemson, South Carolina, 29634-0919 USA*

**Abstract.** Biological nutrient removal (BNR) systems remove carbon, nitrogen, and phosphorus from wastewaters through biodegradation of organic compounds, oxidation of ammonia-N to nitrate-N, reduction of nitrate-N to $N_2$ gas, and sequestration of phosphorus as polyphosphate. The microbial community in such systems is complex because it must contain heterotrophic bacteria capable of aerobic respiration, anaerobic respiration, and fermentation; specialized heterotrophic bacteria that can store polyphosphate; and autotrophic nitrifying bacteria that can withstand long periods without oxygen. Although the basic design principles for BNR systems are reasonably well established, it is becoming apparent that a greater understanding of the microbial interactions involved is required to increase system reliability. For example, although the environment established for the selection of phosphorus accumulating organisms was thought to give them a strong competitive advantage over other heterotrophic bacteria, this has turned out not to be the case. Rather, glycogen accumulating organisms can compete quite effectively in the same environment. Furthermore, BNR systems have suffered from problems with sludge settleability, even though many of the system characteristics are considered to be conducive to the suppression of filamentous bacteria. Finally, the use of molecular techniques has revealed that the autotrophic nitrifying bacteria are different from those that had been considered to be present. This paper reviews the microbial ecology of BNR systems, establishes how molecular biology techniques are changing our understanding of that ecology, and suggests ways in which engineering control can be exerted over community structure, thereby increasing the reliability of BNR systems.

**Keywords:** biological nutrient removal, ecological engineering, filamentous bulking, glycogen accumulating organisms, nitrifying bacteria, phosphorus accumulating organisms

## 1. Introduction

The problems facing wastewater treatment engineers have changed greatly in the past twenty years. Where once the major concern was the removal of biogenic organic matter to prevent the depletion of oxygen in the receiving stream, today the focus is on the removal of specific organic chemicals (typically xenobiotic) to prevent toxic, carcinogenic, or teratogenic effects on organisms living in or consuming the water in the receiving stream, and the removal of nitrogen and phosphorus to prevent eutrophication of the receiving body of water. The engineering community has responded to these challenges with innovativeness and insight, developing new processes that scarcely could have been dreamed of by engineers of the previous generation. One has only to review the proceedings of any conference to find many examples of how designers have applied fundamental

*Water, Air, and Soil Pollution* **123:** 117–132, 2000.

principles to design new and innovative systems for solving specific water quality problems. Truly, innovation is alive and well in the field of wastewater treatment.

Biological wastewater treatment processes have particularly benefited from this innovative spirit. Where once biological processes were designed to accomplish limited objectives, today they are routinely designed to achieve multiple goals with a single microbial community containing numerous populations of diverse character. A key prerequisite for the success of such designs is the ability to engineer the system so that it will foster the development of a microbial community that contains populations capable of achieving the desired outcomes. In other words, engineers must have the ability to create niches within the system that allow the desired microbes to compete favorably with others that do not have the necessary metabolic or physical characteristics.

The act of designing a bioreactor to foster the development of a specific microbial community has been called ecological engineering, or ecoengineering (Thiele and Zeikus, 1988). It requires a basic understanding of the niche occupied by a desired microbial population so that the bioreactor system can be configured in such a way as to create that niche. The term "niche" in this context refers to the types and quantities of resources available, as well as the physicochemical conditions (e.g., pH, temperature, oxygen, etc.) existing (Madigan *et al.*, 1997). Environmental engineers have been consciously practicing ecological engineering for several years now, although not all engineers might recognize it as such. For example, the design of a selector in an activated sludge system creates a niche that allows floc-forming bacteria to compete effectively with filamentous bacteria, thereby generating a floc particle with optimal settling characteristics. Biological nutrient removal (BNR) systems, particularly those for the removal of both nitrogen and phosphorus, are another example. In them, bioreactors containing zones with different electron acceptors are designed to foster the growth of several distinct microbial populations. Anaerobic zones are created to provide a niche for phosphorus accumulating organisms (PAOs), allowing them to sequester phosphorus as polyphosphate granules, which facilitates its removal with the waste biomass. Aerobic zones are provided for growth of the PAOs, as well as nitrifying bacteria, which convert ammonia-N to nitrate-N. Aerobic zones also allow growth of heterotrophic bacteria responsible for the biodegradation of the bulk of the organic matter. Anoxic zones are created to allow enrichment of heterotrophic bacteria capable of using nitrate-N as a terminal electron acceptor, converting it to $N_2$ gas, returning the nitrogen to the atmosphere. Many different system configurations, ranging from fully suspended growth to fully attached growth, have been devised as engineers have attempted to optimize BNR systems (Grady *et al.*, 1999).

As experience has been gained with BNR systems we have learned that our understanding of the ecology of the microbial communities and the niche requirements of the populations involved is far from complete. Part of this awareness has come from studies initiated to increase our understanding of why certain BNR systems have not performed as expected, whereas another part has resulted from application of new molecular biology tools to the resident microbial communities simply for the purpose of gaining a better understanding of their structures. The purpose of this paper is to review three examples of our evolving understanding of the ecological engineering of BNR systems, with the goal of encouraging all involved in the design of innovative wastewater treatment systems to approach their tasks with an appreciation of the need to consider the ecology of that complex microbial community we call activated sludge.

## 2. Enhanced Biological Phosphorus Removal

Enhanced biological phosphorus removal (EBPR) is being applied with increasing frequency for the removal of phosphorus from wastewater. In its simplest configuration, this process requires that the bioreactor be configured as two tanks in series, with the first tank being anaerobic (no significant inputs of dissolved oxygen and nitrate) and the second aerobic, with the entry of the wastewater and the recycled biomass into the first tank, as illustrated in Figure 1 (Grady *et al.*, 1999). The cycling of the biomass between the anaerobic and aerobic zones of the bioreactor provides a niche for the enrichment of PAOs in the microbial community. Although the identity of the bacteria comprising the PAOs remains unknown, their overall biochemistry is reasonably well established (Mino *et al.*, 1998). In the anaerobic zone, PAOs are able to take up short chain fatty acids (SCFAs) and store them as polyhydroxyalkanoates (PHAs), using stored polyphosphate as a source of energy and stored glycogen as a source of reducing power. As a result, soluble phosphate is released to the medium. In the aerobic zone, the PAOs use the stored PHAs as an energy and carbon source for growth, rebuilding the stored glycogen pool and taking up soluble phosphate for replenishment of the stored polyphosphate. If the carbon to phosphorus ratio in the wastewater is adequate, the energy available in the stored PHAs is sufficient to allow essentially complete removal of soluble phosphate.

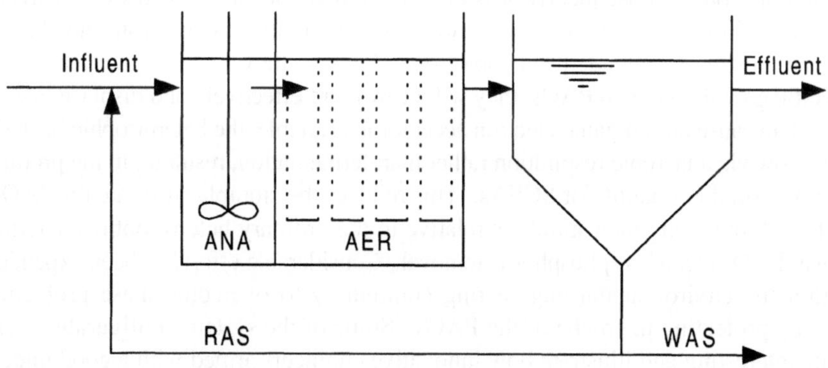


Fig. 1. The simplest bioreactor configuration for enhanced biological phosphorus removal. ANA = anaerobic; AER = aerobic; RAS = recycled activated sludge; WAS = waste activated sludge.

Ecological engineering has been an integral part of EBPR system design since its inception. Because of the ability of PAOs to take up and store SCFAs as PHAs under anaerobic conditions and then to metabolize the PHAs for growth under aerobic conditions, the imposition of sequential anaerobic and aerobic zones, as illustrated in Figure 1, provides an environment in which PAOs can compete effectively with ordinary heterotrophic bacteria. Because no inorganic electron acceptors are available in the anaerobic zone, the growth of ordinary heterotrophic biomass is restricted to that which can occur by fermentation, which is small. As a consequence, a significant amount of the organic matter in the wastewater can be taken up and sequestered by the PAOs as PHAs, making that organic matter unavailable to the normal heterotrophic bacteria, thereby enriching the microbial community with PAOs. In other words, a niche has been created for the PAOs.

As experience was gained with EBPR systems, refinements were made in the system configuration to optimize the PAO niche. For example, if the sewer system is largely anaerobic, then fermentation in the sewer will convert much of the readily biodegradable organic matter into SCFAs by the time the wastewater reaches the treatment system. As a consequence, because the uptake of SCFAs by PAOs is rapid, the size of the anaerobic zone can be small. Conversely, if the sewer system is mainly aerobic, few SCFAs will be produced in it, requiring the anaerobic zone of the EBPR process to be larger so that heterotrophic fermentation, which is a relatively slow process, can produce them.

The occurrence of nitrification in the aerobic zone required even more ecological engineering because of the return of nitrate-N to the anaerobic zone through biomass recycle. Sometimes nitrification is a requirement in the overall system design, but sometimes it is an unavoidable occurrence because of the similarity of the maximum specific growth rates of PAOs and nitrifying bacteria at temperatures above 20°C (Grady *et al.*, 1999). The entrance of nitrate-N into the anaerobic zone damages the PAO niche in three ways. First, some PAOs can use nitrate-N as their terminal electron acceptor, but when they do so they behave metabolically as if they were in an aerobic zone. Consequently, less PHA is formed. Second, because of the availability of nitrate-N as a terminal electron acceptor, denitrifying heterotrophic bacteria can grow in the intended anaerobic zone. Because they have a higher maximum specific growth rate than PAOs, they will grow more effectively and displace PAOs. Third, because an inorganic electron acceptor is available, the heterotrophic bacteria will grow via anaerobic respiration rather than fermentation, resulting in the production of a smaller quantity of SCFAs. This reduces the storable substrate for PAOs, further diminishing their numbers relative to the ordinary heterotrophic bacteria. Fewer PAOs mean less phosphorus removal. Considerable effort has been expended within the environmental engineering community to overcome these problems, thereby protecting the niche of the PAOs. Some of the system configurations are quite interesting and illustrate how innovative engineers armed with a good understanding of the growth requirements of a desired population can apply ecological engineering to ensure its presence in the microbial community. They are discussed in detail elsewhere (Grady *et al.*, 1999).

The premise upon which the ecological engineering of EBPR systems has been based is that the imposition of sequential anaerobic and aerobic zones, with the recycle of biomass around the system, provides a niche that is unique for PAOs. This premise arose from the assumption that polyphosphate provides a unique energy reserve that allows only PAOs to take up SCFAs under anaerobic conditions and sequester them as PHAs. Unfortunately, as more experience has been gained with EBPR systems, we have learned that the assumption, and hence the premise, is false. Rather, another group of bacteria, known collectively as glycogen accumulating non-polyphosphate organisms, or simply as GAOs, can grow quite effectively in the same environment (Mino *et al.*, 1998). In the anaerobic zone they use glycogen for both energy and reducing power, sequestering SCFAs as PHAs, which are then used in the aerobic zone for growth and replenishment of the glycogen. GAOs have also been called G-bacteria (Cech and Hartman, 1993).

The presence of GAOs in EBPR systems has been observed most frequently in laboratory scale systems (Mino *et al.*, 1998), but their presence has also been noted in full-scale systems (Cech and Hartman, 1993; Maszenan *et al.*, 1998). Proliferation of GAOs at the expense of PAOs can lead to loss of the capability for phosphorus removal, and thus a better understanding of the growth and metabolism of both PAOs and GAOs is required to allow engineers to provide the appropriate niche for PAOs while excluding GAOs.

Although uncertainty still exists about the identity of PAOs, several things are known about the identity of GAOs. For example, they are morphologically distinct from PAOs and while they have similar metabolic pathways, those pathways appear to be regulated by different mechanisms (Mino *et al.*, 1998). In other words, GAOs are not simply PAOs that have lost the ability to store phosphorus as polyphosphate. GAOs appear to be wide spread, having been found in activated sludge systems around the world (Maszenan *et al.*, 1998). Finally, GAOs appear to be diverse. Based on their morphological characteristic of growing as cocci arranged in tetrads, Maszenan *et al.* (1998) isolated three gram negative bacteria into pure cultures from activated sludge plants in Australia, Italy, and Macau by using micromanipulation. They subjected them to physiological analysis to confirm their ability to synthesize PHA and to 16S ribosomal DNA (rDNA) sequence analysis for identification. They also studied the original G-bacteria isolated by Cech and Hartman (1990). All four were found to be phylogenetically quite different from any previously identified bacteria and were classified as four species of a new genus, *Amaricoccus*, in the alpha subclass of the *Proteobacteria*. Nielsen *et al.* (1999) studied the composition of a GAO enrichment culture that was developed from a lab-scale EBPR system that lost its ability to remove phosphorus. The lab-scale reactor was started from a full-scale EBPR system in Greenville, South Carolina (USA) that was not achieving full phosphorus removal, so presumably the GAOs originated in that plant. DNA was extracted from biomass from the GAO culture, the 16S rDNA was amplified by

PCR, and separated by denaturing gradient gel electrophoresis to investigate the diversity of the community. Up to eleven different bands were observed, suggesting that up to eleven types of bacteria were present. DNA from the six most prominent bands was further isolated and sequenced for identification. Only one band represented organisms belonging to the alpha subclass of the *Proteobacteria*, but they were not of the genus *Amaricoccus*. Five bands represented organisms belonging to the gamma subclass of the *Proteobacteria*. One of them was associated with the *Legionella* group, but the other four formed a novel group with no close relationship to any previously described species. This latter group represented approximately 35% of the total microbial community. The bacteria in it were coccoid, 3 to 4 μm in diameter, and contained granule inclusions, presumably of PHA. Taken as a whole, these studies suggest that many types of bacteria belong to the loosely defined group of GAOs, further suggesting that studies to define their metabolism must establish something about their identity.

Because of the ability of GAOs to displace PAOs and disrupt EBPR systems, Filipe (1999) studied enrichment cultures of both PAOs and GAOs, growing on acetate as the sole SCFA, with respect to their metabolism and the effects of pH. The organisms identified by Nielsen *et al.* (1999) were from Filipe's GAO culture. Filipe (1999) developed metabolic models for the anaerobic metabolism of both PAOs and GAOs in which the stoichiometry was defined in terms of the well known energetics of the pathways involved and a single parameter representing the energy required to transport acetate into the cells. He also defined the kinetics of acetate uptake under anaerobic conditions and measured the kinetics of growth on the stored PHAs under aerobic conditions. All studies were performed over a range of pH values from 6.5 to 7.5, with some studies going as high as 8.5. Filipe (1999) found that in the anaerobic zone, GAOs were favored over PAOs, both energetically and kinetically at pHs below 7.2 to 7.3, but that PAOs were favored at higher pHs. Furthermore, in the aerobic zone, PAOs were affected very negatively at pH 6.5, whereas pH had little effect on GAOs. His results strongly suggest that while establishment of an anaerobic zone followed by an aerobic zone is a necessary condition for the establishment of a PAO niche, it is not a sufficient condition. Rather, the system must be operated at pH values in excess of 7.3. Only under that condition will GAOs be unable to invade the niche. Because of Filipe's findings, we have routinely used pH control as an additional factor in operation of laboratory scale EBPR systems, observing that when the pH is allowed to drop, phosphorus removal suffers, whereas when the pH is maintained at 7.5, excellent phosphorus removal occurs. Interestingly, Bond *et al.* (1998) reported that when they allowed the pH to rise in a poorly performing laboratory scale EBPR system, they obtained better phosphorus removal. They did not state, however, how high they allowed the pH to rise.

In addition to revealing the importance of pH control as a way to protect the niche for PAOs, the metabolic studies of Filipe (1999) showed that the energetics of GAOs are less favorable under anaerobic starvation conditions than the energetics of PAOs. This suggests that another way to favor the growth of PAOs over GAOs is to hold the community under anaerobic conditions for a longer period than is needed for uptake of SCFAs, thereby forcing the bacteria to use internal storage products to meet maintenance energy needs. PAOs can provide ATP for maintenance energy by cleavage of stored polyphosphate, whereas GAOs must provide ATP through glycolysis of stored glycogen. Because recovery of polyphosphate is easier than recovery of glycogen, increasing the time under anaerobic conditions should have a larger effect on GAOs than PAOs, giving the PAOs a competitive advantage. In support of this concept, Matsuo (1994) observed that when the solids retention time (SRT) was increased in the anaerobic zone of a failing EBPR system, GAOs were lost from the system and effective phosphorus removal was recovered.

In summary, ecological engineering is extremely important to the selection and maintenance of the PAO population in EBPR systems. While it was originally thought that the provision of anaerobic and aerobic bioreactors in series with biomass recycle and wastewater entering the anaerobic zone was both a necessary and sufficient condition for enrichment of PAOs, it is now known that it is not sufficient. Rather, such a condition also provides a niche for GAOs. However, the maintenance of a pH above 7.3 and/or the use of a longer anaerobic SRT both shift the balance, allowing PAOs to outcompete GAOs and to predominate in the community. All of these actions are examples of the application of ecological engineering principles.

## 3. Filamentous Bulking in BNR Systems

While implementation of biological nutrient removal has had positive effects on the environment through reduction of the nutrient load reaching it, the upgrading of secondary wastewater treatment systems to provide BNR capabilities has not been without problems. In particular, many BNR facilities have experienced foaming and filamentous bulking problems, particularly those operated at long SRTs with intermittent aeration or frequent cycling of the biomass between aerobic and anoxic or anaerobic environments (Eikelboom *et al.*, 1998). Even though one might think that the incorporation of anoxic or anaerobic zones into activated sludge systems would minimize such problems by providing niches that are favorable to floc-forming bacteria over foam-causing and filamentous bacteria, this has turned out not to be the case. Furthermore, the configuration of the anoxic zones as selectors has also proven to be ineffective (Ekama *et al.*, 1996). This suggests that new techniques for the control of nuisance organisms must be given special consideration during the ecological engineering of BNR facilities. This, in turn requires knowledge of the major causative organisms and their growth requirements. Consequently, a survey was conducted of more than 200 BNR facilities in Europe, revealing that *Microthrix parvicella* was by far the most important foam-causing and filamentous organism in them (Eikelboom *et al.*, 1998). As a result, several investigators have sought to understand this organism.

Although a number of research groups have studied "*Microthrix parvicella*", it is unlikely that all of the organisms studied were strains of one species, and this has led to apparently contradictory information in the literature about the organism's characteristics. The reason for this is that in most studies the organism was identified by its morphological characteristics as documented in Eikelboom's taxonomy of filamentous microorganisms (long, coiled filaments; Gram positive; Neisser positive granules), rather than by more exact techniques such as 16S rDNA analysis. In fact, Wanner (1994) noted that even though *M. parvicella* has been named, it should be treated as an Eikelboom type, rather than as a true taxonomic species. Recently, however, isolates have been characterized by 16S rDNA sequencing and submitted as true candidate species (Blackall *et al.*, 1996). Only continued work at the molecular level will resolve the true identity and characteristics of the filamentous bacteria commonly associated with BNR systems, and reveal how many identified morphologically as *M. parvicella* are indeed the same. In the mean time, this situation must be kept in mind as the literature about "*Microthrix parvicella*" is considered.

A very important question about *M. parvicella* concerns the types of substrate it prefers because that has an important impact on its ability to grow in BNR systems. Studies by Slijkhuis and Deinema (1982) established that *M. parvicella* cannot use readily biodegradable substrates like simple hexoses, amino acids, and organic acids. Rather, it prefers long chain fatty acids in esterified form, although oleic acid can be used in both esterified and free form. Activated sludge microautoradiographic studies of substrate uptake by filamentous bacteria with the morphological characteristics of *M. parvicella* verified these findings (Andreasen and Nielsen, 1997, 1998). However, pure culture studies with an Italian isolate (RN1) with the same 16S rDNA sequence as the bacteria characterized by Blackall *et al.* (1996) did not totally agree (Tandoi *et al.*, 1998). RN1 was unable to grow on the medium of Slijkhuis and Deinema (1982) or on proteose peptone, glucose, or oleic acid in either acid or esterified form, but it was able to grow on acetate, yeast extract, casaminoacids, pyruvate, sodium oleate, and a mixture of short-chain fatty acids, as well as on olive oil mill effluent containing long chain fatty acids. It should be noted that Tandoi *et al.* (1998) measured growth microscopically as extended filament length, giving them a much more sensitive indicator than had been used previously. During their studies they never observed visible turbidity in their cultures and growth stopped before there had been a measurable decrease in total organic carbon. This also occurred in the studies of Blackall *et al.* (1996). Although the reason for this observation is unknown, it suggests that strain RN1 may require an unidentified growth factor that makes growth in pure culture difficult.

Although there are some inconsistencies in the substrate studies cited above, other evidence verifies that bacteria with the morphological characteristics of *M. parvicella* grow well on oleic acid in either free or esterified form (Mamais *et al.*,

1998; Slijkhuis and Deinema, 1988). Oleic acid and other long chain fatty acids are associated with the slowly biodegradable substrate fraction of domestic wastewater, whereas most of the compounds that *M. parvicella* has been unable to grow on are found in the readily biodegradable fraction. This helps to explain why neither kinetic nor metabolic selectors have been effective in controlling filamentous bulking in BNR systems (Ekama *et al.*, 1996). Both types of selectors provide an advantage to floc-forming bacteria over filamentous bacteria as they compete for readily biodegradable substrate. Since these filamentous bacteria do not grow well on readily biodegradable substrates, control strategies that focus on those substrates cannot be effective. While selectors can control the presence of other types of filamentous, by their very nature they are inappropriate for the control of *M. parvicella* and similar bacteria. This illustrates the importance of understanding the growth characteristics of the microorganism that one wishes to control through ecological engineering techniques.

The types of substrates favored by *M. parvicella* are components of the lipid fraction of domestic wastewater. Since this fraction constitutes 31% of the organic matter in domestic wastewater (Raunkjær, *et al.*, 1994), it is possible that pretreatment to remove some of the lipids might help control *M. parvicella*.

Another factor contributing to the ability of *M. parvicella* to compete effectively with floc-forming bacteria in BNR systems is that they can take up substrates like oleic acid under both anoxic and anaerobic conditions, as well as aerobically (Andreasen and Nielsen, 1998). They can also store oleic acid in esterified form inside the cell (Wanner, 1994), although growth on the storage product has not been established. Nevertheless, this suggests that the environment set up to favor PAOs may also be beneficial to *M. parvicella*. Furthermore, *M. parvicella* can use nitrate-N as an electron acceptor, releasing nitrite, although they cannot denitrify to nitrogen gas. Casey *et al.* (1992) have speculated that the ability to denitrify only to nitrogen gas may give filamentous bacteria an advantage over flocculent denitrifiers in systems in which the biomass is rapidly oscillated between aerobic and anoxic conditions. Flocculent denitrifiers contain a suite of reducing enzymes capable of transferring electrons to nitrate, nitrite, nitric oxide, and nitrous oxide, ultimately yielding nitrogen gas. Little is know about the regulation of either the synthesis or activity of those enzymes when bacteria are rapidly oscillated between aerobic and anoxic environments. However, if that regulation leads to an intracellular buildup of a partially reduced nitrogen product within the flocculent denitrifiers, its presence might inhibit cellular function, preventing the flocculent bacteria from growing as effectively as they might. When one considers the fact that only a small fraction of activated sludge need be filamentous to disrupt activated sludge settling properties (Palm *et al.*, 1980), it is easy to see how just a slight shift in the relative growth rates could allow *M. parvicella* to grow sufficiently to cause problems in BNR systems.

Another characteristic of *M. parvicella* is that it has a very high affinity for dissolved oxygen, allowing it to grow under microaerobic conditions, even in the absence of nitrate as an electron acceptor. When Tandoi *et al.* (1998) studied the effect of dissolved oxygen concentration on the specific growth rate of RN1 on a variety of substrates, they found that it had no effect down to a level of 0.37 mg/L. This suggests that the half-saturation coefficient for dissolved oxygen for *M. parvicella* is lower than that for flocculent bacteria, allowing *M. parvicella* to compete effectively for any oxygen that enters the anoxic or anaerobic zones of BNR systems. Indeed, laboratory studies found that continuous aeration of activated sludge systems infested with *M. parvicella* led to their loss from the sludge (Slijkhuis and Deinema, 1988). While continuous aeration obviously is not possible in BNR systems, it is possible that better exclusion of oxygen from the anoxic and anaerobic bioreactors in such systems could decrease the growth of *M. parvicella.* Since only a small quantity of filamentous bacteria in activated sludge can deteriorate its settling characteristics, perhaps such an action would have a beneficial effect on sludge settleability.

Finally, it should be noted that the morphological characteristics of *M. parvicella* are strongly influenced by the growth conditions (Knoop and Kunst, 1998). At low temperatures and long SRTs, it grows as long coiled filaments that are Gram positive and contain Neisser positive granules (the "classical" morphology). This shape has a strong negative impact on settling. However, when the temperature is raised to 20°C or above, the long filaments break up into short filaments, which no longer interfere with settling. The short filaments also have Gram negative regions. Furthermore, when the SRT is decreased at low temperature, the filaments change into thick, rope-like structures that do not extend beyond the floc, and thus do not interfere with settling. If *M. parvicella* cannot be excluded from BNR systems, perhaps such information will allow the selection of growth conditions that minimize its impact.

In summary, while it is still not certain how BNR systems should be designed to minimize problems from *M. parvicella* and similar filamentous bacteria, a preliminary picture is emerging. System SRTs should be kept as short as possible consistent with the treatment objectives. Primary settling and skimming should be used to minimize the input of complex lipids into the system. Anoxic and anaerobic zones should be well defined and efforts should be made to exclude oxygen. Furthermore, intermittent or inadequate aeration should be avoided as a means of denitrification because it provides an environment conducive to *M. parvicella* growth. Because of the importance of these organisms and their impact on sludge settleability, several labs are actively engaged in research on them. Consequently, we can expect to see the solution to the problem through the application of ecological engineering principles in the near future.

## 4. Nitrifying Bacteria

A key requirement for the development of the appropriate niches in a BNR system is knowledge of the nature and characteristics of the microbial populations for which the niches are intended. In other words, if we don't know the growth requirements of the populations, we have no rational basis for applying ecological engineering to establish the niches. Engineers accept the fact that most activated sludge microbial communities contain a wide variety of heterotrophic bacteria because of the large number of different organic compounds present in the wastewaters being treated. In this case, the organic compounds being degraded help establish the niches required and engineers pay little attention to the identity of the individual heterotrophic bacteria present. Rather, we simply accept the degradation of the organic compounds as evidence that appropriate niches have been created. A different situation exists for the autotrophic nitrifying bacteria responsible for the sequential oxidation of ammonia-N to nitrate-N. Since that oxidation is key to the ultimate removal of the nitrogen through denitrification, it is essential that niches be established for the bacteria involved. Because the reactions are few (oxidation of ammonia-N to nitrite-N, followed by oxidation of the nitrite-N to nitrate-N) and the types of bacteria involved limited, we have acted as if those reactions were performed by bacteria of only two genera and have focused our attention on them. Open almost any textbook on wastewater treatment and you will learn that ammonia oxidation is performed by members of the genus *Nitrosomonas* and nitrite oxidation is performed by members of the genus *Nitrobacter*. Furthermore, much of our knowledge about the physiology of ammonia oxidizing bacteria (AOB) and requirements for their growth has come from pure culture studies performed with a single species, *Nitrosomonas europaea*, primarily because of their availability in culture collections and the ease with which they can be grown (Prosser, 1989). Similarly, much of our knowledge of nitrite oxidizing bacteria (NOB) has come from studies with *Nitrobacter winogradskyi*.

Our knowledge of nitrifying bacteria has evolved rapidly since the development of molecular techniques that allow discrimination among bacteria on the basis of their 16S rDNA sequences. Based on ultrastructural properties, AOB had been divided into five genera, *Nitrosomonas*, *Nitrosococcus*, *Nitrosospira*, *Nitrosovibrio*, and *Nitrosolobus* (Madigan *et al.*, 1997), but 16S rDNA analysis has caused the number of genera of the characterized bacteria to be reduced to three (Schramm *et al.*, 1998; Wagner *et al.*, 1998). *Nitrosococcus mobilis* has now been moved into *Nitrosomonas*, which comprises a coherent group within the beta subclass of the *Proteobacteria*, while *Nitrosospira*, *Nitrosovibrio*, and *Nitrosolobus* have all been reclassified as *Nitrosospira*, which also falls within the beta subclass of the *Proteobacteria* (Schramm *et al.*, 1998). The remaining members of the genus *Nitrosococcus*, which are all marine organisms, constitute a coherent group within the gamma subclass of the *Proteobacteria* (Wagner *et al.*, 1998). The known NOB have been assigned to four genera: *Nitrobacter*, which is a member of the alpha subclass of the *Proteobacteria*; *Nitrococcus*, which is a member of the gamma subclass of the *Proteobacteria*; *Nitrospina*, which is in the delta subclass of the *Proteobacteria*; and *Nitrospira*, which constitutes an independent phylum in the domain *Bacteria* (Juretschko *et al.*, 1998). Known members of the genera *Nitrococcus* and *Nitro-*

*spina* are marine organisms, whereas both freshwater and marine organisms are found within *Nitrobacter* and *Nitrospira* (Madigan *et al.*, 1997).

Because it has generally been assumed by environmental engineers that nitrification in wastewater treatment systems is performed by members of the genera *Nitrosomonas* and *Nitrobacter*, the question arises as to whether this is true. Fortunately, the development of molecular techniques has allowed evaluation of the assumption because those techniques permit detection of specific microbes *in situ* without the need for culturing them, with its associated biases.

As far as the AOB are concerned, it appears that *Nitrosomonas* is indeed the most important genus in activated sludge and trickling filter systems receiving relatively high specific loadings of ammonia-N, whereas *Nitrosospira* is the most important genus in biofilm systems receiving very low ammonia-N concentrations. Wagner *et al.* (1996) analyzed biomass samples from nine nitrifying wastewater treatment plants with probes for *Nitrosomonas* and found it to be present in all of them. *Nitrosospira* was not tested for. In a follow-up study at one of the plants, Juretschko *et al.* (1998) found that *Nitrosococcus* (now classified as *Nitrosomonas*) *mobilis*-like cells were numerically dominant, with small populations of other *Nitrosomonas* species and some novel cells that were not identified. Likewise, when Mobarry *et al.* (1996) probed for AOB in a lab-scale suspended growth bioreactor they found that most of them were identical or closely related to described species of the genus *Nitrosomonas*, although there was great diversity in the species present. Finally, Ballinger *et al.* (1998) found that 75 percent of the AOB 16S rDNA fragments selected at random from their lab-scale activated sludge system were related to *Nitrosomonas* while the other 25 percent were related to *Nitrosospira*. The most important *Nitrosomonas* was related closely to *Nitrosomonas ureae*. In biofilm systems, Schramm *et al.* (1996) probed for *Nitrosomonas* in trickling filter biofilm from an aquiculture facility and found them to be present. However, later studies on the biofilm in a lab-scale fluidized bed bioreactor with a very low ammonia loading revealed that members of the genus *Nitrosospira* were the only AOB present (Schramm *et al.*, 1998). Additional studies will be needed before we can generalize that members of the genus *Nitrosomonas* will always predominate in systems with high ammonia loadings and members of the genus *Nitrosospira* will always predominate in systems with low ammonia loadings, but such a pattern would be consistent with the patterns of these two genera in the environment (Ballinger *et al.*, 1998; Schramm *et al.*, 1998; Stephen *et al.*, 1996). However, based on the results reported, it can be stated that *Nitrosomonas europaea* is not the most important AOB. Rather, other species are important. Consequently, environmental engineers should be cautious in assuming that all AOBs have the characteristics of *Nitrosomonas europaea*. Furthermore, as more is learned about the identity of the predominant AOBs in wastewater treatment systems, studies should be performed to fully characterize their kinetics and growth requirements because that information will allow better application of ecological engineering principles to provide the required niche for effective ammonia oxidation.

As far as the NOB are concerned, it appears that members of the genus *Nitrospira* are the most important in wastewater treatment systems, although *Nitrobacter* has been found. During their analysis of activated sludges from nitrifying full-scale systems, Wagner *et al.* (1996) also used a general probe for *Nitrobacter*. All samples were negative, suggesting that *Nitrobacter* spp. were either absent or present in such low numbers as to be insignificant in the oxidation of nitrite to nitrate. In their follow-up study on biomass from one of the plants, Juretschko *et al.* (1998) found that *Nitrospira*-like cells were the only NOB found by molecular techniques. However, the only NOB isolated by classical isolation techniques was of the genus *Nitrobacter*. This suggests that the concept that *Nitrobacter* are the main NOB in activated sludge systems may have come from the relative ease with which they can be isolated. Studies in Australia also found *Nitrospira* to be the most abundant NOB in samples from a full-scale facility, as well as from an enrichment culture derived from it (Burrell *et al.*, 1998). However, members of the genus *Nitrobacter* were present in the enrichment culture, reinforcing the concept that they are more important in laboratory cultures than in the field. This does not suggest that they are unimportant in full-scale facilities, however. For example, *Nitrobacter* was found in the biofilm from a trickling filter at an aquiculture facility (Schramm *et al.*, 1996), although their relative numbers are unknown because probes for other genera were not used. Finally, *Nitrospira*-like bacteria were identified as the NOB in a lab-scale nitrifying fluidized bed bioreactor receiving a low nitrogen loading (Schramm *et al.*, 1998). Although *Nitrospira* appears to be the most important NOB in engineered systems, their growth characteristics have not been established and thus it is impossible to know whether the knowledge gained from studies on *Nitrobacter* are truly relevant to wastewater treatment systems.

Our ability to apply ecological engineering to the design of our bioreactors is directly related to our knowledge of the interactions among the different microbes involved. In the past, our ability to study microbial interactions has been limited by our ability to distinguish among various microbes *in situ*. That situation changed dramatically with the development of gene probes and their use in fluorescent *in situ* hybridization (FISH). Through the use of FISH, in combination with various microscopic techniques, it is possible to see the spatial relationships between bacterial groups in activated sludge floc and trickling filter biofilms. Juretschko *et al.* (1998) applied these techniques to study the distribution of AOB and NOB in activated sludge floc. They found that the AOB and NOB were distributed in the floc in distinct clusters, with each type of microbe forming microcolonies. The AOB formed microcolonies approximately 60 μm in diameter, whereas the NOB formed microcolonies around 10 to 20 μm in diameter. The surprising finding was that the NOB did not grow around or through the AOB microcolonies, but were distinct from them. Schramm *et al.* (1996) performed similar studies with trickling filter biofilm and found an extremely dense layer of AOB in the upper layer of the biofilm, with only a few cell clusters in the deeper biofilm. As in the activated sludge floc, the AOB grew in microcolonies, which are now thought to be characteristic of microbial growth in biofilms (Costerton *et al.*, 1995). The NOB also grew in microcolonies that were less dense and fewer in number than the AOB microcolonies (Schramm *et al.*, 1996). Interestingly, the NOB microcolonies were more evenly distributed in the biofilm than the AOB microcolonies. Information such as this will help modelers write more realistic models representing microbial growth in wastewater treatment systems, which can then be used in simulation programs to understand how engineering decisions influence the microbial community involved. In other words, better information about microbial interactions will give us better tools with which to practice ecological engineering.

## 5. Closure

Biological nutrient removal (BNR) systems utilize complex microbial communities containing populations of specialized bacteria that have been selected for their ability to perform specific tasks. The success of those systems is dependent on the ability of the design engineer to create the appropriate niches for the required microbial populations, thereby ensuring their maintenance in the community. Such an activity is called ecological engineering in recognition of the exertion of engineering control over the ecology of the bioreactor. Ecological engineering may be as simple as selecting the appropriate SRT, pH, and dissolved oxygen concentration to foster the growth of autotrophic nitrifying bacteria, or as complex as designing a system for simultaneous carbon oxidation, nitrification, denitrification, and phosphorus removal. In all situations, however, the engineer should be aware of the fact that he/she is not just building a physical system, but rather, is exerting control over the microbial community that will reside in that system.

Based on the information provided in this paper, the following points should be stressed:

- The imposition of sequential anaerobic and aerobic zones in an EBPR system is a necessary, but not a sufficient, condition for creating a stable niche for PAOs because GAOs can grow effectively in the same environment, displacing the PAOs.
- Many types of bacteria appear to belong to the loosely defined group called GAOs. More knowledge is needed of their diversity and characteristics.
- Maintenance of the pH above 7.3 in both zones of an EBPR system prevents GAOs from invading the PAO niche, thereby enhancing phosphorus removal.
- The imposition of an anaerobic SRT in excess of the value needed for uptake of SCFAs decreases the competitiveness of GAOs relative to PAOs, providing another mechanism for control of the microbial community in EBPR systems.
- In spite of the presence of anoxic and anaerobic zones, BNR systems have experienced problems from foam-causing and filamentous bacteria. The main causative agent of those problems is "*Microthrix parvicella*".

- In most studies, *M. parvicella* has been characterized by its ultrastructural properties rather than by its genetic characteristics. Consequently, it is unlikely that all studies have been done with strains of the same species, leading to contradictory information in the literature about them.
- *M. parvicella* grows well on slowly biodegradable substrates and is able to reduce nitrate-N to nitrite-N, explaining why neither kinetic nor metabolic selectors have been effective in controlling its growth.
- *M. parvicella* has a very high affinity for dissolved oxygen, allowing it to grow well under microaerobic conditions. This suggests that anoxic and anaerobic zones should be well defined and oxygen should be excluded from them.
- Although members of the genus *Nitrosomonas* are the predominant ammonia oxidizing bacteria (AOB) in activated sludge and trickling filter systems receiving heavy nitrogen loads, *Nitrosomonas europaea* is not the most important species. Others are equally important, although little is known about their physiology and kinetics.
- Although most of our knowledge about nitrite oxidizing bacteria (NOB) has come from studies with bacteria of the genus *Nitrobacter*, they are not the most important NOB in wastewater treatment systems. Rather, members of the genus *Nitrospira* are.
- Studies on the spatial distribution of AOB and NOB in activated sludge floc and biofilms have revealed that each group of organism grows in distinct clusters, with NOB microcolonies separate from, but near, AOB microcolonies. Such knowledge will ultimately be useful in the ecological engineering of BNR systems.

## References


Andreasen, K. and Nielsen, P. H.: 1997, *Applied and Environmental Microbiology.* **63**, 3662-3668.

Andreasen, K. and Nielsen, P. H.: 1998, *Water Science and Technology.* **37**, #4/5, 19-26.

Ballinger, S. J., Head, I. M., Curtis, T. P., and Godley, A. R.: 1998, *Water Science and Technology.* **37**, #4/5, 105-108.

Blackall, L. L., Stratton, H., Bradford, D., Sjörup, C., Del Dot, T., Seviour, E. M., and Seviour, R. J.: 1996, *International Journal of Systematic Bacteriology.* **4**, 344-346.

Bond, P. L., Keller, J., and Blackall, L. L.: 1998, *Water Science and Technology.* **37**, #4/5, 567-571.

Burrell, P. C., Keller, J., and Blackall, L. L.: 1998, *Applied and Environmental Microbiology*, **64**, 1878-1883.

Casey, T. G., Wentzel, M. C., Lowenthal, R. E., Ekama, G. A., and Marais, G. v. R.: 1992, *Water Research.* **26**, 867-869.

Cech, J. S. and Hartman, P.: 1990, *Environmental Technology.* **11**, 651-656.

Cech, J. S. and Hartman, P.: 1993, *Water Research.* **27**, 1219-1225.

Costerton, J. W., Lewandowski, Z., Caldwell, D. E., Korber, D. R., and Lappin-Scott, H. M.: 1995, *Annual Review of Microbiology*, **49**, 711-745.

Eikelboom, D. H., Andreadakis, A., and Andreasen, K.: 1998, *Water Science and Technology.* **37**, #4/5, 281-289.

Ekama, G. A., Wentzel, M. C., Casey, T. G., and Marais, G. V. R.: 1996, *Water SA.* **22**, 147-152.

Filipe, C. D. M.: 1999, *Competition between Phosphorus and Glycogen Accumulating Bacteria: Stoichiometry, Kinetics, and the Effects of pH*, Ph.D. Dissertation, Clemson University, Clemson, SC, USA.

Grady, Jr., C. P. L., Daigger, G. T., and Lim, H. C.: 1999, *Biological Wastewater Treatment: Second Edition, Revised and Expanded*, Marcel Dekker, Inc., New York, NY, pp. 355, 487-560, 949-986.

Juretschko, S., Timmermann, G., Schmid, M., Schleifer, K.-H., Pommerening-Röser, A., Koops, H.-P., and Wagner, M.: 1998, *Applied and Environmental Microbiology*. **64**, 3042-3051.

Knoop, S. and Kunst, S.: 1998, *Water Science and Technology*. **37**, #4/5, 27-35.

Madigan, M. T., Martinko, J. M.., and Parker, J.: 1997, *Brock, Biology of Microorganisms, Eighth Edition*, Prentice Hall, Upper Saddle River, NJ, pp. 535, 660.

Mamais, D., Andreadakis, A., Noutsopoulos, C. and Kalergis, C.: 1998, *Water Science and Technology*. **37**, #4/5, 9-17.

Maszenan, A. M., Seviour, R. J., Patel, B. K. C., Rees, G. N., and McDougall, B.: 1998, *Water Science and Technology*. **37**, #4/5, 65-69.

Matsuo, Y.: 1994, *Water Science and Technology*. **30**, #6, 193-202.

Mino, T., van Loosdrecht, M. C. M., and Heijnen, J. J.: 1998, *Water Research*. **32**, 3193-3207.

Mobarry, B. K., Wagner, M., Urbain, V., Rittmann, B. E., and Stahl, D. A.: 1996, *Applied and Environmental Microbiology*. **62**, 2156-2162.

Nielsen, A. T., Liu, W.-T., Filipe, C., Grady, L. Jr., Molin, S. and Stahl, D. A.: 1999, *Applied and Environmental Microbiology*. **65**, 1251-1258.

Palm, J. C., Jenkins, D., and Parker, D. S.: 1980, *Journal of the Water Pollution Control Federation*. **52**, 2484-2506.

Prosser, J. I.: 1989, *Advances in Microbial Physiology*. **30**, 125-181.

Raunkjær, K., Hvitved-Jacobsen, T., and Nielsen, P. H.: 1994, *Water Research*. **28**, 251-262.

Schramm, A., Larsen, L. H., Revsbech, N. P., Ramsing, N. B., Amann, R., and Schleifer, K.-H.: 1996, *Applied and Environmental Microbiology*. **62**, 4641-4647.

Schramm, A., de Beer, D., Wagner, M., and Amann, R.: 1998, *Applied and Environmental Microbiology*. **64**, 3480-3485.

Slijkhuis, H. and Deinema, M. H.: 1982, in *Bulking of Activated Sludge: Prevention and Remedial Methods*, edited by B. Chambers and E. J. Tomlinson, Ellis Horwood, Ltd., Chichester, UK, pp. 75-89.

Slijkhuis, H. and Deinema, M. H.: 1988, *Water Research*. **22**, 825-828.

Stephen, J. R., McCaig, A. E., Smith, Z., Prosser, J. I., and Embley, T. M.: 1996, *Applied and Environmental Microbiology*. **62**, 4147-4154.

Tandoi, V., Rossetti, S., Blackall, L. L., and Majone, M.: 1998, *Water Science and Technology*. **37**, #4/5, 1-8.

Thiele, J. H. and Zeikus, J. G.: 1988, *Biotechnology and Bioengineering*. **31**, 521-535.

Wagner, M., Rath, G., Koops, H.-P., Flood, J., and Amann, R.: 1996, *Water Science and Technology*. **34**, #1/2, 237-244.

Wagner, M., Noguera, D. R., Juretschko, S., Rath, G., Koops, H.-P., and Schleifer, K.-H.: 1998, *Water Science and Technology*. **37**, #4/5, 441-449.

Wanner, J.: 1994, *Activated Sludge Bulking and Foaming Control*, Technomic Publishing Co., Inc., Lancaster, PA, USA, 170-178.

# BIODEGRADATION OF RECALICTRANT COMPONENTS OF ORGANIC MIXTURES

E. D. SCHROEDER, J. B. EWEIS, D. P. Y. CHANG and J. K. VEIR
*Department of Civil & Environmental Engineering, University of California, Davis*
*One Shields Avenue, Davis, CA USA 95616*

**Abstract.** Biodegradation of two recalcitrant compounds, dichloromethane and methyl tert-butyl ether was investigated individually and in combination with toluene and benzene. A vapor phase biofilter operating at an air flux of 1 $m^3/m^2 \cdot min$ and an empty bed residence time of 1 minute was used as the reaction system. Inlet recalcitrant contaminant concentrations were typically 35 ppm on a volume basis while the aromatic compound concentrations were varied from 8 ppm to 150 ppm. Dichloromethane removals were not impacted by the introduction of toluene. However, a rapid decrease in methyl tert-butyl ether removal resulted from the initial introduction of toluene. Complete removal of both methyl tert-butyl ether and toluene was achieved within a week of operation.

**Keywords:** recalcitrat organics, biodegradation, biofiltration, methyl tert-butyl ether, dichloromethane

## 1. Introduction

The interactions of individual species in mixed microbial populations are of interest in waste treatment for a number of reasons. Most wastes are composed of many contaminants, some of which are degraded by only a few species or groups. Discharge regulations are typically based on lumped or surrogate measures such as biochemical oxygen demand (BOD), chemical oxygen demand (COD), and total organic carbon (TOC). Many discharges contain compounds of specific interest (e.g., benzene in petroleum processing wastes) that are assumed to be removed when the BOD removal is high. If a particular microbial species is required for degradation of the target compound competition for nutrients, oxygen or space may be of significance in the reaction process. For example if the target compound is present in low concentration relative to other, more easily degraded compounds, the target specific microbial species may not be able to compete until more easily degraded compounds have been removed. The concept is illustrated in Figure 1. An example of the effect is shown in the chromatograms of Figure 2 where the data is from a vapor phase biofilter treating gasoline at a soil vapor extraction site. Note that the peaks resulting from the partitioning of the approximately 60 compounds present in gasoline decrease at different rates as they pass through the biofilter. The cause is probably due to differences in interphase transfer rates, compound solubilities, relative hydrophobicity, and the relative growth rates and stoichiometry of

*Water, Air, and Soil Pollution* **123:** 133–146, 2000.

microorganisms metabolizing aliphatic and aromatic hydrocarbons. The complexity of the process makes modeling extremely difficult. Models have been proposed in which spatial distribution of organisms has been attempted. However, the lack of solid information on growth rates, stoichiometry, hydrodynamics and transport coefficients in both phases makes modeling very inexact. For the past decade the biological treatment group at the University of California, Davis has been working with relatively simple mixtures of easy and difficult to degrade compounds in an attempt to address questions of relative removal rates and interactions of microbial groups carrying out noncompetitive degradation processes. To simplify the experimental work we have been using vapor phase bioreactors commonly called biofilters. In these systems, contaminants are introduced in the vapor phase, partition to the liquid phase and are transported into the biofilm where metabolism takes place. The most important feature of the biofilter for the purpose of these studies was that vapor phase organic samples are much easier to analyze than liquid phase sample using gas chromatography.

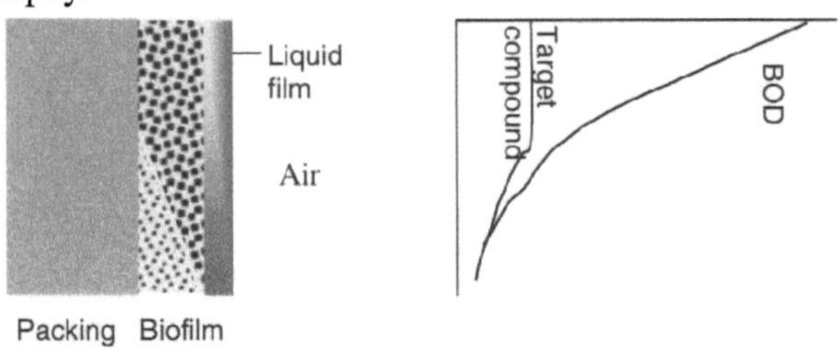


Fig. 1. Conceptual description of delayed degradation of recalcitrant target compound relative to more easily degraded organics.

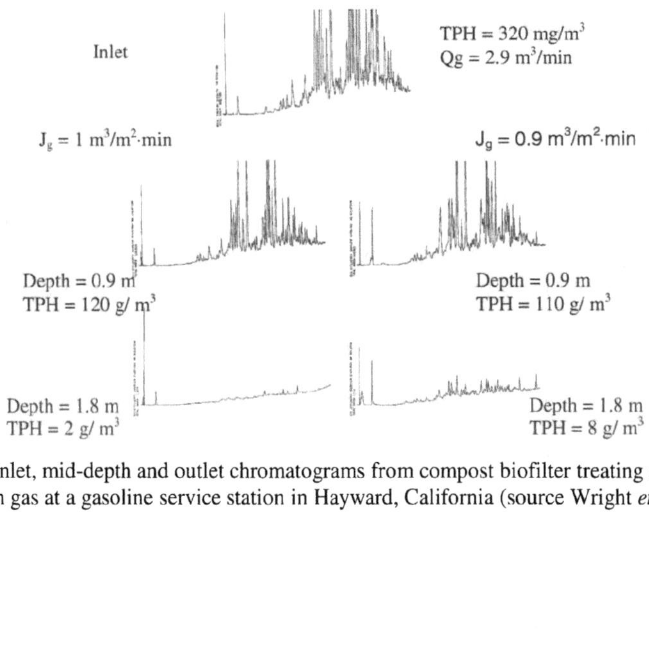


Fig. 2. Inlet, mid-depth and outlet chromatograms from compost biofilter treating soil vapor extraction gas at a gasoline service station in Hayward, California (source Wright *et al.*, 1997)

## 2. Biofilters

Biofilters are packed bed reactors that can be conceptualized as trickling filters without (or at least a minimal) moving liquid phase. Organic loading rates are typically 1 to 5 kg COD/$m^3$·d, about the same as for conventional wastewater trickling filters (0.5 to 2.5 kg COD/$m^3$·d). Two general configurations of biofilters are in use, a conventional system (schematic shown in Figure 3) in which contaminated air is humidified and passed through a packing material on which a microbial community develops. Contaminants and oxygen are transferred into the water phase and then into the biofilm where the degradation reactions take place. Nutrients are supplied from the packing material in some cases or may be added with the humidified air. In the second configuration, usually referred to as a biotrickling filter, the humidification chamber is replaced with a recirculating water stream sprayed over the top of the packing material. Water flows in biotrickling filters are quite low (typically about 1 $m^3/m^2$·d compared to 15 to 35 $m^3/m^2$·d in a high rate trickling filter used for wastewater treatment). Biotrickling filters are most useful in treating contaminant streams having products that change the reaction environment such as sulfides or choro-organic compounds. The ability to continually add nutrients and control the salt balance is a second advantage. However, the thicker liquid film of biotrickling filters may result in mass transport limitations (Hudepohl, 1999; Hudepohl *et al.*, 1999).

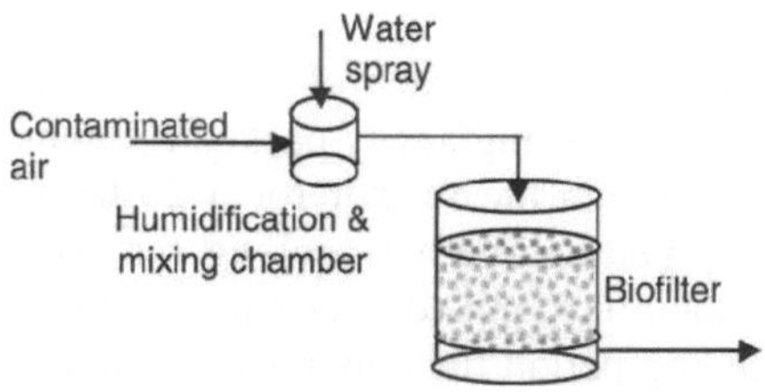


Fig. 3. Schematic diagram of biofilter system

The most commonly used biofilter packing material is a mixture of compost and a bulking agent that prevents consolidation and compaction of the wet mixture. Dry nutrients and buffer (e.g., calcium carbonate) may be included in the mixture. The type of compost does not appear to be important, although it is possible that some sources will provide a better source of microorganisms. Graded pumice has been used successfully in biofilters for both odor and NOx control (Hudepohl *et al.*, 1999). A number of synthetic packings have been used, including granular activated carbon (Devinny *et al.*, 1995), polystyrene pall rings, and diatomaceous earth pellets (Ergas *et al.*, 1995, Sorial *et al.*, 1996).

## 3. Materials and Methods

The biofilters used in this project were constructed of four 150 mm I.D., 300 mm long stainless steel tubes connected by flange couplings. Each of the four sections was packed with 250 mm (5.3 L) of Celite R-635™ porous silicate pellets. The remaining 50 mm of the section was used as a plenum for redistributing flow and sampling. A modified, unpacked section at the bottom of the units served to collect drainage and as an inlet or outlet section, depending on the direction of flow used. A second unpacked section at the top served as an inlet or outlet section, also. Filtered laboratory air was split into two streams controlled by rotameters. Selected contaminants were continuously injected into one stream via a syringe pump (Harvard Apparatus, Syringe Infusion Pump 22). The second air stream was passed through an aerosol generator (Heart ™,Vortran Medical Tech., Inc. Sacramento, CA - nebulizer) containing a solution of inorganic nutrients. The contaminated air stream and the aerosol-containing air stream were then combined in a mixing chamber before entering either the top or bottom of the column. In the manner described above, contaminated air was supplied to the column at a rate of 18 L/min yielding an empty bed residence time (EBRT) of 1 minute. A schematic of the apparatus is shown in Figure 3.

### 3.1. PROCESS START-UP

Two biofilters were used, one for each of the target compounds in this study. Each experiment began by seeding newly packed columns either by circulating a microbial suspension through the reactor for a 24 hour period or soaking the media in microbial suspension before packing the biofilter. The suspension consisted of nutrient solution, a mixed microbial culture, and the target substrate. At the end of the seeding period the column was drained and the contaminated air feed was begun.

### 3.2. TARGET COMPOUNDS

The easily degraded contaminant used throughout the work was toluene, a component of gasoline and a commonly used industrial and commercial solvent. Toluene is metabolized and used as a sole carbon and energy source by a large number of common soil bacteria. Both aerobic and anaerobic degradation is carried out. Aerobic degradation is usually initiated with a monooxygenase attack at the 1-2 bond or a dioxygenase attack at the 2-3 bond. Two difficult to degrade compounds were selected, dicloromethane (DCM) and methyl tert-butyl ether (MTBE). Physical properties of the compounds are given in Table I. Note that based on Henry's law the liquid phase equilibrium concentration of the compounds at 25°C and 50 ppmv (parts per million on a volume basis) would be 0.67 mg/L for toluene, 1.66 mg/L for DCM and 5.45 mg/L for MTBE.

Dichloromethane is a widely used industrial and commercial solvent. As a single carbon compound, only a restricted group of microorganisms are able to use DCM as a sole carbon and energy source. The most common groups reported in laboratory studies are members of the genus Hyphomicrobium (Stucki *et al.*, 1981). The organisms degrading DCM in this research were isolated in pure culture by our colleagues K. M. Scow and Mark Fuller and appeared to be members of the genus *Flavobacterium* based on phospholipid fatty acid analysis.

TABLE I

Properties of the target compounds used in this study, toluene, dichloromethane, and MTBE

| Compound | Molecular formula | Molecular weight | Solubility mg/L | Vp mmHg | H[a] (@25°C) | $\log K_{OW}$ (@25°C) |
|---|---|---|---|---|---|---|
| Toluene | $C_7H_8$ | 92 | 515 | 22 | 0.28 | 2.69 |
| DCM | $CH_2Cl_2$ | 85 | 20,000 | 350 | 0.105 | 1.15 |
| MTBE | $C_4H_9OCH_3$ | 88 | 48,000 | 240 | 0.033 | 0.94-1.16 |

[a]dimensionless Henry's coefficient

Methyl tert-butyl ether is used as an octane booster and combustion enhancer in gasoline. The compound was first added to gasoline when tetraethyl lead was removed and became a major component of gasoline in the United States due to US Environmental Protection Agency requirements that oxygenates be added to fuel in ozone and carbon monoxide nonattainment areas. In California, MTBE comprises between 11 and 15 percent of gasoline by volume. The compound has become extremely controversial as a result of its high water solubility and low sorbability on soil. MTBE travels in groundwater at approximately the same speed as chloride. Consumers are often able to detect the kerosene like odor and taste at concentrations lower than 20 μg/L. Microbial degradation of MTBE was not reported until 1994 (Salinitro, 1994). Since that time, several mixed cultures have been developed in which MTBE is the sole external carbon and energy source and complete MTBE mineralization takes place (Eweis, *et al.*, 1997, Park and Cowen, 1997). The UC Davis culture was developed from samples taken from a compost-based, pilot scale biofilter operating at the Joint Water Pollution Control Plant of the Los Angeles County Sanitation Districts. Off-gas from the activated sludge unit feeding the biofilter contained little organic material other than MTBE, which was typically in concentrations of less than 200 ppbv. Acclimation to MTBE degradation required over 350 days, as shown in Figure 4. Enrichment and evaluation of the culture at UC Davis has reduced start-up times to about 20 days in both attached growth vapor phase and suspended growth liquid phase systems. However, yields remain very low (about 0.2 to 0.4 g solids/g MTBE removed) and the maximum specific growth and saturation coefficient rate are estimated to be about 0.3 $d^{-1}$ and 3 to 4 mg/L, respectively (Eweis, 1999). Two pure cultures capable of mineralizing MTBE have been isolated by our colleagues J. Hanson and K. M. Scow. The morphologically different cultures have identical S16 rRNA sequences but different PFLA make-up. They have been tentatively identified as members of the genus *Sphingomonas*. Other workers have not identified species responsible for MTBE degradation but the physical descriptions supplied have been similar to those of the UC Davis culture.

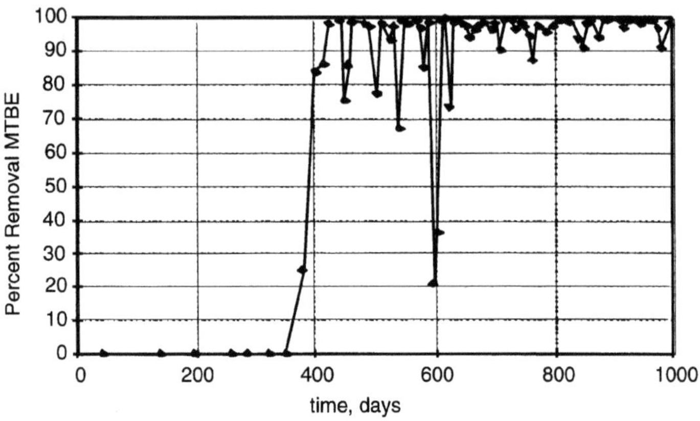


Fig. 4. MTBE removal in a pilot scale compost biofilter at the Joint Water Pollution Control Plant of the Los Angeles County Sanitation Districts. The biofilter received air from the activated sludge discharge channel containing approximately 150 ppbv of MTBE and very little other volatile organic material. EBCRT was 1 minute. (Source Eweis *et al.*, 1997)

## 4. Experimental Program

Three types of experiments have been conducted. All operations were carried out in a controlled temperature room maintained at 23°C. In all of the experiments the air flux and reactor empty bed residence time have been maintained at 1 $m^3/m^2 \cdot min$ and 1 minute, respectively. Each set of experiments has begun with new Celite™ R-635 packing seeded as described above. In the first type of experiment DCM was introduced as the first contaminant. Nominal inlet DCM concentration of 40 ppmv was maintained at that level over the 160 days of the experiment. After 51 days of operation, toluene was introduced into the column at a concentration of 40 ppmv. At 90 days of operation the toluene concentration was increased to 70 ppmv and at 115 days of operation the toluene was increased to 150 ppmv. In the second set of experiments the biofilter was started on MTBE at an inlet concentration of 35 ppmv and run for approximately 9 months with no other source of carbon or energy at an EBRT of 1 minute.

Using an arbitrary time of 0 days, toluene was introduced on day five at a concentration of 8 ppmv. After an additional 8 days the toluene concentration was increased to 25 ppmv and then to 70 ppmv after another week of operation. As will be shown below, removal of both MTBE and toluene decreased shortly after toluene inlet concentration was increased to 70 ppmv (Figure 8). Toluene was removed from the system and nitrogen feed was increased.

In the third set of experiments, the biofilter was operated for eight weeks with MTBE at 35 ppmv as the only organic source. A second arbitrary 0 time was set and performance monitoring begun. After five days of operation, toluene was included in the feed stream at 70 ppmv. After 12.5 days of operation benzene at 70 ppmv was substituted for toluene and continued to be included in the feed stream until 490 hours of operation was reached. From 20 days to 28.5 hours MTBE was again the sole organic fed (Figure 10).

## 5. Results

### 5.1. TOLUENE REMOVAL

Toluene is easily degraded in vapor phase biofilters despite the relative volatility and low solubility of the compound. Start-up times required to establish complete removal of inlet concentrations of up to 200 ppmv (750 mg/m$^3$) are less than three weeks. In the experiments with the DCM column, acclimation was achieved in 18 days at an inlet concentration of 40 ppmv. Subsequent increases in inlet toluene concentration were responded to without a lag. The system was operating on a 1 minute EBRT and toluene loadings were 220, 385, and 825 g toluene/m$^3$·d, respectively. Removals in the first 25 cm section (15 second EBRT) of the column were typically greater than 95 %, as indicated in Figure 5. Over a period of several weeks the performance of the first section deteriorated somewhat. The cause is believed to be channelization due to accumulated biomass. Filling the column with water and allowing the unit to drain under gravity resulted in complete recovery of performance. Breakthrough of toluene beyond the 50 cm mark did not occur.

### 5.2. DCM REMOVAL

Biodegradation of DCM is considerably more difficult than degradation of toluene. Start-up of biofilters from nonacclimated cultures typically takes about 10 weeks. However, when biofilters are seeded with material from acclimated units the start-up time is reduced to about three weeks. Because DCM degradation results in the production of two moles of acid per mole of DCM mineralized, care must be given to maintaining buffer capacity. However, once established the DCM degrading organisms are robust and system operation is not difficult. Although the physical properties of DCM are more favorable than those of toluene in terms of partitioning to the water phase, degradation is slower as shown in Figure 6. Based on the lower rate of removal and considering the physical properties of the two compounds, it appears that the system is reaction limited rather than transfer rate limited. This conclusion is supported by the data shown in Figure 7 where removals in the first 25 cm are consistently slightly less than the inlet concentration.

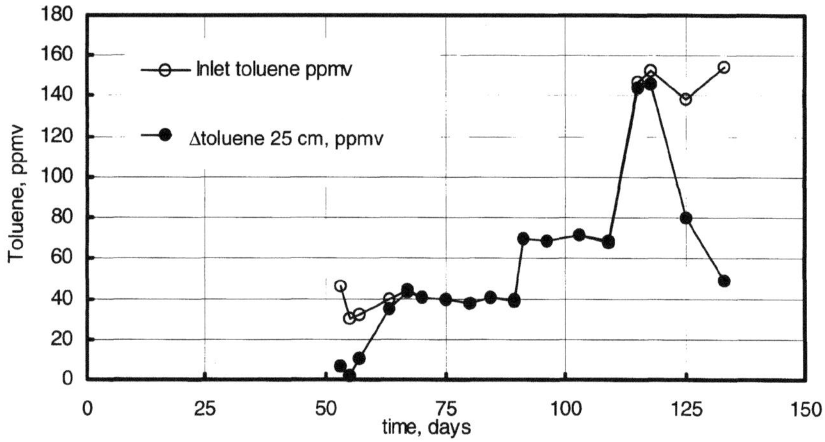


Fig. 5. Toluene removal in first 25 cm of biofilter also receiving 40 ppmv DCM. Decrease in removal in first 25 cm section is believed to have resulted from channelization due to clogging.

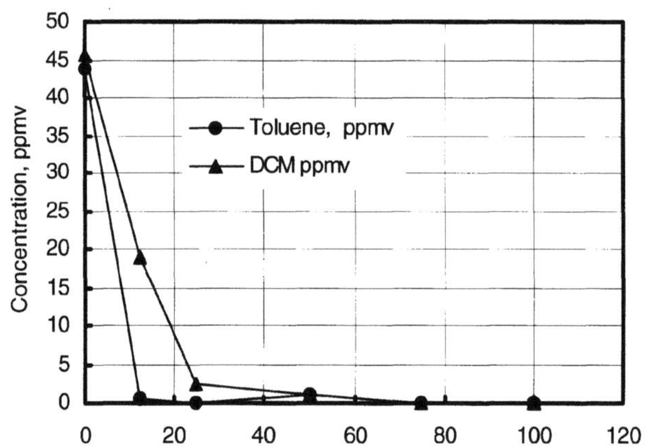


Fig. 6. Removal of DCM and Toluene on day 67 of operation.

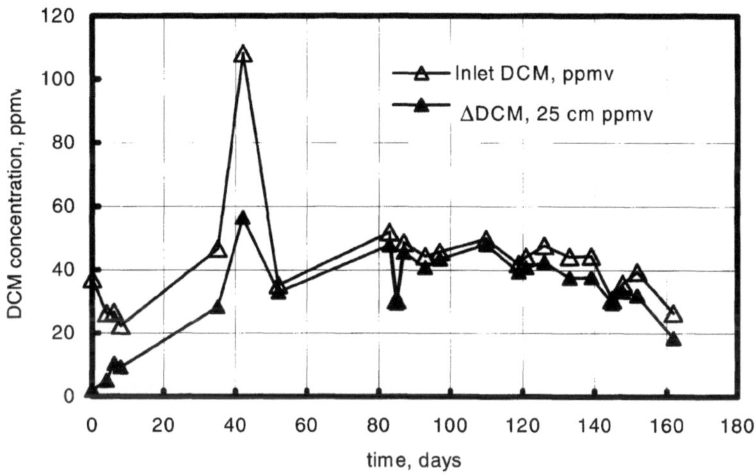


Fig. 7. DCM removal in the first 25 cm of the biofilter. Once acclimated the removals remain close to the inlet concentration value.

## 5.3. MTBE REMOVAL

The biofilter was operated for approximately nine months with MTBE as the sole carbon and energy source prior to the initiation of the experiments reported here. Acclimation of the biofilter to MTBE degradation required several weeks. Over the subsequent nine months of operation at a nominal inlet concentration of 35 ppmv, removal patterns became stabilized. An arbitrary time was set as zero and toluene addition was begun five days later at an inlet concentration of 8 ppmv, as indicated in Figure 8. Toluene degradation began almost immediately and MTBE removal dropped from 100 percent to approximately 95 percent. There was no measurable change in performance of the first section in part due to significant changes in inlet concentration resulting from the operation of the syringe pump. Within six hours MTBE removals had returned to pre-toluene levels. Increasing the inlet toluene concentration to 25 ppmv after 12.5 days of operation resulted in breakthrough of both compounds but the effect lasted only a few hours. When the inlet toluene concentration was increased to 70 ppmv breakthrough occurred quickly and percent removal began to decrease sharply, although total organic removal increased considerably on a mass basis. Toluene was removed from the inlet stream after eight days and the complete MTBE removal was reestablished within 30 minutes, as shown in Figure 8. In subsequent experiments it was determined that breakthrough had resulted from depletion of nitrogen and increasing the nitrogen concentration in the aerosol feed allowed considerable increases in total organic loading without breakthrough (Eweis *et al.*, 1998a).

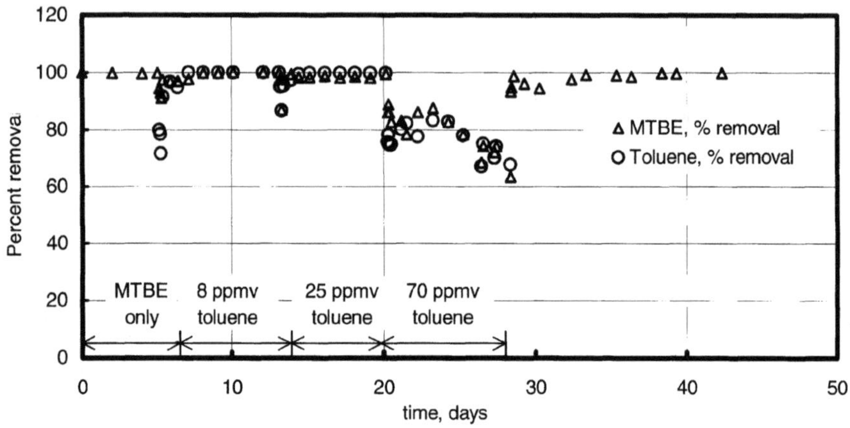


Fig. 8. Removal of MTBE and toluene in biofilter initially acclimated to MTBE for nine months. Decrease in percent removal when toluene inlet concentration reached 70 ppmv was due to nitrogen limitation. Note that actual total carbon removal continued to increase as percent removal decreased.

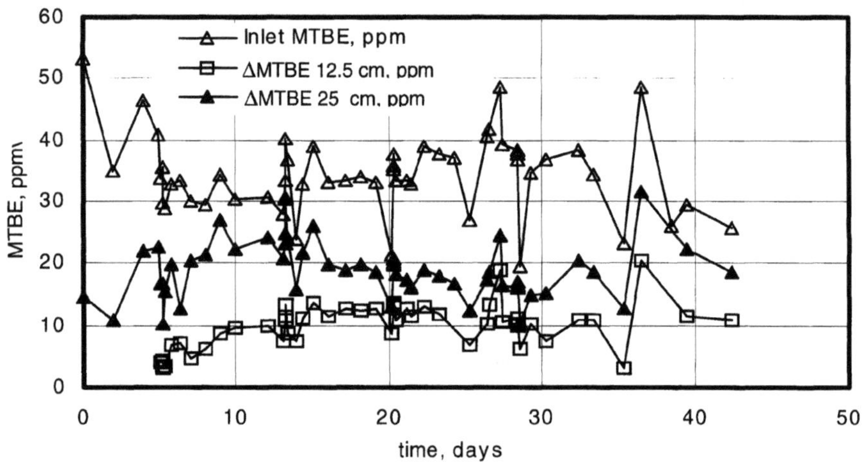


Fig. 9. MTBE removal in first section of biofilter. Note that mass removal decreases from 20 to 28 days of operation, the period of 70 ppmv toluene inlet concentration.

The biofilter was operated for two months with MTBE as the only carbon source and an increased inlet nitrogen concentration. Consequently, a second series of experiments were carried out in which 70 ppmv of toluene, and later benzene, as an alternative substrate was added. A summary of these experiments is shown in Figure 10 where percent removal versus time is plotted.

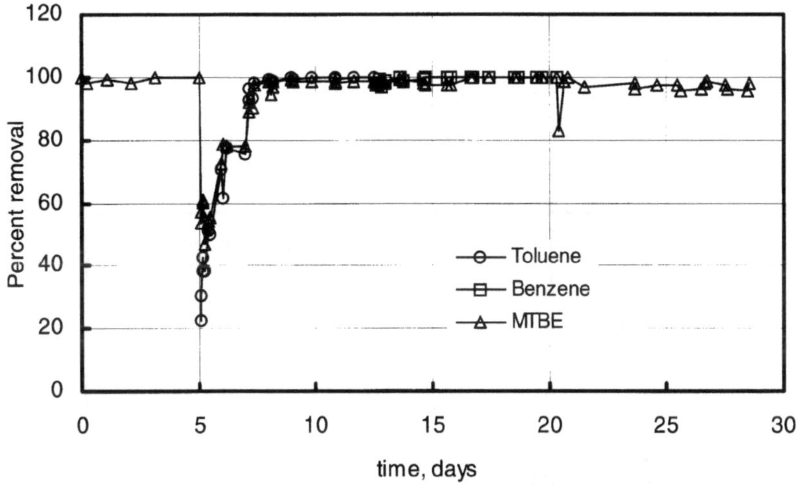


Fig. 10. Removal of benzene, toluene, and MTBE in biofilter initially acclimated to MTBE. Nominal inlet concentrations of toluene, benzene and MTBE were 70, 70, and 35 ppmv, respectively.

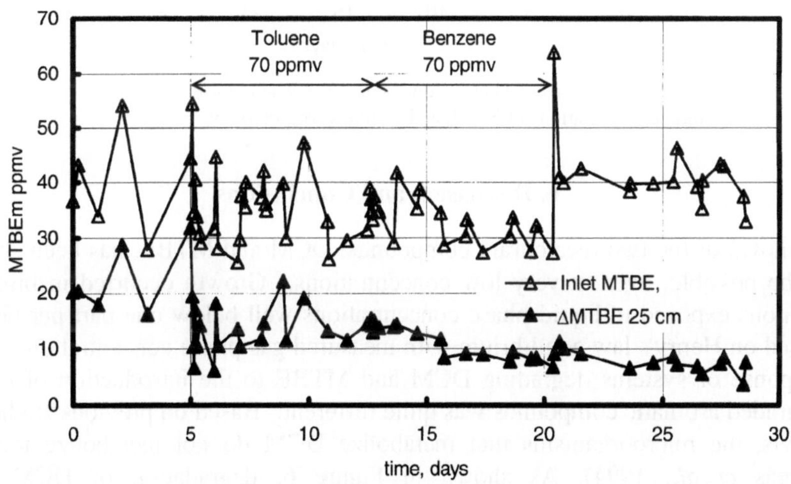


Fig. 11. Removal of MTBE in the first 25 cm of biofilter treating toluene (days 5 to 13) and benzene (days 13 to 20).

The fact that MTBE removal continued to decrease after benzene was removed from the feed suggests that substrate competition or inhibition was not a factor. Long term operation of biofilters results in gradual accumulation of biomass, blockage of pores and apparently channelization (Wright, 1999). However, a significant breakthrough occurred when toluene was first introduced in these experiments, as shown in Figure 12 where outlet concentrations of toluene, benzene and MTBE are plotted. The breakthrough pattern was virtually the same for each of the four biofilter sections and is typical of the response of biofilters to transient loadings. Other plug flow, attached growth processes would be expected to behave in a similar manner.

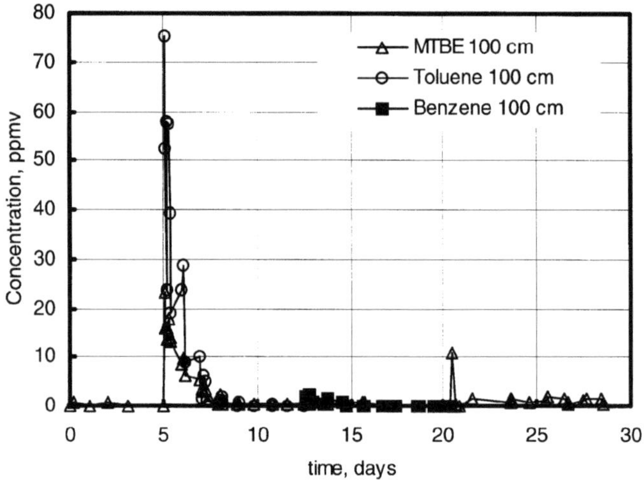


Fig. 12. Outlet concentrations of MTBE, toluene and benzene in biofilter having inlet concentrations of 35, 70 and 70 ppmv, respectively.

## 6. Discussion and Conclusions

Removal of the two recalcitrant compounds, DCM and MTBE, has been shown to be possible, even at very low concentrations. Growth occurred in biofilter sections exposed to liquid phase concentrations well below one part per billion based on Henry's law equilibrium with measured gas phase concentrations. The response of systems degrading DCM and MTBE to the introduction of easily degraded aromatic compounds was quite different. Based on previous studies at Davis, the microorganisms that metabolize DCM do not metabolize toluene (Ergas *et al.*, 1994). As shown in Figure 6, degradation of DCM was considerably slower than degradation of toluene. However, the presence of toluene had no apparent impact on DCM degradation, as shown in Figure 7.

The organisms that metabolize MTBE also metabolize aromatic compounds and the result is that introduction of toluene sharply impacted MTBE removal

(Figures 9 - 12). However, acclimation to combined removal of aromatics and MTBE is relatively rapid as is shown by the removals in the first biofilter section (Figure 9). Clogging of the biofilter with accumualted biomass results in channelization that may be confused with substrate competition. However, there appears to be little impact of multilple substrates on system performance once acclimation has occurred.

The very clear differences in rate of removal of the compounds used in this study demonstates that qualitative make-up of the outlet stream can be very different from that of the inlet stream. Regulations should not be based on the assumption that recalcitrant compounds have been removed when lumped parameters such as TPH or BOD are low.

## Acknowledgments

This research was supported in part by the U. S. Environmental Protection Agency, Environmental Resolutions, Inc., ARCO Corporation, and the American Petroleum Institute.

## References

Morton, R. L. and Caballero, R. C.: 1998, *Proceedings of the USC-TRG Biofiltration Conference,* University of Southern California, October 22-23 1998, pp. 107-114.

Devinny, J. S., Hodge, D. S., Chang, A. N. and Reynolds, Jr., F. E.: 1995, *Inovative Technologies for Site Remediation and Hazardous Waste Management,.* R. D. Vidic and F. R. Pohland (eds), Am. Soc. Civil Engineers, New York, pp. 481-488.

Ergas, S. J., Kinney, K. A., Fuller, M. E. and Scow, K. M.: 1994, *Biotechnology and Bioengineering,* **67,** 1048-1054.

Ergas, S. J., Schroeder, E. D., Chang, D. P. Y., and Scow, K. M.: 1994, *Proceedings of the 87th Annual Meeting of the Air and Waste Management Association,* Cincinnati, Ohio. June 19-24.

Eweis, J. B., Chang, D. P. Y., Schroeder, E. D., Scow, K. M., Morton, R. L. and Caballero, R. C.: 1997, *Proceedings of the 90th Annual Meeting & Exhibition, Air & Waste Management Association*, Toronto, Canada June 8-13.

Eweis, J. B., Schroeder, E. D., Chang, D. P. Y. and Scow, K.M.: 1998b, in G. B. Wickramanayake and R. E. Hinchee (eds), *Natural Attenuation: Chlorinated and Recalcitrant Compounds*, Battelle Press, Columbus, Ohio. pp. 341-346.

Eweis, J. B., Watanabe, N., Schroeder, E. D., Chang, D. P. Y. and Scow, K. M.: 1998, *Proceedings of the National Ground Water Association*, Anaheim, California, June 3-4.

Hudepohl, N. J.: 1999, Department of Civil & Enivronmental Engineering, University of California, Davis.

Hudepohl, N. J., Davidova, Y., du Plessis, C. A., Schroeder, E. D. and Chang, D. P. Y.: 1999, *Biofilter Technology for NOx Control, California Air Resources Board,* Center for Environmental and Water Resources Engineering, University of California, Davis.

Park, K. and Cowan, R.: 1997, *Proceedings of the 213th ACS National meeting, Division of Environmental Chemistry*, San Francisco, California, April 13-17, pp. 421-424.

Sorial, G. S., Smith, F. L., Suidan, M. T. and Biswas, P.: 1995, *J. Air and Waste Management Association,***45,** 801-810.

Steffan, R. J., McClay, K., Vainberg, S., Condee, C. W. and Zhang, D.: 1997, *App. and Environ. Micro.*, **63,** 4216-4222

Stucki, G., Galli, R., Ebersold, H. R. and Leisinger, T.: 1981, *Arch. Micro.* **130**, 366-376.

Wright, W. F., Schroeder, E. D. and Chang, D. P. Y.: 1999, *Proceedings of the 1998 USC-TRG Biofiltration Conference*, University of Southern California, pp. 143-152.

Wright, W. F., Schroeder, E. D., Chang, D. P. Y. and Romstad, K.: 1997, "Performance of a Compost Biofilter Treating Gasoline," *J. Env. Eng*, ASCE, **123,** 547-555.

# APPLICATION OF BIOFILMS AND BIOFILM SUPPORT MATERIALS AS A TEMPORARY SINK AND SOURCE

P. A. WILDERER, P. ARNZ and E. ARNOLD

*Institute of Water Quality Control and Waste Management, Technical University of Munich, Am Coulombwall, D-85748 Garching, Germany, e-mail: wilderer@bv.tum.de*

**Abstract.** Wastewater treatment plants are typically subjected to variable influent loading conditions. Extreme variations of flow, composition and substrate concentration may occur in the influent of industrial wastewater treatment plants, but also at plants serving tourist areas. High rate reactors of low hydraulic buffer capacity are particularly affected. Exploitation of internal equalization capacities are proposed to dampen the fluctuations of biomass loading. In this context, the biomass itself and – in case of biofilm reactors – the biofilm support media may be considered as a sink during peak loading situations, and as a source of substrate as soon as the influent loading drops. Sequencing Batch Biofilm Reactors (SBBR) studies in laboratory and pilot scale were conducted to investigate the capacity of these internal sink and source terms. The reactors were packed with four types of biofilm carrier materials, blasted clay granules, granular activated carbon, zeolite and small size plastic rings (Kaldnes). Temporary storage of substrates was achieved by means of adsorption, ion exchange and absorption processes. As the react phase proceeded, the bulk liquid concentration dropped and desorption processes followed by metabolic reactions became dominant. From the results achieved it can be concluded that thick biofilms, and biofilm support media with sorptive capacities are favorable to counteract peak loading fluctuations, and to keep the effluent concentration from exceeding set discharge levels.

**Keywords:** adsorption, biofilm reactors, desorption, influent loading fluctuations, Sequencing Batch Biofilm Reactor

## 1. Introduction

Traditionally designed wastewater treatment plants are space demanding. To accommodate bioreactors and sedimentation tanks, land of significant size and value must be set aside. Since in many cases, in particular in industry, additional property area is either not available or too expensive or both, the readiness of considering factory operated wastewater purification plants is often rather poor despite of wide spread concerns about impacts of industrial pollution on the environment, and on municipal wastewater treatment plants.

High rate, compact treatment systems are needed to allow implementation of reactors for process wastewater treatment in industrial facilities. During the past years, the development of those systems has made significant progress. Particularly, two specific technologies have received attention: bioreactors coupled with membrane separation units (membrane reactors) and high rate biofilm reactors with inherent filtration capacities (bio-filters). Membrane reactors have their niche where the volumetric loading of the reactors are rather

*Water, Air, and Soil Pollution* **123:** 147–158, 2000.

low but the mass loading is high. On the other hand, biofilm reactors are especially advantageous in cases where the hydraulic loading rate is high. Since in both cases no sedimentation tanks are required, the land demand of those systems is relatively small. In addition, the bioreactor can be kept small when the biomass concentration in the membrane reactor and the specific area of the biofilm support media in bio-filters are enhanced and set higher than in traditional activated sludge tanks and trickling filters.

There is one problem, however, which has often not been properly addressed by the developers of high rate, compact bioreactors: process stability under fluctuating influent conditions. Typically, hydraulic loading as well as mass loading and composition of the wastewater to be treated vary, diurnally, often in a wide range. The variations are the more pronounced the smaller the catchment area of the treatment plant. In industry, changes in flow rate, concentration and composition may be especially high, for instance during working hours when production processes and cleaning cycles are follow each other. The smaller the volume of the bioreactor is - in relation to influent rate - the less is the hydraulic equalization potential of the reactor, and the higher is the risk of breakthrough and subsequent violation of discharge limits. In principle, the problem could be solved by placing an equalization tank in front of the bioreactor. This, however, would require land, and would diminish the advantage of the high rate treatment system applied.

The alternative is to incorporate specific buffer capacities in the bioreactor in the form of a temporary sink for critical wastewater constituents. When the mass loading of the reactor increases, this sink would be activated, and the additionally imported compounds would be stored. When the loading decreases again, the former sink would serve as a source of substrate to support the microorganisms.

Adsorption followed by desorption is a process sequence which satisfies the basic requirements of the temporary sink and source concept. Other options are ion exchange and diffusion of molecules into the void space of biofilms and of the packing of the biofilm support media.

In the following, the potentials of the sink-source concept will be investigated using observations documented in literature and the latest results of experimental trials.

## 2. Literature Review

The idea of using activated carbon as a temporary sink for organic wastewater constituents and as a source of substrate for biofilm microorganisms has been studied by various authors. As a first step, selected substrates were adsorbed to granular activated carbon. As the adsorptive capacity of the activated carbon reached its limits, microorganisms were employed to biologically regenerate the spent carbon granules (Hutchinson *et al.,* 1990). Jaar (1991) and Kolb (1997) did not wait for the exploitation of the adsorptive capacity of the activated carbon but combined the adsorption and regeneration in one process schematic using the Sequencing Batch Reactor (SBR) strategy. A reactor packed with granular activated carbon was loaded with a batch of wastewater and aerated (Figure 1).

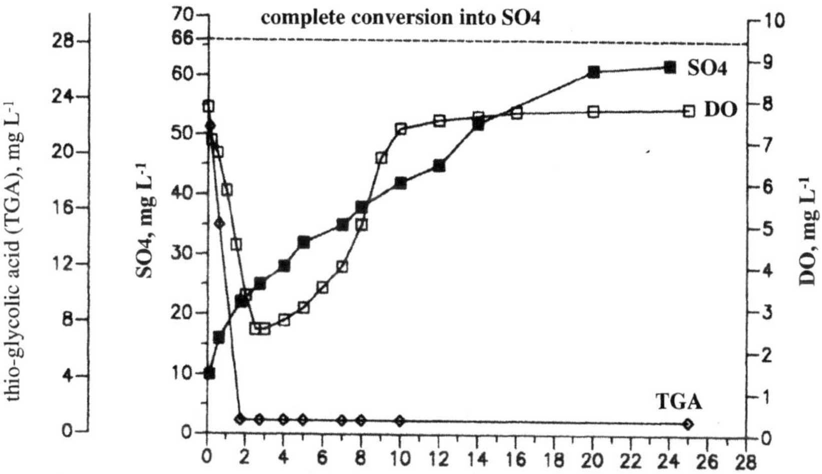


Figure 1 Time course of thio-glycolic acid (TGA) conversion under aerobic conditions in a biofilm reactor packed with granular activated carbon Source: Jaar (1991)

The reactor was inoculated with a mixed culture of microorganisms which were allowed to grow at the surface of the activated carbon granules forming a biofilm there. Since the substrate concentration was high at the beginning of each SBR cycle, adsorption processes dominated during the initial phase of the cycle, and the substrate concentration in the bulk fluid decreased. The biofilm organisms metabolized the substrates as they passed through the biofilm towards the binding sites of the activated carbon. As the adsorptive equilibrium was approached and the bacteria continued to take up substrates from the bulk liquid, desorption was initiated, substrate molecules traveled from the binding sites directly to the biofilm organisms, and the activated carbon was regenerated, eventually. The progress of the regeneration process was observed by measuring chloride ions in case of chlorinated hydrocarbons as test substances, or sulfate ions in case of thio-glycolic acid as a model, respectively.

In case of mixed substrates in the influent the desorption process could not be completed, however, for all of the wastewater constituents applied, and the

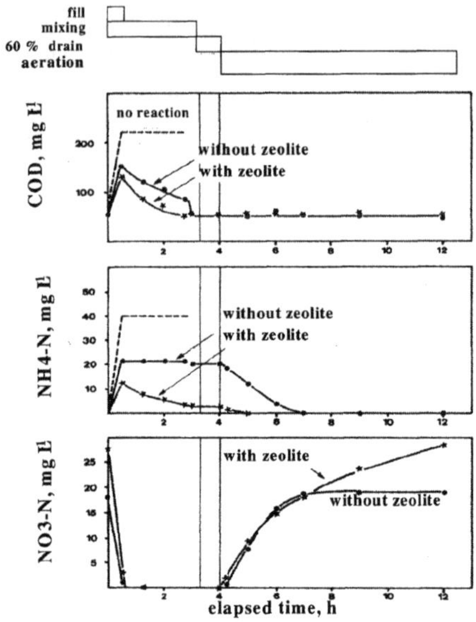


Figure 2:
Time course of material conversion in a biofilm reactor with zeolite granules as biofilm support media.
Source: Kruzic (1986)

activated carbon gradually became loaded with those compounds. Nevertheless, impressive in-service times for the activated carbon columns could be demonstrated, and the observed effluent concentrations were very low over a 165 and 296 days of operation, respectively, despite of occasionally very high influent substrate concentrations applied. For benzene, a regeneration efficiency of 99% was observed. For phenol, 2-chloro-phenol and tri-chloro-ethene the observed regeneration efficiencies were 91%, 43% and 4%, respectively.

Kruzic (1986) conducted similar experiments but used zeolithe as cat-ion exchanger to temporarily store ammonium ions, and provide desorbed ammonium as electron donors later on in the SBR cycle. An ammonium rich wastewater was filled into an activated sludge SBR containing granules of zeolite (400 mg $L^{-1}$).

During an initial, non-aerated phase, ammonium was allowed to get bound to the zeolite, and nitrate generated during the previous cycle to get denitrified. In the following, the activated sludge and the zeolite granules were settled, 60% of the supernatant was removed from the reactor, the aerators were turned on, and the bound ammonium was allowed to get desorbed and nitrified. In Figure 2, the time course of substrate conversion during a SBR cycle is illustrated. In comparison with the control reactor containing no zeolite granules removal of ammonium from the bulk liquid was completed after a relatively short period of time. Generation of nitrate proceeded even when no ammonium was present in the bulk liquid suggesting that previously bound ammonium served as major electron donor during the nitrification phase.

It needs to be mentioned, in this context, that zeolite is a non specific cat-ion exchanger. It may take up not only ammonium but also any other cat-ions from the wastewater. Desorption driven by the progress of nitrification in the adjacent biofilm and subsequent regeneration of the zeolite particles can only be expected for ammonium, however. Consequently, this technology is only applicable as long as the wastewater to be treated is relatively low in concentration of competing cat-ions (for instance condensate of sludge dryers).

Finally, the research of Kaballo (1998) should be mentioned, who conducted experiments with a biofilm reactor loaded with wastewater rich in 4-chlorophenol (4-CP). Blasted clay particles (8 to 16 mm in diameter) were used as biofilm support material. He observed a significant loss of 4-CP in the inlet zone of the reactor during peak loading conditions. Calculations clearly demonstrated that in this case the incoming material (4-CP) diffused into the void space of the biofilm system (pores in the extra cellular polymeric matrix, voids between biofilm carriers) where the substrate concentration was relatively low during normal loading conditions (Figure 3).

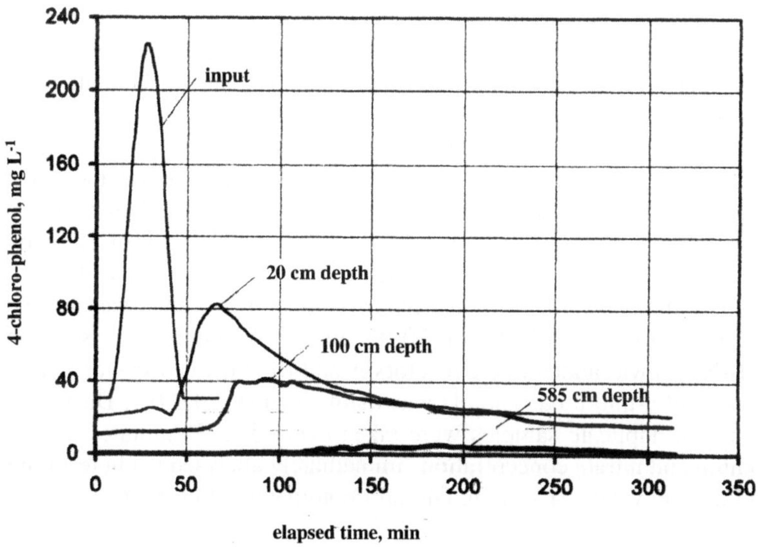


Figure 3 Time course of the concentration of 4-chlorophenole in various depth of a continuous flow fixed bed biofilm reactor, packed with blasted clay balls Source: Kaballo (1998)

The decrease of the substrate concentration in the deeper zones of the reactor was the result of metabolic conversion processes. After the peak loading event the stored material in the biofilm was converted also, and the substrate concentration in the biofilm decreased to normal levels.

In summary it can be concluded that there are a number of possibilities apart from hydraulic equalization units in front of bioreactors to accommodate peak loading condition in high rate systems. Exploitation of the sink and source capacities of the biomass appears to be the most interesting for obvious reasons. As the biomass is capable of accumulating and releasing chlorinated hydrocarbons (Kaballo, 1998) the question was raised whether or not it could also serve as a temporary sink for ammonium. To answer this question a number of experiments were conducted at the Munich laboratories. The results are described and discussed in the following.

## 3. Materials and Methods

### 3.1. BATCH EXPERIMENTS

In order to assess the potential of the internal equalization capacity of a biofilm reactor, batch experiments were conducted in lab scale batch reactors. Fixed bed samples were taken from Sequencing Batch Biofilm Reactors (SBBRs) operated for nitrification and enhanced biological phosphorus removal. The biofilm support materials used were blasted clay balls with a diameter of 4-8 mm covered with a thin biofilm (reactor A), and Kaldnes plastic elements filled with thick biomass aggregates (reactor B), respectively. The fixed bed samples were washed three times to remove any adhering ammonium. The water content of the fixed bed was determined and the weight of the biomass was measured after shaking it off (triplicate samples). The reactors used for the batch experiments had a fixed bed volume of 1 L, a height of 250 mm, a diameter of 105 mm and a water volume of 1.6 L and 1.75 L in reactor A and B, respectively, depending on the support material and its porosity. Mixing of the water was achieved by recirculation from top to bottom (170 mL $min^{-1}$). An ammonium solution (100 mg $NH_4$-N $L^{-1}$) was added to the reactors A and B. Nitrification was inhibited by addition of allyl-thio-urea. The pH was maintained at 7.0 ± 0.1. During 2.5 h of recirculation, duplicate samples were taken every 5 to 30 minutes, filtered and ammonium and nitrate concentrations immediately analysed using test tubes (Dr. Bruno Lange cuvette test system, Digital Photometer ISIS 6000).

### 3.2. ADSORPTION ISOTHERMS

The capacity of the biofilm and the support materials to adsorb ammonium was determined by measuring ammonium adsorption isotherms. The carriers were dried, weighed and placed in an ammonium solution and shaken for three hours. The initial ammonium concentrations ranged from 0 to 300 mg $NH_4$-N $L^{-1}$. For the biomass experiments, sludge from washing the blasted clay from reactor (A) and from washing the Kaldnes plastic carriers from reactor (B) was collected and centrifuged. Total solids concentration was adjusted in the range from 10.7 to 12.9 mg TS $L^{-1}$. Nitrification was inhibited by addition of allyl-thio-urea. Stripping of ammonia was prevented by a lid and by keeping the pH neutral.

Duplicate samples were taken after one and three hours, filtered and immediately analyzed for ammonium, nitrate and nitrite using test tubes (Dr. Bruno Lange cuvette test system, Digital Photometer ISIS 6000). $NH_4^+$ production due to lysis was measured and taken into account for calculations of ammonium adsorption. Also, the volume of the samples was respected in the calculation.

## 4. Results and Discussion

It was found that the blasted clay and the plastic media (Kaldnes) packing exhibit an internal water content of 25.6 ± 0.1 % and 62.3 ± 0.5 %, respectively. Considering the different volumes of the added ammonium solution, it was calculated that the initial ammonium concentration gets reduced to 82.4 % (reactor A) and 81.6 % (reactor B) of the inflow concentration simply by dilution. From Figures 4 and 5 the measured and calculated concentrations of ammonium can be seen.

The measured concentrations in both reactors are well below those calculated. Nitrate was not found in concentrations above 1 mg $NO_3$-N $l^{-1}$; nitrite was not found at all. Nitrate did probably not stem from nitrification but was washed out of the biofilm.

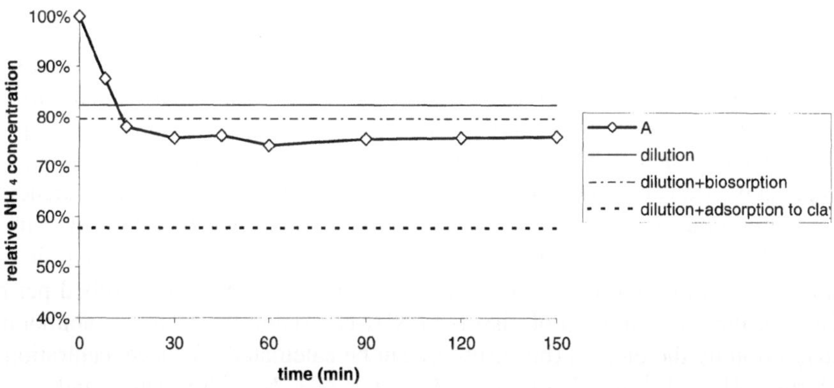


Fig. 4. Measured and calculated ammonium concentrations in reactor A packed with blasted clay particles.

To account for the low ammonium concentrations, adsorption processes are to be considered. The total solids concentrations in the blasted clay and the plastic material reactor were 1.84 ± 0.08 mg TS per liter of the fixed bed, and 11.77 ± 0.83 mg TS $L^{-1}$, respectively.

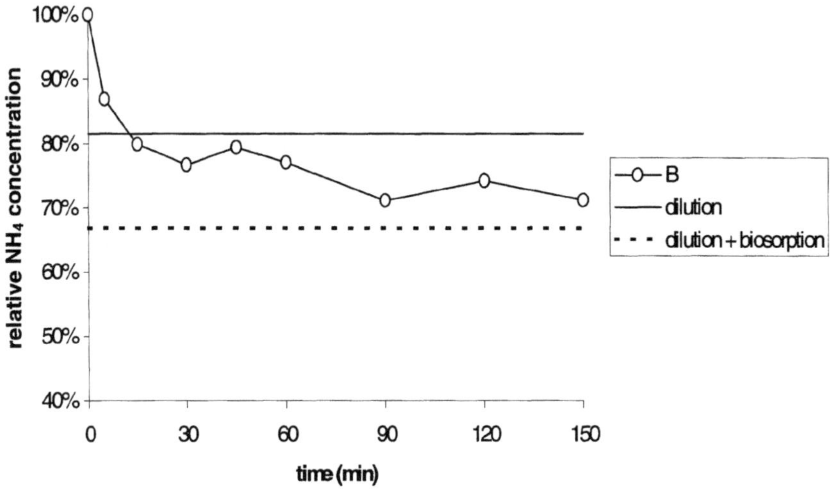


Fig. 5. Measured and calculated ammonium concentrations in reactor B (Kaldnes packing).

The adsorption capacities of the biomass can be seen from Figure 6 (reactor A) and Figure 7 (reactor B). It can be seen that even at equilibrium concentrations above 200 mg $NH_4$-N $L^{-1}$ saturation of the ammonium adsorption is not yet approached. Nitrate and nitrite could be detected only in a few samples in concentrations below 0.3 mg N $L^{-1}$ The results suggest a linear dependency of ammonium adsorption on bulk concentration in the range investigated. Almost the same slope of the linear regression line was found for the two sludges washed from the biofilm carriers (0.028 and 0.030 mg $NH_4$-N adsorbed per mg TS over the concentration of dissolved $NH_4$-N). Using these slopes, ammonium adsorption by the biofilm (biosorption) can be calculated. For a concentration of 15 mg $NH_4$-N $L^{-1}$ an adsorption of 0.45 mg $NH_4$-N $g^{-1}$ TS was found, which comes very close to the value reported by Nielsen (1996). For the reactor packed with clay particles, this corresponds to 1.5 mg N $m^{-2}$ fixed bed, which is a somewhat lower than the value of 2.7 mg N $m^{-2}$ biofilm reported by Wik (1999).

As a binding mechanism, Nielsen (1996) proposed electrostatic interactions as bacteria and extracellular polymeric substances (EPS) have a net negative surface charge. In the experiment summarized above (Figures 4 and 5), the calculated concentrations are now close to those measured. In the reactor packed with the clay particles the true biosorption may be somewhat higher than the one calculated because the biomass was presumably underestimated due to the method of measuring the biomass. By shaking the clay balls in a defined way, low deviations result when repeatedly measuring the biomass, but some of the biomass must be assumed to remain in the pores of the material.

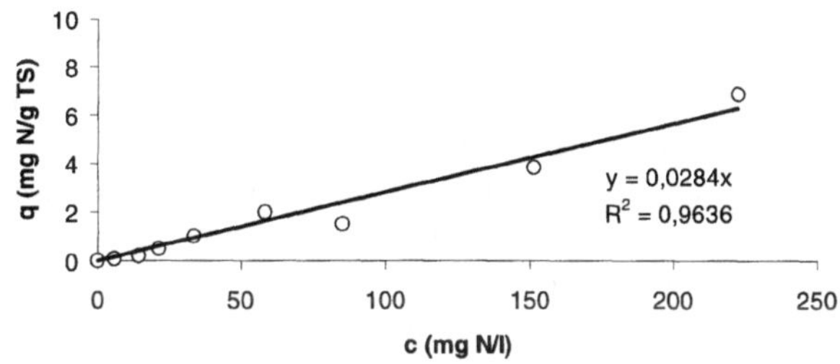


Fig. 6. Ammonium adsorption to sludge from reactor A packed with blasted clay particle.

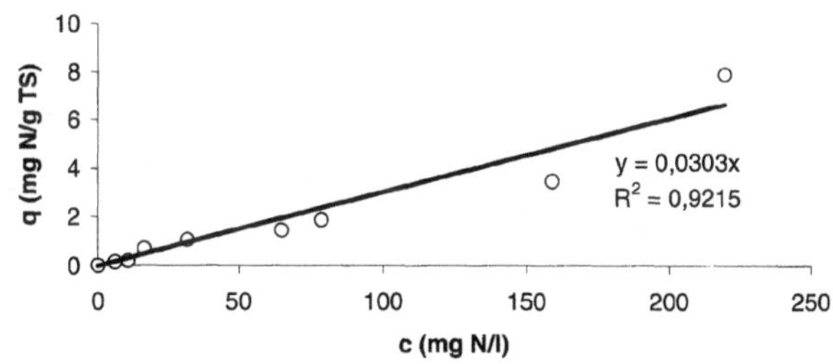


Fig. 7. Ammonium adsorption to sludge from reactor B packed with Kaldnes plastic rings.

No ammonium adsorption was found to naked Kaldnes plastic elements (data not shown), nor was it expected, since polyethylene is a non polar material not apt to adsorb cat-ions. Porous clay, however, has a strong capacity to take up ammonium (Figure 8). An ammonium saturation seems to occur at a load of around 0.1 mg $NH_4$-N $g^{-1}$ clay. The adsorption can be well described by the Langmuir equation. The ammonium concentration that would result if the maximum clay adsorption at the given concentration becomes effective can be seen in Figure 4. The resulting concentration level is far lower than the one measured.

The Langmuir isotherm describes the equilibrium of adsorption and desorption processes. Nevertheless, results from desorption experiments with de-ionized water did not agree with the adsorption experiment. Although the results were not quite consistent, they suggest that almost no desorption of ammonium from clay occurs in de-ionized water. Consequently, once loaded there would be little or no capacity for further adsorption. The clay particles which were used for the experiments were taken from a reactor which received ammonium rich solutions for a considerably long period of time so that complete loading of the particle surface must be considered. Therefore, it is assumed that the observed reduction of ammonium concentrations (Fig. 4 and 5) is due only to dilution and adsorption to biomass. Further evidence for this claim has been obtained by measuring ammonium micro-profiles (Gieseke et. al., 1999) using microelectrodes. During the first non-aerated phase of a SBR cycle, a steep gradient of the ammonium concentration was found within the interfacial boundary layer. The steepness of that gradient decreased quickly and finally leveled off during the subsequent nitrification phase.

Considering the sensitivity against peak loadings of reactors packed with different biofilm carrier materials, it becomes evident that a support material that allows high biomass concentrations to develop is advantageous as it has a higher capacity to adsorb for instance ammonium and act as a temporary sink. Ammonium ad- and absorbed in biomass is certainly accessible to nitrifiers (Nielsen, 1996) and no brine for regeneration is needed.

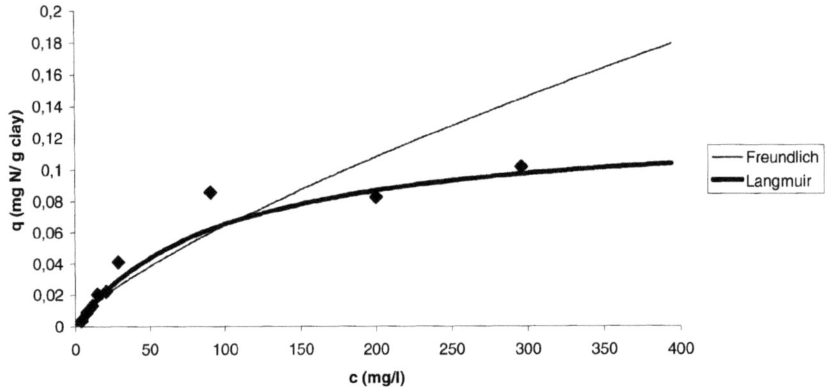


Fig. 8. Ammonium adsorption to porous clay particles applied in reactor A; isotherms according to Freundlich and Langmuir

This statement is supported by the performance data of a SBBRs treating reject waters from sludge dewatering. Despite of occasionally extremely high ammonium concentration for up to 900 mg $L^{-1}$ no inhibition of nitrification was observed (Arnold *et al.*).

In conclusion, high biomass concentrations can constitute a most effective equalization capacity for handling peak loadings of ammonium.

## 5. Conclusions

The results of the studies described and discussed above lead to the following conclusions:

1. Loading variations are typical for municipal and industrial wastewater treatment plants, and must be counteracted by means of equalization, in particular when high rate bioreactors are considered for application.
2. Equalization of variation of flow, mass and composition of wastewater can be achieved not only by positioning of an equalization tank in front of the bioreactor, but also by making use of internal sink and source capacities.
3. Sorption phenomena can be effectively used as internal sink and source terms. Experimental trials conducted by various authors indicate that adsorption to activated carbon, immediately followed by desorption and biodegradation is a very effective method to manage peak loadings of organic substances.
4. Adsorption (ion exchange) of ammonium to zeolite particles followed by desorption and nitrification has also been studied by a number of authors. Positive results have been reported, but competition of cations other than ammonium for the binding sites must be considered.
5. The biomass itself may act as a temporary sink of organic substrates, and as a source of material thereafter.
6. Results presented in this paper clearly demonstrate that ammonium can be effectively stored in the water filling the void space of a packed bed and of the extracellular polymeric substances (EPS) of biofilms and activated sludge flocs. Besides, ammonium gets effectively adsorbed, presumably to the EPS.
7. To exploit the sink and source capabilities of the biomass in an attempt to dampen influent loading variations, thick biofilms appear to be superior over thin ones, despite the fact that only a fraction of thick biofilms can be expected to be penetrated by substrate and oxygen during normal operation of biofilm reactors.
8. Biofilm carrier materials which provide thin biofilms and in addition a significant amount of accumulated biomass, like the Kaldnes carriers do, may be considered as a model for advanced biofilm carrier materials specialized for handling peak loading situations.

## Acknowledgement

The research described above was carried out in the framework of the SFB 411 (Fundamentals of aerobic biological wastewater treatment), and financed by the German Research Foundation (DFG).

## References

Arnold, E., Böhm B. and Wilderer, P. A.: 1999, Application of activated sludge and biofilm SBR technology to treat reject water from sludge dewatering systems: a comparison, Proceedings of 4th IAWQ Specialised Conference on Small Wastewater Treatment Plants, 18-21 April 1999, Stratford-upon-Avon, UK.

Gieseke, A., Arnz, P., Schramm, A., Amann, R. and Wilderer, P. A.: 1999, Nutrient removal with a Sequencing Batch Biofilm Reactor: process parameters and microscale investigations, To be presented at the 4th IAWQ Conference on Biofilm Systems, 17-20 October 1999, New York, USA.

Hutchinson, D. H. and Robinson, C. W.: 1990, A Microbial Regeneration Process for Granular Activated Carbon II: Regeneration Studies, *Wat.Res.* **24**, 1217-1224.

Jaar, M.: 1991, Biologische Regeneration schadstoffbeladener Aktivkohle am Beispiel der Modellsubstanzen 3-Chlorbenzoesäure und Thioglykolsäure (Biological regeneration of spent activated carbon using 3-chlorobezoic acid and thioglycolic acid as model substances), Hamburger Berichte zur Siedlungswasserwirtschaft 9, TU Hamburg-Harburg, Germany.

Kaballo, H. -P.: 1998, Das Sequencing Batch Biofilm Reactor (SBBR)-Verfahren zur Reinigung von chlororganisch belasteten Abwässern im Leistungsvergleich mit einem baugleichen kontinuierlichen Biofilmverfahren (Treatment of chloroorganic substances with a SBBR, in comparison with a continuous flow biofilm reactor), Berichte aus Wassergüte- und Abfallwirtschaft 141, TU Munich, Germany.

Kolb, F. R.: 1997, Biologische Reinigung xenobiotica-haltiger Abwässer in einem Aktivkohle-Festbett-Schlaufenreaktor mit Membran-Stoffübertrager (Biological purification of wastewater containing xenobiotics with the aid of a loop flow reactor packed with activated carbon and equipped with a membrane oxygen transfer unit), Berichte aus Wassergüte- und Abfallwirtschaft 128, TU Munich, Germany.

Kruzic, A. Ph.: 1986, Zwischenspeicherung von Ammonium zur späteren Nitrifikation (temporary storage of ammonia for further nitrification), in Anwendung des Sequencing Batch Reactor (SBR).Verfahrens zur biologischen Abwasserreinigung (Application of SBR technology for biological wastewater treatment), Wilderer, P.A. and Schroeder, E.D. (eds). Hamburger Berichte zur Siedlungswasserwirtschaft 4, TU Hamburg-Harburg, Germany.

Nielsen, P. H.: 1996, "Adsorption of ammonium to activated sludge", *Wat. Res.* **30**, 762-764.

Wik, T.: 1999, "Adsorption and denitrification in nitrifying trickling filters", *Wat. Res.* **33**, 1500-1508.

# EFFECTIVE MANAGEMENT AND OPERATION OF COAGULATION AND FILTRATION

G. S. LOGSDON

*Black & Veatch Corporation, 10250 Alliance Road, Ste. 101, Cincinnati, Ohio 45242 USA*

**Abstract.** The most commonly used water filtration technique involves coagulation and rapid rate filtration, either in conventional plants with flocculation and sedimentation, or in direct filtration plants in which the sedimentation process is omitted. Both versions of coagulation and filtration can be effective for controlling *Giardia* cysts and *Cryptosporidium* oocysts, but research done by several investigators has shown that coagulation and filtration must be operated very carefully to attain the best results. When filtered water turbidity is 0.1 ntu or lower, the process is most effective. Careful control of coagulation chemistry and of filtration rate increases, continuous monitoring of filtered water turbidity, and proper management of backwash water are keys to successful filtration.

**Keywords:** coagulation, *Cryptosporidium*, filtration, *Giardia*, monitoring, turbidity

## 1. Introduction

Coagulation and rapid rate filtration is widely used in water treatment because the process, which has numerous variations, is effective for controlling a wide range of contaminants. These include microorganisms; particulate matter that causes turbidity; color and some forms of natural organic matter (NOM); some inorganic substances such as oxidized iron, manganese, and arsenic; and algae.

The nature and extent of pretreatment have a strong influence on the capabilities of plants employing coagulation and filtration. Conventional plants with sedimentation basins can cope with source water turbidity of 1000 to 2000 nephelometric turbidity units (ntu) or higher, as well as color, NOM, and oxidized inorganics. Direct filtration plants, in which sedimentation is omitted, can treat source waters with turbidity of 10 to 20 ntu and color up to about 40 color units (AWWA Committee Report, 1980). Algae present serious problems to conventional plants because algae do not settle well, if at all, and in direct filtration plants algae can clog the filters. Dissolved air flotation (DAF) is a form of pretreatment that is ideally suited to algae removal, and DAF is very effective for treatment of water with color.

The three above-mentioned variations of coagulation and filtration (direct filtration, DAF/filtration, and conventional treatment) have all proven effective for controlling protozoa in water. Pilot plant investigations have shown, however, that water must be properly conditioned by coagulation in order to attain effective microorganism removal, and filters must be managed with great care to avoid breakthrough of turbidity and microorganisms. This paper reviews research of five groups of investigators and shows how deviations from optimum or near-optimum coagulation and improper management of filtration rate increases can severely deteriorate filtration performance for removal of protozoa. Backwash water recycling can disrupt coagulation if not managed properly. Finally, the need for continuous monitoring of coagulation and filtration as an aid to effective management of the treatment process is discussed. Production of filtered water having a turbidity of 0.1 ntu or lower should be the goal if effective control of cysts and oocysts is to be attained.

## 2. Pilot Plant Testing for Protozoa Removal

Waterborne outbreaks of giardiasis in the 1970s and cryptosporidiosis in the 1980s and early 1990s provided the impetus for numerous pilot plant studies of protozoa removal. Five pilot plant investigations are reviewed in this paper. In these studies, the source water turbidity was generally below 10 ntu, and direct filtration often was used. A common finding in each study has been that when filtered water turbidity is 0.1 ntu or lower, removal of cysts and oocysts is more effective than when turbidity is above 0.1 ntu.

### 2.1. DESCRIPTIONS OF STUDIES

Logsdon *et al.* (1981) used a small direct filtration treatment train with three in-line rapid mixers for mixing first *Giardia* cysts and then coagulation chemicals, two or three 2-L enclosed square jar test jars stirred at 30 rpm for flocculation, and a 3.8-cm diameter filter column with 46 cm of anthracite media over 15 cm of sand. The rate of flow generally was 0.2 L/min, and alum (aluminum sulfate) typically was used for coagulation at a dosage of about 10 mg/L. The filtration rate generally was 10 m/hr, except when effects of rate increases were studied. Operational variations tested included sub-optimal and very inadequate coagulation, rate increases, and turbidity breakthrough with high head loss.

DeWalle *et al.* (1984) used a direct filtration pilot plant with 10.8 cm filter columns having 51 cm of anthracite over 25 cm of sand for research with *Giardia* cysts. A typical filtration rate was 10 m/hr, but rates as low as 4.3 m/hr were used. Alum at a dosage of 12 mg/L was the coagulant used in most filter runs, with some testing of sub-optimal coagulation.

Hendricks *et al.* (1999) operated a 76 L/min pilot plant in an in-line filtration mode, with rapid mixing and filtration. Two filters with square cross-sections (61 cm x 61 cm) were used, one with 94 cm of anthracite monomedium, and the other with 56 cm of anthracite over 33 cm of sand. An alum dosage of 26 mg/L was generally used, with some runs at a sub-optimal dosage of 13 mg/L. Both *Giardia* and *Cryptosporidium* were used.

Ongerth and Pecoraro (1995) studied direct filtration for removal of *Giardia* and *Cryptosporidium* using a 4-L/min pilot plant with flocculation and 15-cm diameter filters. Mixed media was used, with 46 cm of anthracite over 22 cm of sand over 15 cm of garnet. The filtration rate was 12 m/hr, and the optimum alum dosage was about10 mg/L at a pH of 6.4 to 6.6. In a sub-optimal coagulation run, an alum dosage of 5 mg/L was tested.

Patania *et al.* (1995) carried out extensive pilot plant studies involving *Giardia* and *Cryptosporidium.* The portion of their work reported herein was conducted with a conventional process train of rapid mix, flocculation, sedimentation, and filtration at 15 m/hr. Ferric chloride was used as the coagulant, typically at a dosage of 15 mg/L with or without Cat-Floc T at 1 mg/L. They evaluated filtration at optimum coagulation conditions, studying protozoa removal at the beginning of filter runs after backwashing, and later in filter runs when stable filtered water quality had been attained.

The experimental data in the studies reviewed for this paper contained results in which filtered water samples had no detected protozoa, as well as results for filtered water samples with detectable protozoa. With the exception of the Hendricks *et al.*, data, the only filtered water protozoa data used in Tables I and II found later in this manuscript were data for which protozoa were detected.

## 2.2. PILOT PLANT RESULTS

### *2.2.1. Inadequate Coagulation*

Improper coagulation causes poor filter performance, as compared to the results attained when coagulation is done correctly, and removal of cysts and oocysts is erratic and unpredictable when rapid rate granular media filters are used to treat uncoagulated water. *Giardia* removal results of Logsdon *et al.*, DeWalle *et al.*, Ongerth and Pecoraro, and Hendricks *et al.*, are presented in Table I. In every case, the log removals attained by a research team were better for optimized coagulation than the results of the same team when coagulation was sub-optimum. *Cryptosporidium* removal results of Ongerth and Pecoraro and Hendricks *et al.*, also are presented in Table II. In those studies the log removals attained by optimized coagulation were better than the results observed when coagulation was sub-optimum. With optimum coagulation, filtered water turbidity was 0.1 ntu or lower, whereas sub-optimal coagulation resulted in filtered water turbidity above 0.1 ntu and as high as 1 ntu.

### *2.2.2. Returning a Filter to Service after Backwash*

For about 100 years, evidence has been available (Pittsburgh Filtration Commission, 1899) to show that elevated densities of microorganisms are likely to be found in the filtered water produced just after a filter is backwashed and placed into operation. With the passage of time filtered water quality improves. This has been referred to as the ripening period or the initial improvement period. The work of Allen Hazen at Pittsburgh in 1898 was done without use of any disinfectant. Hazen's data in the Pittsburgh Filtration Commission report showed that concentrations of plate count bacteria in filtered water at or shortly after the start of a new filter run were 2-fold to 39-fold higher (geometric mean, 8-fold) than after one hour of operation, in the 14 runs for which data were presented. In several of the filter runs, the concentrations of plate count bacteria in filtered water after one to two hours of operation were similar to concentrations that had been measured during the last hour or the previous filter run, just before the filter was backwashed. Both Logsdon *et al.* (1981) and DeWalle *et al.* (1984) noted higher concentrations of *Giardia* cysts when filter runs started as compared to later in the same runs. The same phenomenon was observed by Patania *et al.*, for *Giardia* cysts (Table I) and *Cryptosporidium* oocysts (Table II). In the work reported by Patania *et al.*, the turbidity of filtered water was higher during ripening than during the period of stable operation after ripening.

TABLE I
Summary of *Giardia* Cyst Removal in Pilot Plant Studies

| Researcher | # of Runs (R) or Samples (S) | Test Condition | Filtered Water | | Log removal, mean & (range) |
|---|---|---|---|---|---|
| | | | Turbidity, mean & (range) | Cyst/L, range | |
| Logsdon | 3 (S) | Optimum Coagulation | 0.08 (0.08 - 0.08) | 40 - 120 | 3.2 (2.6 - 3.7) |
| Logsdon | 6 (S) | Sub-opt. Coagulation | 0.27 (0.24 - 0..32) | 40 - 200 | 2.0 (0.4 - 2.9) |
| Logsdon | 5 (S) | Minimal Coagulation | 0.7 (0.5 - 1.0) | 420 - 4000 | 0.7 (0.1 - 1.2) |
| DeWalle | 6 (R) | Optimum Coagulation | 0.05 (0.02 - 0.19) | About 30 to 300 | 3.0 (2.6 - 3.7) |
| DeWalle | 1 (R) 2 (S) | Sub-opt. Coagulation | 0.44 (0.37 - 0.52) | About 13,000 to 60,000 | 0.8 (0.4 - 1.1) |
| Hendricks | 11 (R) | Optimum Coagulation | 0.08 (0.07 - 0.09) | Not given | 3.2 |
| Hendricks | 3 (R) | Sub-opt. Coagulation | 0.5 - 0.8 | Not given | 1.6 |
| Ongerth | 7(S) | Optimum Coagulation | 0.03 (0.02 - 0.08) | 0.3 - 2 | 3.3 (2.8 - 3.7) |
| Ongerth | 3 (S) | Subopt. Coagulation | 0.36 (0.34 - 0.40) | 36 - 170 | 1.3 (1.0 – 1.7) |
| Patania | 15 (R) | Stable run | 0.04 (0.04 – 0.05) | 0.1 - 18 | 4.5 (2.7 – 5.1) |
| Patania | 17 (R) | Ripening | 0.17 (0.08 – 0.36) | 0.7 - 22 | 3.5 (2.6 – 4.2) |

*2.2.3 Filtration Rate Increases*

An increase in the rate of filtration causes an increase in the shear forces acting on floc particles trapped in a granular media filter bed. Depending on floc strength, increasing the filtration rate may cause some particles to dislodge and be discharged from the filter. Filtration rate increases ranging from 50% to 150% in 10 seconds did not cause filtered water turbidity to increase when floc was strengthened with a nonionic polymer (Logsdon *et al.,* 1981). When alum was used with no polymer, a filtration rate increase from 10 m/hr to 27 m/hr for a period of two minutes caused turbidity to increase from 0.3 to 1.0 ntu, and the *Giardia* cyst concentration increased 25-fold. When the filtration rate was decreased to 10 m/hr, both turbidity and *Giardia* cyst concentration returned to levels observed before the rate increase.

TABLE II

Summary of *Cryptosporidium* Oocyst Removal in Pilot Plant Studies

| Researcher | # of Runs (R) or Samples (S) | Test Condition | Filtered Water | | Log removal, mean & (range) |
|---|---|---|---|---|---|
| | | | Turbidity, mean & (range) | Cyst/L, range | |
| Hendricks | 11 (R) | Optimum Coagulation | 0.08 (0.07 - 0.09) | Not given | 3.4 |
| Hendricks | 3 (R) | Sub-opt. Coagulation | 0.5 - 0.8 | Not given | 2.5 |
| Ongerth | 7(S) | Optimum Coagulation | 0.03 (0.02 - 0.08) | 0.9 - 9 | 2.9 (2.5 - 3.4) |
| Ongerth | 3 (S) | Subopt. Coagulation | 0.36 (0.34 - 0.40) | 7 – 34 | 1.5 (1.2 - 1.9) |
| Patania | 9 (R) | Stable run | 0.04 (0.04 – 0.05) | 0.01 – 0.6 | 5.2 (4.3 - 6.3) |
| Patania | 17 (R) | Ripening | 0.17 (0.08 – 0.36) | 0.1 – 1.2 | 4.5 (3.6 - 5.1) |

*2.2.4. Turbidity Breakthrough*

Turbidity breakthrough at the end of a filter run can be accompanied by a massive discharge of microorganisms even if they are not present in the influent water (Logsdon *et al.,* 1981). After operating their pilot filter for over 24 hours in which *Giardia* cysts were continuously fed, with cysts in influent water ranging from 8011 to 470 cysts/L, a turbidity breakthrough was monitored at 2.5 hours (about 5 theoretical detention times for the treatment train) after cyst feed had stopped. At a filtration rate of 10 m/hr, turbidity was 0.43 ntu and 2900 cysts/L were detected. After the filtration rate was increased to 15 m/hr, turbidity rose to an average of 0.85 ntu and 3600 cysts/L were detected in the filtered water. The source of these cysts was the floc that had been captured earlier in the run when cysts had been added to the influent water. This showed that after a long period of operation in which microorganisms had been removed by a granular media filter, a turbidity breakthrough episode could result in the discharge of large numbers of organisms in a short time.

*2.2.5. Summary of Pilot Plant Findings*

The data of five different studies are consistent in showing that the best removal of protozoa (*Giardia* cysts and *Cryptosporidium* oocysts) occurred when the filtered water turbidity was at or below 0.1 ntu, with a trend of less effective protozoa removal when turbidity exceeded 0.1 ntu. However, no concentration of organisms could be associated with a specific value of filtered water turbidity.

## 3. Recycling Backwash Water

At plants that employ coagulation, recycling backwash water by returning it to the influent raw water has the potential to cause problems by increasing the rate of flow through the plant, by increasing the concentration of suspended solids and protozoa in the influent water, and by upsetting the coagulation chemistry. Return of supernatant liquid from tanks used to settle wash water and sludge was implicated in a cryptosporidiosis outbreak in the U.K. (Dept. of Environment, Dept. of Health, 1990). The U.K. report noted that by settling backwash water and sludge, only 83 percent of oocysts were removed as compared to 99 percent removal by filtration.

Techniques to minimize problems associated with backwash recycling include provision of a flow equalization basin, continuously returning a small stream of backwash water, and properly treating backwash water prior to recycling.

At direct filtration plants, treatment of backwash water is the only mechanism by which cysts and oocysts can be removed from the process flow stream when backwash water is recycled. Treatment options include coagulation and sedimentation, with or without filtration; dissolved air flotation; and membrane filtration. Hess *et al.* (1993) reported on using polymer addition and plate settlers to treat backwash water at a direct filtration plant. The raw water had a low turbidity (<1 ntu) and color of about 20 color units, resulting in backwash solids that were highly organic, with poor settling characteristics. When the rate of flow was reduced and the clarifiers were operated at a loading rate of 0.6 m/hr, turbidity of the recycled flow averaged less than 1.5 ntu. A 5300 $m^3$/day DAF facility has been used to treat backwash water at a 79,000 $m^3$/day water filtration plant (Grubb and Arnold, 1997). They reported that the influent turbidity in the backwash treatment plant ranged from 13 to 19 ntu, and clarified backwash water turbidity ranged from 1.2 to 2.7 ntu, with no addition of coagulant chemicals to the DAF process. The DAF clarifier loading was 5 to 6 m/hr, with a 10 percent recycle. Sludge floated from the DAF unit had a 3 percent solids content. Membrane filtration for backwash water treatment was evaluated by Thompson *et al.* (1995).

They reported that when polymers were used along with inorganic coagulants, the polymer in the floc tended to plug the membrane pores. Treatment was successful when floc produced by iron coagulation (without polymer) was filtered, when the backwash water had been settled before membrane filtration.

## 4. Monitoring and Operation

Process equipment at filtration plants should be designed so that flow of raw and treated water, filter head loss, and chemical feeds can be monitored easily by plant staff. Both clarified water turbidity and filtered water turbidity should be monitored continuously at plants employing a clarification step ahead of filtration.

At plants practicing coagulation and filtration, attaining the proper coagulant dosage and pH is crucial to process optimization. Some plants keep historical data on raw water quality and chemical dosages used for effective treatment. Some use jar tests on a frequent basis. In the 1980s and 1990s, use of streaming current monitors as a guide to controlling coagulant dosage has become more common in the United States, especially at plants where raw water quality can change rapidly. Other plants have used pilot filters, in which a small side-stream of coagulated water is filtered continuously, and the filtered water turbidity is monitored. If turbidity from the pilot filter is acceptable it is likely that turbidity in the filtered water at the full-scale filters will be acceptable also. Some plants that treat low-turbidity water use continuous turbidimeters for raw and filtered water and also continuous particle counting on filtered water. The latest U.S. EPA regulations on surface water treatment will require installation of a continuous turbidimeter for each filter at plants serving over 10,000 persons. Use of continuous turbidimeters at each filter has become a common practice at well-run plants in the U.S. A number of plants use multiple approaches to coagulation control so operators have multiple sources of information for making decisions about pretreatment. This avoids reliance on a single monitoring technique and provides protection against possible failures of one monitoring method.

## 5. Summary

The overall findings of five different pilot plant studies involving *Giardia* and *Cryptosporidium* are all in agreement with respect to the need to attain effective coagulation. Furthermore, the results of Logsdon *et al.*, show that filter rate changes have the potential to cause passage of large numbers of protozoa into filtered water. The findings of Patania *et al.*, with regard to passage of more protozoa during filter ripening as compared to the stable or mature portion of the filter run were foreshadowed by the plate count bacteria results obtained in the 1890s by Allen Hazen.

Filtered water turbidity measurements do not correlate directly with cyst and oocyst concentration in filtered water. Nevertheless production of very low turbidity (at or below 0.1 ntu) has been shown by several researchers to result in high removals of cysts and oocysts, whereas passage of turbidity in the range of 0.3 to 1 ntu has been accompanied by passage of large numbers of cysts and oocysts, as compared to the water quality attained with turbidity below 0.1 ntu.

Recycle of backwash water and other residuals has the potential to disrupt optimum treatment conditions and can lead to serious water quality problems. In areas where water is scarce, backwash water can not be wasted, so effective management of this water resource becomes highly important. At direct filtration plants, backwash water must be effectively treated before it is recycled.

Process monitoring is absolutely necessary for optimized operation of filtration plants. Filtration rates and head loss must be monitored, and rate increases have to be managed with great care. Coagulation must be optimized, and a variety of approaches are available. Use of multiple approaches to coagulation monitoring provides plant operators with a wide range of information on which to base wise decisions about pretreatment. Filtered water turbidity should be monitored at each filter, and particle counting is recommended because of the greater sensitivity of particle counters to incipient turbidity breakthrough. The goal for filtered water turbidity should be 0.1 ntu or below for each filter.

## References


AWWA Committee Report.: 1980, *Jour. AWWA.* **72:7**, 405-411.

Department of Environment, Department of Health.: 1990, *Cryptosporidium in Water Supplies,* Her Majesty's Stationery Office, London, U.K.

DeWalle, F. B., Engeset, J., Lawrence, W.: 1984, *Removal of Giardia lamblia Cysts by Drinking Water Treatment Plants*, EPA-ORD.

Grubb, T. and Arnold, S.: 1997, *Proceedings AWWA Water Quality Technology Conference*, AWWA.

Hendricks, D. W., *et al.*: 1999, *Biological Particle Surrogates for Filtration Performance Evaluation*, AWWA Research Foundation Report (in press).

Hess, A., Affinito, A., Dunn, H. Gaewski, P., Norris, E.: 1993, *Proceedings AWWA/WEF Joint Residuals Management Conference*, AWWA.

Logsdon, G. S., Symons, J. M., Hoye, R. L. Arozarena, M. M.:1981, *Jour. AWWA.* **73:2**, 111-118.

Ongerth, J. E. and Pecoraro, J. P.: 1995, *Jour. AWWA.* **87**:12, 83-89.

Patania, N. L., Jacangelo, J. G., Cummings, L., Wilczak, A., Riley, K., Oppenheimer, J.: 1995, *Optimization of Filtration for Cyst Removal*, AWWA Research Foundation, USA.

Pittsburgh Filtration Commission.: 1899, *Report of the Filtration Commission of the City of Pittsburgh*, 166-169.

Thompson, M. A., Vickers, J. C., Wiesner, M. R., Clancy, J. L.:1995, *Proceedings AWWA Annual Conference, Water Quality*, AWWA.

# WASTEWATER RECLAMATION FOR AGRICULTURAL REUSE IN ISRAEL: TRENDS AND EXPERIMENTAL RESULTS

A. BRENNER[1], S. SHANDALOV[1], R. MESSALEM[2],
A. YAKIREVICH[1], G. ORON[1] and M. REBHUN[3]

[1]*The J. Blaustein Institute for Desert Research, Ben-Gurion University of the Negev, Sede Boker Campus, 84990 Israel.* [2]*The Institutes for Applied Research, Ben-Gurion University of the Negev, Beer Sheva, 84105 Israel.* [3]*Environmental and Water Resources Engineering, Technion-Israel Institute of Technology, Haifa, 32000 Israel.*

**Abstract.** Water shortage and a deterioration in the quality of water resources in Israel have made necessary a national policy recommending reuse of practically all municipal wastewater in order to supply a major part of agricultural water demand. Two pilot-scale systems were operated and studied for several years. The first one consisted of an advanced treatment scheme incorporating a sequencing batch reactor (SBR) system with further deep-bed granular filtration. The second system was an SBR unit, for the purpose of optimizing nitrogen and phosphorus removal and testing further microfiltration of SBR effluents. The SBR process has been shown to be an efficient biological treatment method producing low Biochemical Oxygen Demand (BOD) and Total Suspended Solids (TSS) effluents. SBR effluents, even if loaded with high TSS concentrations, could be further purified in the filtration stage, producing low-turbidity effluents. Granular filtration experiments were carried out using a gravitational single-medium filter composed of uniformly-sieved quartz sand. It was found that most of the suspended solids were removed in the top 10 cm of the filter bed. Influent turbidity was found to be the main parameter affecting the process, while filtration rate had only a minor effect. Microfiltration of SBR effluents showed highly efficient removal of turbidity and pathogens. Advanced mathematical models were developed and calibrated for both the biological process and for the granular filtration process.

**Keywords:** deep-bed filtration, mathematical modeling, microfiltration, reclamation, reuse, sequencing batch reactor, wastewater treatment, water shortage

## 1. Introduction

Water scarcity and deterioration in the quality of water resources in many countries have led to the recognition that water shortage and water pollution control should be solved by a careful water resource management that incorporates advanced technologies. Desalination of seawater and brackish groundwater, as well as reclamation and reuse of municipal wastewater, are the main strategies that have been proposed for investigation and application in Israel.

Wastewater treatment can produce new water resources for various uses and prevent water pollution and health hazards. While pollution prevention is a global necessity, the use of treated wastewater as a new water resource is a management issue particular to semi-arid countries due to the increasing water shortages. In municipal wastewater treatment, there are various reuse

*Water, Air, and Soil Pollution* **123:** 167–182, 2000.

possibilities. However, some require high quality effluents (agricultural crop irrigation, recreational use, aquaculture, groundwater recharge), and advanced treatment technologies must be adopted to provide this. The core of most municipal wastewater treatment plants is the biological process, also known as the secondary treatment stage. The specific processes selected are dictated by the characteristics of the raw sewage (especially the organic and nutrient content) and the permitted destinations for the effluents (either reuse or disposal). In this respect, the activated sludge system is considered a reliable and an efficient biological treatment process, capable of producing low biochemical oxygen demand (BOD) and total suspended solids (TSS) effluents. This process is most prevalent in modern countries where stringent standards for effluent quality exist. It is also considered a versatile process that can be easily modified to meet nitrogen and phosphorus limitations. Secondary effluents may be further refined by a filtration stage, resulting in high quality reclaimed water.

Deep-bed filtration is an efficient means of removing residual particulate matter. Another option for tertiary filtration would be a membrane separation process, such as microfiltration. This process is considered an emerging technology that can remove colloidally dispersed, turbidity-causing particles very efficiently. Its expected benefits include a very considerable improvement in disinfection of the product enabling utilization of the effluent in unrestricted irrigation, including food crop irrigation. With regard to disinfection, organisms present in domestic sewage, such as *Giardia* and *Cryptosporidium*, nowadays constitute a major problem in water and wastewater treatment. Because of limited removal in conventional treatments and their resistance to disinfection (especially *Cryptosporidium*), a tertiary filtration stage may be obvious.

Conventional treatment methods cannot always easily be applied to industrial wastewater. Although application of biological treatment processes to industrial wastes is feasible, this requires judicious design selection. A biological treatment process should accomplish consistent elimination of many specific compounds which are considered hazardous to human health, and meet stringent disposal requirements imposed in most modern countries. Unfortunately, many of the industrial waste materials resist biodegradation, while others may exhibit toxic and inhibitory effects. Therefore, physico-chemical processes such as chemical oxidation and precipitation and carbon adsorption are often required as a pre-treatment or a polishing stage. An alternative approach may be the introduction of source segregation and in-plant control programs (Brenner, 1999). This is a crucial process for industrial wastewater that is sent for biological treatment at publicly-owned wastewater treatment (POWT) plants. This will allow the plant to operate without interference or inhibition. An in-plant control program is considered an economic strategy because the plant could then recycle valuable materials, reduce waste quantities and eliminate the need for large-scale and sophisticated bio-physical treatment processes.

Advanced wastewater treatment methods may eliminate most environmental problems and contribute positively to the net water balance in semi-arid and arid regions. A combined sophisticated treatment sequence, i.e., mechanical biological treatment (such as the activated sludge process) followed by tertiary filtration (deep-bed or membrane separation) and disinfection, will also enable economic flexibility in the selection of agricultural crops to suit market considerations. These concepts have been the driving force for the research described herein, which is aimed at the development of design guidelines for wastewater reclamation plants in Israel and its neighboring countries.

## 2. Water Balance and Wastewater Reclamation in Israel

Basic figures describing present water consumption (year 1995) and water demand forecast (year 2020) in Israel are given in Tables I and II. Two aquifers in Israel are the main sources of fresh water, the coastal aquifer and the mountain aquifer. Their annual production potential is approximately 400 and 300 million cubic meters ($Mm^3/Y$), respectively. Other small local aquifers can add another 300 $Mm^3/Y$. The Sea of Galilee is a surface water source that can supply approximately 450 $Mm^3/Y$. Various runoff catchments and direct utilization of water from the main rivers feeding the Sea of Galilee together add approximately 250 $Mm^3/Y$. There are also various local small aquifers of brackish water, especially in the southern part of Israel (The Negev Desert). This water is partly used in agriculture and industry (see Table I). The brackish water maximum production potential is approximately 300 $Mm^3/Y$. As can be seen in Table I, most of the water is consumed by the agricultural sector, which is gradually converting to the use of marginal water, especially treated effluents. The specific municipal water consumption in 1995 was approximately 100 $m^3$-capita/Y (population of 5.6 M). The 2020 forecast of the specific municipal water consumption is 120 $m^3$-capita/Y, based on a population of 8.2 M.

TABLE I

Water consumption in Israel (1995 balance in $Mm^3/Y$)

| | Fresh | Effluent | Brackish | Total | % of total | % of fresh* |
|---|---|---|---|---|---|---|
| Agriculture | 900 | 250 | 120 | 1270 | 64 | 57 |
| Domestic | 590 | / | / | 590 | 30 | 37 |
| Industry | 90 | / | 30 | 120 | 6 | 7 |
| Total | 1580 | 250 | 150 | 1980 | 100 | 100 |

* % fresh water consumed of total fresh water consumption

The water shortage in Israel is a continuing problem due to over exploitation that has affected the quality of the available water sources. Water shortage, on the one hand, and the concern for the quality of surface and groundwater resources, on the other, have led to the awareness that a national wastewater reclamation program must be developed. Such a program could supply a major part of agricultural water demand and may enable disposal of effluents without any health hazard or environmental nuisance. It has become a national policy to gradually increase the fraction of reclaimed wastewater instead of fresh water for agriculture use. In the year 2020, approximately 50% of agricultural water consumption will be provided by treated wastewater, if agriculture maintains its present size (see Table II). This means that all municipal wastewater should be treated for reuse in agriculture and industry. Since 70-80% of the municipal water consumption results in sewage that can be collected, treated and reused, this is a very reliable water source. In 1995, 93% of the sewage in Israel flowed to the sewer system, 80% of the total wastewater (including municipal and industrial wastewater) was treated and 80% of the municipal wastewater was reused, mainly in agriculture.

TABLE II
Year 2020 water demand forecast ($Mm^3/Y$)

| | Fresh | Effluent | Desalinated | Total |
|---|---|---|---|---|
| Agriculture | 600 | 600 | / | 1200 |
| Domestic | 750 | / | 250 | 1000 |
| Industry | 300 | 100 | / | 400 |
| Total | 1650 | 700 | 250 | 2600 |

Waste stabilization ponds for many years have been the most common method of wastewater treatment. Unfortunately, many of the existing plants were designed long ago, and, therefore, cannot bear the present increased organic load. This results in poor removal of organic matter and suspended solids, odor problems, inefficient and high cost disinfection, and continuing pollution of rivers and groundwater. Moreover, the effluent quality obtained by this treatment is inadequate; therefore, effluent reuse in agriculture is limited to certain crops (Oron and DeMalach, 1987). There is a trend, promoted by new national standards issued by the Ministry of Health, to gradually convert most of the extensive wastewater treatment systems to intensive process systems. The standards require that each municipality larger than 10,000 people must treat its wastewater to the level of 20 mg/L BOD and 30 mg/L TSS. This calls for an activated sludge system, a proven technology which has been applied successfully in Western Europe and the US for many years. Nitrogen removal is also required in some regions to prevent nitrate pollution of the aquifers.

Two pioneering wastewater reclamation plants for agricultural reuse were actually established during the eighties and have proved to be successful during the last ten years. The Dan Region Wastewater Treatment Plant treats the wastewater of Greater Tel-Aviv area having a present capacity of 140 $Mm^3/Y$. The core of this treatment plant is a single-sludge activated sludge system capable of nitrogen removal through nitrification and denitrification. The final effluent is recharged into a local section of the coastal aquifer which serves for an *in-situ* filtration. High quality effluent is pumped after a long storage-and-polishing stage and conveyed by a special pipeline to the Negev Desert area to be utilized for agricultural irrigation. The second large plant is the Kishon Reclamation Plant based on two stabilization reservoirs which receive secondary effluent from the treatment plant of the Greater Haifa area and local runoff water. This system is designed to enable reuse of 20 $Mm^3/Y$ high quality effluent for agricultural irrigation in the Jezreel Valley area.

For several reasons future management is not as simple as may be reflected in Table II. In Israel, as in many dry regions, most of the precipitation occurs during a short season of 4-5 months. Furthermore, there is a steep precipitation gradient from north (600-800 mm rainfall) to south (less than 100 mm rainfall) along a distance of approximately 500 km. This situation requires careful design of water conducts (from north to south) and storage reservoirs (from winter to summer). The storage requirement is necessary also for treated wastewater because it is continuously produced during the entire year, while agricultural demand is mainly during the summer. Storage can be provided in open reservoirs (the most common practice in Israel) or by aquifer recharge. Both strategies affect water quality because of chemical and biological processes occurring during long storage periods.

Another major issue that has to be considered in future management of multi-quality water resources (including fresh water, brackish water and treated effluent) is the salination of soil caused by long-term effluent irrigation, since wastewater becomes more saline during municipal water use. The mixing of different quality water resources and partial desalination of effluent using technologies such as electrodialysis or reverse osmosis are strategies that should be investigated along with the traditional management concepts. There is no doubt, however, that wastewater should be treated properly to prevent environmental nuisances, water pollution and health risks. The level of treatment should be increased to facilitate flexibility in agricultural crop selection according to economic considerations and to enable further application of advanced technologies, such as membrane separation.

## 3. SBR Biological Treatment

The sequencing batch reactor (SBR) activated sludge system has been shown to be a highly-efficient simply-operated process for the treatment of municipal wastewater, and industrial and agricultural wastes (Irvine and Ketchum, 1989). The SBR is a fill-and-draw activated sludge system that offers flexibility of nutrient removal and peak flow buffering. This process can be adapted to the needs of municipalities and settlements in Israel that recently have been required to improve their wastewater treatment systems. The SBR process is an efficient biological treatment stage, producing low BOD and TSS effluents. It is also a versatile process, the operation of which can be easily regulated to achieve N and P removal. SBR effluent may be further refined by a filtration stage, resulting in high-quality effluent.

Two pilot plant systems have been operated and studied for several years. A 280 L reactor, designated SBR I, was built at the Sede Boker Campus as part of a combined system incorporating secondary treatment and tertiary filtration. A schematic description of the entire system including SBR reactor and deep-bed filtration column is shown in Figure 1.

Pilot I was a cylindrical polyethylene vessel with a diameter of 0.7m and height of 0.8m. Domestic wastewater was fed into the reactor via gravity, from a holding tank that received raw sewage pumped by a centrifugal pump from a local sewer. The holding tank served as a short residence time settling chamber in order to prevent passage of rough solids into the biological reactor. The flow of sewage into the reactor was controlled by a pneumatic actuator with an integral solenoid valve controlled by a digital timer. The same arrangement was used for effluent draw and air supply. The air was introduced into the reactor through a ceramic fine-bubble diffuser tube fed from an air compressor. All aeration and discharge functions were controlled with pre-programmed digital timers. The SBR was operated with 4 daily cycles, treating a total of 560 liters. This yielded an hydraulic residence time (HRT) of 12 hours. The organic load (COD/MLSS basis) was 0.4 $day^{-1}$, resulting in low aerobic solids retention time (SRT< 2d) that limited the level of nitrification.

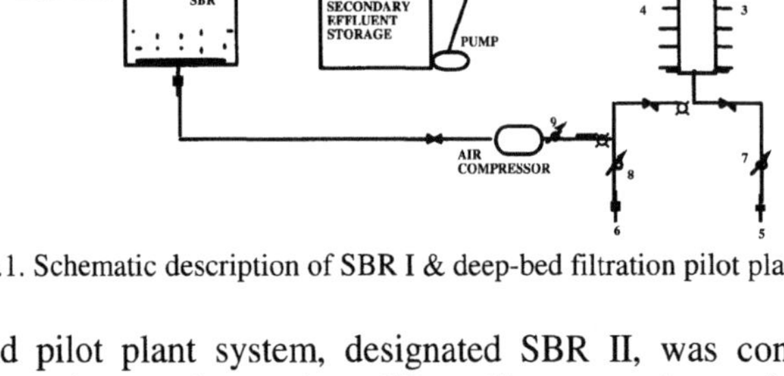


Fig.1. Schematic description of SBR I & deep-bed filtration pilot plant.

The second pilot plant system, designated SBR II, was constructed and operated at the main pumping station of Beer-Sheva, treating typical municipal wastewater. A schematic description of the entire system including SBR reactor and a microfiltration unit is shown in Figure 2. The reactor was a cylindrical fiberglass vessel with a diameter of 1.4m and height of 2.5m. The reactor was fed raw domestic wastewater by a pressure line from the pumping station, controlled by a pneumatic actuator with an integral solenoid valve operated by a timer. The same arrangement was used for effluent draw and excess sludge withdrawal. The aeration of the system was achieved by a unique jet aeration system (Jet Tech Inc.,Kansas, USA), including a mixing pump and a blower which injects mixed liquor and air, respectively, through a jet nozzle into the reactor. The pump and the blower could be operated separately (controlled by separate timers) to achieve only mixing (for denitrification) or aeration.

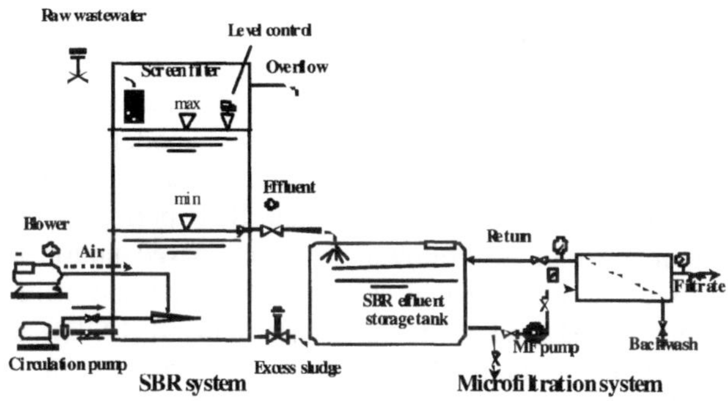


Fig.2. Schematic description of SBR II & microfiltration pilot plant.

SBR biological treatment of the Sede Boker domestic wastewater demonstrated consistent and highly efficient removal of organic matter with a good clarification efficiency. The main performance parameters of the biological process over a 16-month operation period are summarized in Table III.

TABLE III
Performance of SBR I treating Sede Boker domestic sewage

| Parameter | Influent | Effluent |
|---|---|---|
| pH | 8.1 ± 0.2* | 8.0 ± 0.2 |
| Alkalinity, mg/L as CaCO3 | 304.3 ± 54.9 | 240.1 ± 72.7 |
| BOD total, mg/L | 248.2 ± 81.7 | 9.6 ± 5.0 |
| BOD filtered, mg/L | 92.1 ± 27.0 | 4.9 ± 3.0 |
| COD total, mg/L | 483.9 ± 191.4 | 56.9 ± 23.8 |
| COD filtered, mg/L | 164.1 ± 72.6 | 33.1 ± 17.4 |
| NH4, mgN/L | 41.7 ± 23.7 | 16.7 ± 13.5 |
| NOx, mgN/L | 0.1 | 5.4 ± 2.8 |
| PO4, mgP/L | 10.6 ± 3.5 | 3.6 ± 3.4 |
| TSS, mg/L | 218.9 ± 98.4 | 10.2 ± 5.7 |
| Turbidity, NTU | Not determined | 8.1 ± 7.9 |

* Standard deviations

Typical fluctuations of TSS and turbidity of the secondary effluents showed a linear ratio of 1.54 between turbidity and suspended solids (i.e., suspended solids =1.54 x turbidity). Since the SBR effluents were of high-quality, usually with low levels of suspended solids and turbidity, reactor mixed liquor was intentionally added to the effluent storage tank to form a wide range of suspended solids concentrations in the effluents to be further filtered. This was done in order to simulate failure of secondary clarifiers, and to enable testing filtration of a high range of effluent turbidities.

The Beer-Sheva pilot plant (SBR II) was operated in a more sophisticated way, so that various operational strategies for the optimization of nitrogen and phosphorus removal could be investigated. In addition, the study involved development and calibration of a mathematical model (as will be described later). The preferred and most practical operation strategy included five cycles per day, each treating 0.9 $m^3$ of raw sewage. SBR performance was monitored routinely by analysis of the following: reactor mixed liquor suspended solids (MLSS); effluent total suspended solids (TSS); feed and effluent biochemical oxygen demand (BOD) and chemical oxygen demand (COD); and total kjeldahl nitrogen (TKN), ammonia, nitrate, nitrite and phosphorus. Daily measurements were also taken of temperature, pH, and dissolved oxygen in the reactor. Settling characteristics of the sludge were tested by the sludge volume index (SVI). Most of the chemical and microbiological analyses were conducted according to procedures described in *Standard Methods* (APHA, 1992). The main parameters for this operation strategy are detailed in Table IV, and process performance is given in Table V.

TABLE IV
Operation parameters of Beer-Sheva pilot plant (SBR II).

| Parameter | Value |
|---|---|
| Maximum reactor liquid volume, L | 2600 |
| Feed volume per cycle, L | 900 |
| Cycles per day | 5 |
| Reactor MLSS, mg/L | 3265 |
| Hydraulic residence time (HRT), d | 0.58 |
| Organic load, gCOD/gMLSS-d | 0.30 |
| Aerobic Organic load, gCOD/gMLSS-d | 0.63 |
| Solids retention time (SRT), d | 4.3 |
| Aerobic SRT, d | 2.1 |
| Cycle times: | |
| Anoxic Fill, h | 0.5 |
| Aerobic Fill, h | 0 |
| Anoxic React, h | 1 |
| Aerobic React, h | 2.5,2,2,2.5,2.5 |
| Settle, h | 1.0 |
| Draw and Idle, h | 0.5 |
| Total cycle time, h | 5,4.5,4.5,5,5 |
| Cycle aerobic fraction | 0.48 |

SBR II has demonstrated consistent and highly efficient removal of organic matter, as well as excellent clarification. It was also able to achieve complete nitrification and partial denitrification, resulting in approximately 80% removal of N compounds. The average results for a three months period presented in Table V show partial removal of P. With proper operational modifications, P removal could reach higher values, as will be shown later.

TABLE V
Performance of Beer-Sheva pilot plant (SBR II) .

| Parameter | Influent | Effluent |
|---|---|---|
| pH | 7.8 ± 0.2* | 7.9 ± 0.2 |
| BOD total, mg/L | 180.0 ± 3.0 | 12.1 ± 5.1 |
| BOD filtered, mg/L | 64.5 ± 3.0 | 3.3 ± 1.2 |
| COD total, mg/L | 561.5 ± 233.9 | 48.6 ± 26.9 |
| COD filtered, mg/L | 165.0 ± 36.1 | 35.4 ± 13.0 |
| Alkalinity, mg/L as CaCO3 | 348.9 ± 25.1 | 200.0 ± 36.5 |
| NH4, mgN/L | 41.3 ± 3.2 | 0.4 ± 0.4 |
| NO3, mgN/L | 0 | 10.1 ± 4.9 |
| NO2, mgN/L | 0 | 2.1 ± 2.2 |
| PO4, mgP/L | 16.0 ± 5.4 | 10.9 ± 2.9 |
| TSS, mg/L | 259.7 ± 78.9 | 17.6 ± 9.2 |
| SVI, mL/g | / | 79.6 ± 10.6** |

* Standard deviations; ** In the reactor

## 4. Deep-Bed Sand Filtration

The filtration column receiving the secondary effluent of the Sede Boker SBR I unit, which treated domestic sewage, was made of transparent plexiglass, 6.8 cm in diameter and 220 cm in height. The initial bed depth in all experiments was 115 cm. The column was equipped with 24 connections for piezometers at 5 cm intervals starting from the bed surface, and with sampling ports spaced at depths: 3, 13, 23, 33, 43, 53, 63, 73, 83, 93, 103, and 115 cm from the bed surface. The filter column was operated in a downward direction at a constant velocity, controlled by a flow-meter located at the outlet of the column and by a flow adjusting valve located at the outlet of the constant-head feeding tank. The latter, forming a 2m maximum available head, received a continuous flow of secondary effluents from an effluent storage tank by means of a centrifugal pump (see Figure 1). The centrifugal pump also provided continuous mixing of the secondary effluent in the storage tank during filtration tests. The reported filtration experiments were carried out using a single filter medium composed of uniformly-sieved quartz sand. Two sets of experiments using two different filtration media were done. The first set, used a grain size of 1.4-2.0 mm, and the second, 2.0-2.5 mm. Evaluation of filtration efficiency was based on measurements of TSS concentration and turbidity.

As pointed out earlier, the filtration experiments were carried out using two different quartz sand media, each of them serving separately as a single-medium filter bed. A wide spectrum of filtration velocities was employed, ranging from 10 to 25 m/h. TSS and turbidity levels of the secondary effluents being filtered varied widely. Secondary effluents with extreme turbidity levels, as high as 20-40 NTU, were prepared by mixing of SBR effluent with MLSS in order to simulate high-TSS-effluent filtration.

Filtration experiments, although not including the whole range of filtration rates for the 2.0-2.5 mm filter grain size, have indicated that a safe and more efficient design should be based on the 1.4-2.0 mm grain size bed. Therefore, further analysis of filtration results used the results obtained for the 1.4-2.0 mm sand. Turbidity has been shown to be a dominant parameter that significantly affects the length of the filtration cycle. Filtration rate, on the other hand, had only a minor effect on filtration cycle. This observation is presented nicely in Figure 3, which shows the effect of secondary effluent turbidity on filtrate volume for all the experiments carried out with the 1.4-2.0 mm bed filter. Figure 3 includes data for a wide range of filtration rates, showing conclusively that filtration rate has only a minor effect on the filtration cycle.

According to the Israeli recommendations for effluent reuse in agriculture (Shelef, 1990), effluents containing less than 15 mg/L suspended solids and less

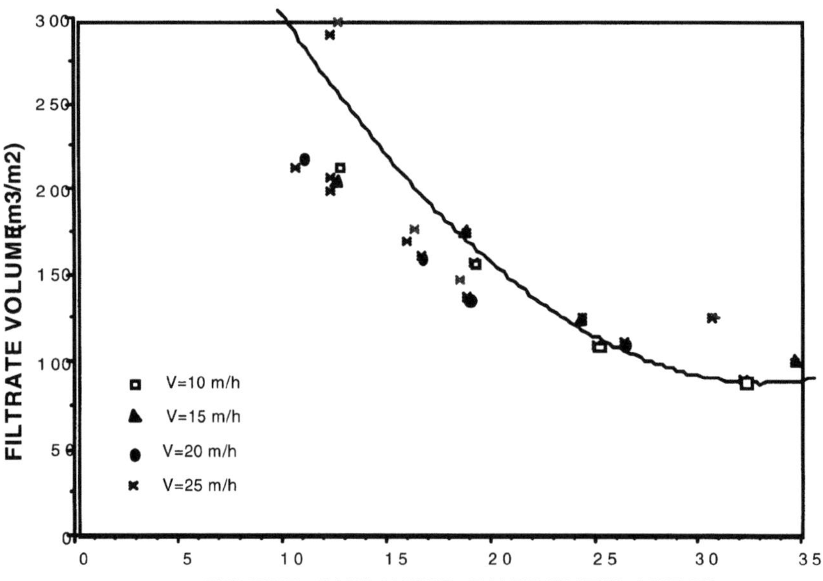


Fig.3. Effect of filter influent turbidity and filtration rate on filtrate volume for the 1.4-2.0 mm bed filter.

than 15 mg/L BOD can be designated for unrestricted irrigation, if they are properly disinfected. According to the correlation obtained in this study, 15 mg/L suspended solids equals 10 NTU. Therefore, the single 1.4-2.0 mm sand bed that was tested with no chemical addition could successfully yield high quality final effluents for a wide range of influent turbidities and filtration rates.

This quality may not meet the effluent turbidity limit of 2 NTU as required, for example, in California. Improvement of final effluent quality in order to meet this 2 NTU standard and to decrease phosphate concentrations can be achieved by the addition of flocculation aids or by an alternative filtration technology such as microfiltration.

## 5. Microfiltration

A self-cleaning, continuous microfiltration system (Memtec, Australia) was installed in the two-stage pilot plant II as shown in Figure 2. Hollow-fiber microporous membranes are the heart of the microfiltration system. They are approximately 0.5 mm in diameter and are encapsulated into a bundle to form a filter module. The nominal filtration area of the unit is 4 $m^2$. The system can treat between 4 and 5 $m^3$ of sewage per day, at a nominal rate of 80 gpd/$ft^2$, i.e., about 500 L/h. A proprietary feature of the Memtec system is its air-operated backwash: raw feed (via the feed pump) and compressed air are used to backwash the unit whenever there is an increase in the resistance to flow, measured by the transmembrane pressure (TMP). Air is introduced into the filtrate side of the system and released through the walls of the hollow fiber. Accumulated solids are flushed from the membrane surface by means of the feed water and gas backwash. The sequence of backwashing involves a pulse of air (1-2s) at 600 kPa across the membrane, followed by a liquid flow on the feed side. The filtration rate and the frequency of backwash depend on the characteristics of the waste stream coming from the SBR. During normal operation, the feed passes from the outside of the membrane (from the module shell) into the center (lumen) and exits as filtrate. The system can be operated either in a cross-flow configuration, or in a dead-end flow mode. In the cross-flow mode, the effluent is recirculated a number of times in the unit, according to the rate of the cross-flow. In the dead-end mode, all the effluent passes through the membrane in a single pass. Typical system feed pressure is 25-35 psi (170 to 240 kPa). The normal operating differential pressure for the membrane is 5-30 psi (35 to 210 kPa), with an average initial differential pressure loss of 5-8 psi (35-55 kPa).

Evaluation of microfiltration efficiency was based mainly on measurements of TSS concentration, turbidity and bacterial counts, including total count, total coliforms and fecal coliforms. Filtrate flow rates, TMP and frequency of backwash, as well as the frequency of chemical cleaning of the membranes, were correlated with the quality of the filtrate and with system performance. Turbidity was measured using a Hach Turbidimeter Model 16800 and expressed as NTU.

The main performance parameters of the entire two-stage secondary and tertiary treatment process over a six-month operation period are summarized in Table VI. The microfiltration unit was operated in the cross-flow mode. Air backwash at 600 kPa across the membrane was carried out automatically every 20-30 minutes, and chemical cleaning with an alkaline detergent solution containing 1-2% NaOH was applied periodically, every two weeks. Filtrate flow rates of 80-180 Lx $m^{-2}$x $h^{-1}$ were consistently obtained during the operating period, when the cross-flows applied varied from 0.10 to 1.25 $m^3xh^{-1}$ per module. These permeate flow rates are higher than those reported for operation in the dead-end flow mode (Sadr Ghayeni *et al.*, 1998). High flow rates, however, cause the transmembrane pressure to increase with time. A decline of the flow rate at high cross-flow (1.25 $m^3xh^{-1}$ per module) after 10 hours of operation was observed, and the experiment was stopped. This could have been caused by compaction of the layer cake that normal backwash could not easily remove. Only after repeated air backwash followed by intensive chemical cleaning, could the original flux be restored.

TABLE VI
Performance of SBR II & microfiltration treatment processes

| | Influent | SBR effluent | MF effluent |
|---|---|---|---|
| BOD total, mg/L | 245±13* | 13.6±7.7 | 3.7±1.5 |
| COD total, mg/L | 800±154 | 77.51±6.5 | 38±6.1 |
| TSS, mg/L | 318±78 | 17.5±9.8 | 0 |
| Turbidity, NTU | >100 | 7.2±3.5 | 0.1±0.03 |
| Total coliforms, cfu/mL** | $25x10^6$ | $8.0x10^5$ | 60 |
| Fecal coliforms, cfu/mL** | $5.4x10^6$ | $2.4x10^5$ | 22 |
| Total count, cfu/mL** | $8.3x10^6$ | $6.0x10^5$ | 387 |

* Standard deviations; ** Average of 5 measurements.

The microfiltration system efficiently removed all suspended matter, thus reducing the total BOD and COD values. Total bacterial counts for MF permeate showed 6-log removal of coliforms and fecal coliforms, and the turbidity decreased from 7-10 to 0.1 NTU. This water quality can ensure an efficient disinfection which will enable unrestricted reuse in agriculture.

## 6. Mathematical Modelling

The terms "modelling or "simulation" refer to a mathematical replication or representation of a process. Mathematical models of activated sludge and other treatment processes have mainly served for the evaluation of system performance for a variety of process configurations, load conditions, and operating strategies. This has enabled improved design of full-scale plants, based on limited (bench-scale or pilot-scale systems) experimental results, but supported by mathematical extrapolations to yield an optimal design and operating strategy (Brenner, 1997).

Most recent models for the activated sludge process are mechanistic and deterministic; that is, all processes are postulated by logical mechanisms which have definite relationships between input and output. In pioneering models of activated sludge systems, steady-state conditions were assumed in which all inputs and outputs were constant over time. The mathematical model which was developed and calibrated for the SBR is a mechanistic-dynamic model, and is a consequence of intensive field and computer studies carried out on SBR II. During initial stages of the study, some of the kinetic and stoichiometric coefficients included in the model were determined experimentally. In addition, detailed characterization of organic carbon (based on the COD measure) was carried out. Most experimental procedures used for coefficient determination and organics characterization made use of advanced respirometric methods.

The model includes 14 processes as follows: aerobic growth of heterotrophs, anoxic growth of heterotrophs, decay of heterotrophs, growth of nitrosomonas, decay of nitrosomonas, growth of nitrobacter, decay of nitrobacter, aerobic growth of polyphosphate accumulating bacteria (PAO), decay of PAO, fermentation of fermentable organics, anaerobic uptake of fermentation products (acetate), hydrolysis of particulate organics, hydrolysis of particulate nitrogen, and hydrolysis of particulate phosphorus. Forty kinetic and stoichiometric coefficients were included in this model. Those that were determined experimentally served as the basis for model calibration/validation. Coefficients that could not be determined were selected from the literature. The calibration was carried out with the aid of the AQUASIM computer program, a universal computer program for the identification and simulation of aquatic systems (Reichert, 1994).

Calibration or parameter estimation was made by comparing theoretical simulated results to actual measurements, using the weighted least squares method, to yield the best-fit set of kinetic and stoichiometric parameters. In other words, model parameters represented by constant variables could be estimated by minimizing the sum of the squares of the weighted derivations between measurements and calculation. Calibration of the SBR model, was based on seven dynamic experiments, each of which contained dynamic data, such as COD, $NH_4$, $NO_3$, $NO_2$, $PO_4$, mixed liquor suspended solids (MLSS), dissolved oxygen (DO), and oxygen uptake rate (OUR). Figure 4 presents representative results which show the comparison of experimental results with calculated values obtained in model calibration for one of the experiments.

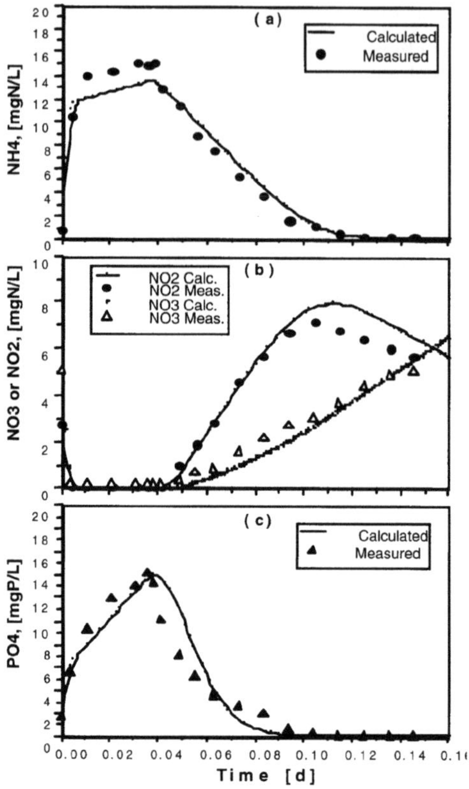


Fig.4. Comparison of calculated and measured N & P transformations in SBR operation.

We have used the concept offered by Adin and Rebhun (1977) for deep-bed filtration modelling in which the attachment and detachment phases are considered as part of the overall process. The detachment mechanism in this model was formulated so that it was independent of the influent concentration. The equation describing the change of the specific deposit as a function of attachment and detachment was solved simultaneously with a general mass balance of solids in the porous filtration media, in order to determine solids concentration and the specific deposit over time and depth (Shandalov *et al.*, 1997). Representative results of a filtration run are shown in Figure 5. Good correspondence between the experimental and simulated breakthrough curves for the filtration run reveals the possibility of employing this model for optimal design and operation of secondary- effluent deep-bed filtration.

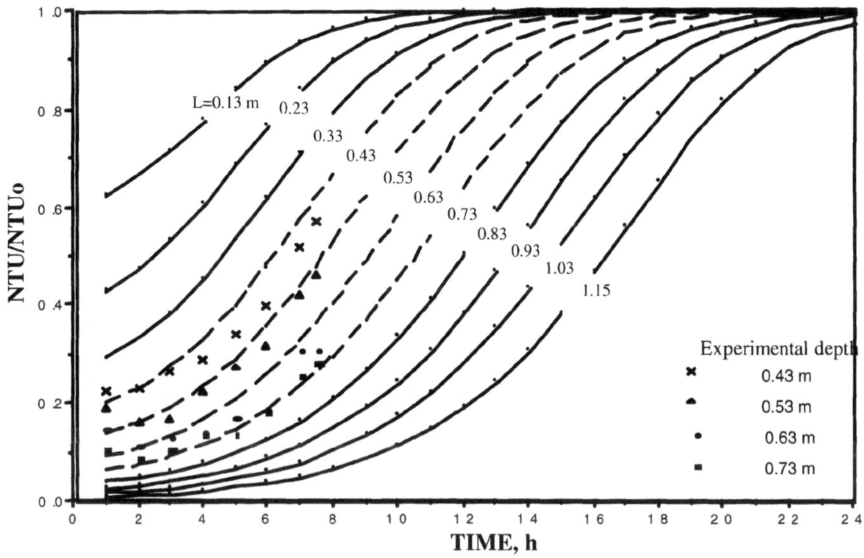


Fig. 5. Simulated (solid and dashed lines) and experimental (dots) breakthrough curves.

## 7. Summary and Conclusions

The overall results of SBR effluent filtration indicate the feasibility of this operation. Tertiary filtration produced desirable filtrate quality with no chemicals added. After disinfection, such effluents can easily meet the Israeli requirements for unlimited crop irrigation. In order to meet an effluent turbidity limit of 2 NTU, improvement of deep-bed filtration efficiency can be simply achieved by the addition of flocculation aids. Membrane separation processes, such as microfiltration, are feasible solutions, too, as evidenced in this study.

The SBR has demonstrated consistent and highly efficient removal of organic matter, as well as excellent clarification. It was also able to achieve complete nitrification and partial denitrification, resulting in approximately 80% removal of N compounds without chemical (external carbon) addition. Long-term average results show partial removal of P; however, this could be optimized to achieve almost complete P removal.

Two single-medium filters composed of uniformly-sieved sand grains in sizes 1.4-2.0 mm and 2.0-2.5 mm were tested at various filtration velocities and at a wide variety of inflow qualities, with no chemical addition. Of the two media, the 1.4-2.0 mm bed demonstrated consistent and highly efficient removal of turbidity.

Filter influent turbidity has been shown to be a dominant parameter that significantly affects the head-loss development profiles and, consequently, the length of the filtration cycle. Filtration rate, on the other hand, had only a minor effect on the head-loss development and filtration cycle. Influent turbidity is also the main parameter affecting the penetration of solids into the depth of the filter medium which consequently determines the quality of the final effluent.

The microfiltration system efficiently removed all suspended matter present in SBR effluent, thus reducing the total BOD and COD values, too. Total bacterial counts for MF permeate showed 6-log removal of coliforms and fecal coliforms, and the turbidity decreased from 7-10 to 0.1 NTU. This water quality ensures efficient enough disinfection to enable unrestricted water reuse for agriculture, both in the US and Israel.

An advanced mathematical model describing the sequencing batch reactor (SBR) system was developed and calibrated, based on operation and testing of a pilot plant system which treated the municipal wastewater of a large city in Israel (Beer-Sheva). This model represents the typical operation of an SBR system, having a solids retention time (SRT) of 5-10 days and which is capable of achieving N and P removal.

The attachment and detachment concepts were utilized as key factors in a mathematical model calibrated for deep-bed filtration of solids. A good correspondence between experimental and simulated breakthrough curves for filtration runs reveals the possibility of using this model for optimal design and operation of deep-bed filtration of activated sludge effluent.

Mathematical models such as those developed and calibrated in this study can be applied in the design of new processes, in the modification of existing systems, and in process analysis and control. They can be further modified and calibrated to fit other operational conditions and wastewater characteristics.

## References


Adin, A. and Rebhun, M.: 1977, *J. American Water Works Association* **70**, 444.

Brenner, A., Shandalov, S., Oron, G., and Rebhun, M.: 1995, *Wat. Sci. Tech.* **30**, 219.

Brenner, A.: 1997, *Wat. Sci. Tech.* **35(1)**, 121.

Brenner, A.: 1999, *Environmental Engineering & Policy.* **4.**

Irvine, R. L., and Ketchum, Jr., L. H.: 1989, *CRC Critical Reviews in Environmental Control* **18**, 255.

Oron, G. and DeMalach, Y.: 1987, *Water Resources Bulletin* **23**, 777.

Reichert, P.: 1994, *Wat. Sci. Tech.* **30(2)**, 21.

Sadr Ghayeni, S. B., Beatson, P. J., Schneider, R. P. and Fane, A. G.: 1998, *Desalination* **116**, 65.

Shandalov, S., Yakirevich, A., Brenner, A., Oron, G. and Rebhun, M.: 1997, *Wat. Sci. Tech.* **36(4)**, 231.

Shelef, G.: 1991, *Wat. Sci. Tech.* **23**, 2081.

# BIOLOGICAL DENITRIFICATION OF GROUNDWATER

M. I. M. SOARES
*Department of Environmental Hydrology & Microbiology, The Jacob Blaustein Institute for Desert Research, Ben-Gurion University of the Negev, Sede Boqer 84990, Israel.*
*e-mail: soares@bgumail.bgu.ac.il*

**Abstract.** Nitrate concentrations in groundwater have increased in many areas of the world. This causes serious concerns because of the link found between nitrate and the blue-baby syndrome, and of the possible formation of carcinogenic compounds in the digestive tract. Biological denitrification, bacteria-mediated reduction of nitrate to nitrogen gas, is a method used in the treatment of nitrate contaminated groundwater. The denitrifying microorganisms require carbon and energy substrates which may be organic or inorganic compounds. Treatment can take place in the aquifer (*in situ* treatment) or in above ground reactors. Numerous biological denitrification processes have been reported; this paper reviews some of this work and studies in progress in the author's laboratory. The choice of a biological denitrification system has to be considered on an individual basis. Although preventive measures are curbing the problem in some developed countries, nitrate pollution is still on the rise in many other countries. Innovative, low-cost biological denitrification processes are specially needed in developing countries.

**Keywords:** bacteria, biological denitrification, groundwater, nitrate removal, water treatment

## 1. Introduction

Widespread pollution of drinking water sources by nitrate is an important environmental problem in many parts of the world. The main causes for increasing levels of nitrate are increased usage of nitrogenous fertilizers, increasing irrigation with domestic wastewater and changes in land-use patterns. Nitrogenous compounds are transformed in soils via microbiologically mediated reactions. As a result, nitrate is formed and, being a very mobile ion, it readily passes through the soils and reaches the aquifer.

Concern over nitrate contamination is due to the link found between nitrate and methaemoglobinaemia, the blue-baby syndrome, in bottle-fed infants (under six months of age), and to the possible formation of n-nitroso compounds which are known to be potent carcinogens in the digestive tract (Mirvish, 1985). For these reasons, standards have been set for nitrate in water for human consumption: the European Community Drinking Water Directive (EC, 1980) incorporated into the legislation of most European countries stipulates a Maximum Admissible Concentration and Guide Level for nitrate-N of 11.30 mg $l^{-1}$ and 5.65 mg $l^{-1}$ respectively, and the same upper limit is also recommended by the World Health Organization (WHO, 1984).

Conventional drinking water treatment does not remove nitrate, and special treatment processes are required for lowering the concentration of nitrate to acceptable levels. The simplest solution, blending with low nitrate water, is often not viable, either because low nitrate water is not available or because of the expense of transferring low nitrate water over long distances. Removal of nitrate, then, has to be carried out, and a number of processes have been developed which can be chemical (chemical reduction), physical (reverse osmosis, electrodialysis), chemical-physical (ion exchange) or biological. Among these, only ion exchange and biological denitrification are feasible on a large scale. Furthermore, the biological process is the most environmentally sound as nitrate is completely eliminated, while ion exchange generates a waste of highly concentrated nitrate (and sulfate in least selective resins) and regenerating chemicals, chloride or bicarbonate. The total cost of biological treatment is usually reported to be lower than that of ion exchange (Hall *et al.*, 1985; Dahab, 1987). Some estimates place the cost of the two processes at the same level or lower for ion exchange, although the high price of disposal of the brine is often not included in the costs indicated for ion exchange (Rogalla *et al.*, 1990; Kapoor and Viraraghavan, 1997).

In view of growing concerns about nitrate contamination, this paper seeks to present and review several biological denitrification processes which have been reported in the literature and are currently conducted in the auther's laboratory.

## 2. General Principles of Biological Denitrification

Biological denitrification occurs naturally when certain bacteria use nitrate as terminal electron acceptor in their respiratory process, in the absence of oxygen. Denitrification consists of a sequence of enzymatic reactions leading to the evolution of nitrogen gas. The process involves the formation of a number of nitrogen intermediates and can be summarized as follows:

$$NO_3^- \rightarrow NO_2^- \rightarrow NO \rightarrow N_2O \rightarrow N_2$$

Denitrifying bacteria are ubiquitous in nature (Gamble *et al.*, 1977; Zumft, 1992), and biological denitrification treatment consists of the provision of suitable carbon and energy sources which may be organic or inorganic compounds. Treatment can take place in the aquifer, or the water may be pumped into above ground reactors.

## 3. Process Systems

Numerous biological denitrification studies have been carried out at laboratory level and several full-scale plants have been operated in Europe. Experience in the U.S. has been limited due to the fact that biological treatment has not yet been sanctioned for the treatment of potable water supplies (Dahab and Sirigina, 1994; Dahab and Woodbury, 1998).

Various configurations have been used in above ground denitrification, with packed-bed and fluidized-bed reactors being the most common. The latter usually affords the highest denitrification rates per reactor volume but high breakthrough of biomass may be a problem. *In situ* treatments are relatively few, and the process usually consists of a central pumping well surround by injection wells through which the substrate is introduced. Advantages of *in situ* treatment are stable temperature in the ground (an important consideration in cold climates), at least partial removal of the biomass in the ground, and low investment. Possible disadvantages are clogging of the ground by biomass and gas formation and difficulty in achieving homogeneous substrate distribution. Furthermore, the site should be carefully selected and the geological conditions well known.

The low solubility of nitrogen gas can lead to clogging of packed-bed reactors (Harremoes *et al.*, 1980; Soares *et al.*, 1989, 1990) and fine-matrix aquifers (Mercado *et al.*, 1988; Ronen *et al.*, 1989). The solubility of $N_2$ in water is approximately 20 mg $l^{-1}$, relative to an atmosphere of pure $N_2$. However, water in equilibrium with the atmosphere contains approximately 16 mg $l^{-1}$, leaving room for the production of only 4 mg $l^{-1}$ $N_2$ before saturation is reached, at atmospheric pressure. Strategies used to control nitrogen gas were application of pressure filtration in closed reactors to increase gas solubility (Philipot *et al.*, 1985), high upflow velocity rates (MacDonald, 1990) and vacuum deaeration preceding the denitrification reactor (van der Hoek *et al.*, 1994).

Biological denitrification systems fall into two types: heteretrophic when an organic compound serves as the source of carbon and energy, and autotrophic when inorganic carbon is supplied for cell synthesis and an inorganic compound serves as energy source. Post treatment of the denitrified water is generally required to remove bacteria and residual organic carbon.

### 3.1. HETEROTROPHIC PROCESSES

Heterotrophic denitrification processes are the most studied and most widely applied in the field. While the nature of the organic compounds may affect the biomass yield, the choice is generally based on economic considerations. Methanol is the least expensive of the simple carbon sources, but its use in the treatment of potable water is not permitted in some countries.

Liessens *et al.* (1993) used methanol in a fluidized sand-bed reactor located at the de Blankaart, Belgium, water production facility. A contact time of 15 min was required to remove 17 mg N $l^{-1}$, and the reactor loading was 2.03 kg N $m^{-3}$ $d^{-1}$, at 3.5°C. Excess biomass was removed from the sand particles using a stirrer and a hydrocyclone. Residual methanol was eliminated by post treatment by trickling filtration and granular activated carbon columns.

In a continuous laboratory system developed by Reising and Schroeder (1996) and Mansell and Shroeder (1999) a membrane separated biomass and methanol from the water to be treated; nitrate ions from the water diffused through the membrane to the biomass. Approximately 18 mg N $l^{-1}$ were removed at a flux of 4 g N $m^{-2}$ $d^{-1}$ of membrane area.

Van der Hoek and Klapwijk (1987) developed a treatment combining ion exchange and biological denitrification. Nitrate was removed from groundwater by ion exchange and regeneration of the resin took place in a closed circuit in an upflow sludge blanket reactor (USB) denitrification reactor with methanol. In laboratory scale studies with a nitrate-selective resin, a reduction of 90% in brine production was obtained, compared to conventional ion exchange (van der Hoek and Klapwijk, 1988).

Using ethanol in USB laboratory reactors Green *et al.*, (1994) obtained loading rates of up to 4 Kg N $m^{-3}$ $d^{-1}$ with a retention time of 8 min but washout biomass could be a problem. In columns of alginate beads with immobilized cells of *Pseudomonas denitrificans* ethanol-dependent denitrification was limited by mass transfer through the alginate matrix, and hampered by the short stability (approximately 2 months) of the alginate (Nilsson and Ohlson, 1982). A specific denitrification rate of 19.2 g N $kg^{-1}$ wet gel $d^{-1}$ was reported.

Several full-scale systems with ethanol have also been reported. In Germany, Roennefahrt (1986) developed the DENIPOR process consisting of fixed-film reactors with buoyant spheres of expanded polystyrene. The concentration of nitrate in the influent was 12.5 to 14.5 mg N $l^{-1}$ and at least 95% of the nitrate was removed at loads of up to 0.7 kg N $m^{-3}$ $d^{-1}$; excess biomass was removed by downwards flushing and postreatment consisted of anthracite/sand filtration and chlorination. In France, Richard (1989) developed the Nitrazur and the Biodenit processes. The first is an upflow process in which acetic acid may also be used, and the second is a downflow process operated under pressure. Both include postreatment with activated carbon filters. A Nitrazur plant located in Champfleur yielded a nitrate removal efficiency of 72% for a continuous flow of 35 $m^3$ $h^{-1}$ and an ethanol to nitrate-N dose of 3.1; the reactor and filter had to be thoroughly backwashed every 48 h to avoid formation of nitrite. A process similar to Nitrazur was the largest full-scale treatment reported so far, located at Guernes/Dennemont. The plant was designed to remove both nitrate and ammonia from contaminated groundwater and operated for a number of years at a capacity of 400 $m^3$ $h^{-1}$ (Rogalla *et al.*, 1990). It consisted of two fixed-bed reactors in series: an anoxic filter where nitrate was removed at filtration rates up to 10 m $h^{-1}$, and an aerobic two-layer filter packed with activated carbon and sand, where the water was polished prior to further disinfection by ozonization. The concentration of nitrate decreased from 9 - 15 to 3 - 4 mg N $l^{-1}$, and an ethanol to nitrogen ratio of 1.2 was used.

In a sand and gravel aquifer in Czechoslovakia, Janda *et al.* (1988) conducted an *in situ* full-scale trial with ethanol. The background concentration of nitrate was about 22.6 mg N $l^{-1}$ and a removal efficiency of approximately 50% was observed with a contact time of one to two days. Their system consisted of one central well and four crosswise injection wells 12 to 15 m from the extraction well; water recycling through the injection wells improved spreading of the substrate.

An *in situ* denitrification system called Nitredox has been in operation in a gravel aquifer in Beesemburg, Austria (Braester and Martinell, 1988). The supply well is surrounded by 16 injection (reduction) wells in an outer ring with a radius of 18 m. Inside this ring is a second ring of 8 (oxidation) wells with a radius of 10 m. A number of inspection wells are placed inside each ring to constantly monitor the redox values of the passing water. Other inspection wells are placed outside the system to allow reference values in the untreated water in the aquifer. Nitrate is reduced in the outer ring, and in the small ring nitrogen is degassed, iron and manganese are oxidized and precipitated in the ground, and any residual nitrite is oxidized. Each injection well is equipped with a small pump and the system is fully automated. The system has a yield of about 215 $m^3$ $h^{-1}$, and the concentration of nitrate is reduced from 22.6 to 5.7 mg N $l^{-1}$.

Hamon and Fustec (1991) tested an *in situ* system consisting of 30 injection wells in a circle 25 m from a central extraction well. Ethanol diluted with recirculated water was injected into the aquifer. A drop of approximately 16 mg N $l^{-1}$ was obtained after 3 days of pumping at a flow of 30 $m^3$ $h^{-1}$. However, rapid clogging of the aquifer followed this; discontinuous operation (1 h pumping, 1 h halt) minimized clogging.

Acetate was used as substrate in Germany by Bockle *et al.* (1986), in a fixed-bed of granulated activated carbon. Water was pumped from the aquifer, amended with acetate, passed in a downflow through the reactor, areated in a cascade and infiltrated into the aquifer. Denitrification rates of up to 3.5 kg N $m^{-3}$ $d^{-1}$ were obtained if a DOC of 1 mg $l^{-1}$ was allowed, and the efficiency decreased at lower DOC breakthrough. The filter was backwashed daily to remove biomass, and after nine months of continuous operation the aquifer showed no signs of clogging. The quality of the treated water was monitored through a well 10 m downstream from the injection point, and it consistently complied with the required standards even when nitrite was eluted from the reactor.

In laboratory experiments aimed at studying problems associated with *in situ* denitrification treatment in sandy aquifers, Soares *et al.* (1988) used sucrose as substrate in sand-packed columns operated in a downflow mode. After a few months of operation, visible accumulation of gas occurred accompanied by marked decrease in the permeability of the bed; vacuum treatment restored permeability to its original level. Further studies with ethanol, formate, acetate and sucrose showed that gas (and not biomass) was the main reason for clogging of the fine matrix (Soares *et al.*, 1989, 1991). It was suggested that *in situ* pulse applications of the carbon source would be preferable to a continuous supply (Soares *et al.*, 1991).

In field studies Mercado *et al.* (1988) used an aqueous solution of sucrose as substrate to treat the aquifer around an unused old well with estimated depth of 90-100 m and nitrate content in the order of 14 mg N $l^{-1}$. A dual-purpose strategy (recharge-pumping) using a single well was only suitable for long cycles with a delay of about one month between the two stages, if filtration and desinfection requirements were to be minimized. The substrate was also injected through three small-diameter injection wells drilled around the extraction well however, due to technical problems and geological characteristics of the site, only one of the injection wells functioned properly. The discharge rate was 50 to 60 $m^3$ $h^{-1}$, and the concentration of nitrate decreased by 10%. The reason for this relatively low effect was that with only one injection well the denitrified water around the injection well was highly diluted by the surrounding nitrate polluted groundwater converging into the pumping well.

Cellulose can also serve as substrate for denitrification of groundwater. Boussaid *et al.* (1988) conducted laboratory and field studies with wheat straw. In their largest field process, part of the water extracted from the well was pumped into three aboveground reactors of approximately 7.5 $m^3$ capacity, situated around the well and packed with a mixture of marl and straw. After passing through the reactors at a rate of 1 $m^3$ $h^{-1}$ (3 h contact time), the water was returned to the aquifer through infiltration pits. A decrease of 2.26 to 4.52 mg nitrate-N $l^{-1}$ was observed in the well water, but the efficiency of the reactors decreased with time, and they developed clogging problems.

Shredded newspapers (Volokita *et al.*, 1996b), wheat straw (Soares and Abeliovich, 1998) and unprocessed cotton (Volokita *et al.*, 1996a) were investigated as possible low cost substrates for denitrification. Laboratory columns were packed with the cellulosic substrate only. The reasons for choosing this configuration were: bulkiness of the substrates, attempts to avoid or at least minimize the entrapment of gases, and the fact that to break down the substrate bacteria have to attach themselves to the cellulose fibers. Cotton was by far the most efficient substrate (Volokita *et al.*, 1996a) and was tested in a 9 $m^3$ field reactor packed with 1200 kg of cotton (Soares *et al.*, 1999). The system could not be operated at full capacity due to serious compression of the bed by high water pressures. The highest rate of denitrification obtained was 0.36 kg N $m^{-3}$ $d^{-1}$, at a feed rate of 6 $m^3$ $h^{-1}$ which could be sustained only for short periods. The system was stable at feed rates of 0.8 and 1.5 $m^3$ $h^{-1}$. It was concluded that a smaller reactor or a system of parallel small-size reactors should be employed; optimization of the process is under investigation. Cotton is completely degraded, and long-term operation of reactors requires only periodic additions of cotton. The cost of removing nitrate with cotton as electron donor is US \$1.48 $Kg^{-1}$ N (Soares *et al.*, 1999).

## 3.2. AUTOTROPHIC PROCESSES

Hydrogen gas and various reduced-sulfur compounds have been used as microbial energy source in denitrification of groundwater. The low cost of these inorganic substrates and low formation of biomass are important advantages although their application can also present some limitations. During denitrification, reduced sulfur compounds are converted into sulfate, and this renders the method unsuitable for the treatment of water containing high levels of endogenous sulfate. High concentrations of sulfate can act as a laxative, especially in combination with magnesium. Sulfate may also affect the taste of water with the effective threshold being different for different salts. A maximum of 400 mg sulfate $l^{-1}$ in drinking water is therefore recommended (WHO, 1984).

Hydrogen gas is an ideal energy substrate for denitrification in the sense that it is completely harmless to potable water, and no further steps are required to remove either excess substrate or its derivatives. However, $H_2$ forms flammable and explosive mixtures with oxygen, and its solubility in water is low (1.6 mg $l^{-1}$ at 20°C).

Studies on the use of sulfur compounds were for a time limited to denitrification of wastewater (Batchelor and Lawrence, 1978; Bisogni and Driscoll, 1978). Denitrification of groundwater was studied in laboratory columns packed with a granular mixture of elemental sulfur and limestone (Martin and Blecon, 1983; Blecon *et al.*, 1985). Limestone served as the source of inorganic carbon for bacterial synthesis and as a pH buffering agent. The reactors were operated in an upflow mode with periodic backwashings to remove accumulated biomass.

In The Netherlands van der Hoek *et al.* (1992) developed the sulfur-limestone process further, and operated a field system with a capacity of 35 $m^3$ $h^{-1}$. It consisted of four unit operations in series: vacuum-deaeration, nitrate removal in a sulfur-limestone filter, aeration (cascade) and soil infiltration as postreatment. The upflow filtration rate was between 0.25 and 0.5 m $h^{-1}$ and volumetric loadings of up to 0.12 kg N $m^{-3}$ $d^{-1}$ were applied. The system had to be operated at oversupply of nitrate to avoid the formation of sulfide which was oxidized to colloidal sulfur during the aeration stage and caused clogging of the subsequent infiltration pond.

A number of denitrification systems have been described in which compressed hydrogen gas is sparged into the water to be treated. Kurt *et al.*

(1987) studied a bench-scale fluidized bed sand reactor in which a residence time of 4.5 h was required for complete denitrification of water containing 25 mg nitrate-N $l^{-1}$, at rates of up to 0.55 kg N $m^{-3}$ $d^{-1}$. Dries *et al.* (1988) tested a two-column system with removal of nitrate in the first column using polyurethane as support medium, and removal of excess hydrogen and oxidation of residual nitrite to nitrate in the second column. Water flowed downwards in the first column while hydrogen entered from the bottom; the water then passed through the second column in a upflow mode. Denitrification rates of 0.5 kg N $m^{-3}$ reactor $d^{-1}$ were obtained, at 20°C.

A full-scale process known as DENITROPUR was developed by various authors and operated in Monchengladbach, Germany (Gros and Treutler, 1986; Gros *et al.*, 1986). The process incorporated a hydrogen saturator, addition of phosphate and carbon dioxide, four packed-bed reactors in series, postaeration, floculant addition, filtration, and UV filtration; the reactor operated at a loading rate of 0.25 kg N $m^{-3}$ $d^{-1}$ and residence times of 1 to 2 h were required to remove 11.29 mg N $l^{-1}$.

More recently, Sakakibara and Kuroda (1993) reported a novel approach which overcomes some of the limitations associated with the use of hydrogen: hydrogen was generated in the denitrification reactor by electrolysis of the water to be treated. They used a batch system with two interconnected reactors, cathodic and anodic. Prior to start up, a biofilm was allowed to develop on the surface of the cathode, by batch cultivation in a rich organic medium. When the biofilm was visible, the electrode was transferred to the reactor and connected to the power supply; the microorganisms immediately took up hydrogen generated at the cathode.

TABLE I
Characteristics of the electrobiochemical systems.

| | | One-reactor System | Two-reactor system |
|---|---|---|---|
| Electrodes active area | Dimensions (cm) | 14 x 7 | 9 x 5 |
| Bed | Dimensions (cm) | 14 x 7 x 0.6 | 4.5(Ø) x 38.5 |
| | Volume ($cm^3$) | 59 | 610 |
| | Void volume | 17 | 270 |
| | GAC (g) | 35 | 550 |

Lately, a bioelectrochemical process has been studied in the author's laboratory (Kiss *et al.*, 1999). Two main process configurations were investigated: a single reactor where both the generation of hydrogen and denitrification took place, and a two-reactor system where water (amended with sodium bicarbonate as carbon source) was first enriched with hydrogen in an electrolysis cell prior to entering a packed-bed bioreactor. In the electrolysis cell a Nafion cation-exchange membrane separated the two electrodes, and prevented oxygen generated at the anode from reaching the cathode. Hypalon rubber spacers intercalated between the membrane and the electrode created cathodic and anodic chambers. The reactors were operated in a continuous mode, and granulated activated carbon (GAC) served as physical support for the biomass. The main characteristics of both systems are summarized in Table I. The denitrification rate of 0.45 kg N $m^{-3}$ $d^{-1}$ was obtained in the single-reactor system, with removal of 18 mg N $l^{-1}$ in 17 min. In the two-reactor system denitrification rates of 0.2 kg N $m^{-3}$ $d^{-1}$ were observed, with 22 mg N $l^{-1}$ being removed in 1.4 h, a much longer contact time but still lower than those reported by Gros *et al.* (1986) and Kurt *et al.* (1987).

The overall performance of the single reactor process appeared to be superior, but formation of scale on and around the cathode became a problem. In the two-reactor system, deposits accumulated on the cathode were not in direct contact with the biomass and could be easily removed by disconnecting the bioreactor from the electrochemical cell for a few minutes while flushing the cathodic compartment with a diluted acid solution. A possible strategy to prevent formation of scale is to periodically disconnect the bioreactor and switch the polarity of the electrodes for a few minutes. The electrochemical cell is the most complex and expensive component, and increasing the volume of its bed will be more costly than using a small electrochemical cell with a larger bioreactor; this makes the two-reactor process more suitable for upscaling. The system has been in continuous operation for the last nine months and is very stable; studies are in progress to optimize its operational parameters.

## 5. Concluding Remarks

For many years groundwater research was focused on supply with little interest being expressed with regard to water quality. Yet, millions of people all over the world are dependent on groundwater for domestic use, and even in some of the most developed countries, large sections of the population utilize untreated groundwater.

Migration of nitrate to the aquifer can be slow and years may lapse before the consequences of past practices are fully manifested. There is today an increasing awareness that the best approach to groundwater contamination by nitrate is prevention, and preventive measures are curbing the problem in some developed countries. However, nitrate pollution is still on the rise and will continue to increase in many developing countries where large-scale use of fertilizers is yet to come, and raw or inadequately treated effluents will continue to be discarded in the environment. Thus, innovative, low-cost biological denitrification processes are still needed New inexpensive substrates have to be tested, and the cost of postreatment minimized. Treatments *in situ* are still relatively few, and further methods for introducing substrate into the aquifer are required as well as studies of the long-term effects of *in situ* treatment on groundwater quality.

The choice of a biological denitrification system has to be considered on an individual basis. A complex aboveground system can be easily integrated in today's large water plants, but a less costly process will be needed for water facilities supplying smaller numbers of consumers. *In situ* treatment may be problematic in fine aquifers but may also be the only practical and affordable solution in a remote village. The lower denitrification rates per volume reactor obtained with inorganic substrates may be compensated by their low cost and postreatment requirements.

## Acknowledgements

The author thanks the financial support from the U.S. - Israel Cooperative Development Research Program, U.S. Agency for International Development, and from the Israel-Federal Republic of Germany Program of the Ministry of Science (MOS) and the Bundesministerium fuer Bildung Wissenschaft, Forschung und Technologie (BMBF).

## References


Batchelor, B. and Lawrence, A. W.: 1978, *J. Water Pollut. Control Fed.* **50**, 1986-2001.

Blecon, G., Martin, G. and Cormier, M.:1985, Proc. Conf. *Nitrates in Water*, Paris, October 1985.

Bocle, R., Rohmann, U. and Wertz, A.: 1986, *Aqua* **5**, 286-287.

Boussaid, F., Martin, G., Morvan, J., Collin, J. J., Landreau, A. and Talbo, H.: 1988, *Environ. Technol. Lett.* **9,** 803-816.

Braester, C. and Martinell, R.: 1988, *Water Sci. Technol.* **20,** 149-163.

Dahab, F. M.: 1987, *J. Environ. Syst.* **17**, 65-74.

Dahab, F. M. and Sirigina, S.: 1994, *Water Sci. Technol.* **30**, 133-139.

Dahab, F. M. and Woodbury, W. L.: 1998, *Proceedings of the AWWA Inorganic Contaminants Workshop*, San Antonio, TX, USA, 22-24 February, 1998.

Dries, D., Liessens, J., Verstrate, W., Stevens, P., de Vos, P. and Ley, J.: 1988, *Water Supply* **6**, 181-192.

Driscoll, C. T. and Bisogni, J. J.: 1978, *J. Water Pollut. Control Fed.* **50**, 569-577.

European Comunity: 1980, *Off. J. Eur. Commun.* ***23,*** L 229, 11-29.

Gamble, T. N., Betlach, M. R. and Tiedje, J. M.: 1977, *Appl. Environ. Microbiol.* **33**, 926-939.

Green, M., Tarre, S., Schnizer, M., Bogdan, B., Armon, R. and Shelef, G.: 1994, *Water Res.* **28,** 631-637.

Gros, H. and Treutler, K.: 1986, *Aqua* **5**, 288-290.

Gros, H., Schnoor, G. and Rutten, P.: 1986, *Water Supply* **4**, 11-21.

Hall, T., Walker, R. A. and Zabel, T. F.: 1985, *Proceedings of the Conference"Nitrates in Water*", Paris, October 1985.

Hamon, M. and Fustec, E.: 1991, *Res. J. Water Pollut. Control Fed.* **63,** 942-949.

Harremoes, P., Jansen, J. la C. and Kristensen, G. H.: 1980, *Prog. Water Technol.* **12,** 253-269.
Janda, V., Rudovsky, J., Wanner, J. and Marha , K.: 1988, *Water Sci. Technol.* **20,** 215-219.
Kiss, I., Szekeres, S., Bejerano, T. T. and Soares, M. I. M.: 1999, *Water Sci. Technol.* (Submitted).
Kurt, M., Dunn, I. J. and Bourne, J. R.: 1987, *Biotechnol. Bioeng.* **29,** 493-501.
Kapoor, A. and Viraraghavan, T.: 1997, *J. Environ. Eng.* **123**, 371-380.
Liessens, J., Germonpre, R., Beernaert, S. and Verstraete, W.: 1993, *J. AWWA* **85,** 144-154.
MacDonald, D. V.: 1990, *J. Water Pollut. Control Fed.* **62**, 796-802.
Mansell, B.O. and Schroeder, E.D.: 1999, *Water Res.* **33**, 1845-1850.
Martin, G. and Blecon, G.: 1983, *Aqua* **32**, 66-67.
Mercado, A., Libhaber, M. and Soares, M. I. M.: 1988, *Water Sci. Technol.* **20**, 197-209.
Mirvish, S. S.: 1985, *Sience* **315,** 461-462.
Nilsson, I. and Ohlson, S.: 1982, *Eur. J. Appl. Microbiol. Biotechnol.* **14**, 86-90.
Philipot, J. M., Chaffange, F. and Pascal, O.: 1985, *Water Supply* **3**, 93-98.
Reising, A. R. and Schroeder, E. D.: 1996, *J. Environ. Eng.*, ASCE **127**, 599-604.
Richard, Y. R.: 1989, *J. Inst. Water Environ. Manage.* **3,** 154-167.
Roennefahrt, K. W.: 1986, *Aqua* 5, 283-285.
Rogalla, F., Ravarini, P., De Larminat, G. and Couttelle, J.: 1990, *J. Inst. Water Environ. Manage.* **4,** 319-329.
Ronen, D., Berkowitz, B. and Magaritz, M.: 1989, *Transport in Porous Media* **4,** 295-306.
Sakakibara , Y. and Kuroda , M.: 1993, *Biotechnol. Bioeng.* **42,** 535-537.
Schlegel, H. G., Kaltwasser, H. and Gottschalk, G.: 1961, *Arch. Mikrobiol.* **38,** 209-222.
Soares, M. I. M. and Abeliovich, A.: 1998, *Water Res.* **32**, 3790-3794.
Soares, M. I. M., Belkin, S. and Abeliovich, A.: 1988, *Water Sci. Technol.* **20**, 189-195.
Soares, M. I. M., Belkin, S. and Abeliovich, A.: 1989, *J. Water Wastewater Res.* **22,** 20-24.
Soares, M. I. M., Braester, C., Belkin, S. and Abeliovich, A.: 1991, *Water Res.* **25**, 325-332.
Soares, M. I. M., Brenner, A., Yevzori, A., Messalem, R., Leroux, Y. and Abeliovich, A.: 1999, *Water Sci. Technol.* (Submitted).
*Standard Methods for the Examination of Water and Wastewater*: 1992, 18th edition, American Public Health Association / American Water Works Association/Water Environmental Federation, Washington, DC, USA.
van der Hoek, J. P., Hijnen, W. A. M., van Bennekom, C. A. and Mijnarends, B. J.: 1992, *Aqua* **41,** 209-218.
van der Hoek, J. P., Kappelhof, J. W. N. M. and Schippers, J. C.: 1994, *Aqua* **43**, 84-94.
van der Hoek, J. P. and Klapwijk, A.: 1987, *Water Res.* **21**, 989-997.
van der Hoek, J. P. and Klapwijk, A.: 1988, *Water Supply* **6**, 57-62.
Volokita, M., Abeliovich, A. and Soares, M. I. M.: 1996a, *Water Sci. Technol.* **34**, 379-385.
Volokita, M., Belkin, S., Abeliovich, A. and Soares, M. I. M.: 1996b, *Water Res.* **30**, 965-971.
WHO: 1984, *Guidelines for drinking water quality*, World Health Organization, Geneva,Vol. 2, pp. 290-292.
Yutaka, S. and Kuroda, M.: 1993, Biotechnol. Bioeng., **42,** 535-537.
Zumft, W.: 1992, 'The dentrifying prokaryotes', in A. Balows, H.G. Truper, M. Dwokin, W. Harder and K.- H. Schleifer (eds), *The Prokaryotes*, Springer-Verlag, New York, USA, pp. 554-582.

# TOWARDS A SCIENCE-BASED INTEGRATED OZONE-FINE PARTICLE CONTROL STRATEGY

R. L.TANNER
*Tennessee Valley Authority, Environmental Research Center, P.O. Box 1010, Muscle Shoals, Alabama 35662-1010 U.S.A.*

**Abstract.** Epidemiology studies relating health effects to ambient levels of ozone and fine particles have led to the modification of standards in the United States for these pollutants (substitution of an 8-h standard for ozone at 80 ppbv, and addition of 24-h and annual standards for fine particles). The interrelationships of these pollutants in the atmosphere suggest the need for an integrated, science-based strategy for their control. Secondary ozone formation has been controlled through emission controls on VOC and $NO_x$ precursors. Fine particles are secondary products largely resulting from the oxidation of precursors ($SO_2$, $NO_x$, and VOCs). The key intermediates in both types of secondary process are free radical species and the photochemically labile compounds that produce them in the atmosphere. However, due to the complex and nonlinear nature of the processes, reductions in precursors may lead to unexpected changes in ozone and fine particle formation rates. For example, reduction in $NO_x$ emissions may reduce ozone and nitric acid levels, but lead also to increased rates of sulfate formation in clouds and increased ammonia availability for neutralization of acidic sulfate aerosols. Reductions of $SO_2$ may reduce aerosol sulfate levels in the summer, but have no effect in other seasons. Reductions in VOCs may reduce ozone levels in urban core areas, but not elsewhere. An integrated, regionally and seasonally specific, emission reduction strategy is needed to cost-effectively reduce both ozone and fine particle levels.

**Keywords:** air quality standards, control strategies, fine particles, ozone

## 1. Introduction

Epidemiological studies appear to implicate ambient levels of ozone and accumulation mode (a.k.a. "fine") particles to a variety of human health effects, mostly related to the respiratory system. The U.S. Environmental Protection Agency's mandate to protect human health has led to the modification of National Ambient Air Quality Standards (NAAQS) in the United States for these pollutants. These changes include the substitution of an 8-h standard for ozone at 80 ppbv for the previous 1-hr standard at 120 ppbv, and the addition of 24-h and annual standards for fine particle mass at 65 and 15 $\mu g/m^3$, respectively. Fine particles are defined as those separated by an aerodynamic sizing device having a 50% upper size cut at 2.5 µm, hence the fine particle NAAQS are commonly referred to as the $PM_{2.5}$ standards.

The interrelationships of the processes forming ozone and fine particles in the atmosphere strongly suggest the need for developing an integrated, science-

*Water, Air, and Soil Pollution* **123:** 195–201, 2000.

based strategy for their joint control. Secondary ozone formation in the atmosphere is due to the interaction of nitrogen oxides, $NO_x = NO + NO_2$, with volatile organic compounds, VOC, in the presence of sunlight. Ambient ozone levels have largely been controlled through limits on emissions of VOC and $NO_x$ precursors. Fine particles are secondary products largely resulting from the oxidation of precursors ($SO_2$, $NO_x$, and VOCs) to form less volatile products—sulfates, nitrates and oxygenated organic species. The key intermediates in the processes forming both types of secondary products are free radical species and the photochemically labile compounds that produce them in the atmosphere.

As a result of these interactions, emission reduction strategies which were designed to control the accumulation of ozone have also affected the processes which form fine particles, and vice versa. Both of these processes are now well enough understood, and the success of emission control strategies for their control sufficiently in doubt, that it is necessary, in this author's view, to develop integrated control strategies for both ozone and fine particles if the desired goal of reducing their concentrations in the ambient atmosphere is to be efficiently and promptly attained. The scientific basis for such a strategy is outlined below.

## 2. Summary of the Science

### 2.1. THE PROCESSES

#### *2.1.1. The Production of Secondary Tropospheric Ozone*

Ozone in the lower troposphere is made through a chain reaction process initiated by the reaction of OH radicals with any of a variety of volatile organic compounds (VOC). Ozone concentrations are maintained by the $O_3$-NO-$NO_2$ photo-stationary state (Finlayson-Pitts and Pitts, 1986), defined by the balance between reaction of NO with ozone and the photolysis of product $NO_2$ during daytime hours. Excess ozone is formed due to oxidation of NO to $NO_2$ by peroxy radicals, $HO_2 + RO_2$. The chain reaction is terminated by several radical-consuming steps, the most important of which is OH + $NO_2$ (Calvert *et al.*, 1985).

#### *2.1.2. Oxidation of $SO_2$ to Sulfate*

Sulfate aerosol is made by reactions in the gas phase which convert $SO_2$ to $SO_4^{=}$ (Eatough *et al.*, 1994), and by reactions in the aqueous phase (hydrometeors, and to a lesser extent, aerosol liquid water) which mainly involve peroxides (Pandis and Seinfeld, 1989; Schwartz, 1984) and which are relatively insensitive to cloud/rain water pH (Martin and Damschen, 1981). Oxidation of S(IV) to sulfate by ozone may also play a role at high solution pHs (Hoffman, 1986).

*2.1.3. Oxidation of Nitrogen Oxides to Inorganic Nitrate*

Nitrate is formed from nitric acid by temperature-dependent reaction with gaseous ammonia (Russell *et al.*, 1983). Ammonia sources are variable in time and space, hence the fraction of nitrate in the particle phase is usually low in Eastern North America, especially when temperatures are high, but high in Western North America (and in some urban areas), especially when temperatures are low. Organic nitrates and peroxyacyl nitrates are also formed by $NO_2$-radical reaction. Their presence in the gas phase is well documented, but documentation of their particulate-phase forms is sparse.

*2.1.4. Photochemical Production of Organic Aerosols*

Volatile organics are converted to less-volatile organics by OH abstraction and $RO_2/HO_2$ addition reactions, leading to oxygenated products (carbonyls, alcohols, and organic acids) (Bowman *et al.*, 1995; Pandis *et al.*, 1991). These are very complicated processes whose rates and extent vary widely from location to location and season to season. The ability to distinguish, through analysis of individual constituents or fraction, the various sources of particulate organics from each other, is still in a primitive state and is in serious need of substantial improvement (Chow, 1995).

2.2. THE INTERCONNECTIONS

Atmospheric photo-oxidation processes do not occur in isolation. All of the major species and especially reactive and free radical intermediate species participate in a host of coupled reactions, such that the overall lifetimes and major sources and sinks of these species vary widely depending on levels of key species. This fact has been known for two decades or more, but has not been incorporated into control strategies for the most part because of gaps in the knowledge base about these coupled processes. More frequently, however, nonscience considerations and the desire to utilize nationally uniform NAAQS for pollution control has led to simplified, one-pollutant-at-a-time approaches. There is a need at this point in time to recognize and act upon the interconnections, outlined below, between ozone and fine-particle formation processes:

- Ozone is made through a chain process initiated by OH + VOC, maintained by the $O_3$-NO-$NO_2$ photostationary state, and terminated by several radical-consuming steps, the most important of which is OH + $NO_2$ to form nitric acid and particulate nitrate. Hence, the processes forming excess ozone and inorganic nitrate are inextricably intertwined.
- The rate and extent of gas-phase oxidation of $SO_2$ to sulfate aerosol depends directly on OH levels. The rate and extent of aqueous phase oxidation of $SO_2$ to sulfate aerosol by peroxides depends nonlinearly on gas-phase $HO_2$

and $RO_2$ levels since their recombination reactions at low $NO_x$ levels are the principal sources of gas-phase precursors. Alternatively, sulfate formed from aqueous oxidation may depend nonlinearly on ozone at high solution pH (Pandis and Seinfeld, 1989; Tanner and Schorran, 1995). In all cases, reactions forming excess tropospheric ozone also contribute simultaneously to $SO_2$-to-sulfate formation.

- Nitrate is formed from nitric acid by reaction with gaseous ammonia, thus fine particulate levels of nitrate depend both on OH levels (Eatough *et al.*, 1995) and *independently* on ambient $NH_3$ levels (Tanner and Harrison, 1992).
- Volatile organics are converted to less-volatile organics by OH abstraction and $RO_2/HO_2$ addition reactions, leading to oxygenated products (carbonyls, alcohols, and organic acids) when $NO_y$ levels are low, but increasingly to nitrate products when $NO_y$ is higher (Finlayson-Pitts and Pitts, 1986). The extent to which organic aerosols are formed in the atmosphere is now becoming clear, but the details of the organic aerosol formation processes are largely unknown.

## 2.3. THE LAW OF UNEXPECTED CONSEQUENCES

The interactions of various photo-oxidation processes in the atmosphere which form ozone and fine particles lead to multiple examples of the "Law of Unintended Consequences", by which we mean that actions designed to achieve one end may affect other actions or decisions in nonintuitive ways, positively or negatively, thereby altering the effectiveness of a particular action, such as a strategy to control the levels of a noxious pollutant in the atmosphere. Some examples of unintended consequences of VOC, $NO_x$ or $SO_x$ reductions are detailed below:

- Reduction in $NO_x$ emissions may reduce ozone and nitric acid levels, but can also lead to increased rates of sulfate formation in clouds because of increases in peroxide levels. Nitric acid levels will also be lower as a result, leading to increased ammonia availability for the neutralization of acidic sulfate aerosols.
- Reductions of $SO_2$ may reduce aerosol sulfate levels in the summer, but have no effect in other seasons. This occurs because gas-phase production of $H_2SO_4$ by the OH + $SO_2$ reaction is always first order in $SO_2$, but aqueous-phase production of sulfate is only first order in $SO_2$ when peroxides are in excess (or at high pH) (Schwartz, 1987). Depending on the location and magnitude of $NO_x$ sources, peroxides are frequently the limiting reagent for aqueous-phase oxidation ($SO_2$ in excess) during periods of lower solar radiation, i.e., in nonsummer months at higher latitudes, leading to nonlinear,

less-than-stoichiometric reduction in sulfate levels following reductions in $SO_2$ emissions.

- Reductions in VOCs may reduce ozone levels in urban core areas, but not in other locations. The consensus drawn from recent science is that ozone formation in most areas of North America is limited by the availability of $NO_x$ ($NO + NO_2$) due to the short lifetimes of these species and the abundance of biogenic emissions during the ozone season in many areas. However, in larger urban areas, the urban core atmosphere is still limited by VOC availability. Thus, VOC emissions reductions may reduce ozone in the urban core but have no observable effects on ozone levels in adjacent suburban and rural areas. This consideration is especially troubling given that the new U.S. ozone standard is based on 8-h rather than one-hour averages, and it is much more likely that rural areas will exceed the 8-h standard than the previous 1-h standard (Parkhurst, 1999).

## 3. Implications for Policymakers

In consideration of the interactions described above, there are several factors which should, in our view, be considered in the further development of ozone and fine-particle control strategies. For example, reduction in VOC emissions may decrease ozone formation rates in some core urban areas and have no effect in other areas, since ozone production is limited, in summer months, by VOC concentrations only in urban cores. In rural areas, ozone production is generally limited by $NO_x$ levels because of the short lifetimes of NO and $NO_2$, but also because of the abundance of natural sources of VOC in most areas during summer months.

Decreases in $NO_x$ emissions may lower regional ozone concentrations but:

- increase the efficiency (per unit $NO_x$) of ozone production (Gillani *et al.*, 1998);
- increase the fraction of $SO_2$ converted to sulfate by increasing the availability of peroxide oxidant species;
- particulate nitrate may stay the same if ammonia emissions remain the same or increase.

Decreases in $SO_2$ emissions may lower $SO_2$ concentrations but not lower aerosol sulfate levels proportionately, because of the increased importance of nonlinear aqueous-phase oxidation processes. Summertime sulfate levels may, however, be reduced if other atmospheric sinks for $SO_2$ are not altered.

The rate of organic aerosol formation, its concentrations, and its atmospheric lifetime may be affected by changes in all of the above: anthropogenic and biogenic VOCs, $NO_x$, $SO_2$, and $NH_3$. Much more information is needed in this area before controls on the precursors to organic particulate matter can intelligently be applied.

## 4. Conclusions

Control of ozone and fine particle levels is best accomplished by an integrated program, which maximizes the reduction in concentrations that occur per unit reduction in precursor emissions. The science summarized herein strongly suggests that regionally and seasonally specific emissions controls may achieve the desired reduction in ozone and fine particle levels much more efficiently than the present national, single-pollutant-at-a-time approach.

## References

Bowman, F. M., Pilinis, C. and Seinfeld, J. H.: 1995, Ozone and aerosol production of reactive organics, *Atmos. Environ.* **29**, 579-590.

Calvert, J. G., Lazrus, A., Kok, G. L., Heikes, B. G., Walega, J. G., Lind, J. and Cantrell, C. A.: 1985, *Chemical mechanisms of acid generation in the troposphere*, *Nature* **317**, 27-35.

Chow, J. C.: 1995, Critical review: measurement methods to determine compliance with ambient air quality standards for suspended particles, *J. Air Waste Manage. Assoc.* **45**, 320-382.

Eatough, D. J., Caka, F. M. and Farber, R. J.: 1995, The conversion of $SO_2$ to sulfate in the atmosphere, *Israel J. Chem.* **34**, 301-314.

Finlayson-Pitts, B. J. and Pitts, Jr., J N.: 1986, *Atmospheric Chemistry-Fundamental and Experimental Techniques,* Wiley and Sons, Ch.

Gillani, N. V., Meagher, J. F., Valente, R. J., Imhoff, R. E., Tanner, R. L. and Luria, M.: 1998, Relative production of ozone and nitrates in urban and rural power plant plumes, 1, Composite results based on data from 10 field measurement days, *J. Geophys. Res.* **103**, 22, 593-22, 602.

Hoffman, M. R.: 1986, On the kinetics and mechanism of oxidation of aquated sulfur dioxide by ozone, *Atmos. Environ.* **20**, 1145-1154.

Martin, L. R. and Damschen, D. E.: 1981, Aqueous oxidation of sulfur dioxide by hydrogen peroxide at low pH. *Atmos. Environ.* **15**, 1651-1656.

Pandis, S. N. and Seinfeld, J. H.: 1989, Sensitivity analysis of a chemical mechanism for aqueous-phase atmospheric chemistry. *J. Geophys. Res.* **94**, 1105-1126.

Pandis, S. N., Paulson, S. E., Seinfeld, J. H. and Flagan, R. C.: 1991, Aerosol formation in the photo-oxidation of isoprene and beta-pinene, *Atmos. Environ.* **25A**, 997-1008.

Parkhurst, W. J.: 1999, unpublished data.

Russell, A. G., McRae, G. J. and Cass, G. R.: 1983, Mathematical modeling of the formation and transport of ammonium nitrate aerosol. *Atmos. Environ.* **17**,49-964.

Schwartz, S. E.: 1984, "Gas-Aqueous Reactions of Sulfur and Nitrogen Oxides in Liquid-Water Clouds", in J. G. Calvert (ed), *$SO_2$, NO, and $NO_2$ Oxidation Mechanisms: Atmospheric Considerations,* Butterworths, Boston, MA., pp. 173-208.

Schwartz, S. E.: 1987, "Aqueous-Phase Reactions in Clouds" in R. W. Johnson, G. E. Gordon (eds), *The Chemistry of Acid Rain: Sources and Atmospheric Processes*, American Chemical Society, Washington, DC, pp. 93-108.

Tanner, R. L., Schorran, D. E.: 1995, Measurements of gaseous peroxides near the Grand Canyon: Implication for summertime visibility impairment from aqueous-phase secondary sulfate formation, *Atmos. Environ.* **29**, 1113-1122.

Tanner, R. L., Harrison, R. M.: 1992, "Acid-Base Equilibria of Aerosols and Gases in the Atmosphere", in J. Buffle, H.P. van Leeuven (eds), *Environmental Particles,* Vol. 1, Lewis, Chelsea, MI, Ch. 3, and references therein.

# AN ASSESSMENT OF THE MOBILE SOURCE CONTRIBUTION TO $PM_{10}$ AND $PM_{2.5}$ IN THE UNITED STATES

A. W. GERTLER, J. A. GILLIES and W. R. PIERSON

*Energy and Environmental Engineering Center, Desert Research Institute, 2215 Raggio Parkway, Reno, Nevada 89512, USA*

**Abstract.** Mobile sources are significant contributors to ambient particulate matter (PM) in the United States. As the emphasis shifts from $PM_{10}$ to $PM_{2.5}$, it becomes particularly important to account for the mobile source contribution to observed particulate levels since these sources may be the major contributor to the fine particle fraction. This is due to the fact that most mobile source mass emissions have an aerodynamic diameter less than 2.5 μm, while the particles of geological origin that tend to dominate the $PM_{10}$ fraction generally have an aerodynamic diameter greater than 2.5 μm. A common approach to assess the relative contributions of sources to observed particulate mass concentrations is the application of source apportionment methods. These methods include material balance, chemical mass balance (CMB), and multivariate receptor models. This paper describes a number recent source attribution studies performed in the United States in order to evaluate the range of the mobile source contribution to observed PM. In addition, a review of the methods used to apportion source contributions to ambient particulate loadings is presented.

**Keywords**: air quality, mobile source emissions, $PM_{10}$, $PM_{2.5}$, receptor modeling, source apportionment

## 1. Introduction

Atmospheric fine particulate matter has been implicated in human health effects. Recent studies have discussed the epidemiology of this relationship (Dockery *et al.*, 1993; Heath *et al.*, 1995; Pope *et al.*, 1995 Schwartz *et al.*, 1996), potential causal mechanisms (Seaton *et al.*, 1995), and the controversy that surrounds the PM and health effects debate (Vedal, 1997). While emission inventories and apportionment studies have shown mobile sources contribute significantly to ambient $PM_{10}$ (particulate matter of aerodynamic diameter less than 10 (μm or less), most of the emissions from mobile sources are in the $PM_{2.5}$ (aerodynamic diameter less than 2.5 μm) fraction. In many urban areas in the US, mobile sources are the dominant source of $PM_{2.5}$. Thus if we are to reduce the health impacts of fine particles, we need to understand the mobile source contribution to the observed particle levels.

Primary particulate emission sources from mobile sources include their exhaust (Mulawa *et al.*, 1997; Sagebiel *et al.*, 1997), the mechanical wear of tires and brakes (Pierson and Brachaczek, 1983), and the injection of particles from the pavement (Nicholson *et al.*, 1989) and unpaved road shoulders

*Water, Air, and Soil Pollution* **123:** 203–214, 2000.

(Moosmüller *et al.*, 1998) by resuspension processes. The products of tire and brake wear, and the resuspended road dust are dominated by particles larger than 10 μm, although a tail extends below this size (Pierson *et al.*, 1976; Mulawa *et al.*, 1997). Weingartner *et al.* (1997) found road dust and tire wear contributions to $PM_3$ to be very small. Particulate matter in the vehicle-exhaust is dominated by particles smaller than $PM_{10}$ (Sawyer and Johnson, 1995; Baumgard and Johnson 1996; Ristovski *et al.*, 1998). Mobile sources also contribute to atmospheric PM through emissions of gaseous precursors of PM from tailpipe and evaporative losses. Areas of research in mobile source PM include the speciation of emissions (e.g., Lowenthal *et al.*, 1994; Watson *et al.*, 1994a; Gertler *et al.*, 1997), the development of emission factors from dynamometers (e.g., Gertler *et al.*, 1996; Mulawa *et al.*, 1997; Cadle *et al.*, 1997; Truex *et al.*, 1998) and on-road measurements (e.g., Ingalls, 1989; Ingalls *et al.*, 1989; Pierson *et al.*, 1996; Gertler and Pierson, 1996, Gertler *et al.*, 1997), and the assessment of the relative contributions of mobile source to total PM in inventory (e.g., Hildemann *et al.*, 1991), and receptor modeling studies (e.g., Lowenthal *et al.*, 1997; Watson *et al.*, 1998).

The primary objective of this paper is to present the results of recent studies apportioning the contribution of mobile sources to observed ambient PM levels in the United States. This will be accomplished through an evaluation of commonly applied methods to attribute source emissions to observed PM levels, the characteristics of chemical profiles of mobile source emissions for use in source attribution studies, and a review of recent large scale source apportionment studies in which apportionment methods have been used to attribute mobile source contributions to ambient PM levels.

## 2. Source Apportionment Techniques

Source apportionment modeling techniques utilize the information contained within the chemical composition of particulate or gaseous phase pollutants in ambient air samples to attribute the relative amounts of pollutants to their respective sources. The relative contributions from each source are estimated based on knowledge of the characteristic chemical species emitted from the sources and the quantities measured in ambient samples. This approach is sometimes referred as the "top-down" approach as opposed to the "bottom-up" or emission inventory approach that estimates relative contributions to ambient levels based upon estimated emission rates, activity levels of emission sources, and dispersion from the sources. Source apportionment models range from the relatively simple, in the case of the material balance approach (Solomon *et al.*, 1989), to the complex mathematical treatments of multivariate receptor models.

A first approximation for apportionment is often done using a material balance approach (Solomon *et al.*, 1989). In this approach, chemical analyses of particulate samples collected at a receptor are used to attribute the components of the total mass to broad categories of aerosol types. These categories can include, but are not limited to, carbonaceous components (from combustion sources), geological material (from fugitive dust emissions), and nitrates and sulfates (secondary aerosols). Secondary aerosols are formed from emitted gases that transform to particles as a result of chemical reactions in the atmosphere.

More detailed source apportionment is accomplished through the application of receptor models that utilize measurements of total and speciated mass concentrations of gases and particulate matter in the ambient air in combination with representative chemically-speciated source profiles for specific emission source categories. Receptor models can provide estimates of contributions to more specific sources than the general categories of the material balance method. For example, the contribution of PM from different combustion sources, such as motor vehicles and wood smoke, can be estimated using receptor models. These sources would be grouped into the carbonaceous particle category using the material balance approach. Receptor models attempt to apportion atmospheric pollutants to sources by relating chemical and physical properties of the source material to the properties observed at a receptor site. There are two main approaches to receptor modeling, chemical mass balance (CMB) and multivariate models. The U.S. EPA sanctions the CMB model version 7.0 (Watson *et al.,* 1990) as its reference method for source apportionment.

## 3. Mobile Source Emission Profiles

Accurate CMB source apportionment results require realistic source profiles. Mobile source particulate emissions are among the most difficult to measure with respect to emission rates and chemical composition. This difficulty arises from: 1) the different mobile source types (e.g., spark-ignition gasoline-fueled cars and trucks, heavy-duty diesel-fueled trucks, diesel buses, 2- and 4-stroke gasoline engines); 2) the inadequate characterization of certain sub-groups within the broader categories in the motor vehicle fleet (e.g., high emitting spark-ignition gasoline-fueled cars); 3) seasonal and annual changes in fuel composition and emission control technology; 4) the large range of emission characteristics of individual emitters within each mobile source category; 5) variation of emissions under different operating conditions; 6) the several emission points on each vehicle (i.e., tailpipe, fuel evaporation, tire wear, brake wear, resuspended dust); 7) contributions from non-road engines including 2- and 4-stroke gasoline engines and diesel engines; and 8) the emission of a mixture of primary particles, semi-volatile organic compounds, and secondary particle precursors.

Recently, it has been recognized that sub-groups within the light-duty category of vehicles (e.g., high-emitters and smokers) are not of equal importance in contributing to ambient particulate levels. However, the predominance of each type of emitter (i.e., smokers versus high-emitters) in terms of their relative contributions to ambient loadings remains to be firmly established. In addition, there are no distinct divisions established that delineate between high emitters and others. There is also a lack of understanding of the contributions made by non-road emissions from engines in agricultural, construction, lawn and garden equipment and recreational marine vehicles. Finally, locomotives, aircraft, and commercial marine vehicles all contribute to ambient particulate levels and must be considered.

Source profiles for mobile vehicles are developed using several methodological approaches. Particulate and gaseous emissions from light- or heavy-duty vehicles can be sampled using a diluted exhaust sampling system coupled with a dynamometer to simulate different vehicle operating conditions (e.g., Lowenthal *et al.,* 1994; Sagebiel *et al.*, 1997; Cadle *et al.,* 1997; Mulawa *et al.*, 1997). An alternative approach is to take roadside samples that represent actual mixtures of vehicles in operation (e.g., Watson *et al.*, 1994b). A similar approach to roadside sampling is the sampling of the aerosol and gaseous composition of the air within the confines of a roadway tunnel (e.g., Pierson *et al.,* 1996; Gertler *et al.*, 1996). In both of these latter situations, samples are obtained where the air is dominated by motor vehicle emissions although stationary source emissions are also present but at lower levels. For roadside sampling the sampler is usually located on the sidewalk or road shoulder as close to the nearest traffic lane as is possible.

In each of these methodological approaches, particulate emissions are collected on substrates that are then submitted to chemical analyses. Typical chemical species include metals and ions (Chow *et al.*, 1997), polycyclic aromatic hydrocarbons (PAHs) (Lowenthal *et al.*, 1994), semi-volatile hydrocarbons (Zeilinska *et al.*, 1996), dioxins and furans (Gertler *et al.*, 1998).

## 4. Examples of Recent Source Apportionment Studies

Source apportionment studies are used to gain understanding of the relative contributions of different sources of pollution to the composition of the observed ambient loadings. Numerous studies have been undertaken over the years at various scales ranging from the local to the regional. In this section the findings of a number of recent source apportionment studies will be reviewed to provide some of the most current information available on the apportionment of particulate matter to its sources. It should be noted that when PM proportions are attributed to mobile sources in the studies reviewed in this section, this refers to the PM directly emitted from motor vehicles.

## 4.1. THE U.S. EPA NATIONAL AIR POLLUTANT EMISSION TRENDS, 1990-1996

In December 1997, the U.S. EPA Office of Air Quality Planning and Standards released the latest edition of the National Air Pollutant Emission Trends, 1990-1996 document (U.S. EPA, 1997). In this document an assessment of regional $PM_{2.5}$ ambient concentrations and patterns of source contributions for the eastern and western U.S. are presented.

Within the area designated as the eastern U.S. (east of the western borders of Minnesota, Iowa, Missouri, Arkansas, and Mississippi) seven monitoring site were chosen to provide a characterization of the regional-scale $PM_{2.5}$. The three non-urban locations show regional background levels of $PM_{2.5}$ of 5.1 μg $m^{-3}$ in northern Minnesota, 9.5 μg $m^{-3}$ in New England, and 11.4 μg $m^{-3}$ in the Appalachian and Mid-Atlantic region. The three sites designated as urban have annual averages of 14.9 μg $m^{-3}$ for Rochester, NY, 16.2 μg $m^{-3}$ for Boston, MA, and 19.2 μg $m^{-3}$ for Washington, DC. The urban site averages are close to or exceed the new annual average $PM_{2.5}$ National Ambient Air Quality Standard (NAAQS) of 15 μg $m^{-3}$ (U.S. EPA, 1998). $PM_{2.5}$ shows relatively consistent composition and is dominated by sulfate of secondary origin and carbonaceous particles emitted directly from combustion processes, including motor vehicles. The high sulfate can be attributed to the higher $SO_2$ emissions throughout much of the east.

In the western U.S. region (defined as the area to the west of the eastern borders of North Dakota, South Dakota, Nebraska, Kansas, Oklahoma, and Texas) eight monitoring sites were examined to provide a measure of regional $PM_{2.5}$ levels and their spatial variability. The four non-urban locations show a range in the annual average $PM_{2.5}$ level between 3.1 μg $m^{-3}$ in the central Rocky Mountains to 4.5 μg $m^{-3}$ in the Sierra Nevada. The regional background for all the sites is 4.1 ±0.7 μg $m^{-3}$, indicating a quite low and stable background level over a very large area in the West. This value is roughly one-half of the overall background value found in the eastern U.S. of 8.7 ±3.2 μg $m^{-3}$ that is an average of the three measurements for northern Minnesota (5.1 μg $m^{-3}$), New England (9.5 μg $m^{-3}$), and the Appalachian and Mid-Atlantic region (11.4 μg $m^{-3}$). The four western sites designated as urban (Spokane, WA, San Joaquin Valley, CA, South Coast (Los Angeles and environs), and West Phoenix, AZ) have annual $PM_{2.5}$ averages of 11.0 μg $m^{-3}$ for Spokane to a high of 30 μg $m^{-3}$ in the San Joaquin Valley, CA. The South Coast area is also high at 28 μg $m^{-3}$, while West Phoenix is similar to Spokane with an annual average of 13.5 μg $m^{-3}$. Both the South Coast region and the San Joaquin Valley areas exceed the new $PM_{2.5}$ NAAQS, while the other two urban areas (Spokane and Phoenix) are close to the standard. It should be mentioned that although the San Joaquin Valley area is designated as "urban" it is in fact a mixture of urban areas inset into a region of intensive agriculture.

The composition of the $PM_{2.5}$ in the west differs significantly from the east with higher ammonium nitrate concentrations in certain areas, especially near cities and in the intensive agricultural region of the San Joaquin Valley, CA. Within the western region Eldred *et al.* (1998) found that carbonaceous particles, primarily attributed to motor vehicles and vegetative burning, are relatively more dominant in the northwest.

## 4.2. NORTHEAST STATES FOR COORDINATED AIR USE MANAGEMENT (NESCAUM)

In September 1988, the Northeast States for Coordinated Air Use Management (NESCAUM), in cooperation with the University of California at Davis, initiated a fine particle monitoring network in eight eastern states (CT, MA, ME, NH, NJ, NY, RI, and VT). The latest data available are for 1995 for five sites including two sites in and near Rochester, NY, one rural site in central Massachusetts, and two sites in and near Boston, MA.

The relatively broad categories that define the material balance cannot be used to specifically identify the contributions from mobile source emissions relative to one another, or relative to other sources of carbonaceous aerosol. The mobile source contributions are aggregated in with the carbonaceous material category as they are direct contributors of carbonaceous components and organic compounds of higher molecular weights. In addition, mobile sources will add some component to the geological material through the addition of resuspended road and shoulder dust. Mobile sources will also impact indirectly as contributors to the secondary aerosol particulate matter through their emissions of gaseous species such as oxides of nitrogen.

The annual averages are close to the new annual average national ambient air quality standard of 15 $\mu g\ m^{-3}$ for $PM_{2.5}$. Salmon *et al.* (1997) report that at the regional background sites in Brockport, NY, and Quabbin Reservoir, MA, the ammonium sulfate and carbonaceous particles are of about equal importance. These two categories each account for approximately 35% of the total fine particle mass. The urban sites show similar levels of sulfate as the background sites, but there is a definite increase in the carbonaceous components in the urban environments. In Kenmore Square in Boston the elemental and organic carbon components account for the majority of the observed fine particle mass concentration. This increase could be attributable to mobile sources in the absence of other obvious industrial combustion sources.

## 4.3. SAN JOAQUIN VALLEY 1995 INTEGRATED MONITORING STUDY (IMS 95)

The IMS 95 study is part of the California Regional Particulate Air Quality Study (CRPAQS), a multiyear effort to understand the causes of elevated suspended particulate concentrations and to evaluate ways to reduce them in central California (Roth and Watson, 1993). The incidence of high PM levels in the Central Valley follows a distinct seasonal trend, with the highest levels recorded in the fall and winter seasons.

Magliano *et al.* (1998) have recently performed a CMB modeling exercise using the IMS 95 database. For the fall 1995 data, Magliano *et al.* (1998) found geological material to be the dominant contributor at all sites accounting for, on average, 60% of the total $PM_{10}$. Secondary ammonium nitrate was the second largest contributor at all sites as well in the fall, comprising on average 16% of the total mass. Magliano *et al.* (1998) attributed 7% of the $PM_{10}$ to mobile sources. However, they also could not successfully apportion excess carbon contributions that suggested the profiles used may not have been adequate to resolve the sources of organic carbon in the study region. In the winter sampling period in the San Joaquin Valley, when $PM_{2.5}$ was generally 70% to 80% of the $PM_{10}$ mass Magliano *et al.* (1998) found the $PM_{2.5}$ sources were very similar to those for $PM_{10}$. Mobile sources were the second largest contributor at urban sites ranging from 18% at Fresno, CA, to 21% at Bakersfield, CA. There was no attempt to separate the diesel versus gasoline contributions in this study. Mobile source contributions were much lower at the rural sites, accounting for 4% to 12% at Kern Wildlife Refuge and Chowchilla, CA, respectively.

## 4.4. NORTHERN FRONT RANGE AIR QUALITY STUDY (NFRAQS), 1996-1997

The Northern Front Range Air Quality Study (NFRAQS), 1996-1997, is one of the most recent studies undertaken to understand aerosol composition ($PM_{10}$ and $PM_{2.5}$) and to apportion the sources of the aerosol to the observed ambient loadings in the city of Denver, CO. Source contributions to the carbonaceous aerosol were estimated by the CMB model using enhanced organic speciation in source emissions and at receptors to apportion motor vehicle, wood burning, and meat-cooking contributions. The enhanced organic compound measurements allowed organic, elemental and total carbon to be apportioned to light-duty and heavy-duty diesel exhaust, and three categories of light-duty gasoline vehicle (LDGV) exhaust: 1) hot stabilized operation; 2) cold starts; and 3) high particle emitters.

The results of this work challenge the conventionally held view that diesel vehicles contribute a proportionally greater amount to the total $PM_{2.5}$ mass than gasoline powered vehicles (e.g., Schauer *et al.*, 1996) for mobile source contributions. Watson *et al.* (1998) reported that exhaust from gasoline powered vehicles and geological materials are the largest direct $PM_{2.5}$ contributors at sites in and near Denver. However, the extension of these results to other areas needs to be tested as the high altitude and winter-time conditions under which the sampling was undertaken may be important contributing factors. On average, CMB model source contribution estimates ascribe 55% of $PM_{2.5}$ in Denver to motor vehicle primary emissions. At non-urban locations in the northern NFRAQS study domain, ammonium nitrate and ammonium sulfate are the largest contributors.

According to Watson *et al.* (1998) vehicle exhaust was the largest $PM_{2.5}$ carbon contributor, constituting ~85% of $PM_{2.5}$ carbon in the Denver metropolitan area and ~75% of $PM_{2.5}$ carbon at more rural sites. The source type that appears to be the greatest contributor to the carbonaceous aerosol, about 60% of $PM_{2.5}$ carbon, are those with emissions similar to light-duty gas vehicles (LDGVs). Pierson and Russell (1979) also reported that the carbon aerosol in Denver appeared to be associated with gasoline powered vehicles. The LDGV source type identified by Watson *et al.* (1998) consists of LDVGs, but also includes any machinery that use gasoline for fuel, for example, gasoline powered lawnmowers. The chemical composition of the PM from these emission sources remains to be characterized at a level that will allow for their separation through application of the CMB. The modeled contribution of $PM_{2.5}$ by LDGVs at urban Denver sites, are 2.5 to 3 times the estimated diesel exhaust contributions. This result challenges the inventory calculations that estimate diesel vehicles contribute a proportionally higher amount than LDGVs to observed PM levels (Hildemann *et al.*, 1991). However, it may be that the inventory values for diesel contributions are accurate, but the actual per cent contributions from this source change as a function of location.

Within the source type identified as LDVG by the NFRAQS study, sub-classifications specified as cold start emissions and high particle emitters contributed 24% (±4%) and 28% (±5%) of $PM_{2.5}$ carbon, respectively, at the Denver urban core site. The addition of $PM_{2.5}$ from LDVG cold start emissions to the ambient levels exhibited a bi-modal distribution with contributions twice as high in the morning and overnight than during the afternoon.

The CMB analysis of the NFRAQS data by Fujita *et al.* (1998) showed that hot stabilized emissions from presumably well-maintained vehicles were minor contributors to $PM_{2.5}$ carbon and total mass and in the range of 3-6%. The $PM_{2.5}$ carbon attributed to a re-suspended road dust component was ~4%. Taking into account the exhaust-generated emissions and the re-suspended road dust, the combined vehicle-related contributions to $PM_{2.5}$ carbon in urban Denver approached 85%. Managing the mobile source contributions to $PM_{2.5}$ appears to be the key to meeting the $PM_{2.5}$ NAAQS in Denver and reducing the intensity of the "brown cloud" that is the highly visible reminder of air quality degradation in this city.

## 5. Conclusions

Mobile sources are major contributors to $PM_{10}$ and $PM_{2.5}$ in the US. Reducing mobile source contributions will be critical for improving air quality and meeting the ambient air quality standards. In order to accomplish this goal, we need to accurately apportion the mobile source contribution to the observed PM levels. However, in many areas of the US the contribution of mobile sources to PM needs to be defined.

The identification of the relative contribution of the mobile source component to ambient $PM_{10}$ and $PM_{2.5}$ through the receptor modeling approach has been improved significantly by research that has identified key chemical components that have been shown to be clearly linked with mobile source emissions. The greatest advance in this respect is the identification and association of certain polycyclic aromatic hydrocarbons (PAHs) with spark ignition and heavy duty diesel emitters. The application of these newly-developed profiles beyond the NFRAQS study would make an important contribution to reconciling the disparity between emission inventory data that suggests diesels are the predominant source and previous CMB studies that support the inventory data (e.g., Schauer *et al.*, 1996). It will hinge on new studies making the appropriate measurements that will allow these source profiles to be used. At this time it would appear that a more definitive apportionment of the mobile sources on a national basis requires data on speciation of PAHs in ambient air samples.

## Acknowledgements

The authors would like to acknowledge the United States Environmental Protection Agency for support of this work under cooperative agreement CX825775-01-0.

## References

Baumgard, K. J. and Johnson, J. H.: 1996, *The effect of fuel and engine design on diesel exhaust particle size distributions*, SAE Technical Paper 960131, SAE International, Warrendale, PA.

Cadle, S. H., Mulawa, P. A., Ball, J., Donase, C., Weibel, A., Sagebiel, J. C., Knapp, K. T. and Snow R.: 1997, Particulate emission rates from in-use high-emitting vehicles recruited in Orange County, California, *Environ. Sci. Technology* **31**, 3405-3412.

Chow, J. C., Egami, R. T., *et al.*: 1997, *San Joaquin Valley 1995 Integrated Monitoring Study Documentation, Evaluation, and Descriptive Data analysis of $PM_{10}$, $PM_{2.5}$ and Precursor Gas Measurements,* Prepared for California Regional Particulate Air Quality Study, Technical Support Division, California Air Resources Board, Sacramento, CA, by Desert Research Institute, Reno, NV, March 3, 1997.

Eldred, R. A., Feeney, P. J. and Wakabayashi, P. H.: 1998, *The major components of $PM_{2.5}$ At remote sites across the United States,* in J. C. Chow and P. Koutrakis (eds), $PM_{2.5}$: A Fine Particle Standard, *Proceedings of the Specialty Conference, Air & Waste Manage. Assoc.*, Pittsburgh, PA, pp. 38-58.

Fujita, E., Watson, J. G., Chow, J. C., Robinson, N. F., Richards, L. W. and Kumar N.: 1998, *Northern Front Range Air Quality Study, Volume C: Source Apportionment and Simulation Methods and Evaluation,* Prepared for Colorado State University, Cooperative Institute for Research in the Atmosphere, by Desert Research Institute, Reno, NV, 1998.

Gertler, A. W., Pierson W. R.: 1996, Recent measurements of mobile source emission factors in North American tunnels, *Sci. Total Environ.* **189/190**, 107-113.

Gertler, A. W., Fujita, E. M., Pierson, W. R. and Wittorff D. N.:1996, Apportionment of NMHCtailpipe vs. non-tailpipe emissions in the Fort McHenry and Tuscarora Mountain Tunnels. *Atmos. Environ.* **30**, 2297-2305.

Gertler, A. W., Sagebiel, J. C. and Pierson, W. R.: 1996, *On-road chassis dynamometer and engine-out dynamometer measurements of heavy-duty vehicle emission factors, Proceedings of the A&WMA Conference on the Emission Inventory: Programs & Progress, RTP, NC,* 11-13 October 1995, Air & Waste Management Association, Pittsburgh, PA.

Gertler, A. W., Sagebiel, J. C., Dippel W. A. and Farina R. J.: 1998, Measurements of dioxin and furan emission factors from heavy-duty diesel vehicles, *J. Air & Waste Manage. Assoc.* **48**, 276-278.

Gertler, A. W., Sagebiel, J. C., Wittorff, D. N., Pierson, W. R., Dippel, W. A., Freeman, D. and Sheetz, L.: 1997, *Vehicle Emissions in Five Urban Tunnels,* Final Report under CRC Project No. E-5, prepared for the Coordinating Research Council, Inc., Auto/Oil Improvement Research Program, National Renewable Energy Laboratory, and South Coast Air Quality Management District Southern Oxidants Study, by Desert Research Institute, Reno, NV, March 1997.

Heath, Jr., C. W., Pope III, C. A., Thun, M. J.: 1995, Particulate air pollution as a predictor or mortality in a prospective study of U.S. adults, *American Journal of Respiratory and Critical Care* **151**, 3, 669-674.

Hildemann, L. M., Markowski, G. R. and Cass, G. R.: 1991, Chemical composition of emissions from urban sources of fine organic aerosol, *Environ. Sci. Technol.* **25**, 744-759.

Ingalls, M. N.: 1989, *On-road vehicle emission factors for measurements in a Los Angeles Area Tunnel,* Paper **89-**137.3, *Proceedings of the 82nd Annual Meeting, Air & Waste Management Association,* Anaheim, CA, June 1989, Air & Waste Management Association, Pittsburgh, PA.

Ingalls, M. N., Smith, L. R. and Kirksey, R. E.: 1989, *Measurements of On-Road Vehicle Emission Factors in the California South Coast Air Basin – Volume I: Regulated Emissions,* Report No. SwRI-1604, prepared by the Southwest Research Institute for the Coordinating Research Council, Inc. Atlanta, GA, June 1989. NTIS Document PB89220925.

Lowenthal, D. H., Wittorff, D., Gertler, A. W. and Sakiyama, S.: 1997, CMB source apportionment during REVEAL, *J. Environ. Engineering* **123**, 80-87.

Lowenthal, D. H., Zielinska, B., Chow, J. C., Watson, J. G., Gautam, M., Ferguson, D. H., Neuroth, G. R. and Stevens, K. D.: 1994, Characterization of heavy-duty diesel vehicle emissions, *Atmos. Environ.* **28**, 731-743.

Magliano, K. L., Ranzieri, A. J. and Solomon, P. A.: 1998, *Chemical Mass Balance modeling of the 1995 Integrated Monitoring Study database,* in J. C. Chow and P. Koutrakis (eds), $PM_{2.5}$: A Fine Particle Standard, *Proceedings of the Specialty Conference, Air & Waste Manage. Assoc.*, Pittsburgh, PA, pp. 824-835.

Moosmüller, H., Gillies, J. A., Rogers, C. F., DuBois, D. W., Chow, J. C., Watson, J. G. and Langston, R.: 1998, Particulate emission rates for unpaved shoulders along paved roads, *J. Air Waste Manage. Assoc.* **48**, 174-185.

Mulawa, P. A., Cadle, S. H., Knapp, K., Zweidinger, R., Snow, R., Lucas, R. and Goldbach, J.: 1997, Effect of ambient temperature and E-10 fuel on primary exhaust particulate matter emissions from light-duty vehicles, *Environ. Sci. Technol.* **31**, 1302-1307.

Nicholson, K. W., Branson, J. R., Geiss, P. and Cannell, R. J.: 1989, The effects of vehicle activity on particle resuspension, *J. Aerosol Sci.* **20**, 1425-1428.

Pierson, W. R., Brachaczek W. W.: 1976, *Particulate matter associated with vehicles on the road,* Society of Automotive Engineers, Inc., Automotive Engineering Congress and Exposition, Detroit, MI, February 23-27, 1976.

Pierson, W. R., Brachaczek W. W.: 1983, *Particulate matter associated with vehicles on the road, II, Aerosol Sci. Technol.* **4**, 1-40.

Pierson, W. R., Gertler, A. W., Robinson, N. F., Sagebiel, J. C., Zielinska, B., Bishop, G. A., Stedman, D. H., Zweidinger, R. B. and Ray, W. D.: 1996, Real world automotive emissions Summary of studies in the Fort McHenry and Tuscarora Mountain tunnels, *Atmos. Environ* **30,** 2233-2256.

Pope III, C. A., Dockery, D. W., Schwartz, J.: 1995, Review of epidemiological evidence of health-effects of particulate air pollution, *Inhalation Toxicology* **7** (1), 1-18.

Ristovski, Z. D., Morawksa, L., Bofinger, N. D. and Hitchins, J.: 1998, Submicrometer and supermicrometer particulate emissions from spark ignition vehicles, *Environ. Sci. Technol.* **32**, 3845-3851.

Roth, P. M., Watson J. G.: 1993, *A Proposed $PM_{10}$ Program for the San Joaquin Valley,* prepared for the San Joaquin Valleywide Air Pollution Study Agency, Fresno, CA, by ENVAIR, San Anselmo, CA, and Desert Research Institute, Reno NV, April 1993.

Sagebiel, J. C., Zielinska, B., Walsh, P. A., Chow, J. C., Cadle, S. H., Mulawa, P. A., Knapp, K.T., Zweidinger, R. B. and Snow, R.: 1997, $PM\text{-}_{10}$ exhaust samples collected during IM-240 dynamometer tests of in-service vehicles in Nevada, *Environ. Sci. Technol.* **31**, 75-83.

Salmon, L. G., Cass, G. R., Pedersen, D. U., Durant, J. L., Gibb, R., Lunts, A. and Utell, M.: 1997, *Determination of fine particle concentration and chemical composition in the northeastern United States, 1995,* Progress Report to Northeast States for Coordinated Air Use Management (NESCAUM), October 1997.

Sawyer, R. F. and Johnson, J. H.: 1995, *Diesel emissions and control technology,* In *Diesel Exhaust, A Critical Analysis of Emissions, Exposure, and Health Effects,* Health Effects Institute, Cambridge, MA.

Schwartz, J., Dockery, D. W. and Neas, L. M.: 1996, Is daily mortality associated specifically with fine particles? *J. Air Waste Manage. Assoc.* **46**, 927-939

Seaton, A., MacNee, W., Donaldson, K., Godden, D.: 1995, Particulate air pollution and acute health effects, *The Lancet* **345** (8943), 176-178.

Solomon. P. A., Fall, T. L., Salmon, Cass, G. R., Gray, H. A., Davidson, A.: 1989, Chemical characteristics of $PM_{10}$ aerosols collected in the Los Angeles area, *JAPCA* **39**, 154-163.

Truex, T. J., Durbin, T. D., Smith, M. R., Norbeck, J.: 1998, *Particulate emissions from in-use,light-duty vehicles in the South Coast Air Quality Management District,* in J.C. Chow and P. Koutrakis (eds), $PM_{2.5}$: A Fine Particle Standard, *Proceedings of the Specialty Conference, Air & Waste Manage. Assoc.,* Pittsburgh, PA, pp. 559-570.

U.S. Environmental Protection Agency: 1997, *National Air Pollutant Emission trends, 1990-1996,* EPA-454/R-97-011, U.S. Environmental Protection Agency, Research Triangle Park, NC.

Vedal, S.: 1997, Ambient particles and health: lines that divide, *Journal of the Air & WasteManagement Association* **47,** 551-581.

Watson, J. G., Chow, J. C., Henry, R. C., Kim, B. M., Pace, T. G., Meyer, E. L. and Nguyen, Q.: 1990, The USEPA/DRI Chemical Mass Balance Receptor Model, CMB 7.0, *Environmental Software* **5**, 38-49.

Watson, J. G., Chow, J. C., Lowenthal, D. H., Pritchett, L. C., Frazier, C. A., Neuroth, G. R. and Robbins, R.: 1994, Difference in the carbon composition of source profiles for diesel- and gasoline-powered vehicles, *Atmos. Environ.* **28**, 2493-2505.

Watson, J. G., Fujita, E. M., Chow, J. C., Zielinska, B., Richards, L. W., Neff, W. and Dietrich, D.: 1998, *Northern Front Range Air Quality Study Final Report,* prepared for Colorado State University, Cooperative Institute for Research in the Atmosphere, by Desert Research Institute, Reno, NV, 1998.

Weingartner, E., Keller, C., Stahel, W. A., Burtscher, H. and Baltensperger, U.: 1997, Aerosol emission in a road tunnel, *Atmos. Environ.* **31** (3), 451-461.

Zielinska, B., Sagebiel, J. C., Harshfield, G., Gertler, A. W., Pierson W. R.: 1996, Volatile organic compounds in the C2 - C20 range emitted from motor vehicles: measurement methods, *Atmos. Environ.* **30**, 2269-2286.

# CHARACTERISATION OF INDIVIDUAL AEROSOL PARTICLES FOR ATMOSPHERIC AND CULTURAL HERITAGE STUDIES

R. VAN GRIEKEN, K. GYSELS, S. HOORNAERT, P. JOOS, J. OSAN, I. SZALOKI and A. WOROBIEC
*Department of Chemistry, University of Antwerp (UIA), Universiteitsplein 1, B-2610 Antwerp, Belgium*

**Abstract.** Microanalysis of individual particles allows straightforward and advanced characterisation of environmental samples. The most obvious technique to study large micro-particle populations is still electron probe X-ray microanalysis (EPXMA). Recently, technical and methodological progress has been made to remedy some of the limitations of conventional EPXMA, as, for example, in the detection of low Z-elements. Recent examples of the use of EPXMA in various environmental fields are presented, namely concerning atmospheric deposition of micropollutants and nutrients to the sea, characterisation of aerosols in the context of their effect on Global Change (remote continental and biogenic aerosols) and aerosol deposition and soiling of paintings in museums.

**Keywords:** aerosols, environment, EPXMA, marine, microanalysis, museums

## 1. Introduction

Characterisation of aerosol samples at the level of individual particles, using micro-analytical techniques, generally permits to obtain more unambiguous and detailed information than bulk analysis, and so it simplifies recognition of the sources of pollution and their processes. Electron probe X-ray microanalysis (EPXMA) is a common micro-analytical technique, which can be used fully automatically. Especially when it is combined with hierarchical cluster analysis, the characteristic properties of the particle groups can be derived straightforwardly. Hence, EPXMA is highly relevant for aerosol characterisation in a number of applied environmental fields.

One of these applications concerns the problem of the deterioration of paintings in museums (Brimblecombe, 1990). Deposition of particles can lead to visual degradation. Chemical reactions with the deposited particles can additionally cause further damage. Camuffo (1998) has already discussed in detail the deposition mechanisms of particles on walls and objects. Several chemical compounds can be considered as threatening for the preservation of works of art. For example, soot can cause significant soiling and visual degradation (Baer and Banks, 1985) and constitutes, together with other organic compounds, a medium for the absorption of damaging gases like $SO_2$. $(NH_4)_2SO_4$ can induce bloom on varnish, while other S-rich material can be responsible for discolouring of the pigments by oxidation to $H_2SO_4$ (De Santis *et*

*Water, Air, and Soil Pollution* **123:** 215–228, 2000.

*al.*, 1992). This process can be catalysed by Fe-rich particles. Detailed knowledge about the concentration and composition of atmospheric particles present in museums can be used to prevent future deterioration of works of art on display. A first example in this work pertains to the field of museum aerosols.

Aerosols in outdoor air are, of course, an even more important application field, in which EPXMA can be used. Next to work environment and urban pollution, a topic of concern, which is touched in this work, is the effect of atmospheric deposition on seawater quality. For example, the North Sea is surrounded by heavily industrialised countries and it has been shown that, for some heavy metals, the supply by the atmosphere exceeds that of all rivers combined (Injuk *et al.,* 1998). The atmospheric supply of nutrients, which potentially cause eutrophication in the North Sea, is nowadays an acute problem in this context.

Also aerosols in remote areas are becoming important again because of the possible climatic effects of aerosols (backscattering and absorption of sunlight and modification of cloud properties might compensate the "greenhouse effect"). In this study, the first preliminary results for a remote continental site in Kazakhstan will be presented. At such a site without industry or significant habitation, aerosol characterisation contributes to the definition of the global "backround" or "baseline" aerosol, i.e., the aerosol which should occur in natural circumstances and on which all pollution is superimposed. In the characterisation of aerosol particles, generally a large amount of soil dust and marine salt particles is found (especially for the coarse fraction), while all other natural and pollution particles are usually numerically unimportant. During winter, however, Kazakhstan is several thousand kilometres remote from soil and sea-salt sources, since the ground is completely covered with snow. Every characterisation of aerosol particles under these circumstances is by definition limited to particles, which are transported over long distances, and to pollution particles. On the other hand, the tropical rain forest is a globally very important ecosystem. It is characterised by intense sources of biogenic gases and aerosols. The primary biogenic aerosol particles consist of different particle-types, including pollen, spores, bacteria, algae, protozoa, fungi, fragments of leaves, excrement and fragments of insects (Simoneit, 1989). A significant fraction of biogenic particles in forested areas comprises secondary aerosol particles formed by gas to particle conversion of organic, nitrogen-or sulphur-related gases. The primary biogenic aerosol can, even in urban areas, represent as much as 50 % in mass, and this shows how the sources of biogenic aerosol particles have been underestimated in the past (Jaenicke and Mathias-Maser, 1992). In the present study, preliminary results for individual particles released to the atmosphere by the Amazon vegetation are presented.

The scope of this article is thus to illustrate the applications of EPXMA in environmental analysis with our recent findings. Moreover, a recent refinement of the EPXMA-technique, namely EPXMA with low-Z element detection using a thin-window detector, will be outlined. Although the thin-window detectors, allowing low-Z element determinations, have been around in X-ray spectrometry for nearly a decade, they have not been applied systematically in single particle analysis (SPA) of aerosols.

## 2. Materials and Methods

### 2.1. SAMPLING OF MUSEUM AEROSOLS

Atmospheric particulate matter has been sampled in three European museums, chosen because of their different architecture and environment. The Correr Museum in Venice, Italy and the Kunsthistorisches Museum in Vienna are both classical museums in historical limestone buildings. They are, however, located in different environments. Venice is influenced by the sea, while in Vienna, no marine influence can be expected at all. The Sainsbury Centre for Visual Arts in Norwich, UK, is a totally different kind of museum. It occupies a very modern building, constructed of glass and aluminium. The museum is located on the campus of the University of East Anglia, in a semi-rural environment.

In all museums, indoor and outdoor size-segregated aerosol samples were collected on polycarbonate by means of cascade impactors for EPXMA. Also, for bulk analysis by energy-dispersive X-ray fluorescence (EDXRF), aerosol samples were collected during the night, using filter units with 0.4 μm pore-size Nuclepore membranes. Assessment of the dry deposition was carried out by attaching filters to the museum walls for a period of six to nine months and subsequent analysis by EDXRF.

### 2.2. SAMPLING OF AEROSOLS IN KAZAKHSTAN

Sampling in Kazakhstan for SPA was performed at a remote astronomical observatory in the Tien Shan mountain region. The station is located on the north slope of the Zailiysky Alatau mountain ridge (43°04' N – 76°58' E) at an altitude of 2760 m above sea level, near the border of Kazakhstan and Kyrgystan. To the south, the Zailiysky Alatau ridge is surrounded by the vast Tien Shan mountains with peaks of more than 4500m. At the north side of the ridge is a vast plain area, about 500 m above sea level, where the city of Almaty is situated and some smaller towns and villages. The influence of these pollution sources on the air quality at the astronomical station is almost non-existent due to the local meteorological conditions. Aerosol samples for SPA were collected on Nuclepore membranes using simple filter units. The filter unit was mounted into a Plexiglass cylinder with a hat-type cover to protect it from any kind of precipitation (rain, snow, etc.). Sampling was interrupted every time the wind direction changed, and the sampling location was under the influence of the main building, and during precipitation (rain or snow) and fog events. The main sampling campaign ran from August 1996 through August 1997. Samples were taken on a weekly basis, with collection times ranging from 2 hours to one day.

## 2.3. SAMPLING OF AEROSOLS OVER THE NORTH SEA

Five sets of marine aerosol samples were collected over the Southern Bight of the North Sea during a cruise of the research ship Belgica, between 28 September and 2 October 1998; contamination by the ship was avoided. The particles were sampled in a nine-stage Berner-impactor on aluminium foil, for later low-Z EPXMA. The air mass trajectories for the sampling period were available.

## 2.4. SAMPLING AEROSOLS IN THE AMAZON BASIN

The sampling was performed during a campaign in March-April 1998 (wet season) as part of the Large-Scale Biosphere-Atmosphere Experiment in Amazonia (LBA). Ground-based aerosol sampling was carried out at Balbina (about 200 km North of Manaus), Brazil. The results that will briefly be discussed below were obtained on cascade impactor samples, using Al-foils as impaction substrates, for later low-Z EPXMA.

## 2.5. AUTOMATED EPXMA MEASUREMENTS

Automated SPA was performed on a JEOL JSM 6300 Scanning Electron Microscope (JEOL, Tokyo, Japan) equipped with backscattered and secondary electron detectors and an EDX detection system. A Si(Li) X-ray detector coupled to a PGT-system (Princeton Gamma Tech, Princeton, NJ, USA) was employed for acquiring the X-ray spectra (accumulation time was 20 s). The typical energy of electron bombarding was 20 keV and the beam current 1 nA. The average diameter of the particles was determined by the backscatter electron (BSE) image. The controlled software localises the particles from the BSE image and performs an X-ray measurement within each particle. The intensities of the characteristic peaks in the spectra are determined by the top-hat filter method. In every sample, 300 to 500 individual particles were analysed. Because of the automated analysis procedure, the resulting data sets are quite large. Therefore, statistical methods are needed for the interpretation of the results. In this work hierarchical cluster analysis was used. The aim of applying cluster analysis was to make a classification of aerosol particles, based upon their chemical composition, into groups of particles with similar chemical composition, to identify the different particle types in the sample.

## 2.6. THIN-WINDOW EPXMA-MEASUREMENTS

The chemical speciation of single aerosol particles requires a quantitative determination of the light elements, like carbon, nitrogen and oxygen. The application of conventional EDX detectors is strongly limited for low-Z element analysis because of the low transmission for low-energy X-rays through the window of EDX detectors. This difficulty can be avoided by the use of windowless or thin-window EDX-detectors. To demonstrate the spectroscopic capability of the thin-window EPXMA, two typical X-ray spectra are shown in Figure 1.

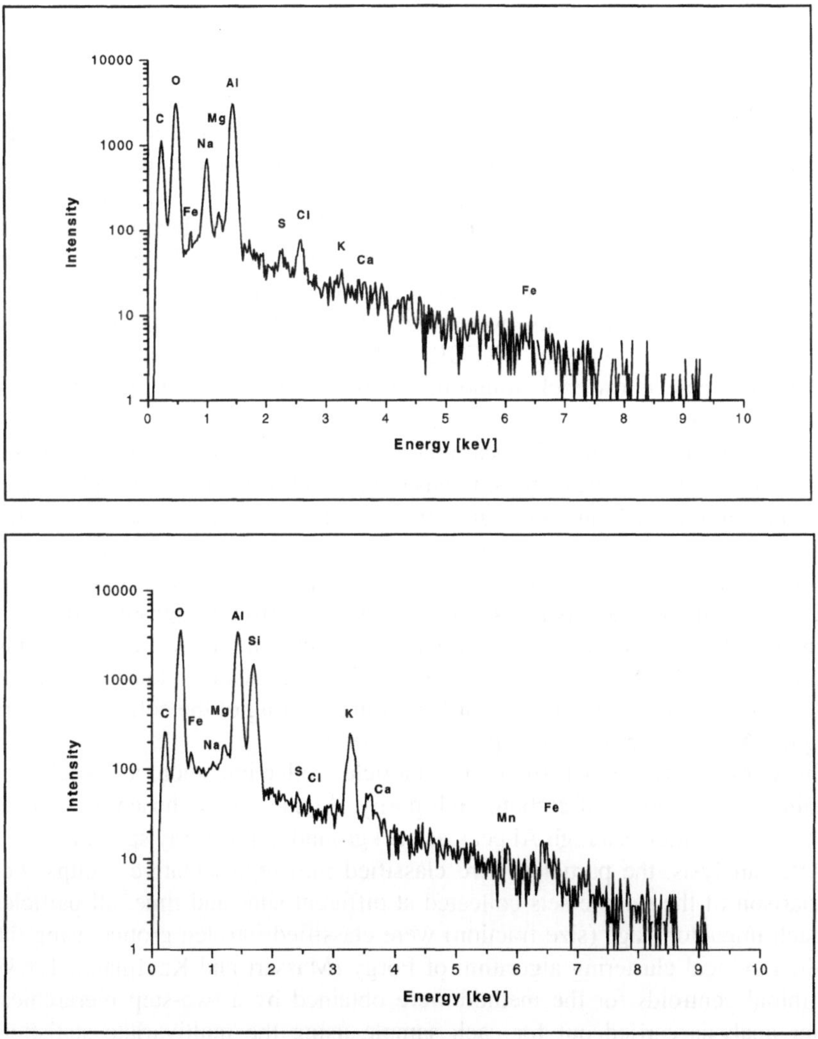


Fig. 1. X-ray spectra recorded by thin-window EPMA equipment, I=0.5 nA, U=10 kV, accumulation time 20 s. (a) organic-type particle with high Na content (b) soil-type particle

The classical quantification methods in EPXMA such as ZAF and $\phi(\rho z)$-based procedures aim to correct for matrix and geometric effects. However, these procedures are strongly limited for light element analysis of individual atmospheric microparticles, because of the strong absorption effect and irregular shape of particles. Also the widely used method of Armstrong (1991), the so-called particle-ZAF algorithm based on the use of bulk standard material, is not ideal for low-Z element analysis of particles.

Therefore, a special quantification algorithm was developed on the basis of Monte Carlo simulation (Gauvin *et al.*, 1995, Hovington, *et al.* 1997) with combination of successive approximations for the mathematical description of the excitation interactions between electrons and atoms, the size and shape effects of the particles and the matrix effects for low-Z elements. The quantification procedure is based on a modified version (Ro *et al.,* 1999) of CASINO Monte Carlo program designed for low-energy beam interactions generating X-ray and electron signals. The simulation code calculates the detected characteristic X-ray intensities of the particle element.

The low-Z measurements were carried out on a JEOL-733 electron microprobe equipped with an OXFORD EDX detector with atmospheric thin window. To achieve low background in the spectra and high sensitivity for light element analysis, a 10 kV accelerating voltage and 0.5 nA beam current were applied. In order to decrease the beam-damage effect on beam-sensitive particles (such as nitrates), each measurement was carried out using a cold sample-holder at the temperature of liquid nitrogen (-193 °C). To reduce the deposition of the oil residues from the vacuum pumps to the sample surface and the detector window, a continuous cleaned $N_2$ gas flow was applied. The size and shape of each individual particle was measured and estimated from a high-magnification secondary electron image. These estimated geometrical data were set as input parameters for the quantification procedure. The net X-ray intensities for the elements were obtained by non-linear least-square fitting of the collected spectra using the AXIL program (Van Espen *et al.,* 1987).

Since the conventional filters for particle collection such as Nuclepore contain a large amount of carbon, high purity, Al sheet was chosen as sample-substrate. This implies a high Al peak as background in the X-ray spectra.

After analysis, the particles were classified into representative groups. For comparison of the sample sets collected at different sites and time, all particles for each impactor stage (size fraction) were classified into ten groups using the non-hierarchical clustering algorithm of Forgy (Massart and Kaufmann, 1983). The initial centroids for the method were obtained by a two-step hierarchical cluster analysis carried out for each sample using the multivariate statistical software package DPP.

## 3. Results and Discussion

### 3.1. AUTOMATED EPXMA TO STUDY ATMOSPHERIC DEPOSITION AND SOILING IN MUSEUMS

In the Correr Museum in Venice, a very high Ca-concentration was detected by EDXRF bulk analysis in all indoor aerosol samples. Also on the filters to measure dry deposition, Ca was the main detectable element. The results of the SPA also indicated that Ca-rich particles and complex Ca-Si-S-Cl particles were the major particle types in the indoor samples. Aluminosilicates constituted a third group. Seasalt, Si-rich, Fe-rich and gypsum particles were identified as minor particle types only. Comparison between the indoor and outdoor abundance of seasalt and aluminosilicates, two particle types of a typical outdoor origin, revealed the indoor seasalt abundance to be lower than the outdoor abundance, while the indoor and outdoor abundances of aluminosilicates are comparable. This suggests another mechanism for aluminosilicate entrance into the museum environment. Soil dust is probably brought in by visitors and later resuspended. A similar comparison can be made for the Ca-rich and the Ca-Si-S-Cl particles. Both particle types have a higher indoor abundance, especially the latter ones. Degradation of wall plaster and cement (and coagulation with other particles) can be considered as important sources of these particle types.

In the Kunsthistorisches Museum in Vienna, Ca was also the main detected element but in much lower concentrations than in the Correr Museum. Ca-rich particles, Ca-Si particles and aluminosilicates were the major particle types identified by SPA. Minor particle types were Fe-rich, Si-rich and $CaSO_4$ particles. Similar particle types were identified in the outdoor environment. At first sight, this indicates a large outdoor influence. However, the similarity between the indoor and outdoor aerosol composition disappears in the smallest size range (<1 μm). This is in contrast with the theory that predicts the indoor/outdoor exchange to be largest for the smallest size range, because of the high mobility and small deposition velocity of small particles. The obtained results indicate that the direct exchange between the outdoor and indoor environment is relatively small. Another mechanism must be responsible for the intrusion of large outdoor particles in the indoor environment. During the sampling campaigns, construction works were carried out on the courtyard of the museum. These can generate many particles, especially those rich in Ca and Si. Visitors could bring them in or they might enter the museum galleries through the air-conditioning shafts giving out in the rooms on the first floor. The Ca dry deposition was highest in those rooms, which have plastered walls. This indicates again the importance of degradation of wall plaster as a source of indoor Ca-rich particles.

At the Sainsbury Centre for Visual Arts, S was the major element present in the bulk aerosol samples. The results of the SPA were totally different for the summer and the winter. In winter, mainly low-Z elemental and S-rich particles were encountered in the smallest size range (<2 μm) and aluminosilicates in the largest size ranges (>2 μm). The indoor summer samples were dominated by aluminosilicates, seasalt and Si-S-Cl-Ca particles. The latter were found to be heterogeneous and probably result from coagulation. The composition of the outdoor samples was very similar to that of the indoor samples. This indicates that the outdoor influence is very large. S was the main element detected in the dry deposition samples. The dry deposition inside display cases was 10 to 100 times lower than on the museum walls.

It can be concluded that the environment and architecture of the selected museums are important factors determining the concentration and composition of the indoor aerosol. In the Correr Museum, degradation of wall plaster constitutes an important source of indoor particulate matter rich in Ca. In the Kunsthistorisches Museum, particles originating from construction works can enter the museum through the air-conditioning system. In the Sainsbury Centre for Visual Arts, practically all the indoor particles originated from outdoors.

From the point of view of conservation, the situation was worst in the Correr museum where the high concentration of Ca-rich material can lead to significant soiling. The concentration of chemically damaging particles was not particularly high in any museum. Moreover, in the Sainsbury Centre, most objects are placed inside display cases, which, as indicated by the dry deposition results, provide protection against potentially harmful particles.

## 3.2. AUTOMATED EPXMA TO STUDY REMOTE AEROSOLS IN KAZAKHSTAN

The preliminary results of a one year sampling campaign at a remote mountain station in the Tien Shan mountain region in Kazakhstan are presented here.

Hierarchical cluster analysis was applied to all weekly samples separately. The relative abundances of the different particle types, averaged over all samples, are shown in Table I.

Aluminosilicates, the dominant particle type, originate naturally from the suspension of soil dust in the air or anthropogenically by the emissions of fly ash from coal combustion. The distinction between the two sources can be made based upon the morphology of the particles. Preliminary results indicate that mainly irregular shaped aluminosilicates are encountered, so that these particles are mainly suspended soil dust. Probably, the Si-rich particles and the different Ca-rich particles can be mainly attributed to the natural suspension of soil dust as well. S-rich particles represent a secondary aerosol, originating from natural sources (soils, volcanoes, biomass burning) or anthropogenic combustion processes. The various K-rich particles are identified as biogenic material such as pollen, spores and fragments of plants. They are mainly observed during the summer months, which supports this assumption.

TABLE I
Aerosol particle types at the Kazakhstan site

| Particle type | Main elements | Average relative abundance (%) |
|---|---|---|
| Aluminosilicates | Si Al Fe | 52 |
| Si-rich | Si | 11 |
| Ca-rich | Ca | 5.1 |
| | Ca S | 5.6 |
| | Ca Si | 1.8 |
| | Ca S Si | 3.1 |
| S-rich | S | 7.2 |
| | S Si | 0.5 |
| | S Zn | 0.1 |
| Low-Z | | 6.3 |
| Fe-rich | Fe | 1.0 |
| | Fe Si | 2.0 |
| | Fe S | 0.2 |
| K-rich | K | 0.2 |
| | K S | 1.9 |
| Ti-rich | Ti | 0.2 |
| | Ti Si | 0.2 |
| Others | Pb | 0.5 |
| | Pb Zn | 0.4 |
| | Cl | 0.4 |

The relative particle type abundance does, however, not give a complete picture; a higher abundance of a specific particle type in a sample does not necessarily imply a higher absolute abundance of this particle type. Due to practical limitations, however, only a limited instrumental set-up could be transported to the sampling site and no information could directly be gathered on the aerosol particle number concentration (number of particles per volume of air). However, to have an idea about the amount of aerosol particles present, an estimate was made of the particle number concentration. This was possible since the samples were simple filter samples, for which the assumption of a homogeneous particle loading on the filter is valid. So knowing the average number of particles per unit area of filter (a value which can be obtained from the output of the automated EPXMA), the total sampled area of the filter and the amount of air that has been sucked through the filter, the absolute number of particles per unit volume of air can be calculated. Of course, this calculation only accounts for particles that could be detected by the instrument (e.g., volatile particles might be lost) and which were larger than the minimum particle diameter which was set to be detected (in this case 0.3 μm).

The particle number concentration, derived in this way for all samples, is shown in Figure 2. Most of the time, the value is less than 2 x $10^6$ particles per $m^3$. In spring, however, an enormous increase in the particle number concentration is observed. This can be explained by the melting of the snow cover. During winter, the soil in the sampling region is covered by a thick snow layer. The aerosols present during this period therefore originate mainly from long-range transport of aerosols produced at distant sources. When the snow layer disappears, the soil remains uncovered, almost without any vegetation, and is therefore subject to wind erosion. Once the vegetation starts to grow, this effect is reduced and somewhat lower particle number concentrations are observed (Figure 2).

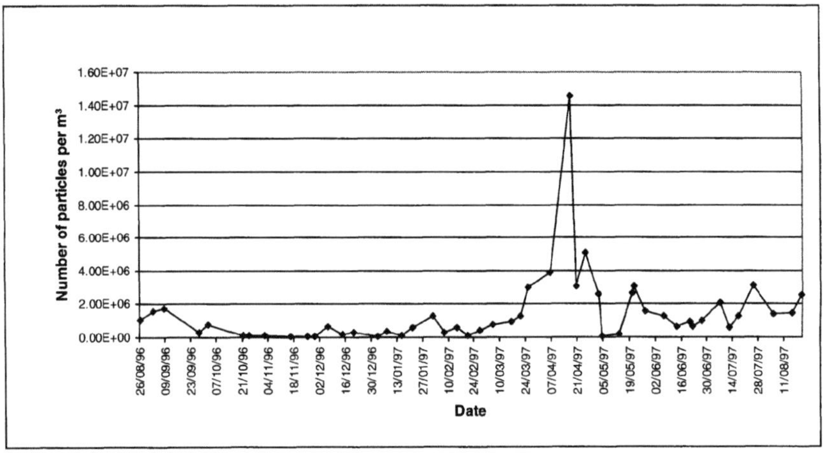


Fig. 2. Particle number concentration (number of particles/m³) for the Kazakhstan site

Knowing the particle number concentration and the relative abundance of the different particle types for each sample, the absolute abundance (number of particles per volume of air) can be calculated for each particle type. In this way and, since the sampling was done on a weekly basis during one year, the seasonal variation of the absolute abundance of the particle types can be investigated. To reduce the effect of short-time variations (due to different meteorological conditions, etc.) the data were smoothed using a two-period moving average. In Figure 3, the annual variation of the absolute abundance of the S-rich particles is shown. While it appeared that the aluminosilicate abundance almost spanned three orders of magnitude with the lowest values during the winter months (background values due to long distance-transport of soil particles), the S-rich particles show smaller changes in their absolute abundance (typically around one order of magnitude). Also the annual variation is rather different and does not show a regular pattern. This indicates that those particles mainly originate from distant sources and consequently are better mixed in the air masses.

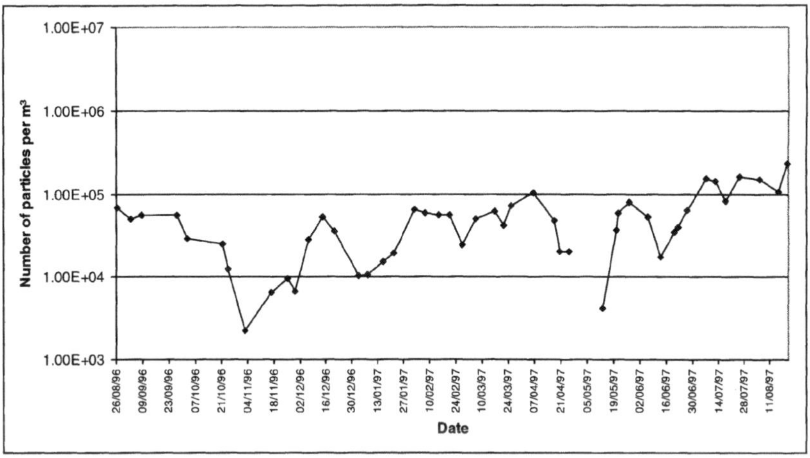


Fig. 3. Seasonal variation of the absolute abundance of the S-rich particles (smoothed data) at the Kazakhstan site

## 3.3. LOW-Z EPXMA OF NORTH SEA AEROSOL PARTICLES

In earlier work, several hundred thousands of individual North Sea aerosol particles were studied by automated EPXMA (Rojas and Van Grieken, 1992) and hundreds of bulk samples were analysed (Injuk *et al.*, 1998) to assess the atmospheric flux of heavy metals to the North Sea and the major sources. Low-Z elements could, however, not be seen with conventional EPXMA. As a preliminary result of low-Z EPXMA, Table IIa shows the average concentrations for seven particle types obtained for the 2-4 μm impactor stage during 5 subsequent sampling episodes; and the average diameter and the abundance variations of the particle classes with time are shown in Table IIb. It appears now that all particle types contain at least 5% of C. Clearly all particles have undergone some coagulation with soot or organic particles, for example. Both seasalt and aluminosilicates incorporate 10% C. N is present as rather pure $NaNO_3$ (overall abundance of 6.7 %) but also, between 0.6 and 11%, in all other particle types. $NaNO_3$ nitrate particles were observed with a relatively higher abundance in the earlier samples. These types of particles could be formed by chemical interaction between drops containing seasalt and gaseous $HNO_3$ in the troposphere (Kerminen *et al.*, 1998; Karlsson and Ljungström, 1998). Most of the $NaNO_3$ type particles had a hemispherical shape, indicating that these particles had been in liquid form during the sampling process. Small $Na_2SO_4$ crystals were often observed connected to the surface of larger $NaNO_3$ particles.

TABLE II
Result of clustering analysis after low-Z EPXMA on North Sea aerosols (2-4μm).

(a) Concentrations of elements

| | Concentration (w%) | | | | | | | | | | | | | | |
|---|---|---|---|---|---|---|---|---|---|---|---|---|---|---|---|
| **Type of particles** | C | N | O | Na | Mg | Si | P | S | Cl | K | Ca | Ti | Cr | Fe | Cu |
| ***Seasalt*** | 9.8 | 0.6 | 9.6 | 36 | 1.0 | 0.0 | 0.0 | 0.9 | 42 | 0.0 | 0.2 | 0.0 | 0.0 | 0.0 | 0.0 |
| ***Alumino-silicates*** | 6.4 | 3.1 | 54 | 1.1 | 1.9 | 24 | 0.1 | 1.0 | 0.1 | 2.3 | 1.0 | 0.3 | 0.0 | 4.7 | 0.1 |
| ***Ammonium+ nitrate*** | 9.7 | 6.1 | 53 | 2.8 | 2.2 | 2.0 | 0.1 | 9.9 | 0.6 | 2.4 | 8.6 | 0.1 | 0.4 | 1.6 | 0.1 |
| ***Organic*** | 37 | 11 | 41 | 2.4 | 1.2 | 0.7 | 1.1 | 2.8 | 0.3 | 1.3 | 0.7 | 0.0 | 0.0 | 0.1 | 0.0 |
| ***Sodium nitrate*** | 5.1 | 11 | 51 | 25 | 1.9 | 0.3 | 0.0 | 4.1 | 0.7 | 0.1 | 0.4 | 0.0 | 0.0 | 0.6 | 0.0 |
| ***Iron oxide*** | 6.4 | 2.1 | 35 | 1.5 | 2.3 | 5.2 | 0.0 | 1.3 | 0.1 | 0.5 | 0.3 | 0.1 | 0.2 | 44 | 0.8 |
| ***Biogenic*** | 59 | 6.4 | 29 | 0.9 | 0.7 | 1.2 | 0.1 | 1.3 | 0.0 | 0.3 | 0.2 | 0.7 | 0.0 | 0.1 | 0.0 |
| ***Sea salt+ organic*** | 15 | 3.8 | 31 | 18 | 2.5 | 1.8 | 0.0 | 2.6 | 20 | 0.6 | 4.2 | 0.0 | 0.0 | 0.2 | 0.0 |

(b) Time-dependence of relative abundances

| | | Relative abundance and sampling time | | | | | |
|---|---|---|---|---|---|---|---|
| **Type of particles** | **Overal abund. (%)** | **d (μm)** | 28.09.98 | 29.09.98 | 30.09.98 | 01.10.98 | 02.10.98 |
| ***Sea salt*** | 36 | 1.6 | 0.0 | 0.7 | 0.0 | 82 | 77 |
| ***Alumino- silicates*** | 22 | 1.9 | 38 | 42 | 39 | 1.3 | 0.5 |
| ***Ammonium+ nitrate*** | 13 | 1.6 | 19 | 17 | 21 | 3.3 | 8.0 |
| ***Organic*** | 12 | 2.3 | 26 | 14 | 15 | 1.3 | 8.5 |
| ***Sodium nitrate*** | 6.7 | 2.0 | 0.0 | 13 | 12 | 6.7 | 2.0 |
| ***Iron oxide*** | 4.8 | 1.4 | 8.0 | 6.0 | 11 | 0.7 | 0.0 |
| ***Biogenic*** | 2.8 | 2.6 | 7.2 | 5.3 | 2.7 | 0.7 | 0.0 |
| ***Sea salt+organic*** | 1.8 | 3.2 | 0.0 | 0.0 | 4.0 | 4.0 | 0.0 |

One of the possible reasons of this effect is the inhomogeneous crystallisation at the impacting surface, since $Na_2SO_4$ crystallises at higher relative humidity (Storms *et al.,* 1984). The highest abundance of sea-salt particles was observed in the last two days. The air mass back trajectories indicated that the samples of 28-30.9.98 originated from continental sources while the1-2.10.98 sampled had indeed a pure marine origin.

## 3.4. LOW-Z EPXMA OF AMAZON BASIN PARTICLES

EPXMA of low-Z element was performed on three series of airborne particles collected by a Batelle-type impactor. Here, the analysis results of the coarse fraction (2-8 μm) are presented as preliminary data. The measuring parameters and general conditions were exactly similar to those applied during the analysis of North Sea samples. In order to compare the elemental compositions of the samples collected at different times, they were classified into ten groups using the Forgy algorithm.

Because the collection of the samples was done during the wet season, organic-type particles could be observed with the highest probability with abundances varying between 40 and 90%, except for the first three days, as shown in Figure 4. The organic particles were divided into three classes by non-hierarchical clustering; this depended on the presence of associated elements such as Na, Mg and K. The most significant difference was observed in the Na-content: only one group of the organic particles contained Na. The

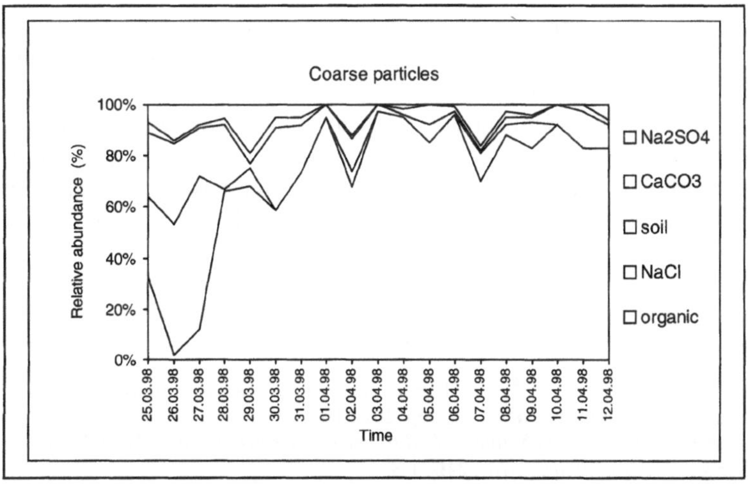


Fig. 4. Percentage variation of the main components (organic and non-organic fractions) of Amazon sample (2-4 μm) versus sampling time, as measured by low-Z EPXMA

Ca, Ti and Fe were seen as related elements. Finally, the last three classes were to be found as separate groups of NaCl, $CaCO_3$ and $Na_2SO_4$ with Mg, Si and S present in low concentrations. The NaCl particles probably originated from the ocean by large-scale transport and their numerical abundance went up to nearly 50% on 26.3.98 when organic particles were relatively unimportant, but the relatively low Cl/Na ratio was indicative for an important reaction with acid compounds underway.

## 4. Conclusions

The presented new applications of automated EPXMA pertain to indoor museum aerosols (where the major sources of particles with a possibly negative effect on painting conservation could be identified) and to aerosols from a very remote site in Kazakhstan. The first results of large-scale low-Z EPXMA are presented. They pertain to North Sea aerosols where an omnipresence of C could be seen and the speciation of the nutrient N could be assessed for continental and marine air flows, and to central Amazon Basin aerosols where, in addition to the expected biogenic aerosols, $Na_2SO_4$ and marine NaCl particles appeared rather unexpectedly. A more thorough interpretation will require the results of the present more continuous and systematic sampling campaigns to be available.

## Acknowledgements

We are grateful to the individuals who provided us with aerosol samples or contributed to the sample campaigns. These were D. Camuffo (CNR, Padua, Italy) and the co-workers in his EU project for the different museums, E. Zakarin (Cosmic Research Institute, Almaty, Kazakhstan) for the Kazakhstan sampling site and W. Maenhaut and his team (Ghent University, Belgium) for the Amazon samples. This work was partially supported by the Belgium Prime Minister's Office of Scientific, Technical and Cultural Affairs (Programme on Sustainable Management of the North Sea, contract MM/DD/10) and by the EU under contract ENV4-CT95-0088.

## References

Armstrong, J. T.: 1991, in K. F. J. Heinrich and D. E. Newbury (eds), *Electron Probe Quantitation,* Plenum Press, New York , pp. 261-316.

Baer, N. S. and Banks, P. N.: 1985, *Intern. J. Museum Management Curatorship* **4**, 9-20.

Brimblecombe, P.: 1990, *Atmos. Envir.*, **24B**, 1-8.

Camuffo, D.: 1998, *Microclimate for Cultural Heritage*, Elsevier, Amsterdam, 235-292.

De Santis, F., Di Paolo, V. and Allegrini, I.: 1992, *Sci. Total Envir.* **127**, 211-223.

Gauvin, R., Hovington, P. and Drouin, D.: 1995, *Scanning* **17**, 202-219.

Hovington P., Drouin D, and Gauvin R.: 1997, *Scanning* **19**, 1-14.

Injuk, J., Van Grieken, R. and de Leeuw, G.: 1998, *Atmos. Environ.* **32**, 3011-3025.

Jaenicke, R, and Mathias-Maser, S.: 1992, Natural source of atmospheric aerosol particles, in S. E. Schwartz and W. G. N. Slinn (eds), *Precipitation Scavenging and Atmosphere-Surface Exchange,* Hemisphere, Washington, DC., pp. 1617-1639.

Karlsson, R. and Ljungström, E.: 1998, *Water Air Soil Pollut.* **103**, 55-70.

Kerminen, V. M., Teinilä, K., Hillamo, R. and Pakkanen T.: 1998, *J. Aerosol Sci.* **29**, 929-942.

Massart, D. and Kaufmann, L.: 1983, *The Interpretation of Analytical Chemical Data by the Use of Cluster Analysis*, Wiley, New York.

Ro, C.-U., Osán, J. and Van Grieken, R.: 1999, *Anal. Chem.* **71**,1521-1528.

Rojas, C. M. and Van Grieken, R. E.: 1992, *Atmos. Environ.* **26A**; 1231-1237.

Simoneit, B. R. T.: 1989, *J. Atmos. Chem.* **8**, 251-275.

Storms, H., Van Dyck, P., Van Grieken, R. and Maenhaut, W.: 1984, *J. Trace Microprobe Techn.* **2**,103-117.

Van Espen, P., Janssens, K. and Nobels, J.: 1987, *Chemom. Lab.* **1**, 109-115

# REACTIVE HALOGEN SPECIES IN THE MID-LATITUDE TROPOSPHERE – RECENT DISCOVERIES

U. PLATT

*Institut für Umweltphysik, University of Heidelberg, INF 229, D-69120 Heidelberg*

**Abstract**. While the role of reactive halogen species (e.g., Cl, Br) in the destruction of the stratospheric ozone layer is well known, up to now it was assumed that these tropospheric halogen events were confined to the polar regions during springtime. However, during the last few years, significant amounts of BrO and Cl-atoms were also found in the Arctic and Antarctic boundary layer. Recently, even higher BrO mixing ratios (up to 90 ppt) were detected by optical absorption spectroscopy (DOAS) in the Dead Sea basin during summer. In addition, evidence is accumulating that BrO (at levels around 1-2 ppt) is also occurring in the free troposphere (in polar regions as well as at mid- latitudes).

In contrast to the stratosphere, where halogens are released from species which are very long lived in the troposphere, likely sources of boundary layer Br and Cl are oxidation of sea-salt halides, while precursors of free tropospheric BrO probably are short-lived organo-halogen species. In addition, it is well possible that boundary layer halogens, in particular bromine, may 'leak out' to the free troposphere and thus could have a regional or even more widespread effect.

At the levels suggested by the available measurements, reactive halogen species have a profound effect on tropospheric chemistry: In the boundary layer during 'halogen events' ozone is usually completely lost within hours or days – the 'Polar Tropospheric Ozone Hole'. In the free troposphere the effective $O_3$- losses due to halogens could be comparable to the known photochemical $O_3$ destruction. Further interesting consequences include the increase of OH levels and (at low $NO_X$) the decrease of the $HO_2$/OH ratio in the free troposphere.

**Keywords**: bromine, chlorine, halogen species, spectroscopy, tropospheric chemistry

## 1. Introduction

The role of reactive halogen species in the destruction of stratospheric ozone is well known and largely understood (e.g., Solomon 1990). Since the suggestion of the BrO – ClO cycle by McElroy *et al.* (1986) and Yung *et al.* (1980) and the subsequent detection of the BrO by Brune and Anderson (1986) and Solomon *et al.* (1989), it is clear that reactive halogen species (RHS = X, $X_2$, XY, XO, HOX, where X, Y denotes a halogen atom) contribute considerably to the loss of stratospheric ozone.

More recently, in the tropospheric boundary layer (BL), significant amounts of BrO and IO could be directly determined by Differential Optical Absorption Spectroscopy (DOAS), and indirect evidence for Cl- and Br atoms was also found under certain conditions. These observations were made at a variety of sites (Table I):

*Water, Air, and Soil Pollution* **123:** 229–244, 2000.

TABLE I

Observation of reactive halogen species in the troposphere and their probable source mechanism

| Species | Found at | Technique | Probable source | Conc. Level, Typ. Rate of $O_3$ Destruction |
|---|---|---|---|---|
| HOCl (?) | Marine BL[1] | Mist Chamber | ? | ? |
| BrO | Arctic and Antarctic BL[6,7] | DOAS (ground based and satellite) | Auto-catalytic Release | Up to 30 ppt<br>1-2 ppb/h |
| BrO | Dead Sea Valley[3] | DOAS | Auto- catalytic Release<br>Possible contribution from $NO_Y$+$Br^-$ | Up to 100 ppt<br>10-20 ppb/h |
| Br | Arctic BL[8,9] | Hydrocarbon Clock | Auto- catalytic Release | $(1\text{-}10)\times10^7 cm^{-3}$ |
| BrO | Mid-Lat. Free Troposphere[4] | DOAS (difference) | $CH_3Br$, stratosphere | 1-2 ppt<br>≈0.05 ppb/h |
| BrO | Polar free Troposphere[10] | Airborne DOAS | Boundary Layer | |
| Cl | Arctic BL[8,9] | Hydrocarbon Clock | ? | $(1\text{-}10)\times10^4$ $cm^{-3}$<br>? |
| Cl | Remote Marine BL[2] | Hydrocarbon Clock | ? | $(1\text{-}15)\times10^3$ $cm^{-3}$<br>? |
| IO | Coastal Areas[5] | DOAS | Degradation of organo-halogens | Up to 6 ppt<br>≈0.2 ppb/h |

[1]Pszenny *et al.* (1993), [2]Wingenter *et al.* (1996)
[3]Hebestreit *et al.* (1999) [4] Frieß *et al.* (1999)
[5]Alicke *et al.* (1999), (J. Plane, privat comm. 1999)
[6]Wagner and Platt (1998), Hausmann and Platt, (1994 Tuckermann *et al.*(1997)
[7]Richter *et al.* (1998) [8]Jobson *et al.* (1994)
[9]Ramacher *et al.* (1997, 1999) [10]McElroy *et al.* (1999)

1. Bromine monoxide was found in the Arctic- (Hausmann and Platt, 1994, Tuckermann *et al.,* 1997) and Antarctic troposphere (Kreher *et al.,* 1997) by ground based and satellite observations (Wagner and Platt, 1998; Richter *et al.,* 1998). These episodes of BL BrO were only found in springtime (Platt and Lehrer, 1995; Lehrer *et al.*, 1997). It could be shown that BrO (at levels up to 30 ppt) is always connected to episodes of BL ozone destruction.
2. Measurements by chemical amplification (Perner *et al.*, 1999), DOAS (Tuckermann *et al.*, 1997), and Hydrocarbon Clock data (Jobson *et al.*, 1994; Solberg *et al.,* 1996; Ramacher *et al.,* 1997, 1999) suggest ClO levels in the ppt range in the Arctic BL.
3. In addition, IO was found at levels reaching several ppt at coastal sites in Ireland (Alicke *et al.,* 1999; Stutz *et al.,* 1999) and at the Canary Islands (J. Plane, personal communication 1999). Model calculations suggest that ozone

destruction due to these IO levels is comparable to the classic photochemical ozone loss in the tropospheric boundary layer.

4. Even higher BrO mixing ratios (up to 90 ppt) were found in the Dead Sea basin (Hebestreit *et al.,* 1999), where also complete BL ozone destruction is associated with episodes of high BrO.
5. Recently, evidence is accumulating that there are significant amounts of BrO (of the order of 1-2 ppt, if uniform distribution over the entire troposphere is assumed) also in the free troposphere (McElroy *et al.,* 1999; Ferlemann *et al.,* 1999; Friess *et al.,* 1999; Platt *et al.,* 1999).

This paper presents the tropospheric regions where reactive halogen species are found, inspects their likely sources and reviews the tropospheric cycles of inorganic halogen species. It is suggested that reactive halogen species have a profound effect on tropospheric chemistry.

## 2. Tropospheric Sources of Inorganic Halogens Species

Inorganic halogen species (in the following denoted as InX = X, $X_2$, XY, XO, HOX, $XONO_2$, HX, where X = Cl, Br, I) are released to the atmosphere either by degradation of organic halogen compounds or by oxidation of sea salt halogenides ($X^-$).

In contrast, the dominating source of InX in the stratosphere is the photochemical degradation of fully halogenated compounds (like $CF_2Cl_2$ or $CF_2ClBr$). In the troposphere InX can only be released from less stable partially halogenated organic compounds (e.g., methyl bromide, $CH_3Br$) or polyhalogenated species such as $CHBr_3$, $CH_2Br_2$, $CH_2I_2$, or $CH_2BrI$ (Cicerone, 1981; Schall and Heumann, 1993; Khalil *et al.,* 1993; Schauffler *et al.,* 1998; Carpenter *et al.,* 1999). Some of these species (e.g., $CH_3Br$) originate in about 50% from anthropogenic sources (the other 50% are emitted by biological processes). Others – in particular polybrominated species – are only emitted from biological sources, for instance from algae in the ocean or in coastal areas. The lifetime of $CH_3Br$ is of the order of one year, $CHBr_3$, has several days lifetime, while $CH_2I_2$ is photodegraded in minutes. (Wayne *et al.,* 1995; Yvon *et al.,* 1996; Davis *et al.,* 1996; Carpenter *et al.,* 1999). In addition, even less stable halocarbon species may also be emitted, which – probably due to their instability – escape detection (Carpenter *et al.,* 1999). Typical atmospheric bromoform levels are of the order of a few ppt (e.g., Cicerone *et al.,* 1988; Dvortsov *et al.,* 1999).

Another source of importance only to the troposphere is release of InX from sea salt (e.g., sea salt aerosol or sea salt deposits) which appears to proceed by three main pathways:

1. Strong acids can release HX from sea-salt halides. Under certain conditions (see below), HX can be heterogeneously converted to reactive halogen species (RHS).
2. Oxidising agents may convert $Br^-$ or $Cl^-$ to $Br_2$ or BrCl. In particular, HOX (i.e., HOBr and HOCl) is likely to be such an oxidator. Thus, the halogen release mechanisms are autocatalytic (see below) as proposed by Tang and McConnel (1996) and Vogt *et al.* (1996). Since the development of a radical-chain and the chain branching mechanism is similar to that of chemical explosions, these processes became known as 'Bromine Explosion' (Platt and Lehrer, 1995; Platt and Janssen, 1996; Wennberg, 1999). However, direct photochemical (Oum *et al.*, 1998a) or non-photochemical oxidation of halides by $O_3$ may also occur (Oum *et al.*, 1998b; Hirokawa *et al.*, 1998), though probably quite slowly.
3. Oxidised nitrogen species, in particular $N_2O_5$ and $NO_3$, and perhaps even $NO_2$, can react with sea-salt bromide or chloride to release either HBr, HCl or photolabile species, in particular $BrNO_2$ (Finlayson-Pitts and Johnson 1988; Finlayson-Pitts *et al.*, 1990; Behnke *et al.*, 1993, 1997; Rudich *et al.*, 1996; Schweitzer *et al.*, 1999; Gershenzon *et al.*, 1999).

## 3. Tropospheric Cycles of Inorganic Halogen Species

In the following we review the reaction cycles of inorganic halogen species in the troposphere (see Figure 1). More information can be found, *inter alia*, in the review of Wayne *et al.*, (1995), Platt and Janssen (1996), Platt and Lehrer (1995), and Lary *et al.* (1996).

Following release by the mechanisms described above, inorganic halogen species will rapidly be photolysed to form halogen atoms, which – in turn – are most likely to react with ozone:

$$X + O_3 \rightarrow XO + O_2 \qquad \text{(R1)}$$

Typical conversion time constants via R1 for Cl are around 0.1 s and for Br and I of the order of 1 s at tropospheric background $O_3$ levels. Halogen atoms are regenerated in a series of reactions including photolysis of XO, which is of importance for X = I, Br and to a minor extent Cl:

$$XO + h\nu \rightarrow X + O \qquad \text{(J2)}$$

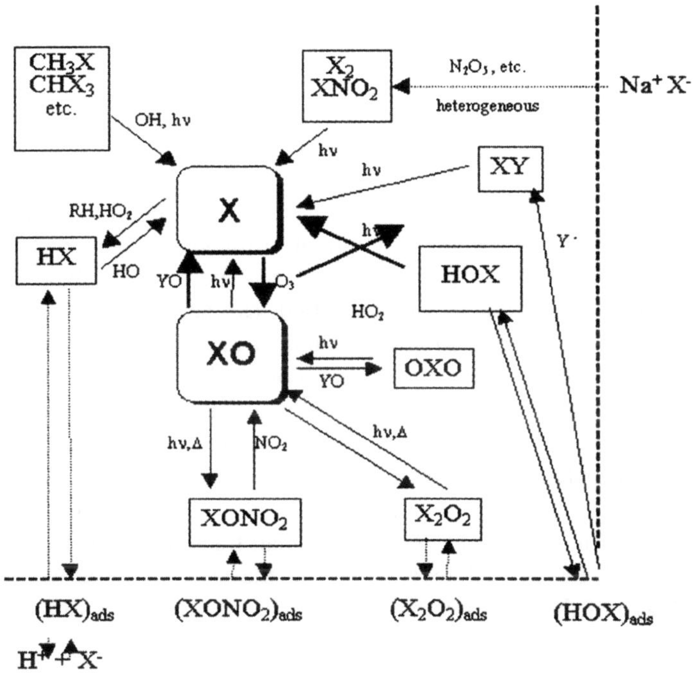


Sea-Salt (Snow Pack or Aerosol Surface)

Fig. 1.Simplified scheme of the inorganic halogen reactions in the boundary layer (X = Cl, Br, I) after Wayne *et al.* (1995) and Platt and Janssen (1996). Sources are released from sea-salt (aerosol or deposits) or photochemical degradation of organohalogen species. Important questions concern the XO/X interconversion and recycling from hydrogen halides (HX). Fat lines: ozone destruction cycles.

Where $J_2 \approx 3\times10^{-5}\ s^{-1}$, $4\times10^{-2}\ s^{-1}$, $0.2\ s^{-1}$ for X = Cl, Br, I, respectively. In the parts of the troposphere affected by pollution reaction with NO is another source of halogen atoms:

$$XO + NO \quad \rightarrow \quad X + NO_2 \qquad (R3)$$

Where $k_3 = 1.7\times10^{-11}$ $cm^3molec.^{-1}s^{-1}$ (X = Cl) and $k_3 = 2.1\times10^{-11}$ $cm^3molec.^{-1}s^{-1}$ (X = Br, I). In addition, the self reactions of XO (or reaction with another halogen oxide YO) also play a role, like:

$$XO + YO \quad \rightarrow \quad X + Y + O_2 \ \ (\text{or } XY + O_2) \qquad (R4)$$

In the case of BrO, the rate constant k = $3.210^{-12}cm^3molec.^{-1}s^{-1}$ (overall rate constant of the self-reaction of BrO including the channel to $Br_2$, accounting for ≈20% at 250 K). Reactions of ClO + BrO (LeBras and Platt, 1995) and, in particular, reactions involving IO are even faster than the above rate constant. Reaction 4 is the rate determining step in an ozone-destruction cycle involving R1, R4, and possibly photolysis of $X_2$ or XY - species. This type of halogen catalysed ozone destruction was identified as the prime cause for polar boundary layer ozone destruction (Oltmanns *et al.*, 1986; Barrie *et al.*, 1988; Barrie and Platt, 1997). While molecular products (XY, e.g., BrCl, $Br_2$, ICl) are assumed to photolyse rapidly into halogen atoms and thus are effectively equivalent to direct production of halogen atoms, the channels leading to OXO would not contribute to the ozone loss.

Ultimately the XO/X ratio is therefore largely determined by reactions R1, R3, R4, and (during daytime) $J_2$. Overall, a small fraction (less than 0.1% for Cl, about 0.1 - 1 % for Br and several 10% in the case of I, see e.g., Platt and Janssen, 1996) of the inorganic halogen will be present in the form of halogen atoms.

In addition, halogen atoms can react with saturated- (Cl) or unsaturated hydrocarbons (Cl, Br) to form hydrogen halides, e.g.,

$$RH + Cl \quad \rightarrow \quad R + HCl \qquad (R5)$$

Here R denotes an organic radical. An alternative is the reaction of halogen atoms with formaldehyde (or higher aldehydes) or $HO_2$ also leading to conversion of halogen atoms to hydrogen halides:

$$X + HO_2 \quad \rightarrow \quad HX + O_2 \qquad (R6)$$

$$X + HCHO \quad \rightarrow \quad HX + CHO \qquad (R7)$$

The former reaction occurs for X = Cl, Br, I, the latter only for X = Cl, Br. In addition, higher aldehydes may play a role. Bromine atoms can add to the C=C double bond of olefins leading to organic Br species, but at least partly – also to HBr. While in the stratosphere with its low hydrocarbon- formaldehyde- and $HO_2$ concentration the fraction of Br atoms being converted to BrO by reaction with ozone (R1) is close to 100%, in the troposphere roughly 99% of the Br atoms react with $O_3$ (Wayne *et al.,* 1995; Platt and Janssen 1996). While HI is hardly formed in the atmosphere, combined with the short lifetime (below one minute) of BrO reaction 7 leads to conversion of reactive Br into HBr within a few hours (e.g., Platt and Lehrer, 1995). In the case of chlorine around 50% of the Cl atoms react with hydrocarbone (e.g., $CH_4$) thus the conversion of reactive Cl to HCl is even faster.

In the lower troposphere, hydrogen halides may be irreversibly lost. Since HX is highly water soluble it is likely to be irreversibly removed from the atmosphere by wet or dry deposition. In the free troposphere, recycling of Br to the $BrO_X$ reservoir is more likely, as will be discussed below. The only relevant gas phase 'reactivation' mechanism of HX is reaction with OH:

$$HX + OH \rightarrow X + H_2O \quad \text{(R8)}$$

Assuming OH levels typical for the free troposphere and gas-phase chemistry only (HBr)/(BrO) ratios should typically reach 10 – 20, thus making HBr the most abundant InBr species there.

Other important reservoirs for inorganic halogen species besides HX are HOX and $XONO_2$, where HOX is formed via the reaction of peroxy radicals with XO:

$$XO + HO_2 \rightarrow HOX + O_2 \quad \text{(R9)}$$

(with $k_9$ is several times $10^{-11}$ $cm^3s^{-1}$, (e.g., Canosa-Mas *et al.*, 1999)), an analogous reaction involving $CH_3O_2$ is also likely to (ultimately) produce HOX (Aranda *et al.*, 1997). Formation of HOX is followed by its photolysis:

$$HOX + h\nu \rightarrow X + OH \quad \text{(R10)}$$

Thus, in the sunlit atmosphere a photo-stationary state between XO and HOX is established with (HOX)/(XO) ranging from roughly 1 to 10 (if heterogeneous reactions are neglected), depending on the oxidation capacity of the atmosphere. Important Side effects of HOX formation are $O_3$ destruction (e.g., Barrie *et al.*, 1988) as well as a reduction of the $(HO_2)/(OH)$ ratio as will be discussed below.

An interesting reaction cycle involving HOBr (Fan and Jacob 1992;Tang and McConnel 1996; Vogt *et al.,* 1996) is the liberation of gaseous bromine species (and to a lesser extent chlorine species) from (sea-salt) halides by the Bromine Explosion mechanism:

$$HOBr + (Br^-)_{Surface} + H^+ \rightarrow Br_2 + H_2O \quad \text{(R11)}$$

The required $H^+$ (the reaction appears to occur at appreciable rates only at pH < 6.5 (Fickert *et al.*, 1999)) could be supplied by strong acids, such as $H_2SO_4$ and $HNO_3$ originating from man made or natural sources. Bromide containing aerosol could be sufficiently acidified by HBr (self-acidification).

Finally, the formation of halogen nitrate, $XONO_2$:

$$XO + NO_2 \xleftrightarrow{M} XONO_2 \qquad (R12)$$

depends on the $NO_2$ concentration. Bromine nitrate is assumed to be quite stable against thermal decay, but is readily photolysed and may be converted to HOX by heterogeneous hydrolysis:

$$XONO_2 + H_2O \xrightarrow{Surface} HNO_3 + HOX \qquad (R13)$$

or to $Br_2$ or BrCl heterogeneous reaction with HY:

$$XONO_2 + HY \xrightarrow{Surface} HNO_3 + XY \qquad (R14)$$

Overall, $XONO_2$ is probably of minor importance at the low $NO_X$ levels typically found in the free troposphere, but can play an important role in polluted air (for instance at polluted coastlines). Under conditions of high $NO_X$ halogen release can occur via the reactions:

$$N_2O_5(g) + NaX(s) \rightarrow NaNO_3(s) + XNO_2(g) \qquad (R15)$$

(e.g., Finlayson-Pitts *et al.*, 1989). The $XNO_2$ formed in the above reaction may photolyse to release a halogen atom or possibly further react with sea salt (Schweitzer *et al.*, 1999):

$$XNO_2(g) + NaX(s) \rightarrow NaNO_2(s) + X_2(g) \qquad (R16)$$

The above reaction sequence would constitute a dark source of halogen molecules.

$$NO_3\,(g) + NaX(s) \rightarrow NaNO_3(s) + X(s) \qquad (R17)$$

(Seisel *et al.*, 1997; Gershenzon *et al.*, 1999). The uptake coefficient of $NO_3$ by aqueous solutions of NaBr and NaCl was found by Rudich *et al.* (1996) to be near 0.01 for sea water, while Seisel *et al.*(1997) found up to 0.05 for dry NaCl.

Recent laboratory investigations have shown that in the troposphere heterogeneous reactions (on aerosol or cloud particle surfaces, as originally suggested by Fan and Jacob (1992)) can efficiently recycle $BrO_X$ (e.g.,

Kirchner *et al.,* 1997). As pointed out by Abbat (1994, 1995) and Abbat and Nowak (1997) an important reaction in this context is (similar to reaction 11):

$$HOBr + HX \xrightarrow{Surface} H_2O + BrX \quad (R18)$$

Where HX denotes HCl or HBr (see also reaction 11). While HBr and HOBr are the most likely forms of InBr in the troposphere as outlined above, HCl is probably more abundant than total Br in the upper troposphere with levels up to 100 ppt reported (e.g., Vierkorn-Rudolph *et al.,* 1984). The $Br_2$ or BrCl formed in reaction 18 will be rapidly photolysed to release Br atoms, which then react as described above.

## 4. Observations of Reactive Halogen Species in the Lower Troposphere

As pointed out above reactive halogen species, in particular Cl- and Br- atoms, BrO, and ClO could be detected by several direct or indirect techniques (see Table I) in the troposphere, both, at polar and mid-latitude sites:

- Differential Optical Absorption Spectroscopy (DOAS) of BrO and IO (and upper limits for ClO). This technique is beeing applied in ground based measurements (e.g., Tuckermann *et al.,* 1997) as well as from aircraft (McElroy *et al.,* 1999), balloons (Harder *et al.*, 1998), and from satellite (Richter *et al.,* 1998; Wagner and Platt 1998).
- 'Hydrocarbon Clock' observations of time - integrated amounts of Cl- and Br- atoms, where the term Hydrocarbon Clock refers to the observation of hydrocarbon ratios in airmasses of known chemical age (e.g., Jobson *et al.*, 1994; Solberg *et al.,* 1994; Ramacher *et al.,* 1997, 1999).
- Detection of ClO by chemical amplification (Perner *et al.,* 1999).
- Sampling (mist chamber) and wet chemical detection of reactive chlorine (Pszenny *et al.,* 1993).

## 5. Observations Suggesting BrO in the Free Troposphere

Recently it became clear that measurements of the BrO column density from the ground (e.g., Fish *et al.,* 1995, 1997; Eisinger *et al.*, 1997; Kreher *et al.*, 1997; Otten *et al.,* 1998; Frieß *et al.*, 1999) and from satellite (Richter *et al.*, 1998; Wagner and Platt 1998) consistently yield higher values than obtained by integrating stratospheric *in-situ* profiles or profiles from balloon-borne spectroscopic measurements (e.g., Harder *et al.,* 1999; Ferlemann *et al.,* 1999).

Additional evidence for a noticeable tropospheric BrO column comes from the analysis of diurnal profiles of the BrO – 'slant column' density. As pointed out by Friess *et al.* (1999) BrO slant column data measured by ground based

ZSL-DOAS in Kiruna, Sweden fit model calculations of the slant columns much better if a tropospheric (vertical) BrO column density of $(0.5 - 2) \cdot 10^{13}$ $cm^{-2}$ is assumed. Similar discrepancies requiring even larger tropospheric BrO columns were found in DOAS observations from the Alps (Jungfraujoch), (D. Perner private communication).

The difference amounts to $(2 - 3.5) \cdot 10^{13}$ $BrO/cm^2$. In view of the several measurement techniques employed, it is unlikely that experimental errors are responsible for the difference. The discrepancy could by far best be explained by assuming an average BrO mixing ratio of 1-2 ppt throughout the troposphere. In view of the remarkable similarity in the magnitude of the tropospheric residual seen at various measurement sites (though to date only in the northern hemisphere) it is likely that BrO could be a rather ubiquitous constituent of the global troposphere.

## 6. The Overall Picture of Inorganic Halogen Species in the Troposphere

From Table II, which summarises the various regions of the troposphere, where reactive halogen species are found as well as their likely sources, it becomes clear that the occurrence of noticeable levels of RHS is a widespread phenomenon in many parts of the troposphere.

Particularly notable is the likely presence of ppt levels of BrO in the free troposphere, which can have a profound effect on the global budget of tropospheric ozone and other chemical key parameters of the troposphere (see below).

TABLE II

Sources of Reactive Halogen Species found in various parts of the Troposphere

| Species, Site | Likely Source Mechanism |
|---|---|
| $ClO_X$ in the Polar Boundary Layer | By-Product of the ‘Bromine Explosion’ |
| $BrO_X$ in the Polar Boundary Layer | Autocatalytic release from Sea Salt on ice (‘Bromine Explosion’ mechanism) |
| $BrO_X$ in the Dead Sea Basin | Bromine Explosion (Salt Pans) |
| $BrO_X$ in the Free Troposphere | (1) Photo – Degradation of hydrogen-containing organo halogen species (e.g., $CH_3Br$)<br>(2) "Spill-out" from the Boundary Layer<br>(3) Transport from Stratosphere (?) |
| $IO_X$ in the Marine Boundary Layer | Photo – Degradation of short-lived organo-halogen species (e.g., $CH_2I_2$) |

Overall, we can now describe the following scenario for reactive bromine species in the free troposphere: It is now clear that methyl bromide is ubiquitous in the troposphere at mixing ratios of 11 – 12 ppt in the northern hemisphere and 8 – 10 ppt in the southern hemisphere (see e.g., Lee-Taylor *et al.,*1998 and references therein). Degradation by OH in the troposphere is the dominant (>95 %) loss of $CH_3Br$. Estimates for the lifetime of this process range from 10 to 18 months (e.g., Mellouki *et al.,* 1992). Assuming a residence time of about one month of air masses in the upper troposphere, a level of inorganic bromine approaching roughly 10% of the $CH_3Br$ mixing ratio – or slightly more than 1 ppt-could be expected. According to model calculations including heterogeneous chemistry and the observations in the polar boundary layer, a large fraction (around 50%) of this inorganic Br would be present in the form of BrO. Assuming this BrO layer to extend from above the BL (roughly 800 mBar) to the tropopause (roughly 200 mBar), it would constitute a $Br_Y$ column density of $1.2{\cdot}10^{13}cm^{-2}$ or roughly half the observed value (if $(BrO)/(Br_Y)=1.0$ is assumed).

In addtition to $CH_3Br$, other – shorter lived – organic bromine species like $CH_2Br_2$ or $CHBr_3$ might also contribute (e.g., Dvortsov *et al.,* 1999). In addition, stratospheric air being transported to the upper troposphere will contain 15 ppt of total inorganic bromine (e.g., Harder *et al.,* 1999).

At the levels suggested by the available measurements tropospheric, BrO can have a noticeable effect on several aspects of tropospheric chemistry, these include:

(1) As already discussed by several authors (Solomon *et al.,* 1994; Platt and Janssen, 1996; Davis *et al.,* 1996), reactive halogen species, and in particular reactive bromine and iodine, can readily destroy tropospheric ozone. While the halogen oxide self- and cross reactions are of negligible influence at 1-2 ppt BrO the XO – $HO_2$ cycle - reactions 9,10 followed by:

$$OH + CO \text{ (or VOC)} \rightarrow HO_2 + CO_2 \text{ (or other products)} \quad (R19)$$

with the net result:

$$2\,O_3 \rightarrow 3\,O_2 \quad (R20)$$

will lead to a rate of ozone loss comparable to photochemical $O_3$ destruction associated with $HO_X$ reactions (due to $O(^1D) + H_2O$ and $HO_2 + O_3$). In addition, the reaction of BrO – $CH_3O_2$ (rate and products of the reaction IO – $CH_3O_2$ is presently unknown) may have a similar effect on $O_3$.

(2) Reactions 9, 10 will lead to the conversion of $HO_2$ to OH and thus reduce the $HO_2/OH$ ratio, in particular at low $NO_X$ levels. At a given photochemical situation ($O_3$-, $H_2O$ levels, insolation) the presence of BrO or IO will therefore increase the OH concentration. It is interesting to note that both effects have been observed in the upper troposphere (e.g., Wennberg *et al.*, 1998).

(3) The reaction of XO with NO, (reaction 3) leads to conversion of NO to $NO_2$ and thus to an increase of the Leighton ratio (L = $(NO_2)/(NO)$). An increase in L is thus usually regarded as an indicator for photochemical ozone production (due to the presence of $RO_2$ (R = organic radical)) in the troposphere. However, as already noted by Platt and Janssen (1996) an increase in L due to halogen oxides will not lead to $O_3$ production.
(4) A more subtle consequence of reactive halogen species for the ozone levels in the free troposphere is due to the combination of the above effects (2) and (3) as pointed out by Stutz *et al.* (1999): Photochemical ozone production in the troposphere is limited by the reaction

$$NO + HO_2 \rightarrow NO_2 + OH \qquad (R21)$$

However, the presence of reactive bromine will reduce the concentrations of both educts of reaction 21 and thus reduce the $NO_X$ – catalysed ozone production.
(5) Heterogeneous reactions of the type reaction 18 with HX = HCl would lead to BrCl, which is readily photolysed to release Cl-atoms. This process would constitute a Br catalysed chlorine activation, which directly enhances the oxidation capacity of the troposphere.
(6) In addition, tropospheric reactive bromine species could be responsible or contribute to the observed cases of upper tropospheric ozone loss (Kley *et al.*, 1996; Reichardt *et al.*, 1996; Davies *et al.*, 1998).
(7) Gas-phase iodine species (like IO or HOI) may facilitate transport of I from the coast to inland areas and thus contribute to our iodine supply (Cauer 1939).

Clearly, at present more measurements are needed to ascertain the distribution and levels of free tropospheric reactive bromine. However, it is already clear that inorganic halogen species are likely to play a major – and yet unrecognised – role in tropospheric chemistry.

## References


Abbat, J. P. D.: 1994, Heterogeneous reaction of HOBr with HBr and HCl on ice surfaces at 228 K, *Geophys. Res. Lett.* **21**, 665-668.

Abbat, J. P. D.: 1995, Interactions of HBr, HCl and HOBr with supercooled sulfuric acid solutions of stratospheric composition, *J. Geophys. Res.* **100**, 14009-14017.

Abbat, J. P. D. and Nowak J. B.: 1997, Heterogeneous interactions of HBr and HOCl with cold sulfuric acid solutions: Implications for arctic boundary layer bromine chemistry, *J. Phys. Chem. A* **101**, 131-2137.

Alicke, B., Hebestreit, K., Stutz, J. and Platt, U.: 1999, Detection of Iodine Oxide in the Marine Boundary Layer, *Scientific Correspondence to Nature* **397,** 572-573.

Aranda, A., LeBras, G., LaVerdet, G. and Poulet, G.: 1997, The BrO + $CH_3O_2$ reaction: Kinetics and role in the tropospheric ozone budget, *Geophys. Res. Lett.* **24**, 2745-2748.

Barrie, L. A., Bottenheim, J. W., Schnell, R. C., Crutzen, P. J. and Rasmussen, R. A.: 1988, Ozone destruction and photochemical reactions at polar sunrise in the lower Arctic atmosphere, *Nature* **334,** 138-141.

Barrie, L. A. and Platt, U.: 1997, Arctic tropospheric chemistry*: overview to Tellus special issue* **49B**, 450-454.

Behnke, W., Scheer, V. and Zetzsch, C.: 1993, Formation of $ClNO_2$ and $HNO_3$ in the presence of $N_2O_5$ and wet pure NaCl- and wet mixed $NaCl/Na_2SO_4$-aerosol, *J. Aerosol Sci.* **24**, 115-116.

Behnke, W., George, C., Scheer, V. and Zetzsch, C.: 1997, Production and decay of $ClNO_2$ from the reaction of gaseous $N_2O_5$ with NaCl solution: bulk and aerosol experiments, *J. Geophys. Res.* **102D**, 3795-3804.

Brune, W. H. and Anderson, J. G.: 1986, In-situ observations of midlatitudestratospheric ClO and BrO, Geophys. Res. Lett. **13**, 1391-1394.

Canosa-Mas, C. E., Flugge, M. L., Shah, D., Vipond, A. and Wayne, R. P.: 1999, Kinetics of the reactions of IO with $HO_2$ and $O(^3P)$, *J. Atmos.* Chem., in press.

Carpenter, L. J., Sturges, W. T., Penkett, S. A., Liss, P. S., Alicke, B., Hebestreit, K. and Platt, U.: 1999, Short-lived alkyl iodides and bromides at Mace Head, Ireland: Links to biogenic sources and halogen oxide production, *J. Geophys. Res.***104,** 1679-1689.

Cauer, H.: 1939, Schwankungen der Jodmenge der Luft in Mitteleuropa, deren Ursachen und deren Bedeutung für den Jodgehalt unserer Nahrung (Auszug), *Angewandte Chemie* **52,** Nr. 11, 625-628.

Cicerone, R. J.: 1981, Halogens in the Atmosphere, *Rev. Geophys. and Space Phys.* **19**, 123-139.

Cicerone, R. J., Heidt, L. E. and Pollock, W. H.: 1988, Measurements of atmospheric methyl bromide and bromoform, *J. Geophys. Res.* **93**, 3745-3749.

Davies, W. E., Vaughan, G. and O'Connor, F. M.: 1998, Observations of near-zero ozone concentrations in the upper troposphere at mid-latitudes, *Geophys. Res. Lett.* **25,** 1173-1176.

Davis, D., Crawford, J., Liu, S., McKeen, S., Bandy, A., Thornton, D., Rowland, F. and Blake, D.: 1996, Potential impact of iodine on tropospheric levels of ozone and other critical oxidants, *J. Geophys. Res.,* **101**, 2135-2147.

Dvortsov, V. L., Geller, M. A., Solomon, S., Schauffler, S. M., Atlas, E. L. and Blake, D. R.: 1999, Rethinking reactive halogen budgets in the midlatitude lower stratosphere, *Geophys. Res. Lett.* **26**, 1699-1702.

Eisinger, M., Richter, A., Ladstätter-Weißenmayer, A. and Burrows, J. P.: 1997, DOAS Zenith sky observations: 1.BrO measurements over Bremen (53°N) 1993-1994, *J.Atmos. Chem.* **26**, 93-108.

Fan, S.-M. and Jacob, D. J.: 1992, Surface ozone depletion in the Arctic springsustained by bromine reactions on aerosols, *Nature* **359**, 522 - 524.

Ferlemann, F., Camy-Peyret, C., Harder, H., Fitzenberger, R., Hawat, T., Osterkamp, H., Perner, D., Platt, U., Schneider, M., Vradelis, P. and Pfeilsticker, K.: 1998, Stratospheric BrO Profile measured at Different Latitudes and Seasons. Measurement Technique, *Geophys. Res. Lett.* **25**, 3847-3850.

Fickert, S., Adams, J. W. and Crowley J. N.: 1999, Activation of $Br_2$ and BrCl via uptake of HOBr onto aqueous salt solutions, *J. Geophys. Res.*, in press.

Finlayson-Pitts, B. J.: 1993, Indications of Photochemical Histories of Pacific Air Masses from Measurements of Atmospheric Trace Species at Point-Arena, California - Comment, *J. Geophys. Res.* **98**, 14991-14993.

Finlayson-Pitts, B. J., Ezell, M. J. and Pitts, J. N.: 1989, Formation of chemically active chlorine compounds by reactions of atmospheric NaCl particles with gaseous $N_2O_5$ and $ClONO_2$, *Nature* **337**, 241-244.

Finlayson-Pitts, B. J. and Johnson, S. N.: 1988, The reaction of $NO_2$ with NaBr: Possible source of BrNO in polluted marine atmospheres, *Atmos. Environ.* **22**, 1107 - 1112.

Fish, D. J., Jones, R. L. and Strong, E. K.: 1995, Midlatitude observations of the diurnal variation of stratospheric BrO:2, interpretation and modelling study, *Geophys. Res. Lett.* **24**, 1199-1202.

Fish, D. J., Aliwell, S. R. and Jones, R. L.: 1997, Mid-latitude observations of the seasonal variation of BrO: 2. interpretation and modelling study, *Geophys. Res. Lett.* **24**, 1199-1202.

Frieß, U., Otten, C., Chipperfield, M.,Wagner, T., Pfeilsticker, K. and Platt, U.: 1999, Intercomparison of measured and modelled BrO slant column amounts for the Arctic winter and spring 1994/95, *Geophys. Res. Lett.*, in press.

Gershenzon, M. Y., Il'in, S., Fedetov, N. G. and Gershenzon, Y. M.: 1999, The mechanism of reactive $NO_3$ uptake on dry NaX (X=Cl, Br), *J. Atmos. Chem.*, in press.

Harder, H., Camy-Peyret, C., Ferlemann, F., Fitzenberger, R., Hawat, T., Osterkamp, H., Perner, D., Platt, U., Schneider, M., Vradelis, P. and Pfeilsticker, K.: 1998, Stratospheric BrO profile measured at different latitudes and seasons: Atmospheric observations, *Geophys. Res. Lett.* **25**, 3843-3846.

Harder, H., Camy-Peyret, C., Chipperfield, M., Perner, D., Platt, U., Pyle, J. and Pfeilsticker, K.: 1999, Comparison of measured and modelled stratospheric BrO: Constraints on the stratospheric bromine photochemistry at mid-latitudes and the total amount of $Br_y$, *Geophys. Res. Lett.*, in press.

Hausmann, M. and Platt, U.: 1994, Spectroscopic measurement of bromine oxide and ozone in the high Arctic during Polar Sunrise Experiment 1992, J. Geophys. Res. **99,** 25,399-25,414.

Hebestreit K., Stutz J., Rosen D., Matveev V, Peleg M., Luria M., and Platt U.: 1999, First DOAS Measurements of Tropospheric BrO in Mid Latitudes, *Science* **283**, 55-57.

Hirokawa, J., Onaka, K., Kajii, Y. and Akimoto, H.: 1998, Heterogeneous processes involving sodium halide particles and ozone: Molecular bromine release in the marine boundary layer in the absence of nitrogen oxides, *Geophys. Res. Lett.* **25**, 2449-2452.

Jobson, B. T., Niki, H., Yokouchi, Y., Bottenheim, J., Hopper, F. and Leaitch, R.: 1994, Measurements of $C_2$-$C_6$ hydrocarbons during Polar Sunrise Experiment 1992, *J.Geophys. Res.* **99**, 25,355-25,368.

Khalil, M. A. K., Rasmussen, R. A. and Gundwardena, A.: 1993, Atmospheric methyl bromide: Trends and global mass balance, *J. Geophys. Res.* **98**, 2887-2896.

Kirchner, U., Benter, Th. and Schindler, R. N.: 1997, Experimental verification of gas phase bromine enrichment in reactions of HOBr with sea salt doped ice surfaces, *Ber. Bunsenges. Phys. Chem.* **101**, 975-977.

Kley, D.: 1997, Tropospheric chemistry and transport, *Science* **276**, 1043-1045.

Kley, D., Smit, H. G. J., Crutzen, P. J., Vömel, H., Oltmans, S., Grassl, H. and Ramanathan, V.: 1996, Observation of near-zero ozone concentrations over the convective Pacific: effects on air chemistry, *Science* **274**, 230-233.

Kreher, K., Johnston, P. V., Wood, S. W. and Platt, U.: 1997, Ground-based measurements of tropospheric and stratospheric BrO at Arrival Heights (78°S), Antarctica, *Geophys. Res. Lett.* **24**, 3021-3024.

Langendörfer, U., Lehrer, E., Wagenbach, D. and Platt, U.: 1999, Observation of filterable bromine variabilities during Arctic tropospheric ozone depletion events at high (1 hour) time resolution, *J. Atmos. Chem.*, in press.

Lary, D. J., Chipperfield, M. P., Toumi, R. and Lenton, T.: 1996, Heterogeneous atmospheric bromine chemistry, *J. Geophys. Res.* **101**, 1489-1504.

LeBras, G. and Platt, U.: 1995, A Possible mechanism for combined chlorine and bromine catalysed destruction of tropospheric ozone in the Arctic, *Geophys. Res. Lett.***22**, 599-602.

Lee-Taylor, J. M., Doney, S. C., Brasseur, G. P. and Müller, J.-F.: 1998, A global three-dimensional atmosphere-ocean model of methyl bromide distribution, *J. Geophys*. Res. **103**, 16039-16057.

Lehrer, E., Wagenbach, D. and Platt, U.: 1997, Chemical composition of the Aerosol during Arctic Spring in Svalbard, *Tellus* **49B**, 486-495.

Manö, S. and Andreae, M. O.: 1994, Emission of methyl bromide from biomass burning, *Science* **263,** 1255-1256.

McElroy, M. B., Salawitch, R. J., Wofsy, S. C. and Logan, J. A.: 1986, Reductions of Antarctic ozone due to synergistic interactions of chlorine and bromine, *Nature* **321**, 759-762.

McElroy, C. T., McLinden, C. A. and MCConnell, J. C.: 1999, Evidence for bromine monoxide in the free troposphere during Arctic polar sunrise, *Nature* **397,** 338-340.

Mc Kinney, K. A., Pierson, J. M. and Toohey, D. W.: 1997, A wintertime in-situ profile of BrO between 17 and 27 km in the Arctic vortex, *Geophys.Res. Lett.* **24**, 853-856.

Mellouki, A., Talukdar, R. K., Schmoltner, A.-M., Gierczak, T., Mills, M. J., Solomon, S. and Ravishankara, A. R.: 1992, Atmospheric lifetimes and ozone depletion potentials of methyl bromide ($CH_3Br$) and dibrommethane ($CH_2Br_2$), *Geophys. Res. Lett.* **19,** 2059-2062.

Oltmans, S. J. and Komhyr, W. D.: 1986, Surface ozone distributions and variations from 1973-1984 measurements at the NOAA Geophysical Monitoring for Climate Change baseline observatories, *J. Geophys. Res.* **91**, 5229-5236.

Otten, C., Ferlemann, F., Platt, U., Wagner, T. and Pfeilsticker, K.: 1998, Groundbased DOAS UV/visible measurements at Kiruna (Sweden) during the SESAME winters 1993/94 and 1994/95, *J. Atmos. Chem.* **30**, 141-162.

Oum, K. W., Lakin, M. J., DeHaan, D. O., Brauers, T. and Finlayson-Pitts, B. J.: 1998a, Formation of molecular chlorine from the photolysis of ozone and aqueous Sea-Salt particles, *Science* **279**, 74-77.

Oum, K. W., Lakin, M. J. and Finlayson-Pitts, B. J.: 1998b, Bromine activation in the troposphere by the dark reaction of $O_3$ with seawater ice, *Geophys. Res. Lett.* **25**, 3923-3926.

Perner, D., Arnold, T., Crowley, J., Klüpfel, T., Martinez, M. and Seuwen, R.: 1999, The measurements of active chlorine in the atmosphere by chemical amplification, *J. Atmos. Chem.*, in press.

Platt, U., Harder, H., Perner, D., Stutz, J., Wagner, T. and Pfeilsticker, K.: 1999, BrO in the Free Troposphere, in prep.

Platt, U. and Hausmann, M.: 1994, Spectroscopic measurement of the freeradicals $NO_3$, BrO, IO, and OH in the troposphere, *Res. Chem. Intermed.* **20**, 557-578.

Platt, U. and Lehrer, E.,: 1995, *ARCTOC, Final Report to the European Union.*

Platt, U. and Janssen, C.: 1996, Observation and role of the free radicals $NO_3$, ClO, BrO and IO in the Troposphere, *Faraday Discuss.* **100**,175-198.

Pszenny, A. A. P., Keene, W. C., Jacob, D. J., Fan, S., Maben, J. R., Zetwo, M. P., Springeryoung, M. and Galloway, J. N.: 1993, Evidence of Inorganic Chlorine Gases other than Hydrogen Chloride in Marine Surface Air, *Geophys. Res. Lett.* **20**, 699-702.

Ramacher, B., Rudolph, J. and Koppmann, R.: 1997, Hydrocarbon measurements in the spring arctic troposphere during the ARCTOC 95 campaign, *Tellus, Ser. B*, 466-485.

Ramacher, B., Rudolph, J. and Koppmann, R.: 1999, Hydrocarbon measurements during tropospheric ozone depletion events, *J. Geophys. Res.* **C104,** 3633-3653.

Reichardt, J., Ansmann, A., Serwazi, M., Weitkamp, C. and Michaelis, W.: 1996, Unexpectedly low ozone concentration in midlatitude tropospherericice clouds: a case study, *Geophys. Res. Lett.* **23**, 1929-1932.

Richter, A., Wittrock, F., Eisinger, M. and Burrows J. P.: 1998, GOME observations of tropospheric BrO in northern hemispheric spring and summer 1997, *Geophys. Res. Lett.* **25**, 2683-2686.

Rudich, Y., Talukdar, R. and Ravishankara, A. R.: 1996, Reactive uptake of $NO_3$ on pure water and ionic solutions, *J.Geophys. Res.* **D101**, 21023-21031.

Sander, R. and Crutzen, P. J.: 1996, Model study indicating halogen activation and ozone destruction in polluted air masses transported to the sea, *J. Geophys. Res.* **D101**, 9121-9138.

Schall, C. and Heumann, K. G.: 1993, GC determination of volatile organoiodine and organobromine compounds in seawater and air samples, *Fresenius J. Anal. Chem.* **346**, 717-722.

Schauffler, S. M., Atlas, E. L., Flocke, F., Lueb, R. A., Stroud, V. and Travnicek, W.: 1998, Measurement of bromine-containing organic compounds at the tropical tropopause, *Geophys. Res. Lett.* **25**, 317-320.

Schweitzer, F., Mirabel, P. and George, C.: 1999, Heterogeneous chemistry of nitryl halides in relation to tropospheric halogen activation, *J. Atmos. Chem.*, in press

Seisel, S., Caloz, F., Fenter, F. F., Van den Bergh , H. and Rossi, M. J.: 1997, The heterogeneous reaction of $NO_3$ with NaCl and KBr: a nonphotolytic source of halogen atoms, *Geophys. Res. Lett.* **24,** 2757-2760.

Solberg S., Schmidtbauer, N., Semb, A. and Stordal, F.: 1996, Boundary-Layer Ozone Depletion as Seen in the Norwegian Arctic in Spring, *J. Atmos. Chem.* **23,** 301-332.

Solomon, S.: 1990, Progress towards a quantitative understanding of Antarctic ozone depletion, *Nature* **347,** 347-354.

Solomon, S., Sanders, R. W., Carroll, M. A. and Schmeltekopf, A. L.: 1989, Visible and near-ultraviolet spectroscopy at McMurdo station, Antarctica, 5, Diurnal varations of OClO and BrO, J. Geophys. Res. **94**, 11393-11403.

Solomon, S., Garcia, R. R., Ravishankara, A. R. : 1994, On the role of iodine in ozone depletion, *J. Geophys. Res.* **99**, 20491-20499.

Stutz, J., Hebestreit, K., Alicke, B. and Platt, U.: 1999, Chemistry of halogen oxides in the troposphere: comparison of model calculations with recent field data, *J. Atmos. Chem.*, in press.

Tang, T. and McConnel, J. C.: 1996, Autocatalytic release of bromine from arctic snow pack during polar sunrise, *Geophys. Res. Lett.* **23**, 2633-2636.

Tuckermann, M., Ackermann, R., Gölz, C., Lorenzen-Schmidt, H., Senne, T., Stutz, J., Trost, B., Unold, W. and Platt, U.: 1997, DOAS-Observation of Halogen Radical- catalysed Arctic Boundary Layer Ozone Destruction During the ARCTOC-campaigns 1995 and 1996 in Ny-Alesund, Spitsbergen, *Tellus* **49B,** 533-555.

Vierkorn-Rudolph, B., Bächmann, K., Schwarz, B. and Meixner, F. X.: 1984, Vertical profiles of hydrogen chloride in the troposphere, *J. Atmos. Chem.* **2**, 47-63.

Vogt, R., Crutzen, P. J. and Sander, R.: 1996, A mechanism for halogen release from sea-salt aerosol in the remote marine boundary layer, *Nature* **383**, 327-330.

Wagner, T. and Platt, U.: 1998, Observation of Tropospheric BrO from the GOME Satellite, *Nature* **395**, 486-490.

Wayne, R. P., Poulet, G., Biggs, P., Burrows, J. P., Cox, R. A., Crutzen, P. J., Haymann, G. D., Jenkin, M. E., LeBras, G., Moortgat, G. K., Platt, U. and Schindler, R. N.: 1995, Halogen oxides: radicals, sources and reservoirs in the laboratory and in the atmosphere, *Atmosph. Environ.* **29**, 2675-2884.

Wennberg, P. O. *et al.*: 1998, Hydrogen radicals, nitrogen radicals, and the production of $O_3$ in the upper troposphere, *Science* **279**, 49-53.

Wennberg, P. O.: 1999, Bromine Explosion, *Nature* **397**, 299-301.

Wingenter, O. W., Kubo, M. K., Blake, N. J., Smith, T. W., Blake, D. R. and Rowland, F. S.: 1996, Hydrocarbon and halocarbon measurements as photochemical and dynamical indicators of atmospheric hydroxyl, atomic chlorine, and vertical mixing obtained during Lagrangian flights, *J. Geophys. Res.* **101**, 4331-4340.

Yvon, S. A and Butler, J. H.: 1996, An improved estimate of the oceanic lifetime of atmospheric $CH_3Br$, *Geophys. Res. Lett.* **23,** 53-56.

Yung, Y. L., Pinto, J. P., Watson, R. T. and Sander, P.: 1980, Atmospheric bromine and ozone perturbations in the lower stratosphere, *J. Atmos. Sci.* **37**, 339-353.

# HYDROGEN SULFIDE AND ODOR CONTROL IN İZMIR BAY

A. MÜEZZINOĞLU[1], D. SPONZA [1], I. KÖKEN [1], N. ALPARSLAN [2],
A. AKYARLI [3] and N. ÖZTÜRE[3]
[1]*Dokuz Eylül Univ., Depart. of Env. Eng., İzmir, Turkey,* [2] *Dokuz Eylül Univ., Env. Research Center, İzmir, Turkey,* [3]*Akoks Environmental Technologies, Şehit Nevres Bulvarı, Kızılay İşhanı, Kat 7 Alsancak, İzmir, Turkey*

**Abstract**. The city of İzmir on the Aegean Sea shoreline is suffering from rotten odors emitted by anoxic river mouths. Anaerobic conditions in the shallowest portion of İzmir Bay due to industrial and domestic wastewaters as well as eutrophication products in this very calm part of the Bay are responsible for this. The inner section of the Bay is becoming shallower with sediments rich in organic matter. Aerobic digestion of organic pollutants is limited by the oxygen input and the warm climate leads to an optimal medium for anaerobic processes when anoxic conditions take over. Anaerobic digestion products are odorous gases among which $H_2S$ with a characteristic pungent odor is most effective in this case. Sulfur containing gases are formed from sulfides and sulfates in the sediment-water interface and are released into the air. Airborne $H_2S$ concentrations are variable as they depend on factors such as high atmospheric diffusion coefficients under changing wind direction and speed, as well as variable such as water depths, organic loadings from rivers, air and water temperatures, sulfate concentrations in sediment and water phases, pH, and Eh.

This study aimed to control noxious smells by inhibiting the anaerobic sulfate-reducing bacteria at the sediment surface. To achieve this aim, pH of the sediment-water interface was increased by spreading hydrated lime. Successful lime doses were investigated in laboratory models first and were then tested and applied in the estuary and downstream segment of one of the creeks. Rates of $H_2S$ gas emissions and aqueous $H_2S$ levels were measured. It was noted that waterborne $H_2S$ levels at river mouths were reduced. Odor control efficiencies of 80-96% even after 10 days of lime addition in the field experiments were obtained. Therefore, the results of this study indicated that lime application to the sediment surface is a recommended method in odor control programs.

**Keywords:** anaerobic activity at sediment-water interface, anoxia in estuaries, environmental uses of hydrated lime, estuarine pollution, hydrogen sulfide emissions, odor control

## 1. Introduction

The presence of high sulfate levels in the marine environment and estuaries accepting industrial and domestic wastewaters is a source of rotten smells from polluted waters if they have turned anoxic. These odors are due to sulfurous gases desorbed from the water phase into the atmosphere at lower pH. Gaseous sulfur compounds appear as intermediate products of a diverse series of biological and chemical processes in the natural sulfur cycle. Odorous biogenic sulfurous gases evolve an unbearable odor which reaches the city of İzmir. Seasonal fish kills are also recorded.

*Water, Air, and Soil Pollution* **123:** 245–257, 2000.

There are costly projects for collection/treatment of urban wastewaters and for dredging the bottom sediments in the shallow part of the Bay. However, this odor control project is proposed for the transient period of time until these projects are fully underway for a complete solution.

The aim of this study was to investigate the reasons and mechanism of this problem and find an easy and cheap solution to apply. This presentation covers a literature survey in this direction, laboratory bench scale model test series, field tests and full-scale applications of hydrated lime addition for control of rotten smells.

## 2. Earlier Studies

Sulfate-reducing microorganisms are capable of using sulfate as a sulfur source for growth, can reduce it to $H_2S$ intracellularly, and can replace the hydroxyl groups, serine and homoserine with sulfydryl groups. These reductions are named assimilatory as virtually all of the $H_2S$ produced is incorporated into aminoacids. Reduction of sulfate to sulfide by microorganisms can be seen in any anoxic habitat containing organic matter and sulfate. Dissimilatory sulfate reduction is the process in which sulfate is used as the electron acceptor for the oxidation of organic matter. The end product of this oxidation is $H_2S$, which is excreted by the cells as opposed to being assimilated.

The activities of sulfate-reducing bacteria, such as *Desulfovibrio, Desulfomonas,* and *Desulfotomaculum*, are geochemically and environmentally important in the global sulfur cycle. The $H_2S$ produced in sediments rapidly combines with metal ions to form metal sulfides. In organic-rich or polluted waters, when $H_2S$ is produced in excess of the iron and other metal ions it can diffuse out of the sediments and cause fish kills and odor problems (Nriagu *et al.*, 1978). Such fish kills coming after anoxia were noted in the Danish Sea between 1981-1983 (Spiro and Stigliani, 1996).

$H_2S$ is not the only product of breakdown of sulfate or organic sulfur compounds. Depending on the conditions, sulfur-containing aminoacids can be broken down to sulfides, $H_2S$ and volatile organic sulfur compounds. Anaerobic production of $H_2S$ from organic sulfur compounds such as cysteine, is well known and especially important in fresh waters. The role of methionine is less well known and many sources note that methyl mercaptan is the product of its reduction. Dimethyl sulfide (DMS) is often formed after the breakdown of sulfonium compounds (Nriagu, *et al.*, 1978). $H_2S$, DMS (dimethyl sulfide), OCS (carbonyl sulfide), $CS_2$ (carbon disulfide), and $CH_3SH$ (methyl mercaptan) are the most important biogenic sulfurous gases generated from the marine environments (Jorgensen, 1985).

In stagnant waters with a lot of organic sedimentary materials, anoxia is frequently seen at deep layers. The Black Sea, the largest deep stagnant sea in the world, is a good example of sulfide rich deep-sea environments. Other examples of sulfate reduction are seen in the sediments of bays, river deltas, fjords, the continental shelf and inland marginal waters (Nriagu *et al.*, 1978).

Sulfur-bearing organic matters are decomposed by the action of anaerobic bacteria and sulfate ions are reduced to $H_2S$ by microorganisms in the lagoon which lies between the breakwater and the eastern coastal bank of İzmir Bay. Odors created can be recognized from all over the city and can be traced with $H_2S$ measurements in the air. Earlier tests in airborne studies indicated about 5 $mg.L^{-1}$ of $H_2S$ in winter and 100-200 $mg.L^{-1}$ of $H_2S$ in the summer period in the problem area (Özel, 1991; Ermiş and Köse, 1990, Öztürk, 1994).

Several cases of lime applications to the bottom sediments are noted in the world aiming at overcoming such problems. In some studies lime precipitation led to the hypothesis that the addition of lime to lakes can also reduce eutrophication. Lime has been added to several lakes and dugouts in Western Canada to improve water quality. Recent studies have shown that the use of hydrated lime is a more complete and longer lasting method of dugout water treatment. The benefits of dugout liming include better control of algae and submerged rooted plants, as well as phosphorus removal and silt precipitation, to have a clean and clear dugout. For example, in lakes with valuable fisheries, alternative approaches to enhance apatite formation could include hypolimnetic injection of $Ca(OH)_2$ or larger surface applications of $CaCO_3$ (Mandaville, 1997). Hydrated lime is mixed into the dugout water to provide an excellent source of good quality water and reaction products are allowed to settle. Lime acts as coagulant to settle down the algae, silt and phosphorus to the bottom of the dugout. A similar application by Bunchanan, *et al.* (1996) was made in dugout water to overcome heavy growths of algae and aquatic plants. The proper dosage of hydrated lime was determined to be in the range of 120g to 240g per $m^3$ of dugout. Hydrated lime is mixed with water to make a slurry and sprayed evenly over the entire water surface of the dugout (Williamson, 1994).

In Mikawa Bay applications, it was found that when lime was added to the mud it rapidly precipitates over the neferoid layer in which $H_2S$, phosphorus dissolution and oxygen consumption occurs, and the water has clarified (Japan Lime Association, 1995). This treatment was done in seawater samples. More interestingly, improvements were held even after one month. Different types of lime were used in these experiments; calcium oxide, calcium hydroxide, and marine cleaner (CaO pellets). $Ca(OH)_2$ is prepared as a solution with 10% of water and sprayed into the sea. If it can be sprayed close to sea bottom, it makes a good mixture and helps speed improvements of seawater environments. These have shown that if the lime is added to the mud, this process prevents $H_2S$ formation, stabilizes phosphorus, oils and heavy metals in bottom dredgings and also prevents them from mixing with soils.

Several creeks causing heavy environmental pollution discharge to Inner İzmir Bay at the port region are the main contributors of pollution such as

Melez, Arap, Bornova, Manda and Laka. In Figure 1 a sketch of the problem area is given. Sedimentary materials carried by the creeks as well as the high rate of rapid eutrophication made İzmir Bay shallower for the last one or two decades. Estuarine water movements are slowed down due to a jetty built to protect the Port, and a lagoon has formed. There are many industries which directly or indirectly discharge their wastes into these streams. Although these industries have pretreatment facilities frequently inspected by the city administration, during this transition period at which the main sewer leading to the municipal treatment plant is under construction, this problem persists. In the estuary of combined Melez and Arap mouths, brackish waters and sludge contain high concentrations of sulfate, sulfides, and sulfur containing organic compounds. A study of sulfate and sulfide chemistry in the sediments of Melez estuary was made and gaseous $H_2S$ in the water and air phases were measured between sediments and water column as well as air-sea interface (Taşdemir, 1989).

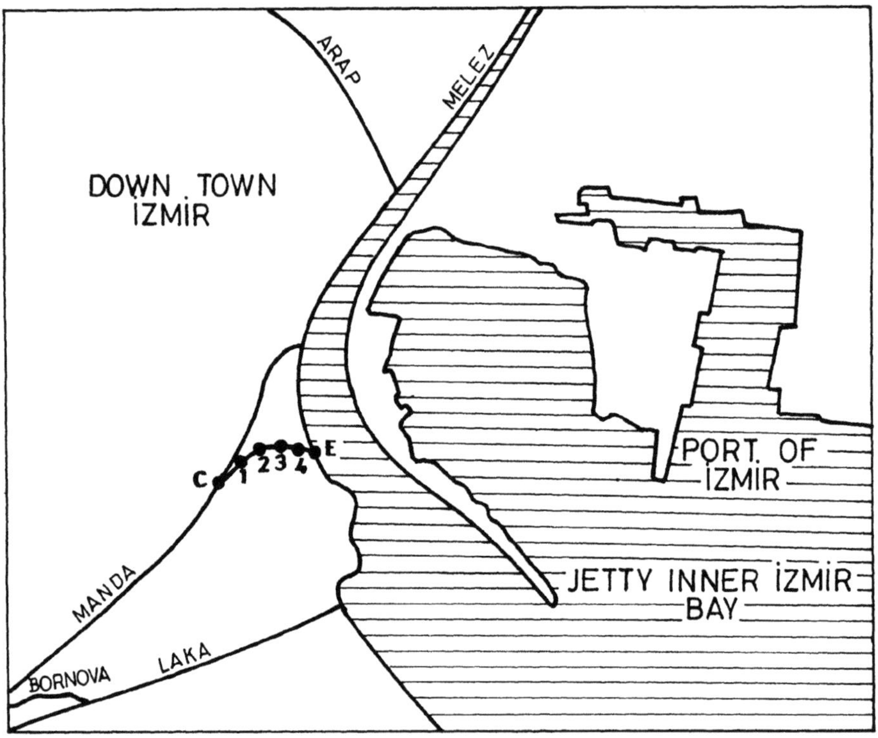


Fig. 1. Problem area in the city of İzmir and field study points

## 2. Materials and Methods

### 2.1. LABORATORY MODEL RUNS

Sediment and water samples were taken from Melez lagoon (Figure 1) and analyzed. Three anaerobic reactors of 2 L capacity with 1.76 x $10^{-2} m^2$ surface area each were set up in the laboratory and studies conducted in triplicate. Sediment samples were placed into the reactors and enough water was added to cover them. In order to simulate summer conditions the reactors were held at 30°C. The gaseous products of anaerobic reactions were kept in telescopic gas storage units. Free $H_2S$ in gas and water phases were measured periodically. Total bacteria, anaerobic bacteria, sulfate-reducing bacteria populations in the sediments were also counted. Sulfate concentrations in the sediment and water phases were measured throughout the tests. Different doses of hydrated lime were added into anaerobic reactors for controlling the activity of sulfate reducers. After stabilization of the model system for sufficient anaerobic process yields, different doses of hydrated lime were added into the reactors to suppress $H_2S$ formation. The first test was done on February 1998 but was unsuccessful as the sediment was already mineralized before it was taken from nature. Therefore sulfur-bearing aminoacids were added into the system to start $H_2S$ evolution. Therefore this first model run was discarded and tests were repeated two more times on samples taken on 28 April and 26 May 1998 to obtain reliable performance.

### 2.2. FIELD STUDIES

Preliminary field investigations before a full-scale lime application were carried out during the summer months of 1998. Odor reduction was followed with respect to time in Arap Creek estuary by measuring free $H_2S$ (aq.) after adding certain amounts of lime into 400 ml capped bottles containing mud and water samples at site.

The full-scale field study was performed with the permission of the Municipality of İzmir on the Manda Creek. Manda Creek collects eight domestic sewers and some additional direct industrial wastewater discharges. A segment of 1 km river length between an estuarine site (E) in the sea and an upstream control point ( C ) without lime addition were chosen. In between, four sampling points from 1 to 4 were determined on the stream with ascending numbers as one goes downstream (Figure 1). Application began with a dose of 200 $g.m^{-2}$ hydrated lime spread over the bottom sediments. $H_2S$, pH and temperature were measured and recorded during 10 days after liming.

## 2.3. METHODS OF MEASUREMENTS

Instantaneous airborne and waterborne free $H_2S$ measurements were made by the Dräger-Liquid-Extraction method (German Env. Mon. Institute, 1990) in both laboratory model runs and field tests. This technique consists of sucking in a known volume of air through the Dräger-tube using a hand pump. A chemical reaction takes place between the $H_2S$ gas and the dry sorbent reagent in the Dräger-tube which can be followed by a color change. The colored length of the tube is a direct indication of the amount of free $H_2S$. For waterborne $H_2S$, a 200 ml liquid sample was poured into the washing bottle and tightly capped by a cover containing two gas connection tubings. Inlet tubing was equipped with a carbon filter to free the purge-air from pollutants to go into the reaction bottle. This air created bubbles in the bottle and purged out the $H_2S$ gas from the water, Then $H_2S$ containing outlet gas stream was passed through the detector tube placed on the outlet tubing. The detector tub was placed on the train after both ends were broken by a glass cutter. Measurement was completed when bubbles stopped rising in the washing bottle and the established color stain length in the detector tube was read. A suitable scale on the tube allowed a direct reading of the $H_2S$ concentrations purged out from the water sample. Parameters in the laboratory and field programs and the methods used to measure them are indicated in Table I.

TABLE I

Measured parameters and their test methods

| Parameter | Method |
|---|---|
| $H_2S$ (both water and air borne) | Dräger tube method. For waterborne $H_2S$ added by an extraction system |
| DO | DO-meter (Jenway 9071) |
| Total suspended particles | APHA/AWWA/WPCF, 1985 |
| Total nitrogen | Spectroquant kit method |
| Total phosphorus | APHA/AWWA/WPCF, 1985 |
| PH, temperature | Portable pH meter and probe (Hanna, HI 8314 membrane) |
| Salinity and conductivity | Portable salinity and conductivitymeter (YSI) |
| Total bacteria | Dilution plate count (nutrient agar (Pelezar, *et al.*,1972) |
| Ammonium producing bacteria | MPN from Mc Cardy Table (mineral medium containing organic nitrogen source, Pelezar, *et al.*, 1980) |
| Anaerobic bacteria | MPN from Mc Cardy Table (anaerobic broth containing Na-thioglicollate) (Shank, 1988) |
| Sulfate reducing bacteria | MPN from Mc Cardy Table (mineral medium containing sulfate source and lead acetate paper sheet (Postgate, 1980) |

## 3. Results

### 3.1. LABORATORY MODEL TESTS

When the February 1998 unsuccessful attempt was abandoned, laboratory model runs for sludge and water samples were taken two times, on April 28 and May 26, 1998. The February tests were unsuccessful as anaerobic sulfate reduction process with gas evolution did not begin. This is due to completed natural mineralization of sulfur compounds in the sample prior to sampling. Therefore some sulfur containing synthetic organic and inorganic chemicals were added to start the reactions. It is also noted that February coincides with the minimum sulfide emissions. Results of tests are given in Table II for the next two successful model runs.

Analytical results showed that the sulfate concentrations as well as all other constituents in samples of April and May 1998 are high enough to start biological sulfate reduction. It is noted that COD concentrations are also high and pollution is due to high organic matter and low inorganic matter content of sediment samples. The number of bacteria in the environment increased at the order of ten thousand from April to May.

TABLE II

Characteristics of sediment-water samples taken from estuary of Arap Creek

| Sampling date | 28/4/1998 | 26/5/1998 |
|---|---|---|
| PH | 6.91 | 7.28 |
| Temperature °C | 18 | 26.4 |
| Dissolved Oxygen ($g.m^{-3}$) | 0 | 0 |
| Salinity $^{o}/_{oo}$ | 2 | 109 |
| Conductivity (μmhos) | 1800 | 1650 |
| Chemical Oxygen Demand (COD, $g.m^{-3}$) | 1200 | 420 |
| Total Nitrogen ($g.m^{-3}$) | 32 | 29 |
| Total Phosphorous ($g.m^{-3}$) | 9 | 7 |
| Sulfate ( $SO_4^=$, $g.m^{-3}$) | 127 | 187 |
| $H_2S$ ($g.m^{-3}$in gas ) | 0 | 0 |
| $H_2S$ ($g.m^{-3}$in liquid ) | 160 | 420 |
| Total Solids ($g.m^{-3}$)) | 224,500 | 79,960 |
| Total Solids ( %) | 22.45 | 7.99 |
| Water Content ( %) | 77.53 | 80.04 |
| Organic Solid Matter ($g.m^{-3}$) | 200,600 | 57,740 |
| Organic Solid Matter ( %) | 89.35 | 71.25 |
| Inorganic matter ( %) | 10.65 | 28.75 |
| Total Bacteria (pcs/ml) | $255x10^6$ | $230x10^{10}$ |
| Anaerobic Bacteria (pcs/ml) | $140x10^5$ | $60x10^{10}$ |
| Sulfate Reducing Bacteria (pcs/ml) | $140x10^5$ | $45x10^9$ |

The gaseous products formed in anaerobic reactors were tested for $H_2S$ concentration and total gas evolution. $H_2S$ gas concentration in the gas reached a maximum of 220 ppm under controlled conditions. This represents the steady state mode. Cumulative gas production rate in the model was measured as 20 ml.day$^{-1}$. After steady state gas production was reached, predetermined doses of hydrated lime were added into the reactor for controlling production of $H_2S$. Waterborne $H_2S$ concentrations, sulfate-reducing and total bacteria counts were measured and results were correlated with different doses of hydrated lime. Laboratory model results (Sponza *et al.*, 1998; Erol, 1999) are given in Figure 2. In this figure absisca shows the hydrated lime doses in g.m$^{-2}$, one ordinate $H_2S$

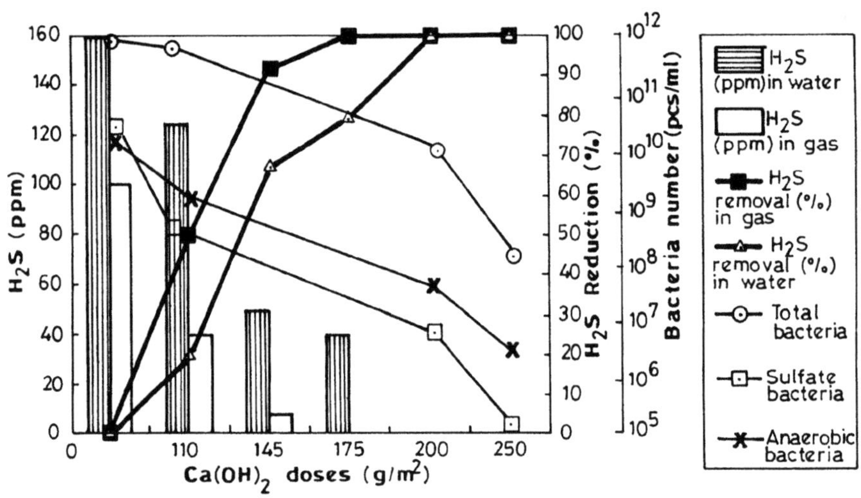


Fig. 2. Variations of control parameters versus $Ca(OH)_2$ doses in laboratory scale reactor

TABLE III

Total, sulfate-reducing, anaerobic bacteria numbers and percentage of killed bacteria with respect to several $Ca(OH)_2$ Doses

| $Ca(OH)_2$ Doses (g.m$^{-2}$) | Total Bacterial Count (pcs.ml$^{-1}$) | % of Killed Total Bacteria | Sulfate Reducing Bacteria (pcs.ml$^{-1}$) | % of Killed Sulfate-Reducing Bacteria | Anaerobic Bacterial Count (pcs/ml) | % of Killed Anaerobic Bacteria |
|---|---|---|---|---|---|---|
| 0 | $2.2x10^{12}$ | 0 | $4.5x10^{9}$ | 0 | $6x10^{9}$ | 0 |
| 114 | $1x10^{12}$ | 54.54 | $5x10^{7}$ | 98.88 | $2x10^{9}$ | 66 |
| 198 | $5x10^{10}$ | 97.72 | $1x10^{6}$ | 99.99 | $1x10^{7}$ | 99.83 |
| 255 | $1x10^{8}$ | 99.99 | $1x10^{5}$ | 99.99 | $1x10^{6}$ | 99.98 |

(aq.) concentrations, another ordinate percent reduction and a third ordinate bacterial counts for each dose level. Optimum lime dose for full reduction of $H_2S$ in water and therefore in emissions is determined as 200 $g.m^{-2}$ under laboratory conditions.

Total, anaerobic and sulfate-reducing bacterial counts and their percentage reductions at several hydrated lime doses are given in Table III.

## 3.2. PRELIMINARY FIELD TESTS

After determination of the optimal dose at laboratory, capped bottle tests for dose control were carried out three times on two different days (12 June and 15 June 1998) at site. Sediment-water mixed samples taken from Arap estuary were tested for $H_2S$ concentrations against lime doses with respect to time elapsed after dosing. These applications were preliminary field studies applied before full-scale tests. By addition of certain $Ca(OH)_2$ doses, variations of pH were also measured. Results containing the preliminary field test results are shown in Table IV.

TABLE IV

Results of bottle tests with hydrated lime application at Arap Creek estuary

| Parameter | First run (12 June 1998) | Second run (15 June 1998) | Third run (15 June 1998) |
|---|---|---|---|
| Initial Ph | 6.89 | 6.91 | 6.91 |
| DO $mg.L^{-1}$ | 0 | 0 | 0 |
| Temperature, °C | 24.5 | 26.5 | 26.5 |
| $H_2S$ (aq), ppm | 600 | 240 | 240 |
| Dosage of $Ca(OH)_2$ $g.m^{-2}$ | 250 | 125 | 189 |
| Time elapsed, min | 15 minutes | 15 minutes | 30 minutes |
| $H_2S$(aq.),ppm | 0 | 16 | 0 |
| Percent Reduction of $H_2S$ | 100 | 93 | 100 |
| Final pH | 9.50 | 9.05 | 8.00 |

## 3.3. FIELD LIME APPLICATION STUDIES

Full-scale lime application was tested in the Manda Creek on 17 October 1998 to control odor and $H_2S$ concentration. This creek was more advantageous because of ease of approach for liming practice. Applications were made by 200 $g.m^{-2}$ of hydrated lime slurry over the sediment-water interface between sampling points. Waterborne free $H_2S$, pH, and temperature readings were taken for 10 consecutive days in the six points. Field test results are given in Table V.

It was shown that the $H_2S$ concentration decreased to zero after 5 minutes of application. A range of 80-96% $H_2S$ reduction was obtained within 10 days of

TABLE V

Data obtained from field liming tests at Arap Creek segment

| Station | $H_2S$( ppm) | PH | T ( $^0C$) | % $H_2S$ REDUCTION |
|---|---|---|---|---|
| Control | | | | |
| Start( t=0 ) | 140 | 7,04 | 19,9 | |
| 7$^{th}$ day | 280 | 7 | 23 | -- |
| Station 1 | | | | |
| Start (t=0) | 80 | 6,8 | 19,7 | 0 |
| After t=10 minutes | 0,5 | 7,3 | 21,8 | 99,4 |
| 1$^{st}$ day | 10 | 7 | 22 | 87,5 |
| 3$^{th}$ day | 14 | 7 | 24 | 82,5 |
| 4$^{th}$ day | 23 | 6,3 | 23 | 71,3 |
| 5$^{th}$ day | 62 | 6,5 | 22 | 22,5 |
| 8$^{th}$ day | 52 | 6,5 | 21 | 35 |
| Station 2 | | | | |
| Start (t=0) | 50 | 6,7 | 21 | 0 |
| After t=10 minutes | 10 | 7,4 | 22,9 | 80 |
| 1$^{st}$ day | 30 | 6,7 | 21,9 | 40 |
| 2$^{nd}$ day | 23 | 6 | 20 | 54 |
| Station 3 | | | | |
| Start (t=0) | 240 | 6,7 | 19,2 | 0 |
| After t=10 minutes | 24 | 7 | 22,7 | 90 |
| 1$^{st}$ day | 56 | 6,9 | 21,7 | 76,7 |
| 3$^{th}$ day | 48 | 6,5 | 24 | 80 |
| 4$^{th}$ day | 50 | 6,5 | 23 | 79,2 |
| 5$^{th}$ day | 52 | 7 | 23 | 78,3 |
| 8$^{th}$ day | 120 | 6,3 | 22 | 50 |
| Station 4 | | | | |
| Start(t=0) | 320 | 6,9 | 19 | 0 |
| After t=10 minutes | 34 | 6,8 | 21,8 | 89,4 |
| 1$^{st}$ day | 42 | 6,7 | 21,8 | 86,9 |
| 3$^{nd}$ day | 49 | 6,8 | 20 | 84,7 |
| 4$^{th}$ day | 150 | 6 | 20 | 53,1 |
| 5$^{th}$ day | 65 | 6,5 | 23,5 | 79,7 |
| 8$^{th}$ day | 140 | 7 | 22 | 56,3 |
| Estuarine | | | | |
| Start (t=0) | 280 | 6,7 | 19,1 | 0 |
| After t=10 minutes | 41 | 7,1 | 22,1 | 85,4 |
| 2$^{nd}$ day | 80 | 7 | 22,5 | 71,4 |
| 4$^{th}$ day | 115 | 6 | 22 | 58,9 |
| 6$^{th}$ day | 108 | 7 | 22 | 61,4 |
| 7$^{th}$ day | 120 | 6 | 23 | 57,1 |
| 8$^{th}$ day | 150 | 6,5 | 22 | 46,4 |

the experimental period. At the tenth day, the tests were stopped as a heavy rain destroyed the sediment structure. Table VI shows the first and last day $H_2S$, pH, temperature and $H_2S$ reduction at the end of 10 days.

TABLE VI

Measurements of $H_2S$, pH, temperature on the 1[st] and 10[th] days and calculated $H_2S$ reductions

| Point | $H_2S$ (ppm) | | PH | | Temperature ($^0C$) | | % Reduction $H_2S$ (aq.) |
|---|---|---|---|---|---|---|---|
| | First | Last | First | Last | First | Last | |
| C | 140 | 280 | 7.0 | 7.0 | 19.9 | 23 | - |
| 1 | 80 | 3 | 6.8 | 6.5 | 19.7 | 20 | 96.3 |
| 2 | 50 | 23 | 6.7 | 6.0 | 21.0 | 20 | 54.0 |
| 3 | 240 | 120 | 6.7 | 6.3 | 19.2 | 22 | 50.0 |
| 4 | 320 | 22 | 6.9 | 6.5 | 19.0 | 20 | 93.1 |
| E | 280 | 150 | 6.7 | 6.5 | 19.1 | 22 | 47.0 |

## 4. Discussion of Results and Conclusions

From Table V it may be noted that waterborne $H_2S$ levels at 6 study points decreased from 50-320 ppm down to 0.5-41 ppm after 10 minutes of hydrated lime dosing. This corresponded to 80-99+% reduction in free aqueous $H_2S$. From the data in Tables V and VI on riverine versus estuarine points, it can be seen that the brackish waters had the lowest reduction efficiency between 85-47% with respect to time. This is normal when one considers the presence of widespread sulfate sources in the marine environment and their renewal possibilities. Varying efficiencies between points of measurement are noted and they are due to direct industrial wastewater inputs.

The variation of remaining $H_2S$ concentrations within the next week was rather low indicating that they originate from anaerobic activities that were blocked by lime addition. Tests continued until the 10[th] day during which a strong rain with a lot of sedimentary material input into the river destroyed this blocking effect.

From Table VI it can be noted that although the $H_2S$ concentration in the control point had doubled on the 10[th] day, 47-96% of $H_2S$ reduction at measurement points was still possible after addition of hydrated lime. Such high efficiencies must be in parallel with the destruction of sulfate-reducing bacteria populations in the water-sediment interface at these lime doses (Figure 2). It is also known from earlier laboratory experience that a new active culture can only evolve within two or more weeks under favorable conditions.

Experimental results are in agreement with success obtained elsewhere such as the applications in bays, lagoons, dugouts, lakes and estuaries. A literature survey shows the success of lime-based chemicals in alleviating $H_2S$ originated odor problems and eutrophication risks. After such treatments fish populations came back to these environments.

By adding a calculated dose of hydrated lime to the bottom sediments sulfate-reducing bacteria were killed very efficiently. Odor problems created by organic pollution and anoxia in the shallowest and most stagnant part of İzmir Bay continuously fed by polluted inputs of the rivers, are evident during the long warm and dry season. In this study, it was found that when hydrated lime is added to the bottom sediments, gaseous sulfides are considerably decreased and are odors are prevented in the area. It is also expected that the lime applications improve the structure of the ecosystem in the most polluted part of of İzmir Bay. That will play a complementary role in the success of the municipal treatment plant of İzmir Bay before it starts operation.

## Acknowledgement

This study was started in accordance with an agreement between Dokuz Eylül University Research Centre for Environmental Studies (ÇEVMER) and AKOKS Environmental Engineering Company in İzmir. Sponsorship by AKOKS should also be acknowledged. Any know-how produced in field applications belongs to these organisations.

## References

American Public Health Association, American Water Works Association, Water Pollution Control Federation (APHA-AWWA-WPCF): 1985, *Standard Methods for the Examination of Water and Wastewater*, 15th Ed., Washington, DC.

Bunchanan, B., Kenzie, O. and Williamson, V.: 1996, *Dugout Maintenance*, Adapted from Agdex FS 716 (B 34), Revised July 1995.

Erol, A.: 1999, Hydrogen Sulfide Pollution in İzmir Bay, Diploma Project (in Turkish), Supervised by Sponza, D, DEU, Dept. of Env. Eng., İzmir.

Ermir, T. and Köse, S.: 1990, *Research on the atmospheric levels of $H_2S$ around Melez Creek*, diploma study (in Turkish) supervised by A.Müezzinoğlu, DEU, Dept. of Env. Eng., İzmir.

Japan Lime Association: 1995, *Report on water pollution prevention*, (in Japanese).

Jørgensen, B. B., and Okholm-Hansen, B.: 1985, *Atmos.Env.* **19**,11,1737-1749.

Köken, İ., Sponza, D. and Müezzinoğlu,A.: 1998, *A Biochemical sulfur cycle with a view to odorous sulfide gas emissions in İzmir Bay*, 1st International Workshop on environmental quality and environmental engineering in the middle east region, S.U. Env. Eng. Dept., Konya, p. 54-66.

Mandaville, S. N.: 1997, *Soil and Water Conservation Society of Metro Halifax Restoration* (Summary of in-lake methodology for both culturally and naturally eutrophic lakes, the Canadian experience), Dortmouth, Canada.

Nriagu, J. O. and Hem, J. D.: 1978, *Sulfur in the Environment*, J.O. Nriagu (ed.), John Willey and Sons, New York, 448-450.

Özel, Ç.: 1991, *Update of Research on the Atmospheric Levels of $H_2S$ Around Melez Creek*, diploma study supervised by A.Müezzinoğlu, DEU, Dept. of Env. Eng., İzmir.

Öztürk, Z.: 1994, *Update of research on the atmospheric levels of $H_2S$ around Melez Creek*, diploma study supervised by A.Müezzinoğlu, DEU, Dept. of Env. Eng., İzmir.

Pelezar, M. T., and Chan, E.C.S.: 1972, *Laboratory Experiences in Microbiology*, Third edition, Mc Graw-Hill Book Comp.

Postgate, J. R.: 1980, *Laboratory Practices* **15**, 1239-1244.

Shank, J. L.: 1988, *Bacteriology J.* **2**, 95-100

Spiro, T. G. and Stigliani, W. M.: 1996, "*Chemistry of the Environment*", Prentice Hall, New Jersey, pp. 234-236.

Sponza, D. T., Müezzinoğlu, A., Alpaslan, N, Dölgen, D., Yılmaz, Z.: 1998, *A world symposium of 4*4"*, An Oral Presentation at the Symposium" AKOKS Env. Eng. Comp., İzmir and Ankara.

Standard Methods for DLE-Kit: 1990, *Measurement of contaminants in liquids with DLE-Kit*, German Environmental Monitoring Institute.

Taşdemir, Y.: 1989, *Research on Sulfate and Sulfide Parameters in Melez Creek*, Diploma study supervised by A.Müezzinoğlu, DEU, Dept. of Env.Eng.

Uslu, O., Müezzinoğu, A., Özdağlar, D.: 1998, *adı* Proceedings of ENV' 88, Dokuz Eylül Univ. Department of Env. Eng, İzmir.

Williamson, K.: 1996, *Hydrated Lime for Algae Control in Dugouts*, adapted from Agdex FS 716 (B 37).

# INVENTORY OF EMISSIONS OF GREENHOUSE GASES IN ISRAEL

J. KOCH[1], U. DAYAN[2] and A. MEY-MAROM[1]
[1] *Soreq Nuclear Research Center, 81800 Yavne, Israel.* [2] *Department of Geography, the Hebrew University of Jerusalem, 91905 Jerusalem, Israel*

**Abstract.** As a Party to the United Nations Framework Convention on Climate Change, Israel is committed to develop a national inventory of anthropogenic emissions and removals of greenhouse gases. This paper presents the national inventory, which was developed according to the guidelines of the Intergovernmental Panel on Climate Change (IPCC). The inventory includes the following sectors: energy, industrial processes, agriculture, forestry and waste. In this paper, only the inventory of the direct greenhouse gases ($CO_2$, $CH_4$ and $N_2O$) is presented. Emissions of these gases were converted to $CO_2$ equivalent emissions by means of their Global Warming Potentials (a measure of the radiative effects of the different gases relatively to $CO_2$). $CO_2$ emissions from burning fossil fuels to produce energy are by far the largest source (50 million tons in 1996). The contribution of methane emissions from decomposition of landfilled municipal solid waste is second in importance (8 million tons of $CO_2$ equivalent). Industrial processes emit about 2 million tons $CO_2$ equivalent, the most important process being cement production. Agricultural emissions amount to about 2 million tons $CO_2$ equivalent and are due to soil emissions of nitrous oxide, methane emissions from enteric fermentation in domestic livestock and $N_2O$ and $CH_4$ emissions from animal waste management. Although most forests in Israel are in a growing stage and atmospheric $CO_2$ is therefore removed to form biomass, this removal amounts to 0.4 million tons only and is very small as compared to emissions from other sectors. On a per capita basis, Israel's emissions of $CO_2$ from fuel combustion are not far behind those of some of the most developed countries.

**Keywords**: climate change, greenhouse gases, greenhouse gases in Israel, inventory of emissions of greenhouse gases

## 1. Introduction

The world community, concerned about the harmful effects of global climate change, reached a landmark agreement at the United Nations Conference on Environment and Development (UNCED) held in Rio De Janeiro in 1992. Most countries joined the UN Framework Convention on Climate Change (UNFCCC), the ultimate objective of which is "stabilization of greenhouse gas concentrations in the atmosphere at a level that would prevent dangerous anthropogenic interference with the climate system" (UNFCCC, 1992).

Two categories of countries were defined in the Convention: developed and industrialized countries (Annex I countries) and developing countries (all other countries not included in Annex I).

According to the Convention and the ensuing Kyoto Protocol agreed upon in December 1997, only Annex I countries are committed to reduce their emissions

*Water, Air, and Soil Pollution* **123:** 259–271, 2000.

of greenhouse gases (GHGs) and reduction targets are defined as amounts and periods of implementation.

As a Party to the UNFCCC not included in Annex I, Israel is only committed to: 1) develop, update periodically, and publish a national inventory of anthropogenic emissions and removals of greenhouse gases; 2) formulate and implement a national programme containing measures to mitigate climate change.

This paper presents the national inventory compiled for the year 1996, according to the Guidelines for National Greenhouse Gas Inventories of the Intergovernmental Panel on Climate Change (IPCC, 1995) and the Revised 1996 Guidelines (IPCC, 1997).

According to the IPCC Guidelines, the inventory includes five sectors which will be treated in the following sections: energy, industrial processes, agriculture, forestry and waste.

Both direct GHGs [carbon dioxide ($CO_2$), methane ($CH_4$) and nitrous oxide ($N_2O$)] and indirect GHGs which are precursors of tropospheric ozone [carbon monoxide (CO), oxides of nitrogen ($NO_x$) and non-methane volatile organic compounds (NMVOCs)] were estimated. Chlorofluorocarbons (CFCs) were not included, since they are already reported under the Montreal Protocol. Hydrofluorocarbons (HFCs), perfluorocarbons (PFCs) and sulphur hexafluoride ($SF_6$), which are all potent greenhouse gases, were not estimated since it was impossible to obtain reliable data concerning their use. Although $SO_2$ is not a GHG, it is an aerosol precursor and as such, has a cooling effect on climate and its emissions were thus estimated.

The complete national inventory is presented in two reports (Koch and Dayan, 1997; Mey-Marom and Koch, 1998). In this paper, we report only on the inventory of the direct GHGs, the contribution of which can be quantitatively estimated as $CO_2$ equivalent.

## 2. Energy

In 1996, about 50 million tons of $CO_2$ were emitted by the oxidation of carbon when fossil fuels are burned to produce energy, making it by far the largest source of $CO_2$ emissions. This estimate was made based on amounts of fuels used and the carbon content of fuels.

We adopted the breakdown of $CO_2$ emissions by the source categories defined by IPCC (IPCC, 1997). These categories are: energy industries, manufacturing industries and construction, transport, residential/ commercial/ institutional sector and agriculture.

A breakdown of the $CO_2$ emissions from fuel combustion by fuel and source category is presented in Figure 1.

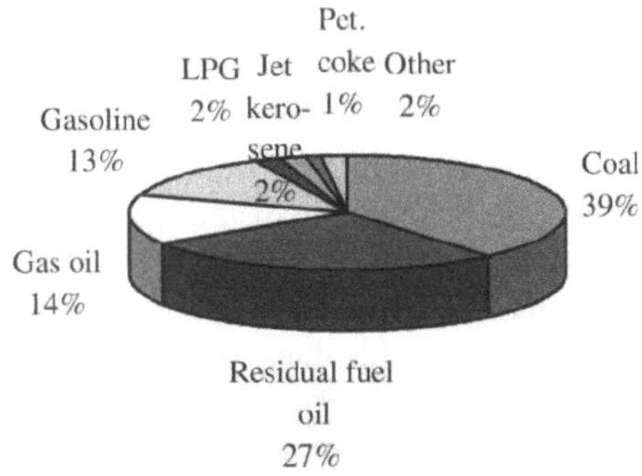


(a)

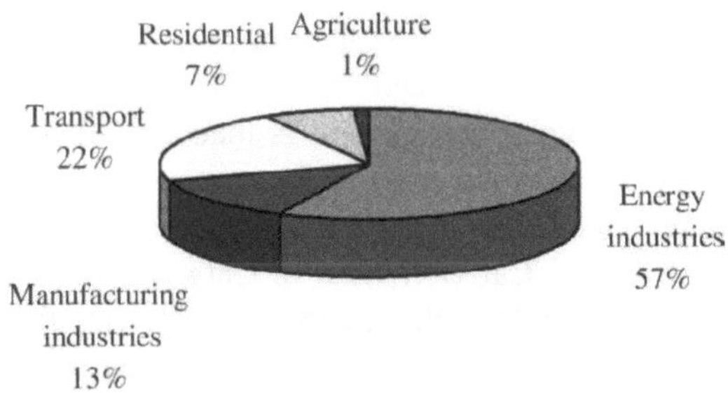


(b)

Fig. 1. A breakdown of the $CO_2$ emissions from fuel combustion by fuel (a) and source category (b).

Regarding the contribution of the different source categories, it can be seen that the energy industries (power plants and oil refineries) are by far the largest source of emissions of $CO_2$ (about 60%), followed by transport (about 20%).

When $CO_2$ emissions are assigned to the different fuels, coal contributes about 40%, whereas the contribution of residual fuel oil is 27%. Gas oil and gasoline contribute about 13% each.

The contribution of fuel combustion to emissions of $CH_4$ and $N_2O$ is minor. Methane is produced in small quantities from fuel combustion due to incomplete combustion of hydrocarbons in fuel, whereas nitrous oxide is emitted for combustion temperatures around 1000°K.

## 3. Industrial Processes

Greenhouse gases are emitted from a large variety of industrial processes not related to energy production or consumption. The main emission sources are industrial production processes which chemically or physically transform materials. Cement production is a notable example of an industrial process that releases a significant amount of $CO_2$.

We estimated emissions of $CO_2$, $CH_4$ and $N_2O$ for the relevant processes existing in Israel and for which reliable data could be found.

The general methodology used to estimate emissions associated with each industrial process involves the product of the amount of material produced/consumed by an emission factor per unit of production/consumption.

### 3.1. $CO_2$ EMISSIONS FROM CEMENT PRODUCTION

Cement production is the most important non-energy industrial process emitting $CO_2$. Carbon dioxide is produced during the production of clinker, an intermediate product from which cement is made. In the cement kiln, calcium carbonate from limestone is calcined to form lime (calcium oxide) and carbon dioxide, according to the following reaction:

$$CaCO_3 + \text{Heat} \rightarrow CaO + CO_2$$

Lime combines in the kiln with clay to form calcium silicates, which are the major compounds in clinker. "Portland" cement is then produced by pulverising the clinker and adding a small amount of gypsum.

$CO_2$ emissions are estimated by adopting an emission factor of 0.5071 t of $CO_2$ per t of clinker produced (IPCC, 1997). In 1996 the production of clinker was 3.3 x $10^6$ t and a total amount of about 1,700 kt $CO_2$ was therefore emitted.

### 3.2. $CO_2$ EMISSIONS FROM LIME PRODUCTION

Calcined limestone (quicklime) is formed by heating limestone to decompose the carbonates. This is usually done at high temperatures in a rotary kiln and the process releases $CO_2$ according to the reaction mentioned in cement production.

Dolomite may also be processed at high temperature to obtain dolomitic lime and release $CO_2$.

We adopted an emission factor of 0.79 t $CO_2$ per t lime produced (IPCC 1997). 135 kt of quicklime were produced, leading to emissions of about 100 kt $CO_2$.

### 3.3. $CO_2$ EMISSIONS FROM AMMONIA PRODUCTION

In most instances, anhydrous ammonia is produced by a catalytically assisted reaction of natural gas or other fossil fuels in the presence of steam. Natural gas is used as the feedstock in most plants, while other fuels (e.g., heavy oils) may be used with the partial oxidation process. Hydrogen is chemically separated from the fuel and combined with nitrogen to produce ammonia. The remaining carbon is emitted as $CO_2$.

Production of ammonia was 61 kt. We adopted an emission factor of 1.5 t $CO_2$ per t $NH_3$ produced (IPCC, 1997). About 100 kt of $CO_2$ were therefore emitted.

### 3.4. $N_2O$ EMISSIONS FROM NITRIC ACID PRODUCTION

The production of nitric acid ($HNO_3$) generates nitrous oxide as a by-product of the high temperature catalytic oxidation of ammonia. Nitric acid is mainly used as a raw material in the manufacture of nitrogenous fertilizers.

Emission factors of $N_2O$ from $HNO_3$ production plants depend upon technology and operating conditions and we adopted relevant values according to the process pressure in the different plants (IPCC, 1997). A total amount of 1.73 kt $N_2O$ is estimated to be emitted.

## 4. Forestry

Vegetation withdraws carbon dioxide from the atmosphere through the process of photosynthesis. Carbon dioxide is returned to the atmosphere by the (autotrophic) respiration of the vegetation and the decay (heterotrophic respiration) of organic matter in soils and litter. Changes in forest and other woody biomass stocks affect these fluxes and their balance, and may therefore cause either emissions or removals of carbon dioxide. In Israel, afforestation programmes have been implemented during the last decades and are still under way, causing a net removal of $CO_2$ from the atmosphere.

The net change in forest and other woody biomass stocks is the balance between the annual biomass growth and the annual harvest. We adopted the conservative assumption stating that all carbon removed in wood and other biomass from forests is oxidised in the year of removal (IPCC, 1997).

As shown in Table I, the annual biomass growth is calculated for each tree species or forest category by multiplying the area of forest land by the average annual growth increment (mass of dry matter) per unit area. The annual biomass loss is given by the commercial harvest data. The net annual $CO_2$ removal from the atmosphere is calculated by subtracting the amount of biomass removed

TABLE I

Calculation of $CO_2$ removal by forests

| Forest category | | Area (kha) | Growth rate (t d.m./ha/yr) | Commercial harvest (kt d.m./yr) | $CO_2$ removal (kt) |
|---|---|---|---|---|---|
| | Conifers | 57 | 4 | 85 | 236 |
| Plantations | Eucalyptus | 9 | 7 | 21 | 69 |
| | Broad-leaved | 14 | 1 | - | 17 |
| Natural Woodlands | | 40 | 0.8 | 4 | 53 |
| Total | | 120 | | 110 | 375 |

from forest stocks from the annual growth in these stocks. The increase in biomass stocks is converted to an increase in carbon by using a value of 0.45 t C per t of dry matter (d.m.) (IPCC, 1997).

It is noteworthy to mention that, although most forests in Israel are composed of conifers and broad-leaved trees, the relatively small area planted with eucalyptuses contributes about 20% of the $CO_2$ removals.

## 5. Agriculture

The agricultural sector in Israel emits both methane and nitrous oxide.

### 5.1. METHANE EMISSIONS

Methane is produced by two sources associated with domestic livestock husbandry: enteric fermentation and manure management.

Methane from enteric fermentation is produced in herbivores as a by-product of the digestive process by which carbohydrates are broken down by micro-organisms into simple molecules for absorption into the blood-stream. Ruminants (cattle, sheep) are the largest source. The amount of methane that is released depends upon the type, age and weight of the animal, as well as upon the quantity and quality of the feed consumed.

Methane from the management of animal manure is produced as the result of its decomposition under anaerobic conditions, which often occur when a large number of animals are managed in a confined area (e.g., dairy farms, beef feedlots, poultry farms).

Methane emissions were calculated by multiplying the number of heads of each livestock type by the respective emission factor for both sources: enteric fermentation and manure management.

Annual methane emissions amount to 42 kt. A breakdown of the emissions by the two sources and the main livestock types reveals that about three quarters of the total emissions are contributed by enteric fermentation, out of which more than 50% are from dairy cattle and 90% from all cattle.

The elevated contribution of the dairy cattle is explained by a high emission factor of 150 kg $CH_4$ /head/yr in Israel. The emission factor takes into account the feed energy requirements for the different physiological functions (i.e., maintenance, feeding, growth, lactation and pregnancy). Its high value is mainly due to a record milk production of 30 kg/head/day.

With regard to emissions from manure management (one quarter of total emissions), the contribution of poultry slightly exceeds that of dairy cattle, owing to the fact that in Israel manure from dairy cattle is mainly managed in dry systems, which do not favor anaerobic conditions.

## 5.2. NITROUS OXIDE EMISSIONS

### *5.2.1. Nitrous oxide direct emissions from agricultural soils*

$N_2O$ production in the soil is biogenic and it results primarily from the nitrification and denitrification processes. $N_2O$ is a gaseous intermediate in the reaction sequences of both processes and it leaks from microbial cells into the soil atmosphere. In most agricultural soils, biogenic formation is enhanced by an increase in available mineral nitrogen, which in turn increases nitrification and denitrification rates. Therefore, addition of fertilizer nitrogen directly results in extra $N_2O$ formation (emissions from unfertilized fields are considered background emissions).

The different sources of $N_2O$ which are considered are synthetic fertilizers, animal excreta nitrogen used as fertilizer, biological nitrogen fixation in N-fixing crops, nitrogen from crop residue mineralization and cultivation of high organic content soils.

Direct emissions from agricultural soils amount to 1.65 kt $N_2O$ per yr; synthetic fertilizers and animal waste contribute 66% and 20% of that amount respectively.

### *5.2.2. Nitrous oxide emissions related to animal production*

$N_2O$ is emitted from two sources related to animal production: animal waste management and animal grazing.

During storage of manure, some manure nitrogen is converted to $N_2O$ by nitrification or denitrification. The amount of $N_2O$ released depends on the system and duration of waste management. Only emissions taking place during storage or handling of manure, i.e., before the manure is added to soils, are taken into account.

$N_2O$ emissions from the relevant animal waste management systems amount to 0.80 kt per yr, of which emissions from solid storage contribute 82%.

Waste produced by animals grazing on pasture ranges induces direct soil emissions of 0.40 kt $N_2O$ per yr.

*5.2.3. Nitrous oxide emissions indirectly induced by agricultural activities*

The following pathways give rise to indirect emissions from nitrogen used in agriculture:

- volatilization and subsequent atmospheric deposition of $NH_3$ and $NO_x$ (originating from the application of fertilizers and livestock nitrogen excretion).
- nitrogen leaching and runoff.

In the first pathway, nitrogen compounds such as nitrogen oxides ($NO_x$) and ammonium (from $NH_3$), which are deposited from the atmosphere, fertilize soils and as such enhance biogenic $N_2O$ formation. The resulting $N_2O$ emissions were estimated as 0.28 kt per yr.

In the second pathway, a considerable amount of fertilizer and manure nitrogen is lost from agricultural soils through leaching and surface runoff. We assumed that the emission of $N_2O$ occurs from groundwater and surface water draining agricultural lands. It was assessed as 0.68 kt per yr.

Indirect emissions from agriculture therefore add up to 0.96 kt $N_2O$ per yr.

*5.2.4. Total $N_2O$ emissions from agriculture*

Total soil emissions are therefore the sum of direct emissions from agricultural soils, soil emissions from grazing animals and indirect emissions from agriculture, adding up to 3.01 kt $N_2O$ per yr. Total $N_2O$ emissions from agriculture are 3.81 kt per yr, after including the emissions due to animal waste management.

## 6. Waste

Methane is the main greenhouse gas emitted from the waste sector in Israel. It is produced by the anaerobic decomposition of man-made waste, when methanogenic bacteria break down organic matter in the waste. The two major sources are landfilling of municipal solid waste and treatment of domestic and industrial wastewater.

### 6.1. LANDFILLING OF MUNICIPAL SOLID WASTE

Landfilling represents the major form of Municipal Solid Waste (MSW) disposal in the industrialised world. In Israel, 90% of MSW is landfilled and the remainder is recycled. Waste management practices encourage the development and maintenance of anaerobic conditions within the landfill: the waste is compacted to minimise void space and it is covered with a soil layer. However, there are no provisions in Israel for collection of the biogas generated.

Methane emissions from landfills were calculated using a mass balance approach, which is conservative and assumes that all methane is released during the year of landfilling (IPCC, 1997). The methodology uses the degradable organic carbon (DOC) content of MSW, i.e. the organic carbon available for biochemical decomposition. It is also conservative, since it does not account for oxidation of methane within the landfill.

In 1996, the total MSW generated in Israel was $4x10^3$ kt, including domestic, commercial and yard waste. Based on a 20% fraction of DOC in MSW, the total methane emissions were estimated as 370 kt.

## 6.2. TREATMENT OF DOMESTIC AND INDUSTRIAL WASTEWATER

Wastewater can produce methane if it is treated anaerobically. Anaerobic methods are used to treat wastewater from municipal sewage and from food processing and other industrial facilities, particularly in developing countries. In contrast, developed countries typically use aerobic processes for municipal wastewater treatment or anaerobic processes in enclosed systems where methane can be recovered and utilised. In Israel a 15% fraction of the wastewater is anaerobically treated.

The main factor which determines the methane generation potential of wastewater is the amount of organic material in the wastewater stream, expressed in terms of Biological Oxygen Demand (BOD).

Methane production is calculated separately for domestic wastewater and for industrial wastewater.

### *6.2.1. Domestic wastewater*

In Israel, the average $BOD_5$ value (5-day test) of wastewater is 60 g/person/day. This value is higher than the value for domestic wastewater itself and takes into account that about 20% of the industrial wastewater streams are discharged with domestic wastewater.

The total annual methane emissions amount to 4 kt.

The annual wastewater outflow of these industries is calculated by multiplying the water consumption per unit of product by the annual product output.

The total methane released was assessed as 15 kt, out of which 9 kt are recovered (mainly in the beer breweries), resulting in net emissions of 6 kt.

TABLE II

Emissions and removals of $CO_2$, $CH_4$ and $N_2O$ from the different sectors in 1996

| Sector | $CO_2$ (kt) | $CH_4$ (kt) | $N_20$ (kt) | $CO_2$ equivalent (100 yr) (kt) |
|---|---|---|---|---|
| Energy | 50,344 | 3.6 | 0.58 | 50,599 |
| Industry | 1,889 | | 1.73 | 2,425 |
| Cement production | 1,673 | | | 1,673 |
| Lime production | 107 | | | 107 |
| Soda ash use | 17 | | | 17 |
| Ammonia production | 92 | | | 92 |
| Nitric acid production | | | 1.73 | 536 |
| Agriculture | | 42.4 | 3.81 | 2,071 |
| Domestic livestock | | 32.4 | | 680 |
| Manure management | | 10.0 | 0.80 | 458 |
| Soil emissions | | | 3.01 | 933 |
| Forest growth | -370 | | | -370 |
| Waste | | 380 | | 7,980 |
| MSW disposal | | 370 | | 7,770 |
| Wastewater treatment | | 10 | | 210 |
| Total | 51,863 | 426 | 6.12 | 62,705 |

## 7. Conclusions

### 7.1. SUMMARY OF THE INVENTORY

Table II summarizes the emissions and removals of $CO_2$, $CH_4$ and $N_2O$ from the different sectors, as they were estimated for the year 1996. The methane and nitrous oxide emissions are converted to $CO_2$ equivalent, by means of the Global Warming Potential (GWP) which is a measure of the radiative effects of the different greenhouse gases relatively to $CO_2$. For a time horizon of 100 years, the GWPs of methane and nitrous dioxide are 21 and 310 respectively (IPCC, 1996).

The contribution of the different sectors and of the three direct GHGs is presented as two pie-charts in Figure 2.

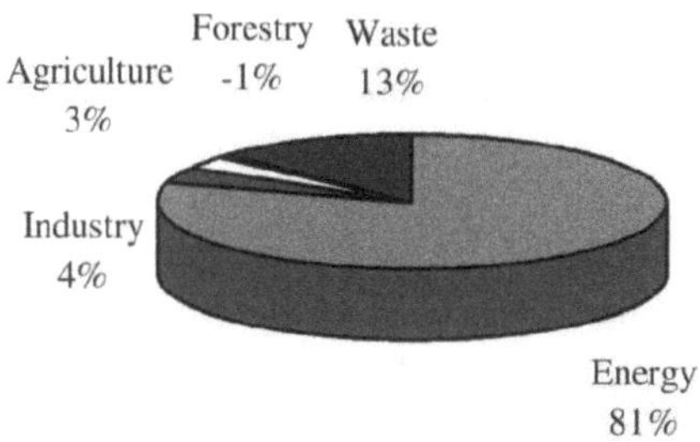


(a)

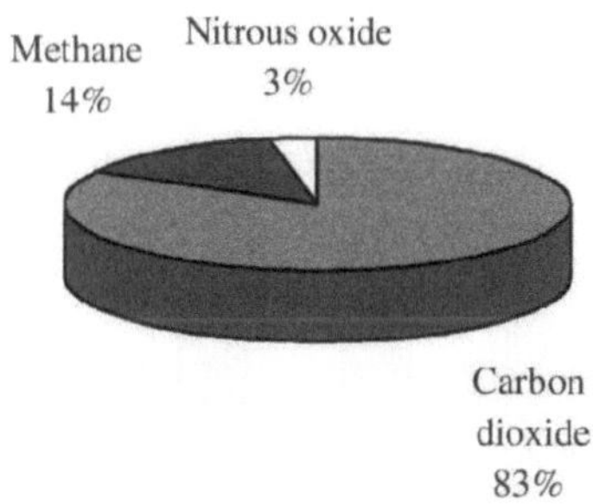


(b)

Fig. 2. The contribution of the different sectors (a) and of the different GHGs (b) to total $CO_2$ equivalent emissions

It can be seen from Table II and Figure 2a that the contribution of methane emissions from decomposition of the solid waste is significant (13% of all emissions for a time horizon of 100 years). It is second in importance to the contribution of $CO_2$ emissions from energy production. This relatively high contribution is explained by the large production of solid waste on one hand, and the fact that only a small fraction is recycled and none is incinerated on the other hand. The very small extent of $CO_2$ removals by forest biomass increase is also evident.

Among the different gases, the contribution of carbon dioxide is very dominant, whereas the contribution of nitrous dioxide is only marginal (Figure 2b).

7.2. ISRAEL AS COMPARED TO THE WORLD

It is also interesting to compare - on a per capita basis - the emissions of Israel with the emissions of other countries and the world average. This comparison was held for $CO_2$ emissions from fuel combustion, which account for 80% of the total radiative forcing of GHGs. As can be seen in Figure 3, the world average stands at 1.1 t carbon (C) per person and per year, but the distribution is very uneven. On the one hand, the U.S. as one of the most energy-consuming countries among the developed ones, emits fivefold the world average. On the other hand, large developing countries such as China, are approaching the world average, while India's emissions are only one fifth the world average. Israel, with its emissions of 2.4 t C, is not far behind some of the more developed

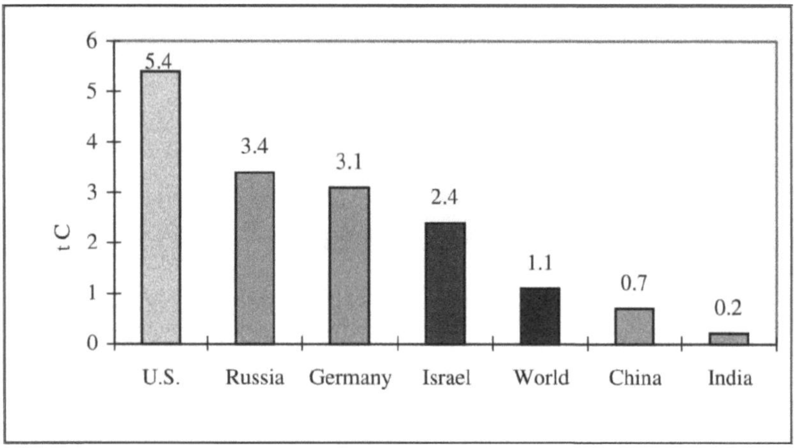


Fig. 3. Comparison of $CO_2$ emissions from fuel combustion in different countries

Annex I countries, such as Germany. The implication of this last finding is that recently-developed countries like Israel could consider voluntary restrictions on the growth of their emissions, as advocated by Koch *et al.* (1999).

**Acknowledgments**

This study was supported by the Ministry of the Environment. We thank the following ministries, institutions and industrial companies for providing us the relevant data and information: Ministry of National Infrastructures, Electricity Authority and Fuel Authority; Israel Electric Company; Ministry of Industry and Trade; Nesher Israel Cement Enterprises Ltd.; Fertilizers and Chemicals Ltd.; Haifa Chemicals Ltd.; Land Development Authority, Forest Department; Central Bureau of Statistics, Agriculture Department; Ministry of Agriculture,

Extension Service, Cattle Department and Mechanisation and Technology Department; Ministry of the Environment, Division of Solid Waste Management and Coordinator for Industrial Wastewater; Biotech Environmental Company Ltd.; Ministry of Agriculture, Water Commission.

## References


IPCC: 1995, *IPCC Guidelines for National Greenhouse Gas Inventories.* United Nations Environment Programme (UNEP), the Organisation for Economic Co-operation and Development (OECD), the International Energy Agency (IEA) and the International Panel on Climate Change (IPCC).

IPCC: 1996, *Climate Change 1995. The Science of Climate Change*, J.T. Houghton *et al.*(ed), Cambridge University Press, Cambridge, UK.

IPCC: 1997, *Revised 1996 IPCC Guidelines for National Greenhouse Gas Inventories,* United Nations Environment Programme (UNEP), the Organisation for Economic Co-operation and Development (OECD), the International Energy Agency (IEA) and the International Panel on Climate Change (IPCC).

Koch, J. and Dayan, U.: 1997, *Inventory of Emissions and Removals of Greenhouse Gases in Israel. Part A: Carbon Dioxide and Methane*, Report SNRC-2784, Soreq Nuclear Research Center, Yavne, Israel, 32 pages.

Koch, J., Dayan, U. and Amir, S.: 1999, Stabilization of atmospheric concentration of greenhouse gases. A proposal for "recently-developed countries to participate in the global effort, submitted to *Global Environmental Change*.

Mey-Marom, A. and Koch, J.: 1998, *Inventory of Emissions and Removals of Greenhouse Gases in Israel. Part B: Nitrous Oxide and Precursors of Ozone and Aerosols. Part C: Reporting the National Inventory*, Report SNRC-2865, Soreq Nuclear Research Center, Yavne, Israel, 72 pages.

UNFCCC: 1992, *United Nations Framework Convention on Climate Change*, UNEP/IUC, Geneva, Switzerland.

# ANALYSIS OF UNCERTAINTY AND VARIABILITY IN EXPOSURE TO CHARACTERIZE RISK: CASE STUDY INVOLVING TRICHLOROETHYLENE GROUNDWATER CONTAMINATION AT BEALE AIR FORCE BASE IN CALIFORNIA

J. I. DANIELS, K. T. BOGEN and L. C. HALL
*Health & Ecological Assessment Division, Earth & Environmental Sciences Directorate*
*Lawrence Livermore National Laboratory*
*P.O. Box 808, L–396, Livermore, CA 94551-0808*

**Abstract.** Quantitative assessments of potential human-health consequences from contaminants in environmental media routinely involve conservative deterministic, screening-level calculations of exposure and risk. Because these calculations generally are based on multiple upper-bound point estimates of input parameters, particularly for exposure attributes, they can yield results for decision makers that actually overstate the need for costly remediation. Alternatively, quantifying uncertainty and variability in exposure can provide a more informative and quantitative characterization of health risk. To illustrate, uncertainty and variability in exposure were analyzed for a hypothetical population at a specific site in California where there is trichloroethylene (TCE) contaminated ground water and a potential for its residential use. When uncertainty and variability in exposure were addressed jointly, the 95th-percentile upper-bound value of individual excess lifetime cancer risk was a factor approaching 10 lower than the most conservative deterministic estimate. Also, the probability of more than zero additional cases of cancer can be estimated, and in this case study it is less than 0.5 for a prospective residential population of up to 26,900 individuals present for any 7.6-y interval of a 70-y time period. Clearly, this probabilistic approach can provide reasonable and equitable risk-acceptability criteria for contaminated sites.

**Keywords:** exposure, ground water, joint uncertainty and variability, probability, risk, trichloroethylene

## 1. Introduction

Quantitative assessments of the potential human health risks from contaminants present at hazardous-waste sites typically involve conservative deterministic, screening-level calculations of exposure and risk, often based on multiple upper-bound point estimates of input parameters. Because inherent conservatism in such estimates may result in highly inefficient strategies for site cleanup, there is growing interest in obtaining more informative and quantitative characterizations of human-health risk (NRC, 1994). Such assessments require quantitative methods to characterize joint uncertainty and interindividual variability (JUV) in estimated risk, based on both uncertainty and/or interindividual variability reflected in each input parameter (Bogen and Spear, 1987; NRC, 1994; Bogen, 1995). Uncertainty here refers to an absence

*Water, Air, and Soil Pollution* **123:** 273–298, 2000.

of measurement data or incomplete knowledge; interindividual variability (or "variability") here refers to true differences or heterogeneity in an empirical, risk-related characteristic (e.g., physiological differences) among individuals in a population (Bogen and Spear, 1987). Such probabilistic assessments can be resource intensive, but are generally appropriate at sites for which deterministic upper-bound calculations of risk overstate the need for costly remediation efforts.

This paper provides a site-specific illustration of how JUV in exposure may be used to characterize risk. Results from such analyses provide an improved understanding of risk for decision makers, including estimates of the upper-bound risk to the average person in the population, the risk to an individual at the upper-bound of exposure, and the likelihood of additional cases of cancer in a population exposed to low-level site contaminants. The case study addresses inactive Landfill Site LF-13 on Beale Air Force Base in California, where groundwater contaminated with trichloroethylene (TCE) has moved beyond the site boundary and so soil-vapor extraction and air-stripping treatment of groundwater have been undertaken at Site LF-13 to remediate this situation (URSGWC, 1998). These actions are designed to reduce to low-levels the concentrations of TCE (and other volatile organic compounds) in the ground water beneath Site LF-13. This is especially important because in this currently rural area of the Sacramento Valley of California, groundwater wells are the principle source of domestic-water supplies and elevated levels of TCE contamination, particularly, would prevent this water from being used for this purpose. Accordingly, the potential risks considered are those attributable to a scenario involving possible future domestic, residential uses of groundwater containing residual, low-level concentrations of TCE that theoretically could be extracted and distributed from Site LF-13 for hypothetical residential populations of different size that might eventually occupy lands adjacent to the site. Those TCE-concentration measurements obtained in 1997 from the groundwater monitoring well near the possible location of such groundwater extraction on Site LF-13 (Purrier, 1997) were used for this analysis. A characterization of JUV in risk is performed, and corresponding estimates of the expected number of additional cancer cases and the probability of greater than zero additional cases for specified populations are obtained. Finally, risk estimators from the JUV approach are compared to those obtained using the traditional framework for computing risk deterministically.

## 2. Methods

The procedures utilized here are ones designed to address JUV in the context of risk characterization (Bogen and Spear, 1987; NRC, 1994; Bogen, 1995). For TCE in groundwater at Site LF-13 on Beale Air Force Base in California, total risk, *R,* is defined as the increased lifetime probability of cancer for an individual attributable to TCE exposure from three pathways: direct ingestion, $E_{\mathrm{Ing}}$, of TCE-contaminated groundwater; dermal absorption, $E_{\mathrm{Derm}}$, of TCE while showering or bathing; and inhalation, $E_{\mathrm{Inh}}$, of TCE volatilized from water to household air. For volatile organic compounds such as TCE, these three pathways typically are the most significant contributors to total dose.

Consistent with previous JUV notation an overbar (i.e., $\overline{\ \ }$) is used below to denote expectation with respect to heterogeneous parameters only, and angle brackets $\left(\text{i.e., } \langle\ \rangle\right)$ to denote expectation with respect to uncertain parameters only (Bogen and Spear, 1987; NRC, 1994; Bogen, 1995). A tilde (i.e., ~ ) appearing above a term denotes a sample mean of empirical values.

## 2.1. EXPOSURE-PATHWAY MODELS

The equations used to model the three most important human exposure pathways for TCE in ground water are described below. These models are consistent with those described by USEPA (1989) and also CalEPA/DTSC (1994) for these pathways.

Exposure to TCE from direct ingestion of groundwater was calculated using Eq. 1.

$$E_{\mathrm{Ing}} = Ing \times ED \times \frac{EF}{AT} \times C_{\mathrm{w}} \ , \tag{1}$$

where

$E_{\mathrm{Ing}}$ = TCE-exposure (intake) resulting from direct ingestion of contaminated ground water [mg/(kg d)];

$Ing$ = daily water ingestion rate per unit body weight [L/(kg d)];

$ED$ = exposure duration (y);

$EF$ = exposure frequency (d/y);

$AT$ = averaging time corresponding to a 70-y lifetime of exposure (d); and

$C_{\mathrm{w}}$ = TCE concentration in ground water (mg/L).

TCE can volatilize to indoor air from water used in showering, bathing, and by the use of toilets, dishwashers, washing machines, and cooking. Inhalation exposure to TCE was calculated by the procedure of McKone and Bogen (1992) for estimating uptake of a volatile contaminant in tap water for a hypothetical four-occupant household. That approach utilizes contaminant water-to-air transfer factors in conjunction with the model developed by Fisk *et al.* (1987) to estimate household-compartment concentrations of volatile contaminants in air. Therefore, the resulting exposure to TCE in indoor air was derived using Eq. 2.

$$E_{\mathrm{Inh}} = Inh \times \frac{1}{D} \times \left[ \left( \frac{W_{\mathrm{sh}} \times \phi_{\mathrm{TCE\text{-}sh}}}{AE_{\mathrm{sh}}} \times ET_{\mathrm{sh}} \right) + \left( \frac{W_{\mathrm{b}} \times \phi_{\mathrm{TCE\text{-}sh}}}{AE_{\mathrm{b}}} \times ET_{\mathrm{b}} \right) + \left( \frac{W_{\mathrm{h}} \times \phi_{\mathrm{TCE\text{-}h}}}{AE_{\mathrm{h}}} \times ET_{\mathrm{h}} \right) \right] \times ED \times \frac{EF}{AT} \times C_{\mathrm{w}}, \tag{2}$$

where

$E_{\mathrm{Inh}}$ = TCE-exposure (intake) resulting from inhalation of TCE volatilized into indoor air from contaminated ground water used for domestic purposes [mg/(kg d)];

$Inh$ = daily inhalation rate per unit body weight [$\mathrm{m^3}$/(kg d)];

$D$ = averaging time for daily water use (24 h/d);

$W_{\mathrm{sh}}$ = water-useage rate(s) for shower and also for bathroom, $W_{\mathrm{b}}$, (L/h);

$W_{\mathrm{h}}$ = water-useage rate for all household activities (L/h);

$\phi_{\mathrm{TCE\text{-}sh}}$ = water-to-air transfer efficiency of TCE in the shower (dimensionless);

$\phi_{\mathrm{TCE\text{-}h}}$ = water-to-air transfer efficiency of TCE in the house (dimensionless), and equal to $\phi_{\mathrm{TCE\text{-}sh}} \times \frac{0.54}{0.70}$ (where the fraction is the ratio of radon transfer in the shower to radon transfer in the house as reported by McKone and Bogen, 1992), with $\phi_{\mathrm{TCE\text{-}h}}$ modeled as statistically independent of $\phi_{\mathrm{TCE\text{-}sh}}$;

$AE_{\mathrm{sh}}$ = air-exchange rate in the shower or bath stall ($\mathrm{m^3}$/h);

$AE_{\mathrm{b}}$ = air-exchange rate in the bathroom ($\mathrm{m^3}$/h);

$AE_{\mathrm{h}}$ = air-exchange rate in the house ($\mathrm{m^3}$/h);

$ET_{\mathrm{sh}}$ = exposure time in showering or bathing (h/d);

$ET_{\mathrm{b}}$ = exposure time in bathroom (h/d);

$ET_{\mathrm{h}}$ = exposure time in house (h/d);

$ED$ = exposure duration (y);

$EF$ = exposure frequency (d/y);

$AT$ = averaging time corresponding to a 70-y lifetime of exposure (d); and

$C_w$ = TCE concentration in ground water (mg/L).

Dermal uptake of TCE while showering or bathing is based on the model of Brown *et al.* (1984) and was calculated from the relationship shown in Eq. 3.

$$E_{\text{Derm}} = A \times f_s \times k_p \times ET_{\text{sh}} \times \text{cf} \times ED \times \frac{EF}{AT} \times \left[ C_w \times \left( 1 - \frac{\phi_{\text{TCE-sh}}}{2} \right) \right], \quad (3)$$

where

$E_{\text{Derm}}$ = TCE-exposure (intake) resulting from dermal uptake of TCE while showering or bathing [mg/(kg d)];

$A$ = surface area of skin per unit body weight ($cm^2$/kg);

$f_s$ = fraction of total skin surface that is in contact with water during showering or bathing (dimensionless);

$k_p$ = dermal permeability rate of TCE from dilute aqueous solutions (cm/h);

$ET_{\text{sh}}$ = time spent showering or bathing (h/d);

cf = conversion factor ($10^{-3}$ L/$cm^3$);

$ED$ = exposure duration (y);

$EF$ = exposure frequency (d/y);

$AT$ = averaging time corresponding to a 70-y lifetime of exposure (d);

$C_w$ = TCE concentration in ground water (mg/L); and

$\phi_{\text{TCE-sh}}$ = water-to-air transfer efficiency of TCE in the shower (dimensionless).

Three concentration measurements of TCE were obtained in 1997 from a monitoring well at Site LF-13 on Beale Air Force Base (Purrier, 1997). This monitoring well is used for evaluating remediation efforts and is located in the immediate vicinity of the site of an extraction well that hypothetically could eventually supply ground water for domestic purposes to possible future residences in the surrounding area. Because soil-vapor extraction and air-stripping treatment of the ground water have been taking place at Site LF-13 to reduce the concentration of TCE to low-levels in the ground water (URSGWC, 1998), it is assumed that there are now no real differences between the three reported sample measurements and that the TCE concentration in the ground water is unlikely to be changing in time. On the basis of these assumptions (which are made for purposes of this illustration and require validation) and because there will be mixing and blending of the ground water during its extraction and distribution, a hypothetical resident using such ground water domestically is likely to be exposed to the mean concentration. Accordingly, the uncertain mean TCE concentration in ground water was modeled as

$$C_w = \left[ \frac{e^{\tilde{\sigma}_{\log \tilde{c}_w} \times T_2}}{E\left( e^{\tilde{\sigma}_{\log \tilde{c}_w} \times T_2} \right)} \right] \times \tilde{c}_w \ , \tag{4}$$

where

$C_w$ = mean TCE concentration (mg/L), where uncertainty in $\log \tilde{c}_w$ is assumed to be T-distributed with two degrees of freedom;

$\langle C_w \rangle \equiv \tilde{c}_w$ = the sample mean of the three $c_w$ measures (Purrier, 1997);

$\log \tilde{c}_w$ = sample mean of the three $\log c_w$ measures;

$\tilde{\sigma}_{\log c_w}$ = sample standard deviation of the three $\log c_w$ measures;

$\tilde{\sigma}_{\log \tilde{c}_w}$ = standard deviation of the sample mean of $\log \tilde{c}_w$, where $\tilde{\sigma}_{\log \tilde{c}_w} = \dfrac{\tilde{\sigma}_{\log c_w}}{\sqrt{3}} = 0.1295$; and

$T_2$ = variate distributed as Student's T with two degrees of freedom.

The expected-value term in Eq. 4, $E\left( e^{\tilde{\sigma}_{\log \tilde{c}_w} \times T_2} \right)$, was determined to be 1.0812, based on a Monte-Carlo simulation involving 2000 trials. The bracketed term in Eq. 4 thus reflects a log-$T_2$-distributed variate normalized to have an expected value equal to one.

Inter-household variability in water-to-air transfer efficiency of TCE in shower water ($\phi_{\text{TCE-sh}}$) was modeled based on 14 experimental measures involving showers running water at ≥ 30 °C summarized by Corsi and Howard (1998). It was assumed that these measures reflect the effects on TCE transfer of variable conditions that may pertain to each household at risk over the course of any residential duration. Effective residential TCE water-to-air transfer efficiency, $\phi_{\text{TCE-sh}}$, was therefore estimated as the mean value of the reported measures (0.76), and variability in $\phi_{\text{TCE-sh}}$ was modeled by the relation

$$\phi_{\text{TCE-sh}} = 0.76 + \left( 0.029 \times T_{\phi_{\text{TCE-sh}}} \right) , \tag{5}$$

where 0.029 is the standard deviation of the mean of the measured values and $T_{\phi_{\text{TCE-sh}}}$ has a Student's T distribution with 13 degrees of freedom.

The term $\left[ C_w \times \left( 1 - \frac{\phi_{\text{TCE-sh}}}{2} \right) \right]$ in Eq. 3 estimates the concentration of TCE in the water contacting the skin during showering, based on the assumption that TCE volatilization is approximately linearly proportional to the vertical distance water has fallen from the showerhead to the floor (Giardino *et al.*, 1992) , and that during showering the body contacts the water about half the distance between the showerhead and the floor. This is a conservative assumption with respect to bathing, because all TCE volatilization would occur during the bathtub filling prior to bathing.

Table I presents the input parameters identified or implied in Eqs. 1–3, but does not include the regulatory default values for such inputs, which appear in Table II. In Table I, distributions for parameters are identified as representing either uncertainty or variability (heterogeneity) and corresponding distribution types are also listed. The exposure-model parameters treated as constants in this assessment are *EF* and *AT*. Other input variates were assumed to be distributed as summarized in Table I and as further described below. As indicated in Table I, all distributed input variates were assumed to be heterogeneous (i.e., to reflect interindividual variability), except the concentration $C_w$ and the fraction $f_m$ variates as also described below.

TABLE I
Inputs (not including regulatory default values; see Table II) for obtaining cancer risk-related estimators (see Table III) for multiple-pathway exposure to trichloroethylene (TCE) concentrations in ground water at Beale Air Force Base in California.

| | | Distribution | | Range | | Arithmetic | | Geometric | | Percentile | | |
|---|---|---|---|---|---|---|---|---|---|---|---|---|
| Variate (units) | Sym-bol | Repre-sents | Type[a] | Min. | Max. | Mean | Stnd. Dev. | Mean | Stnd. Dev. | 5th | 95th | Source of data |
| Mean TCE concentration in water (mg/L) | $C_w$ | Uncertainty | log-$T_2$ | | | 0.0223 | | | | | 0.0301 | Purrier (1997) |
| Fraction of emigrant residents moving out of a local water-supply district in western US (dimensionless) | $f_m$ | Uncertainty | Tri | $\frac{1}{3}$ | 1 | $\frac{2}{3}$ | $\sqrt{\frac{1}{54}}$ | | | 0.439 $\left(=f_m^*\right)$ | | USCB (1997) |
| Cumulative distribution function for total residence time ($ED$) in the western US $\leq t$ (y) | $1-R(t)$ | Variability | E | | | 3.49 | | | | | | Israeli and Nelson (1992) |
| Approximate upper-bound residence duration (used to calculate $\hat{R}_{High}$) | $\hat{ED}$ $\left[=1-R(t)^{f_m^*}\right]$ | Variability | E | | | | | | | | 55.3 | See Eqs. 8 and 11 |

TABLE I (continued)

| | | Distribution | | Range | | Arithmetic | | Geometric | | Percentile | | |
|---|---|---|---|---|---|---|---|---|---|---|---|---|
| Variate (units) | Symbol | Represents | Type[a] | Min. | Max. | Mean | Stnd. Dev. | Mean | Stnd. Dev. | 5th | 95th | Source of data |
| Ingestion rate for western region of US [L/(kg d)] | *Ing* | Variability | LN | | | 0.0242 | 0.0170 | 0.0198 | 1.88 | | 0.0399 | Ershow and Cantor (1989)[b] |
| Inhalation rate [$m^3$/(kg d)] | *Inh* | Variability | E | | | 0.264 | | | | | 0.363 | OEHHA (1996); Marty (1998); USCB (1998)[b] |
| Shower (and also bathroom) water-use rate(s) (L/h) | $W_{sh}$ (and $W_b$) | Variability | LN | | | 480 | 160 | 455 | 1.38 | | 777 | McKone and Bogen (1992), based on James and Knuiman (1987) |
| Household water-use rate (L/h) | $W_h$ | Variability | LN | | | 42.0 | 15.0 | 40.0 | 1.41 | | 69.9 | McKone and Bogen (1992), based on James and Knuiman (1987) |

TABLE I (continued)

| | | Distribution | | Range | | Arithmetic | | Geometric | | Percentile | | |
|---|---|---|---|---|---|---|---|---|---|---|---|---|
| **Variate (units)** | **Sym-Bol** | **Repre-sents** | **Type[a]** | **Min.** | **Max.** | **Mean** | **Stnd. Dev.** | **Mean** | **Stnd. Dev.** | **5th** | **95th** | **Source of data** |
| Normalized mean water-to-air transfer efficiency for TCE (dimensionless) | $^{T}\phi_{TCE\text{-}sh}$ | Variability | $T_{13}$ | | | 0 | | | | | 1.771 | Corsi and Howard (1998) |
| Air-exchange rate for shower ($m^3/h$) | $AE_{sh}$ | Variability | U | 4.0 | 20.0 | 9.94[c] | | | | 4.82[c] | | McKone and Bogen (1992) |
| Air-exchange rate for bathroom ($m^3/h$) | $AE_b$ | Variability | U | 10.0 | 100.0 | 39.1[c] | | | | 14.6[c] | | McKone and Bogen (1992) |
| Air-exchange rate for house ($m^3/h$) | $AE_h$ | Variability | U | 300 | 1200 | 649[c] | | | | 344[c] | | McKone and Bogen (1992) |
| Exposure time in shower (h/d) | $ET_{sh}$ | Variability | LN | | | 0.129 | 0.052 | 0.120 | 1.47 | | 0.226 | Burmaster (1998) |
| Exposure time in bathroom (h/d) | $ET_b$ | Variability | LN | | | 0.330 | 0.220 | 0.274 | 1.83 | | 0.744 | McKone and Bogen (1992) |

TABLE I (continued)

| Variate (units) | Sym-bol | Distribution | | Range | | Arithmetic | | Geometric | | Percentile | | Source of data |
|---|---|---|---|---|---|---|---|---|---|---|---|---|
| | | Repre-sents | Type[a] | Min. | Max. | Mean | Stnd. Dev. | Mean | Stnd. Dev. | 5th | 95th | |
| Exposure time in house (h/d) | $ET_h$ | Variability | U | 8.0 | 20.0 | 14.0 | | | | | 19.4 | McKone and Bogen (1992) |
| Surface area per unit body weight ($cm^2/kg$) | $A$ | Variability | E | | | 326 | | | | | 373 | Phillips *et al*. (1993); USCB (1998)[b] |
| Fraction of skin exposed in shower or bath (dimensionless) | $f_s$ | Variability | U | 0.40 | 0.90 | 0.65 | | | | | 0.875 | McKone and Bogen (1992) |
| Skin-permeability coefficient (cm/h) | $k_p$ | Variability | N | | | 0.263 | 0.018 | | | | 0.293 | Bogen *et al*. (1998) |
| Cancer slope factor applicable to both ingestion and dermal exposures {R/[mg/(kg d)]} | $CSF_{Ing}$ | Not applicable | C | | | | | | | | 0.015 | CalEPA (1996) |
| Cancer slope factor applicable to inhalation exposure {R/[mg/(kg d)]} | $CSF_{Inh}$ | Not applicable | C | | | | | | | | 0.010 | CalEPA (1996) |

TABLE I (continued)

| Variate (units) | Symbol | Distribution | | Range | | Arithmetic | | Geometric | | Percentile | | Source of data |
|---|---|---|---|---|---|---|---|---|---|---|---|---|
| | | Represents | Type[a] | Min. | Max. | Mean | Stnd. Dev. | Mean | Stnd. Dev. | 5th | 95th | |
| Averaging time for 70-y lifespan (d) | *AT* | Not applicable | C | | 25,550 | | | | | | | USEPA Region 9 (1998); USEPA (1989) |
| Exposure frequency (upper-bound value; d/y) | *EF* | Not applicable | C | | 350 | | | | | | | USEPA Region 9 (1998); USEPA (1991) |

[a] Distribution types: C = constant; E = empirical (or fitted); LN = lognormal; N = normal; $T_{df}$ = Student's T with df equal to degrees of freedom; log-$T_{df}$ = exponentiated $T_{df}$ distribution; Tri = triangular, U = uniform.

[b] Upper-bound (95[th] percentile) values for lifetime, time-weighted-average quantities calculated using information from the cited references (see Methods).

[c] Mean and corresponding 5%-tile values associated with each air-exchange rate were obtained from the inverse-uniform distribution (1/U) that was constructed from a Monte-Carlo simulation, involving 2,000 trials, of the uniform distribution. Thus, values reported in units of the data are the harmonic mean and the inverse of the 95%-tile of 1/U. This was done so that expected values of risk-related estimators could be calculated using the corresponding exact expressions (which include *AE* values appearing as denominators—see Eq. 2).

TABLE II

Inputs and corresponding regulatory default values applicable to a deterministic calculation of excess-lifetime cancer risk for a "reasonably maximum exposed" person ($\hat{R}_{\text{RME}}$).[a]

| Variate (units) | Value | Reference |
|---|---|---|
| Ingestion rate (L/d) | 2.0 | USEPA Region 9 (1998); USEPA (1989) |
| Body weight (kg) | 70.0 | USEPA Region 9 (1998); USEPA (1989) |
| Inhalation rate ($m^3$/d) | 20.0 | USEPA Region 9 (1998); USEPA (1989) |
| Exposure time in house (h/d) | 16.4 | USEPA Region 9 (1998); Tsang and Klepeis (1996) |
| Shower duration (h/d) | 0.13 | USEPA (1997); James and Knuiman (1987) |
| Skin-surface area ($cm^2$) | 23,000.0 | CalEPA/DTSC (1994) |
| Residential-exposure duration (y) | 30.0 | USEPA Region 9 (1998); USEPA (1989) |
| Residential-exposure frequency (y) | 350.0 | USEPA Region 9 (1998); USEPA (1991) |
| Averaging time (d) | 25,550.0 | USEPA Region 9 (1998); USEPA (1989) |
| Ingestion (and used for dermal) cancer-slope factor ($CSF_{\text{Ing}}$) {Risk/[mg/(kg d)]} | 0.015 | CalEPA (1996) |
| Inhalation cancer-slope factor ($CSF_{\text{Inh}}$) {Risk/[mg/(kg d)]} | 0.01 | CalEPA (1996) |

[a] Characterizing risk for the "reasonable maximum exposure" case involves combining upper-bound and mid-range factors so that a conservative estimate (i.e., above the average) results that is within the range of reasonable possibilities, and is not the worst-possible case (USEPA, 1989 and 1991). The inputs to the $\hat{R}_{\text{RME}}$ identified here are consistent with this goal. Specifically, the inputs and corresponding regulatory default values shown are used. Where default values are not given (and cannot be obtained from those shown) for *uncertain* variates (e.g., TCE concentration in water) the expected value for that input is used (see Table I). Similarly, in the absence of default values for *heterogeneous* variates (e.g., water-use rates) the 95%-tile value for that input is used (see Table I); unless the heterogeneous variate was in the denominator of an equation (e.g., air-exchange rates), and then the 5%-tile value is used (see Table I).

The exposure duration *(ED)* term, in Eqs. 1–3 denotes household residence time in the area that would be supplied with the contaminated ground water for domestic purposes. Because *ED* should account for households moving into and out of the water-supply area, it is modeled to reflect nonlinear JUV. The procedure used to obtain $\overline{ED}$ and $\langle ED \rangle$ distributions adapts the Israeli and Nelson (1992) model of the variability in the time of residence for households living in the Western Region of the US.

Specifically, this model defines the fraction *R*(*t*) of households living in the same residence for a total of *t* years or more for the Western Region [see Eq. 12 and the corresponding parameter values in Table II of Israeli and Nelson (1992)].* According to this model,

$$l(s) = \frac{-\,\mathrm{d}[\log R(s)]}{\mathrm{d}s}, \tag{6}$$

where $0 \le s \le t$ and $l(s)$ is the rate of household moves, implying that $R(t)$ is modeled as a single "compartment" with loss rate $l(s)$ for $0 \le s \le t$, i.e., as

$$R(t) = R_0 \mathrm{e}^{-\int_0^t l(s)\,\mathrm{d}s}, \tag{7}$$

where $R_0 = R(0) = 1$, and $R(\infty) = 0$. Now, let $f_m$ be the fraction of household moves that are "effective", because they involve moves out of an area of concern (in this case study, a hypothetical future water-supply district). Thus,

$$R_{f_m}(t) = \mathrm{e}^{-\int_0^t l(s)\times f_m \mathrm{d}s} = [R(t)]^{f_m}, \tag{8}$$

where $R(t)$ is heterogeneous and $f_m$ is uncertain. Based on geographic mobility data reported by the US Census Bureau (1997), about $\frac{2}{3}$ of all US moves are within the same county. We assume that these moves include an uncertain fraction $(1 - f_m)$ that are within the same water-supply district, and that $(1 - f_m)$ is triangularly distributed between 0 and $\frac{2}{3}$ with a mode at $\frac{1}{3}$, which is consistent with data indicating that many households move small distances within corresponding local areas (ARC, 1999; and Duke-Williams, 1999). Thus, as indicated in Table I we assume that $f_m$ is triangularly distributed between $\frac{1}{3}$ and 1 with a mode at $\frac{2}{3}$.

---

* Note that we retain here the Israeli and Nelson (1992) notation for the fraction *R*(*t*) as a function of time, which should not be confused with risk, *R*, defined (independent of time) in Eq. 13 of this report.

The population-average value of total residence time, $\overline{ED}$, with respect to variability in *ED,* is defined by Israeli and Nelson (1992) as

$$\overline{ED} = \int_0^\infty R(t)\mathrm{d}t \ , \tag{9}$$

(i.e., conditional on $f_\mathrm{m} = 1$). It follows that for any value of $f_\mathrm{m}$, the corresponding population-average value of uncertain total residence time is specified by

$$\overline{ED} = \int_0^\infty R_{f_\mathrm{m}}(t)\mathrm{d}t = \int_0^\infty [R(t)]^{f_\mathrm{m}}\mathrm{d}t \ , \tag{10}$$

in which uncertainty in $f_\mathrm{m}$ was discussed above.

The cumulative probability distribution reflecting variability in total time of residence, *t,* is defined as $1 - R(t)$, in the model of Israeli and Nelson (1992; see their Eq. 4). The corresponding definition of variability in expected *ED,* conditional on $f_\mathrm{m}$, is given by

$$\langle ED \rangle = 1 - R_{f_\mathrm{m}}(t) \ , \tag{11}$$

which, in view of the nonlinear relationship between uncertainty and heterogeneity in $R_{f_\mathrm{m}}(t)$, was approximated as follows, using a second-order estimate of $\langle R_{f_\mathrm{m}}(t) \rangle$ (see Bogen and Spear, 1987):

$$\langle ED \rangle = \langle 1 - [R(t)]^{f_\mathrm{m}} \rangle \approx 1 - \left\{ [R(t)]^{\langle f_\mathrm{m} \rangle} + \left( \frac{1}{2} \times \frac{\partial^2 R_{f_\mathrm{m}}(t)}{\partial f_\mathrm{m}^2} \times \sigma_{f_\mathrm{m}}^2 \right) \right\} \approx$$

$$1 - \left\{ [R(t)]^{\langle f_\mathrm{m} \rangle} \times \left( 1 + \frac{\{\ln[R(t)]\}^2 \times \sigma_{f_\mathrm{m}}^2}{2} \right) \right\} . \tag{12}$$

For further details concerning this derivation, and the procedures for obtaining $\overline{ED}$ and $\langle ED \rangle$ distributions, see Daniels *et al.* (1999).

## 2.2 . CANCER-RISK MODEL

Because hypothetical residential low-dose exposure to TCE, such as might occur as a result of groundwater contamination at Site LF-13 on Beale Air

Force Base, is assumed to have a positive, nearly constant slope at doses small enough to ensure $R$<<1, $R$ can be estimated by Eq. 13:

$$R \cong (E_{\mathrm{Ing}} \times CSF_{\mathrm{Ing}}) + (E_{\mathrm{Derm}} \times CSF_{\mathrm{Ing}}) + (E_{\mathrm{Inh}} \times CSF_{\mathrm{Inh}}), \tag{13}$$

where $CSF_{\mathrm{Ing}}$ is the oral cancer slope factor ($CSF$) for TCE [assumed to apply to both ingestion and dermal exposures (CalEPA/DTSC, 1994; USEPA Region 9, 1998)], and $CSF_{\mathrm{Inh}}$ is the inhalation $CSF$ for TCE that are reported by CalEPA (1996). Each $CSF$ represents an upper-bound estimate of the probability of cancer per unit intake of TCE and unit body weight over a lifetime $\left[\text{i.e., } \frac{\text{Risk}}{\text{mg/(kg d)}}\right]$. The $CSF_{\mathrm{Ing}}$ and $CSF_{\mathrm{Inh}}$ parameters are treated as constants (Table I).

The more traditional approach for arriving at estimators of risk can involve substituting into Eq. 13 those values for $E_{\mathrm{Ing}}$, $E_{\mathrm{Inh}}$, and $E_{\mathrm{Derm}}$ that were all obtained using input parameters either at (1) means only, (2) regulatory defaults, in combination with mean values for uncertain parameters and upper bounds (e.g., 95$^{\mathrm{th}}$ percentiles, or where applicable, 5$^{\mathrm{th}}$ percentiles; see footnote c in Table I) for those parameters without default values (USEPA, 1989 and 1991) or (3) upper bounds (e.g., 95$^{\mathrm{th}}$ percentiles, or where applicable, 5$^{\mathrm{th}}$ percentiles; see footnote c in Table I) exclusively. In the first case, the value of $R$ equates to a "best" estimate, $\hat{R}_{\mathrm{E}}$. In the second case, the value of $R$ is considered to be for a "reasonably maximum exposed" person, $\hat{R}_{\mathrm{RME}}$. In the third case, the value of $R$ corresponds to an upper "conservative" bound, $\hat{R}_{\mathrm{High}}$. All three of these types of risk-related estimators were calculated so these traditional-type estimators could be compared to analogous risk estimators that are more explicitly defined regarding uncertainty and/or variability. The input means used for calculating $\hat{R}_{\mathrm{E}}$ and the input upper bounds (e.g., 95$^{\mathrm{th}}$ percentiles, or where applicable, 5$^{\mathrm{th}}$ percentiles; see footnote c in Table I) used for calculating $\hat{R}_{\mathrm{High}}$ all appear in Table I. The default inputs used to calculate $\hat{R}_{\mathrm{RME}}$ appear in Table II, with the upper bounds (or where applicable, 5$^{\mathrm{th}}$-percentile values) for parameters without default values appearing in Table I for those uncertain and heterogeneous variates, respectively, for which default values are not given. Thus, the $\hat{R}_{\mathrm{RME}}$ is considered to be a conservative estimate of risk (i.e., above the average) that is within the range of reasonable possibilities, and is not the worst-possible case (USEPA, 1989 and 1991).

Risk-related estimators explicitly defined regarding uncertainty and/or variability involve the conditional expectations $\overline{R}$ and $\langle R \rangle$ (Bogen and Spear, 1987; NRC, 1994; Bogen, 1995). $\overline{R}$-type estimators of risk involve $\overline{R}$, which represents uncertain lifetime cancer risk to a (hypothetical) person at a population-average level of risk relative to others. The symbols $\overline{R}_{.05}$ and $\overline{R}_{.95}$ are used to represent the two-tailed lower and upper 90% confidence limits on the cumulative distribution function (cdf) of $\overline{R}$; and $\langle\overline{R}\rangle$ denotes the expected value (i.e., expectation with respect to uncertainty) of $\overline{R}$. $\langle R\rangle$-type estimators of risk involve $\langle R\rangle$, which denotes the set of expected values (with respect to uncertainty) of all the (potentially) different ("heterogeneous") cancer risks incurred within the population at risk. Thus, $\langle R\rangle_{.05}$ and $\langle R\rangle_{.95}$ represent the two-tailed lower and upper 90% confidence limits on the cdf $\langle R\rangle$; and $\overline{\langle R\rangle}$ is the population-average value of $\langle R\rangle$. (Note that the "population average" or arithmetic-mean value of a heterogeneous variate is just the expected value of that variate within a defined population.) Expectations of risk $R$ with respect to variability (i.e., $\overline{R}$) and uncertainty (i.e., $\langle R\rangle$) are defined by

$$\overline{R} = \left(\overline{E}_{\mathrm{Ing}} \times CSF_{\mathrm{Ing}}\right) + \left(\overline{E}_{\mathrm{Inh}} \times CSF_{\mathrm{Inh}}\right) + \left(\overline{E}_{\mathrm{Derm}} \times CSF_{\mathrm{Ing}}\right), \text{ and} \tag{14}$$

$$\langle R\rangle = \left(\langle E_{\mathrm{Ing}}\rangle \times CSF_{\mathrm{Ing}}\right) + \left(\langle E_{\mathrm{Inh}}\rangle \times CSF_{\mathrm{Inh}}\right) + \left(\langle E_{\mathrm{Derm}}\rangle \times CSF_{\mathrm{Ing}}\right), \tag{15}$$

where the terms $E_{\mathrm{Ing}}$, $E_{\mathrm{Inh}}$, and $E_{\mathrm{Derm}}$ are defined in Equations 1–3, and $CSF_{\mathrm{Ing}}$ and $CSF_{\mathrm{Inh}}$ are defined in Eq. 13.

The term $R^*$ denotes an upper-bound estimate with respect to JUV in risk. Specifically, $R^*_{.95}$ denotes the risk to an individual who is at a 95$^{\mathrm{th}}$-percentile level of risk relative to those risks incurred by others in the population at risk. Alternative first-order approximations of this upper-bound JUV estimator (see Bogen, 1995) are given by

$$R^*_{.95} = \overline{R}_{.95} \times \rho_{.95} \qquad \text{or} \qquad R^*_{.95} = \langle R\rangle_{.95} \times \rho'_{.95}\ , \tag{16}$$

where the terms $\rho_{.95}$ and $\rho'_{.95}$ denote "dispersion" ratios between upper-bound risk and expected individual risk; that is,

$$\rho_{.95} = \frac{\langle R\rangle_{.95}}{\overline{\langle R\rangle}} \qquad \text{and} \qquad \rho'_{.95} = \frac{\overline{R}_{.95}}{\langle\overline{R}\rangle}\ . \tag{17}$$

Note that $\rho_{.95}$ may be interpreted as an index of the "inequity" reflected in the distribution of individual risks incurred by a population at risk, insofar as this ratio is proportional to the variance (which measures interindividual differences in) that population. Similarly, $\rho'_{.95}$ represents an index of uncertainty associated with individual risk.

For a population of size $n_T$, $N$ is used to denote the uncertain number of additional cancer cases due to $R$, where expected number of cases is defined as $\langle N \rangle = n_T \times \langle \overline{R} \rangle$. Of specific interest to stakeholders and decision makers may be the probability, $1 - P_0$, that for a given population $n_T$, there will be one or more additional cases of cancer (i.e., the probability that $N \geq 1$).

The value of $P_0$ can be well approximated generally (see Bogen and Spear, 1987; NRC, 1994; Bogen, 1995) by the integral of the conditional Poisson likelihood function:

$$P_o \approx \int_0^1 e^{-n_T \overline{R}} f_{\overline{R}}(\overline{R}) \, \mathrm{d}\overline{R} \ , \tag{18}$$

where the compound-Poisson variate, $n_T\overline{R}$, incorporates the uncertain parameter $\overline{R}$ defined in Eq. 14.

## 2.3. CALCULATIONS

Variabilities in route-specific intake quantities (*Ing, Inh, A*) were calculated using corresponding demographic and exposure-related data cited in Table I. Note that variability in each quantity necessarily depends on the duration of exposure (*ED*) experienced by the population at risk. If all people were exposed for an entire lifetime, then this variability is properly characterized as the distribution of the lifetime, time-weighted-average (TWA) value of the corresponding quantity (*Ing, Inh,* or *A*). In contrast, if exposure duration were in all cases very brief, then this variability for each quantity would better be characterized as the composite distribution reflecting the weighted (or "age-adjusted") average of the age-specific distributions of that quantity, using age-specific population fractions as weights.

For the present analysis, all calculations of $\hat{R}_{\mathrm{High}}$ used upper-bound (95$^{\mathrm{th}}$ percentile) values (listed in Table I) of the lifetime TWA distributions of variability in *Inh, Ing,* and *A* ($\hat{R}_{\mathrm{E}}$ and $\hat{R}_{\mathrm{RME}}$ used means listed in Table I and default values listed in Table II, respectively, for these same three inputs). All other output risk-characterization quantities were calculated using corresponding composite, "age-adjusted" distributions reflecting people of all ages within the modeled exposed population. The latter procedure used is necessarily "conservative", in the sense that for each quantity the composite distribution (which is a weighted mixture of age-specific distributions) is necessarily more broad (i.e., has a larger variance) than the corresponding lifetime TWA distribution (which is the distribution of a weighted sum of random independent variate values). Thus, the larger *ED,* the more likely exposure will involve more than one of the age ranges used to construct the composite distribution, and hence the relevant quantity would more accurately be calculated as a TWA value involving the age ranges involved.

Ideally, computation would involve sampling a value of ED as well as a starting age, and then calculating (or, via a lookup method employing pre-calculated distributions, selecting) the relevant variability distributions for *Ing, Inh,* and *A*. This procedure is numerically taxing, however, so the alternative, simpler, albeit somewhat conservative, approach described above was used instead. This approach implies only very little conservatism in the case of risk characterizations involving $\langle R \rangle$, because the $\langle ED \rangle$ distribution was highly skewed (with a median value of only approximately 2 y), due to the highly skewed nature of residential turnover $R(t)$. Somewhat greater conservatism is implied for risk characterizations involving $\overline{R}$, because $\overline{ED}$, not nearly as skewed, has a median value of approximately 7 y.

For the reasons discussed above, the calculation of $E_{\text{Ing}}$ was based on the Ershow and Cantor (1989) lognormal approximation of the composite distribution reflecting variability in tap-water ingestion per kg body weight by people of all ages in the Western Region of the US. The corresponding lifetime TWA distribution was calculated assuming a 70-y lifespan and using the age-specific intakes reported by Ershow and Cantor (1989). The mean for both distributions was nearly the same.

The calculation of $E_{\text{Inh}}$ was based on age-specific rates of total inhalation per kg body weight for California youth and adults [data collected by Adams (1993) and Wiley *et al.* (1991a,b) were reevaluated and presented by OEHHA (1996), according to discussion with Marty (1998)]. From these data a corresponding composite distribution was calculated using youth and adult population weights derived from national census data (USCB 1998), and a corresponding lifetime TWA distribution was calculated using $\frac{12}{70}$ and $\frac{58}{70}$ as exposure-duration weights for youth and adults, respectively. The mean for both distributions was the same.

The calculation of $E_{\text{Derm}}$ was based on age-specific estimates of body surface area per kg body weight, *A*, reported by Phillips *et al.* (1993). From these data a corresponding composite distribution was calculated using infant/toddler, youth, and adult population weights derived from national census data (USCB, 1998), and a corresponding lifetime TWA distribution was calculated using $\frac{2}{70}$, $\frac{16}{70}$, and $\frac{52}{70}$ as exposure-duration weights for the respective age groups. The mean for both distributions was the same.

Calculations of derived input-variate distributions, the output $\overline{R}$ and $\langle R \rangle$ distributions, and related estimators were performed by Monte-Carlo simulations using *Crystal Ball*®, version 4.0 (Decisioneering, Inc., 1996), and/or *Mathematica*®, version 3.0 (Wolfram, 1996).

## 3. Results

Table III summarizes the lifetime excess cancer risk-related estimates for hypothetical residents theoretically supplied ground water with low-level concentrations of TCE from beneath Site LF-13 at Beale Air Force Base. The traditional risk-related estimator approach yields values of $\hat{R}_{\mathrm{E}}$, $\hat{R}_{\mathrm{RME}}$, and $\hat{R}_{\mathrm{High}}$ equal to $3.1 \times 10^{-6}$, $6.1 \times 10^{-5}$, and $2.4 \times 10^{-4}$, respectively. The risk-related estimator approach that is explicit regarding uncertainty in population-average risk, $\overline{R}$, produces a value for $\langle \overline{R} \rangle$ equal to $3.1 \times 10^{-6}$, and two-tailed lower and upper 90% confidence limits on the cdf of $\overline{R}$ equal to $1.4 \times 10^{-6}$ for $\overline{R}_{.05}$, and $5.5 \times 10^{-6}$ for $\overline{R}_{.95}$. The risk-related estimator approach that is explicit regarding variability in expected risk (with respect to uncertainty), $\langle R \rangle$, produces a value for $\overline{\langle R \rangle}$ equal to $3.1 \times 10^{-6}$, and two-tailed lower and upper 90% confidence limits on the cdf of $\langle R \rangle$ equal to $3.6 \times 10^{-8}$ for $\langle R \rangle_{.05}$, and $1.4 \times 10^{-5}$ for $\langle R \rangle_{.95}$. The index of "inequity" in expected risk (or "dispersion" ratio), $\rho_{.95}$, equals 4.7, which is not substantial (i.e., less than a factor of 10) and therefore indicates there is not a great amount of interindividual variability within the population in this situation. The upper JUV bound ($R^*_{.95}$), which is approximated by the product of $\overline{R}_{.95}$ and $\rho_{.95}$ equals $2.6 \times 10^{-5}$. (The result was nearly identical using $\rho'_{.95}$ to estimate $R^*_{.95}$—see Eq. 16.) Both the upper-bound population-average risk estimator, $\overline{R}_{.95}$, and the upper-JUV-bound risk estimator, $R^*_{.95}$, have values less than $10^{-4}$ and within the range of acceptability (i.e., $10^{-4}$ to $\leq 10^{-6}$) with respect to generally followed regulatory guidance (USEPA, 1990).

Table IV contains the results of the analysis of population risk associated with multipathway exposures to the TCE-contaminated ground water. These results reveal that the probability $(1 - P_0)$ of greater than zero additional cases of cancer for local on-ground, exposed populations of up to several hundred (i.e., corresponding to $n$ individuals; which is the reasonably foreseeable short-term scenario), is less than 0.01. Even for $n$ up to 26,900, the probability of more than zero cases remains below 0.5. In fact, the expected value of the total number of additional cancer cases remains less than 0.01 for $n$ up to several hundred and does not exceed 0.8 until $n$ exceeds 26,900. Even then, it is not clear that the extraction well would be capable of supporting such a large on-ground population, or even the comparable total exposed population over 70 y (i.e., $n_{\mathrm{T}}$) that is equal to 247,767.

The applicable CVM% values shown in Tables III and IV are all small (i.e., < 2%). This result indicates Monte-Carlo quality-control is assured and corresponding estimates are highly reliable.

TABLE III
Lifetime excess cancer risk-related estimates for hypothetical residents at Beale Air Force Base in California, based on multiple-pathway (i.e., ingestion, inhalation, and dermal) exposures to TCE-contaminated ground water.

| **Risk-related estimator approach**[a]<br>Type of estimator | **Symbol**[a] | **Value** | **CVM (%)**[b] |
|---|---|---|---|
| **Traditional** | | | |
| "Best" estimate (using input means) | $\hat{R}_{\mathrm{E}}$ | $3.1 \times 10^{-6}$ | NA |
| Risk to "reasonably maximum exposed" person | $\hat{R}_{\mathrm{RME}}$ | $6.1 \times 10^{-5}$ | NA |
| Upper "conservative" bound | $\hat{R}_{\mathrm{High}}$ | $2.4 \times 10^{-4}$ | NA |
| **Explicit regarding uncertainty and/or variability in:** | | | |
| ➤ **population-average risk, $\bar{R}$** | | | |
| Expectation (with respect to uncertainty) | $\langle \bar{R} \rangle$ | $3.1 \times 10^{-6}$ | NA |
| Lower uncertainty bound | $\bar{R}_{.05}$ | $1.4 \times 10^{-6}$ | 0.46 |
| Upper uncertainty bound | $\bar{R}_{.95}$ | $5.5 \times 10^{-6}$ | 0.45 |
| ➤ **expected risk (with respect to uncertainty), $\langle R \rangle$** | | | |
| Population average | $\overline{\langle R \rangle}$ | $3.1 \times 10^{-6}$ | NA |
| Lower variability bound | $\langle R \rangle_{.05}$ | $3.6 \times 10^{-8}$ | 0.97 |
| Upper variability bound | $\langle R \rangle_{.95}$ | $1.4 \times 10^{-5}$ | 1.2 |
| ➤ **jointly uncertain and heterogeneous risk** | | | |
| Index of "inequity" in expected risk | $\rho_{.95}$ | 4.7 | 0.58 |
| Upper JUV bound | $R^{*}_{.95}$ | $2.6 \times 10^{-5}$ | 0.62 |

[a] Note that $\hat{R}_{\mathrm{E}}$, $\langle \bar{R} \rangle$, and $\overline{\langle R \rangle}$ denote three closely related estimators of mean risk; that $\hat{R}_{\mathrm{RME}}$ and $\hat{R}_{\mathrm{High}}$ are crude and typically conservative approximations of $R^{*}_{.95}$; and that JUV refers to joint uncertainty and variability.

[b] Coefficient of variation of the mean (expressed in percent), $\mathrm{CVM\%} = 100\% \times \frac{\tilde{\sigma}}{\tilde{X} \times \sqrt{m}}$, where $\tilde{\sigma}$ is the sample standard deviation and $\tilde{X}$ is, for each estimator, the sample arithmetic mean obtained from $m$ equal to 10 repeated Monte-Carlo simulations each involving 2,000 trials. Small CVM% values (i.e., < 2%) indicate the estimates obtained are highly reliable, despite Monte-Carlo sampling error. A CVM% value is not applicable (NA), when value of risk-related estimator was not estimated by simulation, but rather was calculated using the corresponding exact expression.

TABLE IV
Population risk associated with multipathway exposures to TCE-contaminated ground water at Beale Air Force Base in California.

| Total exposed population over 70 y, $n_T$ | Exposed population during 7.6 y, $n$[a] | Probability of > 0 additional cases of cancer, $1 - P_0$[b] | CVM[c] | Expected value of the total number of additional cancer cases, $\langle N \rangle = n_T \times \langle \overline{R} \rangle$ |
|---|---|---|---|---|
| 100 | 11 | 0.0003 | 0.00030% | 0.00031 |
| 2,000 | 217 | 0.0063 | 0.0053% | 0.0062 |
| 30,000 | 3,257 | 0.0879 | 0.022% | 0.094 |
| 247,766.9 | 26,900 | 0.5000 | 0.031% | 0.77 |

[a] Here $n$ denotes the number of individuals residing at the impacted site within any 7.6-y time interval (i.e., the expected value, $\langle \overline{ED} \rangle$, of uncertain exposure duration, $\overline{ED}$, for the average exposed person) during the total 70-y time period considered (i.e., $n = n_T \times \frac{7.6 \text{ y}}{70 \text{ y}}$). Note that $n$ is not used to compute $P_0$, and is shown rounded to the nearest integer.

[b] Each value listed is the mean of 10 estimates obtained using the $\overline{R}$ distributions generated by 10 corresponding Monte-Carlo simulations, each involving 2,000 trials. For further details concerning this procedure, see Daniels *et al.* (1999).

[c] Coefficient of variation of the mean was derived as explained in Table III, footnote b.

## 4. Discussion and Conclusions

Bogen (1995) has shown that upper-bound estimators of JUV in risk may be approximated using calculations involving cumulative distribution functions (cdfs) that reflect only uncertainty and only interindividual variability, thus avoiding relatively tedious "nested" Monte-Carlo techniques that are otherwise required to obtain estimators of JUV in risk. That procedure was successfully employed for this analysis of uncertainty and variability in exposure to characterize risk from TCE-contaminated ground water at Site LF-13 on Beale Air Force Base in California.

Comparing the results obtained from this analysis procedure to the more traditional ones shows that the risk-estimators computed more traditionally overestimate the risk to an upper-bound individual $R^*_{.95}$, when JUV in the population is addressed explicitly. Furthermore, it can be seen from the results in Table III that $\hat{R}_E$, $\langle \overline{R} \rangle$, and $\overline{\langle R \rangle}$ all represent expected risk to the average individual. The equality of $\langle \overline{R} \rangle$ and $\overline{\langle R \rangle}$ (and hence the consistency between the alternative $R^*_{.95}$ estimates) suggests that the first-order approximation approach is reliable in this case. More accurate $R^*_{.95}$ estimation would require numerically intensive nested Monte-Carlo methods.

Results presented in Table IV indicate that the greater the exposed population, the less likely will be the chance that there will be zero observed cases. However in this analysis, for $n$ in the hundreds, there is a probability of less than 0.01 that the number of additional cases will be greater than zero. Even for $n$ up to 26,900, the probability of more than zero cases remains below 0.5. Clearly, this information is more valuable than that provided by a single, point estimate of $\langle N \rangle$ alone, especially for large populations (e.g., $n_T$ is 247,767) where $\langle N \rangle$ approaches but does not exceed 1, and there is really no measure of confidence (or uncertainty) associated with just that expected value.

On the basis of this analysis for Site LF-13 at Beale Air Force Base, and as pointed out by Bogen (1995), specific risk estimators might provide the bases for risk-acceptability criteria for a site, along with a specified value for $1 - P_0$. For example, risk-acceptability criteria might take the form of a joint requirement that $\overline{R}_{.95}$ be at least within the range of generally followed regulatory guidance $\leq 10^{-6}$ to $10^{-4}$ (USEPA, 1990); and $\rho_{95} < 10^2$; $R^*_{.95} \leq 10^{-4}$; and $1 - P_0 < 0.5$. Under such conditions, the upper-bound population-average risk, $\overline{R}_{.95}$, is low and within generally accepted regulatory limits; there does not appear to be a great amount of interindividual variability within the population, because the index of "inequity" in expected risk, $\rho_{95}$, is not substantial and so special susceptible groups do not need consideration; relatively highly exposed people in the population are not incurring inordinate risk as $R^*_{.95}$ is even less than $10^{-4}$; and the probability of 1 or more cases of cancer is less than 0.5 for a reasonably foreseeable population equal to $n$ and an expected exposure duration of 7.6 y. Clearly, the results presented here for TCE-contamination of ground water at Site LF-13 on Beale Air Force Base meet these requirements and such risk criteria for this site can ensure that individual lifetime risks are both *de minimis* and equitable.

The more traditional estimates of risk in this case all overestimate the level of risk to the upper-bound individual, including $R_{\text{RME}}$. Therefore, while providing an expedient and standardized assessment tool for screening risk levels at a particular site, such traditional approaches to estimating risk will always overestimate upper-bound individual risk, and may lead regulatory agencies to impose more stringent and costly remediation standards than might otherwise be appropriate. The approach illustrated in this report for TCE-contaminated ground water at Site LF-13 on Beale Air Force Base demonstrates a systematic mechanism for deriving risk-acceptability criteria that can help convince decision makers and stakeholders that money and resources being dedicated to remediation might better be spent on other public health measures that might be more cost effective.

The results of this work reinforce the importance of considering variability and uncertainty in estimates of risk. They also demonstrate that the calculations can be both simple and convenient, while yielding information of sufficient detail to establish reasonable and equitable site-specific risk-acceptability criteria.

## Acknowledgment

This work was performed under the auspices of the U.S. Department of Energy at Lawrence Livermore National Laboratory under contract W-7405-Eng-48. We are especially grateful to the Institute for Environment, Safety, and Occupational Health Risk Analysis (IERA), Environmental Science Branch (RSRE) of the 311th Human System Wing of the U.S. Air Force at Brooks Air Force Base, TX, USA, for supporting this work and providing important guidance. We also extend special thanks to the State of California Environmental Protection Agency, Department of Toxic Substances Control, for furnishing technical advice, and to the Chief for Environmental Restoration at Beale Air Force Base in California, and the Headquarters of the Air Combat Command at Langley Air Force Base in Virginia, for providing and permitting us to use data from Beale Air Force Base.

## References

Adams, W.: 1993, *Measurement of Breathing Rate and Volume in Routinely Performed Daily Activities.* Prepared for California Air Resources Board, Research Division, Sacramento, CA.

American Relocation Center (ARC): 1999, *All About Buyer Representation.* American Relocation Center, Administrative Offices, Shawnee Mission, KS, available online at *http://www.buyingrealestate.com/ pages/article/article2.html* (May 1999).

Bogen, K. T.: 1995, *Risk Analysis* **15**, 411–419.

Bogen, K. T. and Spear, R. C.: 1987, *Risk Analysis* **7**, 427-436.

Bogen K. T., Keating, G. A., Meissner, S. and Vogel, J. S.: 1998, *J. Exposure Analysis Environ. Epidemiol.* **8**, 253–271.

Brown, A. H., Bishop, D. R. and Rowan C. R.: 1984, *Am. J. Publ. Health* **74**, 479–484.

Burmaster, D. E.: 1998, *Risk Analysis* **18**, 33–35.

California Environmental Protection Agency (CalEPA): 1996, *California Cancer Potency Factors: Update—Criteria for Carcinogens.* Memorandum to Cal/EPA Departments, Boards, and Offices from Standards and Criteria Work Group, California Environmental Protection Agency, Sacramento, CA, April 1, 1996 (originally prepared in November 1994), available online at *http://www.oehha.org/docs/ covrltrb.htm* (May 1999).

California Environmental Protection Agency, Department of Toxic Substances Control (CalEPA/DTSC): 1994, *Preliminary Endangerment Assessment (PEA) Guidance Manual (A Guidance Manual for Evaluating Hazardous Substances Release Sites).* California Environmental Protection Agency, Department of Toxic Substances Control, Sacramento, CA (January 1994).

Corsi, R. L. and Howard, C.: 1998, *Shower Database.* In *Volatilization Rates From Water to Indoor Air: Phase II.* Prepared for United States Environmental Protection Agency, Office of Research and Development, The National Center for Environmental Assessment, Washington, DC (Technical Merit Review Draft, released for citation by J. Moya, USEPA Project Officer), Grant Number CR 824228–01, pp. 167–193.

Daniels, J. I., Bogen, K. T. and Hall, L. C.: 1999, *Procedures for Addressing Uncertainty and Variability in Exposure to Characterize Potential Health Risk From Trichloroethylene Contaminated Ground Water at Beale Air Force Base in California,* Lawrence Livermore National Laboratory, Livermore, CA, UCRL–ID–135784, Rev. 1 (Technical Report prepared for U.S. Air Force).

Decisioneering, Inc.: 1996, *Crystal Ball®, Forecasting and Risk Analysis for Spreadsheet Users.* Version 4.0, Decisioneering, Inc., Denver, CO.

Duke-Williams, O.: 1999, *Special Migration Statistics.* Center for Computational Geography, School of Geography, University of Leeds, Leeds, UK, available online at *http://www.geog.leeds.ac.uk/staff/o.duke-williams/sms.html* (May 1999) [see also conference publication by Rees, P. and Duke-Williams, O.: 1995, *The Special Migration Statistics: A Vital Resource for Research Into British Migration. Annual Conference of the Institute of British Geographers,* University of Northumbria at Newcastle, January 1995; cited online at *http://www.geog.leeds.ac.uk/staff/ o.duke-williams/pubs.html* (May 1999)].

Ershow, A. G. and Cantor, K. P.: 1989, *Total Water and Tapwater Intake in The United States: Population-Based Estimates of Quantities and Sources.* Prepared under National Cancer Institute Order #263-MD-810264 by the Life Sciences Research Office, Federation of American Societies for Experimental Biology, Bethesda, MD.

Fisk, W. J., Spencer, R. K., Grimsrud, D. T., Offermann, F. J., Pedersen, B. and Sextro, R.: 1987, *Indoor Air Quality Control Techniques, Radon, Formaldehyde, Combustion Products.* Pollution Technology Review No. 44, Noyes Data Corp., Park Ridge, NJ.

Giardino, N. J., Esmen, N. A. and Andelman, J. B.: 1992, *Environ. Sci. Technol.* **26,** 1602–1606.

James, I. R. and Knuiman, M. W.: 1987, *J. Am. Stat. Assoc.* **82**, 705–711.

Israeli, M. and Nelson, C. B.: 1992, *Risk Analysis* **12**, 65–72.

Marty, M.: 1998, *Personal Communication Regarding the Methodology Used by OEHHA Staff to Develop Breathing Rate Per Unit Body Weight Distribution From Data of Adams (1993) and Wiley et al. (1991a,b) and Presented in OEHHA (1996).* Office of Environmental Health Hazard Assessment (OEHHA), California Environmental Protection Agency, Berkeley, CA.

McKone, T. E. and Bogen, K. T.: 1992, *Regul. Toxicol. Pharmacol.* **15,** 86–103.

National Research Council (NRC), Committee on Risk Assessment of Hazardous Air Pollutants: 1994, *Science and Judgment in Risk Assessment.* National Academy Press, Washington, DC.

Office of Environmental Health Hazard Assessment (OEHHA), California Environmental Protection Agency: 1996, *Exposure Assessment and Stochastic Analysis, Air Toxics Hot Spots Program, Risk Assessment Guidelines Part IV, Technical Support Document, Public Review Draft.* Office of Environmental Health Hazard Assessment, California Environmental Protection Agency, Berkeley, CA.

Phillips, L. J., Fares, R. J. and Schweer, L. G.: 1993, *J. Exposure Analysis Environ. Epidem.* **3**, 331–338.

Purrier, W.: 1997, *Personal Communication Regarding Detailed Data For TCE Groundwater Concentration at Site LF-13 of Beale Air Force Base, California.* Law Engineering and Environmental Services, Inc., Sacramento, CA (December 1997).

Tsang, A. and Klepeis, N.: 1996, *Results Tables From a Detailed Analysis of The National Human Activity Pattern Survey (NHAPS) Response, Draft Report.* Prepared for the U.S. Environmental Protection Agency, Washington, DC, by Lockheed Martin Corp., Contract No. A68-W6-001, Delivery Order No. 13.

URS Greiner Woodward Clyde (URSGWC), Omaha, Nebraska: 1998, *Management Action Plan, Beale Air Force Base, California, December 1998.* Prepared for Headquarters Air Combat Command (ACC), Langley Air Force Base, VA, under contract to US Air Force Center for Environmental Excellence (AFCEE), Brooks Air Force Base, TX, Project No. ACCH19987544, available from Chief, Environmental Restoration (M. E. O'Brien), Beale Air Force Base, CA.

United States Census Bureau (USCB): 1997, *Geographic Mobility of People 1 Year Old And Older, By Sex, Between March 1996 and March 1997.* Source: *March 1997 Current Population Survey (CPS), last revised September 29, 1997.* US Bureau of the Census, Washington, DC, available online at *http://www.bls.census.gov/cps/pub/1997/mobility.htm* (May 1999)

United States Census Bureau (USCB): 1998, *Resident Population of The United States: Estimates By Age And Sex.* Source: *United States Population Estimates, By Age, Sex, Race, And Hispanic Origin, 1990 To 1997 (with Associated Updated Tables For Recent Months), last revised December 28, 1998.* Population Division, US Bureau of the Census, Washington, DC, available online at *http://www.census.gov/population/ estimates/nation/intfile2-1.txt* (May 1999).

United States Environmental Protection Agency (USEPA): 1989, *Risk Assessment Guidance for Superfund Volume 1: Human Health Evaluation Manual, Interim Final.* Office of Emergency and Remedial Response, US Environmental Protection Agency, Washington, DC, EPA/540/1-89/002.

United States Environmental Protection Agency (USEPA): 1990, *Fed. Regist.*, **55**(46), 8666–8865; particularly 8715–8719 (March 8, 1990).

United States Environmental Protection Agency (USEPA): 1991, *Risk Assessment Guidance for Superfund Volume 1: Human Health Evaluation Manual; Supplemental Guidance: Standard Default Exposure Factors, Interim Final.* Office of Emergency and Remedial Response, Toxics Integration Branch, US Environmental Protection Agency, Washington, DC, OSWER Directive: 9285.6-03.

United States Environmental Protection Agency (USEPA): 1997, *Exposure Factors Handbook, Volumes I-III, Final.* Office of Research and Development, National Center for Environmental Assessment, US Environmental Protection Agency, Washington, DC, EPA/600/P-95/002F (August 1997).

United States Environmental Protection Agency Region 9 (USEPA Region 9): 1998, *Region 9 Preliminary Remediation Goals (PRGs) 1998.* Technical Support Team, U.S. Environmental Protection Agency Region 9, San Francisco, CA, available online at *http://www.epa.gov/region09/waste/sfund/prg/* (May 1999).

Wiley, J. A., Robinson, J. P., Piazza, T., Garrett, K., Cirksena, K., Cheng, Y. T., and Martin, G.: 1991a, *Activity Patterns of California Residents.* Prepared for California Air Resources Board, Research Division, ARB/R-93/487, Sacramento, CA, available from National Technical Information Service, Springfield, VA.

Wiley, J. A., Robinson, J. P., Cheng, Y. T., Piazza, T., Stork, L. and Pladsen, K.: 1991b, *Study of Children's Activity Patterns.* Prepared for California Air Resources Board, Research Division, Sacramento, CA, ARB/9-93/489, available from National Technical Information Service, Springfield, VA.

Wolfram, S.: 1996, *The Mathematica Book, 3rd Ed.,* Wolfram Media/Cambridge University Press, Cambridge, UK.

# BIOLOGICAL DIVERSITY - AN OVERVIEW

M. WALDMAN[1] and Y. SHEVAH[2]
[1]*Ministry of Science,* [2]*TAHAL Consulting Eng. Ltd, Israel*

**Abstract.** Despite the overwhelming importance of biodiversity and the growing demand for the genes and chemicals, biodiversity is being lost at alarming rates, largely as a result of human action enhancing degradation of biologically rich ecosystems like tropical rain forests, grassland and coral reefs. A positive global change was achieved at the Earth Summit in Rio, 1992, requiring all nations to take four basic steps: develop national strategies for conservation, establish a system of protected areas, begin to rehabilitate damaged ecosystems and integrate conservation policy into national decision making. The conservation of biological diversity, sustainable use, technology transfer, intellectual property rights, provision of financing and the principle of equitable sharing of benefits were also promoted.

On the local scene, Israel is known for its rich natural vegetation and diversified species which contribute to the biological diversity and plant species population. Beyond the intrinsic value, the diversified plant population is an extremely valuable genetic source for improvement of agricultural crops and extraction of new drugs. The Israeli Gene Bank (IGB) was established to preserve the natural endowment, and is responsible for *in-situ* conservation, formation of gene banks and other related activities with emphasis on regional and international cooperation.

**Keywords**: biodiversity, conservation, ecosystems, gene banks, Israel

## 1. Introduction

Biodiversity or biological diversity, is defined as the conservation of the variety of species and the variability of living organisms, their habitats and the biological ecosystems, encompassing the ecological and evolutionary processes of the natural environment. Biodiversity also reflects the uniqueness and the universality of the biological world and the organizational and hierarchical integration of the molecular, the community and the social and cultural diversity (Di Castri, 1991). The wider definition of biodiversity refers to the variability among living organisms from all sources including *inter alia*: terrestrial, marine and other aquatic ecosystems and the ecological complexes of which they are part, including diversity within species, between species and of ecosystems (Convention on Biological Diversity, 1992).

Thus, biodiversity is not restricted to the preservation of particular animal and plant species of a supposed high aesthetic value or strongly charismatic nature, such as the panda, sequoya tree or orchids, nor is biodiversity about the conservation of threated ecosystems (tropical rain forest, coral reefs, etc.) or its applications in biotechnology, pharmaceutical industry and crop improvement. Biotechnology is already an example of the pervasive introduction of elements

*Water, Air, and Soil Pollution* **123:** 299–310, 2000.

of biodiversity into almost all the production sectors (Di Castry, 1991). Biodiversity is therefore the total sum of all life forms that exist on earth, encompassing: genes, species and ecosystem diversity and their interaction in space and time.

Biodiversity is the wealth of species ecosystems and ecological processes that help make possible the economic and environmental systems. Mankind depends on earth's biodiversity for food, fiber, medicines and new products from the rapidly expanding field of biotechnology. The active ingredients in more than 25% of medical prescriptions derive from natural plants, while the demand for genes and chemicals that remained untapped within the world remaining wild life is fast growing.

Despite the overwhelming and continuous demand for new genes and chemicals, biodiversity is directly and indirectly affected by human activities which leads to unintended reductions in diversity of the live organisms. Biodiversity is being lost at alarming rates, faster today than at any time, largely as a result of human action enhancing degradation of biologically rich ecosystems. Although some decrease of biodiversity, particularly as the result of the conversion of land into areas of intensive agriculture and urbanization is absolutely inevitable and in many ways it is compatible with the functioning of ecosystems and the biosphere, the extinction of species is some order of magnitude higher than during previous centuries and is occurring over an extremely short period of the evolutionary scale, exceeding enormously the current rate of speciation.

Thus, crop genetic variability is decreasing and ecological diversity suffers the consequences of deforestation, desertification and the destruction of traditional landscapes. Species extinction is also the consequence of the common practices of fragmenting landscapes and ecosystems for intensive agriculture or forestry, of destroying habitats or urbanization, transport (railway, roads, etc.) or of drying up wetlands for tourism. The extraordinary rate of loss of biological diversity that is experienced worldwide carries real risk for the future generations who will be deprived of the economic, health, nutritional, aesthetic and other benefits derived from our planet's rich biological inheritance.

This presentation sets out to present and review both local and international efforts to preserve biodiversity. Special attention is accorded to activities in the area of collection, conservation, documentation, evaluation and exploitation of genetic resources and measures to halt the disappearance of genetic resources, using local and international cooperation.

## 2. The Global Action

A positive global change in this down road direction was taken at the 1992 Rio Earth Summit, where the Convention on Biological Diversity was first negotiated providing an international framework for the conservation and sustainable use of biodiversity. Conservation of biodiversity encompasses previous legal acts which protect wild species of wild flora and fauna and limit world trade in rare and endangered species. The new treaty was the first step in recognizing the importance of biological diversity, requiring all the nations to take the following four basic steps:

- Develop national strategies for conservation;
- Establish a system of protected areas;
- Begin to rehabilitate damaged ecosystems;
- Integrate conservation policy into national decision making.

The international convention, which entered into force in December 1993, also put forward the need for the following activities:

- conservation of biological diversity;
- sustainable use; and
- the principle of equitable sharing of benefits so that developing countries should benefit from the use of their biological resources.

The convention also recognized that benefits stemming from the productive use of the genetic flow of resources would flow back to those nations that act to conserve biological diversity and provide access to their genetic resources, as an incentive for countries to protect their resources. These activities are financed by the Global Environment Facility (GEF), which is the institutional mechanism for funding global climate change, biodiversity, ocean management and other biodiversity conservation activities.

The preservation of diversity, particularly of endangered plants, landraces and historic varieties is no longer the domain of large national and international programs. Community seed banks, NGOs and others are becoming more and more visible in the global conservation of diversity.

Among the various agencies, the International Plant Genetic Resources Institute (IPGRI) is a leading actor in the formulation of conservation policies. It serves as the coordinator in the global conservation and exchange of genetic resources among breeder-demand-oriented gene banks and supporting national and international gene banks.

## 3. Conservation Processes and Activities

The process of conservation is conceived as a concerted global effort and a centralized action, allowing a universal comparison of findings, in contrast to a country driven approach in which inventory and monitoring may be carried out, using different taxonomic groups and different sampling methods, making it very difficult to understand the dynamics of global biodiversity. Therefore, the implementation of an acceptable global framework, using uniform methodology and design and allowing the exchange of comparable results is recommended.

## 3.1. POPULATION DYNAMICS AND GENETIC STRUCTURE

Managing genetic resources requires an understanding of the biological dynamics of the population in question. In particular, knowledge of the diversity and distribution of genes makes it possible to predict the likelihood of gene loss in a population and to develop strategies to prevent such deletion. The concept of minimum viable population size, below which the population in a given habitat cannot persist, is used to determine the right size for a stand conservation, although, the minimum population size is likely to differ among species and among habitats for the same species. Understanding genetic structure makes it possible to collect the genetic diversity and to ensure its conservation. Evolutionary responsibility includes conservation and continued evolution, according to the biodiversity laws.

## 3.2. CONSERVATION AND PRESERVATION

The conservation system is a dynamic process in which the species still has the genetic and ecological potential to pursue its evolutionary pathways. Preservation, on the other hand, provides the conditions for the maintenance of the individual or groups, similar to the "fix it management" of a botanical garden or a zoo. Conservation within and outside protected areas should be considered as an essential dimension of overall land use planning. Introduction of environmental concerns in the early stages of development can minimize, through preventive and low cost measures, the risk of decreasing biodiversity. The curative approach is too often prohibitively expensive and incidentally impractical like in the case of conservation of tropical rain forests, grassland and coral reefs. Lack of land management may ultimately be the reason for devastating fires, erosion and loss of open areas for grazing animals and their predators.

Conservation and preservation combine *in-situ* and *ex-situ* conservation methods. The two methods are treated as complementary strategies and the choice of each method depends on:

- size of the population under consideration;
- its genetic history;
- its isolation; and
- regeneration possibility and long term survival.

## 3.3. *IN-SITU* SYSTEMS

*In-situ* conservation includes the designation of protected reserve areas in the form of natural parks in which native plants are protected together with the associated animals, birds and other biological resources, in a manner that the whole ecosystem is strictly preserved. *In-situ* conservation is an essential stage in the development of a coordinated world genetic conservation program, assuming that gene pools of local varieties should be considered from the dynamic point of view, in which the conserved population is under a continuous evolution. *In-situ* conservation has significance for nature conservation (Safriel *et al.*, 1997), but it is more than simply establishing protected natural areas which contain random or fragmentary populations of species.

For the successful management of the natural stands intended for *in-situ* conservation. morphological variation or changes in gene frequency due to environmental or ecological events have to be monitored. This will require a knowledge of ecosystem dynamics including:

- species composition;
- the number of species and their occurrence;
- the soil's physical and chemical properties; and
- temporal and spatial genetic variation.

All this knowledge is vital for successful management of the natural stands intended for *in-situ* conservation.

Because of the high costs involved and the limited number of crop germplasm that can be conserved *in-situ*, the optimal size, the viable population and other scientific and social limitations as compared to *ex-situ* conservation, it has been argued, that *in-situ* conservation should be reserved for wild relatives to preserve genetic polymorphism and to provide opportunity for recombination, or where conservation for an infinite period is required. The Amiad Project, Israel, in which the diversity and degree of adaptation of populations of wild crop relatives is done in their natural habitat conforms with this policy. However, for the conservation of particular species for economic purposes, *ex-situ* preservation may be more economical (Frankel and Soul, 1981). *In-situ* conservation of 28 priority crops, eco-geographic surveying for gene pools of 20 mandate crops and the *in-situ* conservation of another 17 gene pools which could lead to proposals was recommended (IBPGR Task Group,1985).

## 3.4. *EX-SITU* SYSTEMS

*Ex-situ* systems have become the primary tool for the conservation of crop plants, offering gene banks which are easily observed, studied and used, although reflect a preservation process rather than the dynamic and adaptive principles of conservation. Advantages of *ex-situ* conservation include:

- long term conservation;
- easy access to a diversified quantity and accession of seeds;
- wider viability to breeders.

Drawbacks of *ex-situ* conservation include:

- loss of germination rate over time;
- loss of resemblance to the original parents due rejuvenation (genetic drift);
- subject to political and financial implications;
- complementary *in-situ*, including on-farm conservation is essential to meet the growing demand for material;
- fear of overstocking, resulting in bad management;
- little interest by breeders in large gene banks.

*Ex-situ* systems include the collection of seeds and other propagation material, such as progeny gene conservation stands and seed storage. *Ex-situ* progeny gene conservation stands are one example of the duplication protection method for rare and endangered species gene conservation.

Seed storage or seed banks, the mainstay of germplasm conservation of agriculturally important crop species, refers to storage of intact seeds in a controlled environment, under controlled temperature and moisture conditions. The stored seeds can remain viable for decades, but appropriate storage conditions such as temperature, humidity and moisture content should be adapted and monitored for each species.

Tissue culture is another important technique which makes it possible to store large numbers of genotypes of rare and threatened species in a relatively small area. However, methods have yet to be developed to produce plantlets at a reasonable cost.

## 3.5. *IN-SITU* AND *EX-SITU*

Combining existing local farming practices with *in-situ* conservation strategies has been pursued by IBPGR, now IPGRI, despite resistance from the establishment.

## 3.6. TESTS, MARKERS AND INDICATORS

Provenance testing as well as isozyme studies and DNA markers contribute to the knowledge of the genetic variation of a selected population, being indicators of morphological variation or for monitoring changes in gene frequency due to environmental or ecological events. The use of DNA markers to measure genetic diversity is on the increase replacing non molecular indicators.

## 4. Biodiversity Conservation Activity in Israel

On the local scene, Israel is known for its richness in natural vegetation and diversified species of Mediterranean, Irano-Turanian and Saharo-Arabian origin (Zohari, 1973, Pavlicek, *et al.*, 1999, Friedman *et a*l., 1999). As part of the Fertile Crescent, Israel is located on the border between deserts and temperate regions, where domestication of Old World crops began. Beyond the intrinsic value, the diversified plant population has been an extremely valuable genetic source for improvement of agricultural crops and extraction of specific drugs.

Currently, with the rapid increase in population and urbanization and the related infrastructure development, such as the Trans-Israel Highway, the genetic resources of native species are under a serious threat of extinction (Kaplan and Kadmon, 1999 and Yarom, 1999). The genetic resources will most likely be affected in the medium and long term in a way that may significantly reduce genetic variation. Protective measures to counter current influences, including rescue and conservation of the wild species as well as of the remnants of the genetically broad-based landraces and primitive crop varieties have to be immediately taken (Frankel *et al.,* 1995, Dayan, 1999).

### 4.1. ISRAEL GENE BANK (IGB)

To avoid the disappearance and the extinction of endangered species and, on the other hand, to beneficially exploit the natural endowment and respond to the International Convention On Conservation and Sustainable Use of Biological Diversity, Agenda 21, a national gene bank was established. The Israel Gene Bank (IGB), under the Ministries of Science and Agriculture, operates from its headquarters at the Agricultural Research Organization (ARO) at Bet Dagan, and is responsible for the implementation of a strategy for national genetic resources conservation, taking into consideration the specific conditions and characteristics of Israel, including: size, geography, topography, ecology, infrastructure and the administrative and organizational structure (Diwan *et al.*, 1999).

Accordingly, the IGB has responsibility for:

- search for plants potentially suitable for extraction of beneficial substance;
- collection, preservation, documentation and evaluation of genetic resources of crop plants and their relatives, including landraces and primitive cultivars in gene banks across Israel;
- development of *in-situ* and *ex-situ* conservation techniques;
- documentation of all the genetic resources in gene banks throughout Israel and the maintenance of the database and the Information Network;
- promotion of an integrated phenotypic and molecular research;

- promotion of regional and international exchange of plant material and cooperation in preservation of species and biological diversity, through workshops, conferences and training activities.

Within this framework, the IGB has begun exploring, conserving and evaluating the natural stands of some economically important species with regards to their regional patterns of adaptation and their genetic diversity. These efforts are not, however, systematic as yet and their emphasis is mostly on securing breeding material rather than on conserving genetic diversity. Currently, a strategy is being pursued with an overview of genetic conservation, conceptualizing measures and a plan of implementation which will contribute to gene resource conservation, including:

- Identification of target species for conservation;
- Identification of immediate and future threats to the gene pool of target species.

The strategy combines *in-situ* and *ex-situ* conservation methods.

To date, IGB has set aside several natural stands for in situ gene conservation of the target species, including wild wheat gene pools in two nature reserves. Currently, the work plan covers a total of 174 indigenous species divided into 4 groups, although some of the species overlap 2-3 groups:

- Group I – rare and threatened species (wild oats);
- Group II – economically important species (crop's gene pool, untapped potential species, breeding material species;
- Group III – introduced species; and
- Group IV – protected (by law) species.

For *ex-situ* conservation, seeds and other propagation material are collected. These include:

- National cloning repository of deciduous fruit tree landraces and rootstocks, ARO – Matityahu Experimental Station;
- International collection of short day *allium* taxa, Hebrew University of Jerusalem (HUJ), Rehovot;
- Spice and medicinal plants collection, ARO Neve-Yaar;
- Wild wheat field laboratory, Amiad;
- Several short term seed collections of cereals, legumes, cucurbits and other plants, including wild cereals collection, Haifa University;
- *Triticeae* collection, Weizmann Institute of Science;
- Legumes seed collection, HUJ;
- Botanical Garden, HUJ;
- National Herbarium, HUJ;
- ARO Long Term Seed Bank, Bet Dagan, which holds some 16,000 accessions of indigenous species, landraces and crop varieties (Table I).

*Ex-situ* methods include also clones and collections of material for evaluation and screening as well as collections for breeding. Cultivar collections and botanical gardens were also established as part of *ex-situ* conservation.

TABLE I
Types and accessions stored at the volcani center seed bank

| Type | Israel Origin | Worldwide Origin |
|---|---|---|
| Botanical species, excluding endemic | 410 | 284 |
| Local endemic | 1870 | |
| Wild relative of crops | 2200 | 285 |
| Landraces | 1015 | 810 |
| breeding lines, hybrid and botanic garden accessions | 7345 | |
| Modern cultivars, including forage spp | | 1775 |
| Total | 16,000 | |

Future plans for activity and development at the IGB include a survey and rescue collection of gene resources of economic value which are under immediate danger due to development. This will include:

- collection and mapping of genetic variation of selected species;
- development of guidelines and conservation protocols;
- exploration and evaluation of target species with regard to their ecological traits and genetic diversity;
- regeneration of old stored material according to their unique value and storage age;
- supervision and inspection of the existing *in-situ* and *ex-situ* stands;
- promotion of field mission for the selection and collection of reproductive material for further expansion of *ex-situ* stands and seed banks;
- development of conservation sites for important wild relatives of crop plants, including: wild emmer wheat, the *Sitopsis* group of *Aegilops*, *Pisum fulvum*, *P. elatius* and *Carthamus glaucus*.

## 5. International Cooperation

The dominant general lack of information and knowledge regarding biological diversity emphasizes the urgent need to develop scientific, technical and institutional capacities to provide the basic understanding upon which to plan and implement appropriate measures. The adoption of the Convention on Biological Diversity has provided for the implementation of large scale international activities. Following is a short description of some of the salient activities.

The Global Biodiversity Assessment, undertaken by UNEP and financed by GEF, involves more than 1100 specialists from some 80 countries.

The Seville Conference (1995) on biosphere reserves, organized by UNESCO, in which the Seville Strategy was adopted, provides a statutory framework for the World Network of Biosphere Reserves (325 reserves over 83 countries). This enables the conservation of biological diversity linked with sustainable development of ecosystems resources, while supporting research, monitoring and training activities.

UNESCO, jointly with ICSU has launched a major international research initiative on the origin and composition, functioning and conservation of biodiversity, known as DIVERSITAS. The program provides scientific and technical advice to Governments and international institutions for the implementation of Agenda 21 and other articles of the Convention. DIVERSTAS includes 9 complementary elements:

- origins, maintenance and loss of biodiversity;
- ecosystem functioning;
- inventory, classification and inter-relationships;
- assessment and monitoring;
- conservation, restoration and sustainable use;
- human dimensions of biodiversity; and
- microbial biodiversity.

Within the framework of the European Cooperative Programme for the Conservation and Exchange of Crop Genetic resources, the IGB has responsibility for the world collection of vegetative propagated short day *Allium* taxa and the ECP/GR Central Crop Data Base for *T. trifolium alexandrinum* and *T. resupinatum.*

Workshops, popular lectures, visit to conservation sites and schools, radio and television programs are an essential part of the conservation process. Training of students and technical staff in the various aspects of conservation is regularly conducted. International courses on exploration of wild genetic resources and principles of collection and sampling are being conducted.

## 6. Summary and Conclusions

Agriculture, biology and ecology are all concerned with conservation. While within the agricultural realm the conservation of plant genetic resources emerged in the early 1960s, ecological and biological initiatives remained limited, because of the difficulties in setting up a gene pool conservation strategy. The inability to link conservation with development has resulted in a tendency whereby most of the conservation was initiated by nature conservation agencies which tend to neglect the human factor, often preventing sustainable use of the environment.

The technical/practical aspects and hence the financial constraints related to *in-situ* work have long been an issue for debate. Lack of funds tends to strengthen arguments for *ex-situ* conservation, while by contrast, public pressure and increased financial support for nature conservation tends to support a greater focus on the *in-situ* conservation effort by GEF, WB, UNDP and UNEP. Most of these funds are allocated for the conservation of highest species diversity, explained by the public call to protect nature from man.

Moreover, there is a widespread misconception that *in-situ* conservation in a few areas will adequately conserve enough wild genotypes for future crop enhancement when wild relatives in other areas will also be valuable. Furthermore, most *in-situ* conservation has been directed toward habitat preservation and has focused on ecological rather than genetic considerations.

Today, it is recognized that the maintenance of the conditions that create and maintain diversity with emphasis on the role of indigenous knowledge in the evolution and cultivation of different genotypes is the more important, not the preservation of specific genotypes. It seems that despite the many efforts in conservation and preservation, expertise on conservation strategies of useful species is still very limited and national and international efforts are still much needed.

## References


Anikster, Y.: 1998, "Ten Years of In-situ research of a wild wheat population at Amiad, Israel," *Proc. Int. Symp. On In-situ Conservation of Plant Genetic Diversity*, Turkey.

Dayan, T.: 1999, "The Status of Natural History Collections in Israel", in *Proc. of the 7th Int. Conf. of the Israel Soc. for Ecology and Env. Quality Sci.*, Jerusalem.

Di Castri, F.: 1991, "Ecosystem evolution and global change," in O. T. Solbrig and G. Nicolis (eds), *Perspectives in Biological Complexity*, IUBS, Paris, pp. 189 - 218.

Diwan, N., Levy, A. and Anyalem, H.: 1999, "The Conservation of Plant Genetic Resources in Israel by the Israel Gene Bank for Agricultural Crops," in *Proc. of the 7th Int. Conf. of the Israel Soc. for Ecology and Env. Quality Sci.*, Jerusalem.

Frankel, O. H. and Soul, J.: 1981, *Conservation and Evolution*, Cambridge Univ. Press, Cambridge.

Frankel, O. H., Brown, A. D. H. and Burdon J. L.: 1995, *The Conservation of Plant Biodiversity*, Cambridge Univ. Press, Cambridge.

Friedman, J., Barazani, O., Ravid, U. and Putievsky, E.: 1999, "Saving Variability of Unique Indigenous Plant Populations in Israel, a Case Report Illustrated by Bitter Fennel (*Foeniculum vulgare Mill*)", in *Proc. of the 7th Int. Conf. of the Israel Soc. for Ecology and Env. Quality Sci.*, Jerusalem.

IBPGR:1985, *Ecogeographical Surveying in In-Situ Conservation*, Report of an IBPGR Task Force, Washington, D.C.

Kaplan, D. and Kadmon, A.: 1999, "Ecodiversity Changes in the Hula Valley, Israel," in *Proc. of the 7th Int. Conf. of the Israel Soc. for Ecology and Env. Quality Sci.*, Jerusalem.

Pavlicek, T., Chikatunov, V. and Nevo, E.: 1999, "Spatiotemporal Microscale Biodiversity Divergence of Insects at 'Evolution Canyon' Lower Nahal Oren, Mt. Carmel," in *Proc. of the 7th Int. Conf. of the Israel Soc. for Ecology and Env. Quality Sci.*, Jerusalem.

Pistorium, R.: 1997, *Scientists, plants and politics, A history of the plant genetic resources movement*, IPGRI Publ., Rome, p. 130.

Safriel, U. N., Anikster, Y. and Waldman, M.: 1997, "Management of Nature Reserves for Conservation of wild relatives and the significance of marginal population," *Boconea* **7**, 233-239.

UN: 1992, *Report of the UN Conference on Env. and Dev.* Rio de Janeiro, UN, NY, 3-14.

Yarom, I.: 1999, "Insects in a Rapidly Changing Environment: A Biodiversity Survey of the Arava Valley in Israel," in *Proc. of the 7th Int. Conf. of the Israel Soc. for Ecology and Env. Quality Sci.*, Jerusalem.

Zohari, M.: 1973, *Geobotanical Foundations of the Middle East*, Gustav Fischer Verlag, Stuttgart.

# RESTORATION OF THE RIVERS IN ISRAEL'S COASTAL PLAIN

Y. BAR-OR
*Department of Water and Rivers, Ministry of the Environment, Israel*

**Abstract.** Israel's chronic water shortage has aggravated problems of river pollution, as almost no dilution of inadequately treated wastewater takes place. Since the early 1990s, large investments were made to build new wastewater treatment plants, designed to produce effluents of secondary quality or better.. Consequently, it became of interest to examine the possibility of using tertiary treated effluents as a water source for rehabilitation of aquatic habitats. A survey of water quality in clean and polluted segments of Israel's coastal rivers was initiated, including analyses of riverbed sediments. Although many segments are still heavily polluted, there are examples of measurable improvement in water quality, which should lead to better sediment quality. It is argued that sediments serve both as a memory bank of past pollution events and as an ongoing source of anaerobic and toxic compounds, such as sulfides and ammonium compounds. Their chemical compositions are, therefore, of much importance for river restoration processes and monitoring.

**Keywords:** river restoration, riverbed sediments, surface water quality

## 1. Introduction

Israel's coastal plain is transected by ten major rivers and their tributaries (Map 1). These are very small in comparison to other Middle Eastern rivers such as the Nile or the Euphrates, with a typical catchment area of a few hundred square kilometers and a length of tens of kilometers. Nevertheless, until the 1940s most of them drew water through all or most of the year and supported diverse ecosystems in which fauna and flora thrived and flourished.

High quality groundwater of the mountain aquifer discharged naturally through the Yarkon and Taninim rivers and supported substantial flow (220 million cubic meters (CM) through the Yarkon and 60 million CM through the Taninim annually) and rich habitats, including endemic fish and plant species.

Since the establishment of the State of Israel in 1948, the country has absorbed, at a rapid pace, massive waves of immigration from European and Middle Eastern countries, and has undergone accelerated urbanization and industrialization. As a result, and due to low environmental awareness, the coastal rivers progressively became receivers of insufficiently treated wastewaters and solid wastes of diverse origins.

In order to supply the needs of the growing population, larger and larger mounts of groundwater were pumped from aquifers, and there was less and less freshwater discharging to the rivers from springs at their sources. Aquatic and wetland habitats subsequently dried up, and some segments of the coastal rivers,

*Water, Air, and Soil Pollution* **123:** 311–321, 2000.

such as the Ayalon and Dalia rivers, were converted to concrete channels designed exclusively to protect cities against floods.

In 1994, the Ministry of the Environment established the River Restoration Administration as a coordinating body for actions taken by various governmental and non-governmental bodies, to restore or at least rehabilitate damaged streams. The declared goals of the Administration have been:

- Restoration of Israel's rivers by prevention of pollution, release of sufficient amounts of good quality water to their channels, and rehabilitation of adjacent landscapes and ecosystems.
- Proper functioning of the rivers for flood control.
- Control of development and land-use in the rivers' environment, so that these will be based on ecological considerations and potential of natural and heritage resources, for purposes of nature preservation, recreation, tourism, education and research.

More specifically, the aims of the River Restoration Administration have been:

- Collection of data and formation of a data base on natural and heritage values in rivers and their physical, chemical and biological condition.
- Formulation of an integrated national policy for preservation and controlled development of the rivers' environment, based on guidelines for keeping a balance between effective land use and preservation of natural ecosystems and landscapes.
- Preparation of masterplans for restoration of the rivers.
- Advancing preparation and execution of detailed plans for river restoration according to a list of priorities.
- Increasing public awareness and involvement in river rehabilitation and landscape preservation.

One of the major obstacles in restoration has been the chronic lack of adequate water flows, due to the need to divert almost all the renewable sources of freshwater for supply of urban, agricultural and industrial purposes. Israel is one the poorest countries in the region, with only 320 CM per capita (and even less when transfers of water to Jordan and the Palestinian Authority, pursuant to the Peace Accords, are taken into account). Thus, non-conventional water sources must be considered, including treated wastewater. It has become necessary to collect data on water quality in the still clean segments of rivers, and adopt a coherent policy on the use of effluents for river restoration. This issue has become even more pressing, as the amounts of wastewater are constantly increasing and the ability of the agricultural sector to re-use even high quality effluents for irrigation remains constant. This paper presents some of the data collected and the conclusions reached based on it.

## 2. Results and Discussion

### 2.1. WATER QUALITY DATA

Most of Israel's coastal rivers are actually wadis, in which water flows only during winter. In some sections, water may flow throughout the year due to discharges from small springs. The most notable exceptions are the Yarkon river and the Taninim river (Map 1).

The Yarkon, transecting the Greater Tel-Aviv area, was until the 1950s a stable river. However, as most of its spring waters were pumped for irrigation and municipal use, it too has nearly dried up. The Taninim still maintains flow (20-30 million CM annually) due to the fact that the water is brackish, with salinity levels reaching 1800 mg/L chlorides, making it unsuitable for drinking and irrigation.

Table I summarizes water quality data in the major coastal rivers, as monitored by sampling during May 1999. All analyses were carried out in an accredited commercial laboratory. These data supposedly represent the influence of the rainy winter season. However, the 1998/99 winter season was one of the driest ever recorded, with precipitates averaging some 50% of yearly values. Thus, the results obtained can be interpreted as typical of early summer, with low natural flows and increased concentrations of pollutants.

Several conclusions can be derived from Table I:

1. In unpolluted springs and clean segments (in the Naaman, Zippori, Tanini and Dalia rivers), B.O.D. levels are very low (0.4-1 mg/L), with a B.O.D/C.O.D. ratio of 7-60. This may indicate the presence of some non-biodegradable organic matter, presumably of little environmental consequence.
2. In the clean segments, the PO4 levels observed were at the range of 0.02-0.5 mg/L. Total Kjeldahl Nitrogen (TKN) levels were between 0.02-1.8 mg/L. The N/P ratio was 0.6 in the Naaman springs, 1.2 in the Zippori, and 10 in the Dalia river. Dissolved Oxygen levels reached 74-97% saturation, and oxygen levels were often higher in polluted segments, probably due to algal photosynthesis or to aeration taking place in shallow streams (often 50 cm deep ).
3. Boron was monitored as a sensitive indicator of detergents, and hence of domestic wastewater. Its background levels in clean springs and segments ranged between 0.16 mg/L at the Dalia springs and 0.43 mg/L at the Taninim springs, reflecting variations in the hydrogeochemical characteristics of the different water sources. Boron concentrations in domestic sewage are usually higher by approximately 0.4 mg/L than in the source tap water (unpublished data), and this is reflected in the increased boron concentrations in the polluted segments.

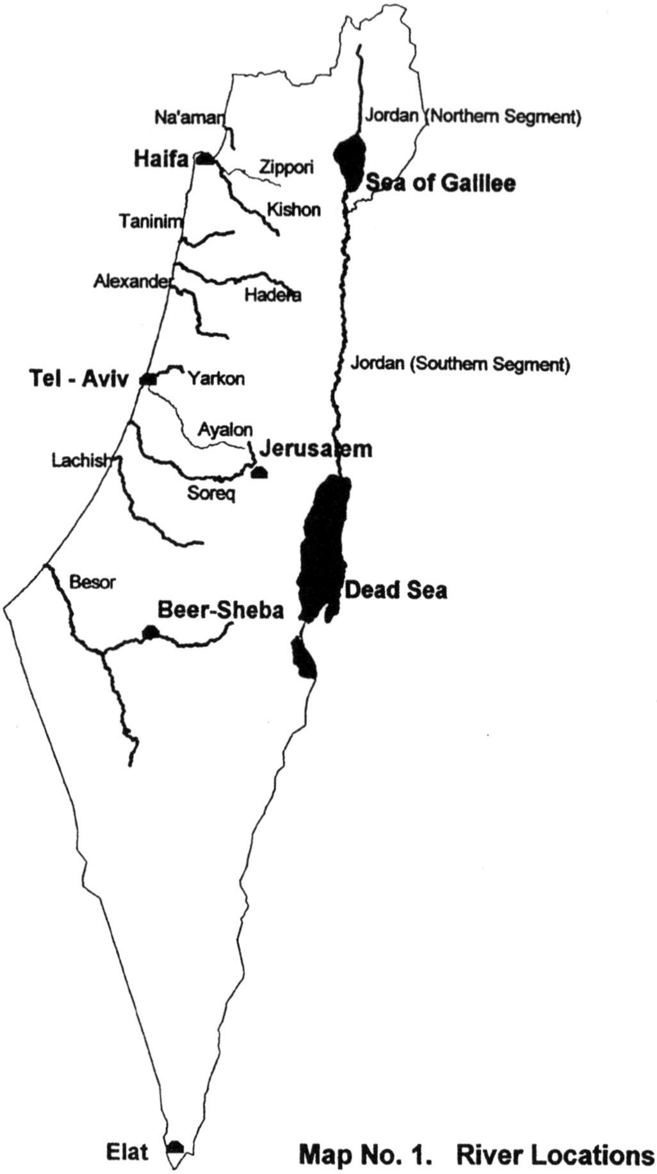


Fig. 1. River locations in Israel

TABLE I

Pollutant concentrations in Israel's coastal rivers, May 1999 (all values in mg/L, except D.O. which is expressed as % saturation at the temperature actually measured):

| River | BOD | COD | PO4 | TKN | Boron | D.O. |
|---|---|---|---|---|---|---|
| Naaman springs | 0.4 | 15 | 0.46 | 0.3 | 0.28 | 90 |
| Naaman downstream (En mifratz) | 22 | 190 | 12.2 | 13.6 | 0.46 | 37 |
| Zippori (springs) | 0.45 | 28 | 0.017 | 0.02 | 0.38 | 74 |
| Zippori downstream (Qaabiyeh) | 19 | 77 | 7.2 | 22.1 | 0.8 | 104 |
| Kishon upstream (Kfar Baruch) | 4 | 54 | 0.3 | 3.3 | 0.7 | 127 |
| Kishon downstream (Mizra) | 32 | 52 | 9.8 | 29.6 | 0.83 | 49 |
| Taninim upstream (Roman bridge) | 1 | 7.4 | 0 | 1.8 | 0.43 | 97.5 |
| Taninim downstream (Ada) | 11 | 40 | 0.84 | 4.1 | 0.52 | 80 |
| Dalia springs | 1 | 12.4 | 0.14 | 1.3 | 0.16 | 77.2 |
| Dalia downstream (estuary) | 3 | 65 | 0.09 | 2.8 | 0.44 | 54 |
| Hadera upstream (L. Haviva) | 3 | 36 | 0.74 | 10.1 | 0.49 | 83 |
| Hadera downstream (coastal road) | 7 | 126 | 2.8 | 10.2 | 0.82 | 83 |
| Alexander (Elyashiv) | 51 | 266 | 6.9 | 19 | 0.6 | 9.1 |
| Soreq (Yesodot) | 4 | 20 | 3.5 | 8.6 | 0.57 | 33 |
| Soreq (Refaim) | 337 | 133 | 9.17 | 43 | 0.56 | 5 |
| Lachish (Ad Halom) | 8.5 | 54 | 0.06 | 2.4 | 0.43 | 142 |

4. The presence of non-point sources of pollution is clearly indicated in the clean parts of the Naaman (high P), Taninim (high TKN), and the Soreq. In the latter case, the relatively low levels of organic matter (due to localized influence of good quality discharges of high groundwater) could not mask the high concentrations of nutrients and boron. These were derived from the wastewater flow coming from the upstream part of the river.

## 2.2. WATER AND EFFLUENT DISCHARGE QUALITY REQUIREMENTS

In the early 1990s it became quite obvious that no substantial dilution of effluents with freshwater could be expected (particularly in summer), and that if all effluent discharges were to be stopped, for the sake of restoration, some of the rivers would be left dry. Consequently, a dilemma arose: is it better to leave the river bed dry, apply for allocation of precious freshwater (almost against hope) or try partial restoration based on the release of treated wastewater of reasonable quality to the river bed. Eventually it was decided that treated effluents can be considered as an acceptable water source for river restoration, based on the following conditions:

- Wastewater treatment needs to be at a tertiary level, including removal of organic matter, nitrogen, phosphorous and ammonia, and include disinfection (preferably by methods other than chlorination) and aeration of the final effluents (Table II).
- Release of the effluents to the river should be after storage in a reservoir for 'polishing' the water quality, with a retention time of 10 days or more.
- The point of discharge should be such as to prevent damage to existing natural habitats and to allow for convenient inspection and monitoring of effluent quality.

Table II shows tentative water quality requirements for release of effluents into river beds included in restoration programs:

TABLE II

Quality requirements for effluents of domestic/urban origin released to rivers:

| | |
|---|---|
| Biological Oxygen Demand | 10 mg/L |
| Total Suspended Solids | 10 mg/L |
| Ammonia | 3 mg/L |
| Total Nitrogen (TKN) | 8 mg/L |
| Total Phosphorous | 0.5-1.0 mg/L |
| Fecal Coliform bacteria | 50 MPN/100mL |
| Dissolved Oxygen | 3 mg/L (minimum) |

## 2.3. SEDIMENT QUALITY DATA

In the summer of 1997, a temporary bridge collapsed into the Yarkon river, and tens of people fell into the water head on. Three of them eventually died as a result of pulmonary and brain damage and complications. This tragedy drew much public attention to the situation of the rivers, and to the need for additional data on the chemical composition of their sediments (even though it was concluded that at least two of the deaths were caused by the fungus *Pseudoallescheria boydii* and by *Pseudomonas* sp., not necessarily related to the prevalent pollution level of the water). Consequently, a survey of sediment quality of some of coastal rivers was initiated. Representative results (Avnimelech, 1998, 1999) of samples taken from the upper layer of the sediments (0-5 cm), during May 1998 (for the Yarkon) and April 1999, are presented in Table III:

TABLE III

Sediment quality in representative samples taken from coastal rivers, spring 1999. The Yarkon results refer to sampling done in May 1998
Sediment layer depth: 0-5 cm; Ambient temperature: 23-25$^0$C; pH values: 7.6-8.4

| River and station | C organic (mg/g) | H2S (ug/g) | NH4 (ug/g) | Hydro-carbons (mg/g) | Cd (ug/g) | Cr (ug/g) | Hg ug/g | V (ug/g) |
|---|---|---|---|---|---|---|---|---|
| Alexander centr. seg. | | 497 | 1131 | 2.06 | <2 | 46.6 | <2 | 45 |
| Alexander west. seg. | 22 | 85 | 448 | 1.77 | <2 | 26.5 | <2 | 24.5 |
| Alexander estuary | 9.4 | 48 | 233 | 0.001 | 2.0 | 49.4 | <2 | 36.7 |
| Soreq west. seg. | 29 | 715 | 683 | 1.86 | 3.5 | 66.4 | <2 | 66.4 |
| Lachish west. seg. | 66 | 1519 | 297 | 1.01 | <2 | 28.0 | <2 | 27.5 |
| Yarkon centr. seg. | 25 | 896 | 121 | 6.5 | <5 | 15.1 | <5 | 6.8 |

The results show heavy pollution in the central segment of the Alexander river, expressed by very high levels of organic matter, sulfides and ammonium, as well as hydrocarbons and vanadium, an indicator of mineral oils and fuels. This part of the river receives raw sewage from the cities of Nablus and Tul-Karm, as well as accidental spills from dairy farms, fish ponds and small villages. The point of sampling is very close to a major road (Highway 4), and this may explain the pollution by hydrocarbons and vanadium. In the western points of sampling the pollution level is progressively alleviated. This could be explained by dilution with the secondary effluents of the city of Netanya (discharged upstream of the “western segment”) and the influence of incoming seawater at the “estuary” sampling point. At present, it is not clear why levels of heavy metals increase at the estuary.

Heavy pollution is also evident in the Yarkon, Soreq and Lachish rivers, due to constant or accidental discharges of untreated wastewaters from coastal towns, and the vicinity of the sampling points to highways.

Conditions in the upper sediment layers are strongly reducing, as evidenced by the high sulfide and ammonium levels, which are toxic even at relatively low levels to many species of fish and invertebrates. This is apparently due to decomposition of organic matter which is continuously contributed from the pollution sources. Settling down of organic particles is facilitated by the very slow flow rate (typically - less than 1 m/sec), characteristic of the coastal plain rivers, where the slope is very shallow.

Analysis of sediment samples is of particular interest because the sediment serves as a kind of memory for pollution events and situations, even when the water column undergoes temporary improvement, such as by floods. It is estimated that the minimum flow, needed to move the upper sediment layers westwards and to the sea, is 100 CM/sec., and such conditions never occurred during the 1998/1999 winter season. Thus, monitoring of indicative chemical parameters in the sediment over several years is supposed to serve as a much more reliable indicator of changes in the overall physico-chemical river environment than mere water analyses, which are often hard to interpret (Topalian *et al.*, 1999).

## 2.4. POLLUTION REDUCTION

Since 1992, some $120 million have been invested annually in a nation-wide campaign to build new wastewater treatment plants (WWTP's) and upgrade outdated ones. Table IV shows the volume and percentages of treated and untreated wastewater in Israel in 1994 and 1999.

TABLE IV
Wastewater treatment in Israel
All figures in million CM

| year | total amount | amount treated (1) | % treated (1) | amount discharged to streams | amount treated (1)& discharged to streams | % treated of amount discharged to streams |
|---|---|---|---|---|---|---|
| 1994 | 390 | 115 (2) | 29.4 | 130 (2) | - | - |
| 1999 | 430 | 181 | 42.1 | 97.4 | 40.8 | 41.8 |

(1) treated to secondary level (B.O.D.<20 mg/L; TSS< 30 mg/L), or better
(2) estimation

The largest project carried out so far was the construction of the Jerusalem WWTP, finished in the summer of 1999 at a cost of $90 million. Currently it handles close to 20 million CM of wastewater annually, discharged to the upper Soreq river.

Most of the new WWTP's were designed to produce effluents with concentrations of B.O.D. and T.S.S. not exceeding 20 and 30 mg/L, respectively, in accordance with requirements set in regulations issued in 1992. However, in a few cases WWTP's were designed and built in conformity with the future requirements (Table II). Of special interest are the treatment plants of Kfar-Saba (KS) and Ramat-Hasharon (RH), discharging to the Yarkon. The KS plant has been producing secondary effluents since 1996, while the RH plant started operation in summer 1999, producing tertiary level effluents.

Table V shows the effect of the upgraded WWTP's on water quality in the river:

TABLE V

Water quality in the Yarkon river
All values in mg/L

| Station | B.O.D. | C.O.D. | T.S.S. | TKN | NH4 | PO4 |
|---|---|---|---|---|---|---|
| Source, 6.98 | 0.3 | 4 | 42 | 0.85 | <0.05 | - |
| Source, 7.99 | 2.0 | 23 | 68 | 2.1 | 0.02 | 1.25 |
| Downstream KH, 6.98 | 15 | 24 | 34 | 35.6 | 29.5 | - |
| Downstream KH, 7.99 | 9.9 | 89 | 29 | - | 30.8 | 30.9 |
| Downstream RH, 6.98 | 22.9 | 57 | 47 | 34.3 | 29.7 | 7.2 |
| Downstream RH, 7.99 | 4.6 | 49.5 | 4 | 23.5 | 20.6 | 24.5 |
| Seven mills (west) 6.98 | 2.8 | 43 | 2 | 31.2 | 25.4 | - |
| Seven mills (west), 7.99 | - | 62 | 7 | - | 16 | 21.4 |

The marked improvement in river water quality as a result of upgrading of the WWTP's can be seen clearly. However, the nutrient removal processes are still inoperative, and this leaves unacceptably high levels of total nitrogen, ammonium and phosphates in the water. This, in turn, is expected to result in undesirable algal growth, which would increase the T.S.S. levels. Sources of nutrient pollution other than wastewater are also expected to be revealed and will need to be addressed (Harremoes, 1998) Further close monitoring is therefore obviously needed.

## 3. Conclusions

River restoration efforts started in Europe in the early 1950s, with the establishment of the International Commission for the Protection of the Rhine against Pollution (Huisman *et al.*, 1998). Since then, many projects of river restoration were initiated in Denmark (Madsen, 1995), the Netherlands (Middelkoop and van Haselen, 1999), England (Wood and Handley, 1997) and other central European countries. The nature of attempts to restore rivers in Israel is fundamentally different than in European countries, as the most pressing issue is the lack of good quality freshwater. This also makes pollution problems more difficult, due to lack of the dilution factor. On the Israeli national level, the River Restoration Administration has prepared so far a preliminary evaluation of water quantities and qualities necessary to rehabilitate the major rivers, and a more detailed study on the Soreq river. This will serve as a basis for requests for water allocations submitted to the Water Commission, according to Israel's Water Law, which regulates water supply for the various consumers, even in the case of protection of endangered species living in wetlands and aquatic habitats is one of the direct results of river restoration efforts. For example, it is estimated that of 297 plant species found in the Taninim river ecosystem in the 1970s, some 100 perished during the 1980s. It is believed that rehabilitation of this major river could result in re-introduction of many of these species, based on experience gained in restored parts of the Yarkon River.

Another expected benefit will be the decrease in pollution of the Mediterranean Sea close to river estuaries. Prevention of uncontrolled discharges to these streams will undoubtedly contribute to protection of the marine environment as well. Israel is a party to the Barcelona Convention for the Protection of the Marine Environment and the Coastal Region of the Mediterranean, the Bonn Convention on the Conservation of Migratory Species of Wild Animals, the Convention of Wetlands of International Importance (the Ramsar Convention), and the Convention on Biological Diversity. The efforts made to significantly reduce pollution loads in rivers (Table IV) will undoubtedly contribute to achieve the goals of these international agreements.

Of particular relevance is the fifth European Community environmental action program (1993), calling for the conservation of nature and natural resources, and the subsequent proposed directive on the ecological quality of water (94/C 222/06). This directive aims at the adoption of “integrated programmes designed to improve the quality of.....surface water”, and “measures .....for the control of pollution of surface water from point sources, sources of diffuse pollution and other anthropogenic factors affecting surface water quality”. In 1997, the European Community adopted a more specific "Water Framework Directive", which calls for all member states to achieve "good status" in all waters by the end of 2010. For surface water, this means adequate quantities and chemical purity (Bloch ,1998) to achieve an acceptable ecological quality.

## Acknowledgement

Data on the Yarkon river water quality was received from the Yarkon River Authority. Water sampling of other rivers mentioned in this paper was done with the help of the Environmental Monitoring Unit at the Nature and National Parks Protection Authority. Thanks are due to all of them.

## References

Avnimelech, A.: 1998, *Monitoring the Yarkon river sediment,* Report submitted to the Ministry of the Environment, Israel, and the Yarkon River Authority, The Environmental Laboratory, Technion, Haifa (in Hebrew).

Avnimelech, A.: 1999, *Sediment analyses in the Alexander, Soreq and Lachish rivers*, Report submitted to the Ministry of the Environment, Israel, The Environmental Laboratory, Technion , Haifa, (in Hebrew).

Bloch, H.: 1998, in M. Landsberg-Uczciwek, M. Adriaanse and R. Enderline (eds), Proc. Internat. Conf. Management of Transboundary Waters in Europe, UN/ECE, 25-30.

De Wit, M. and Behrendt, H.: 1999, "Nitrogen and Phosphorous emissions from soil to surface water in the Rhine and Elbe basins," *Wat. Sci. Tech.* ***39***, 109-116.

Hakstege, A. L., Heynen, J. J. M., Eenhoorn, J. K. and Versteeg, H. P.: 1998, "Strategies for management of contaminated sediments within the Meuse river system", *Wat. Sci. Tech.* **37** 419-424.

Harremoes, P.: 1998, "The challenge of managing water and material balances in relation to eutrophication", *Wat. Sci. Tech. 37,* 9-17

Huisman, P., Wieriks, J. P. and de Jong, J.:1998, *Co-operation on management of transboundary waters: the river Rhine case*, Proc. Internat. Conf. Management of Transboundary Waters in Europe, UN/ECE, 97-111.

Madsen, B. L.: 1995, *Danish watercourses - ten years with the new watercourse act*, Ministry Envir. Energy, Denmark.

Middelkoop, H. and van Haselen, C. O. G.(eds):1999, *Twice a river: Rhine and Meuse in the Netherlands*, National Institute for Inland Water Management and Wastewater Treatment (RIZA).

Topalian, M. L., Rovedatti, M. G., Castane, P. M. and Salibiane, A.: 1999, "Pollution in a lowland river system. A case study: the Reconquista river (Buenos Aires, Argentina)", *Wat. Air Soil Poll.* ***114***, 287-302.

Wood, R. and Handley, J.: 1997, Mersey River Campaign (mid term report), England.

# LAKE KINNERET (ISRAEL) ECOSYSTEM: LONG-TERM INSTABILITY OR RESILIENCY?

M. GOPHEN

*Israel Oceanographic and Limnological Research. The Yigal Allon Kinneret Limnological Laboratory, P.O.Box 345, Tiberias Israel 14102*

*Tel.:972-6-6721444, Fax: 972-6-6724627, E-mail: Gophen@Ocean.Org.Il*

**Abstract.** The Kinneret ecosystem has undergone significant man-made modifications. These include, among others: construction of the south dam (1932); drying of Hula Lake and swamps (1950s); salty springs diversion (1964); lake water salinity decline from 395 (1961) to 210 ppm chlorid; deepening of the Jordan River near Bnot-Yaacov bridge (1971) with consequent high loads of suspended matter; National Water Carrier (NWC) operation (1964); sewage removal and fishpond restriction in the catchment; implementation of the Hula Project (1994-1998); introduction of native and exotic fishes; an amplitude of water level fluctuations of 4.8 m. Natural fluctuations occurred as well: heavy floods with high nutrient loads in 1968/69 and 1991/92 and droughts in 1973, 1989-1991 and 1998-1999; 4.6 m water level fluctuations; low temperatures (12.3$^0$C) and fish kill in 1971/72 and 1991/92; no mixing period in 1984; low (1975, 1996-1997) and high biomass of *Peridinium* (1994, 1998); $N_2$-fixing cyanobacterium bloom (1994, 1995); high stock of Lavnun (*Acanthobrama*) fishes (1990s); zooplankton biomass decline (1969-1993) and increase afterwards. Although man made changes and natural environmental changes were significant, the lake ecosystem did not deviate from the level of resilient fluctuations and the classification of long-term instability in Lake Kinneret is probably an overestimation. In this paper I use long-term (31 years) distribution data of plankton, nutrients, and water level in Lake Kinneret to distinguish between resiliency and instability.

**Keywords:** instability, Lake Kinneret, long-term record, resiliency

## 1. Introduction

Lake Kinneret (the Sea of Galilee), the only natural freshwater lake in Israel, supplies ~30% (600 x $10^6$ $m^3$) of the national water demands in Israel and ~55% (300 x $10^6$ $m^3$) of drinking water requirements. The Kinneret ecosystem (lake: 170 $km^2$; catchment 2730 $km^2$) is located in the northeastern part of the country. Water utilization is carried out as follows: 100 x$10^6$ $m^3$ are removed in the Hula Valley north of the lake (Upper Galilee and Golan Heights); 400 x $10^6$ $m^3$ are delivered through the National Water Carrier (NWC) system; 150 x $10^6$ $m^3$ are taken directly from the lake by local consumers. Needless to say, the water quality of the Kinneret ecosystem is significant and vital state of Israel.

The Kinneret ecosystem has undergone significant man-made modification during the last 70 years: 1) The population in the drainage basin has increased to almost 300,000 inhabitants, and domestic sewage production is approximately 18 x $10^6$ $m^3$ per year of which more than 90% are removed (Gophen *et al.*, 1999; Serruya, 1978); 2) Agricultural management in the Hula Valley was

*Water, Air, and Soil Pollution* **123:** 323–335, 2000.

significantly modified and included new irrigation systems and pesticide, herbicide and fertilizer use which caused changes in nutrient inputs (composition and quantity) (Shacham, 1992; Rom, 1999); 3) The south dam was constructed near Kibbutz Degania, accompanied by deepening of the Jordan outlet between the lake and the dam, to control outflow discharges; 5) The construction of the NWC system (1964) modified outflow regimes and nutrient dynamics: before dam construction the outflow regime was seasonal and after dam operation the outflow moves through the NWC system at a steady rate of ca $10^6$ $m^3$ per day throughout the entire year from a fixed depth of the intake (11-13 m, affected by water level changes) (Serruya, 1978); 6) An annual diversion of approximately 60,000 tons of salt started in 1967 through a newly constructed canal, and lake salinity declined from 395 ppm chloride in early 1960s to 200 ppm in late 1980s and the 1990s (LKDB, 1969-1999). Presently, as a result of a drought, salt diversion is not operated and water salinity has increased to 260 $mgCl^-/L$ (LKDB, 1969-1999); 7) Lake Kinneret is intensively stocked by native and exotic fish species, some of which were found to have a negative effect on water quality and their planting was eliminated (*Oreochromis aureus*, *H. molitrix*-silver carp). The introduction of the native species, *Sarotherodon galilaeus* together with bleak subsidised fishery improved water quality through suppression of *Peridinium* biomass and top-down trophic effect respectively (Gophen *et al.,* 1999); 8) Deepening and widening of a section of upstream Jordan River was followed by heavy flux of suspended matter and nutrients; 9) Restriction of the fishpond area in the Hula Valley from 1700 ha during 1970s to 450 ha presently resulted in a reduction of nutrient inputs (Serruya, 1978; LKDB, 1969-1999); 10) Construction of sewage removal systems with total storage capacity of ca 14 x $10^6$ $m^3$ in the catchment resulted in the decline of nutrient inputs (Rom, 1999); 11) The Hula reclamation project (1994-1998) was implemented including renewal of a hydraulic system and construction of a new shallow lake (Agmon) which functions as a drainage basin within the Valley (Shacham, 1992); 12) Annual diversion of 25-50 x $10^6$ $m^3$ Yarmuk waters (south of Lake Kinneret) into the lake which was recently formulated as part of the peace accord with the State of Jordan (Water Commission, unpublished data).

Alongside man-made modifications in the system, natural events of extreme varied levels of parameters have occurred within the Kinneret system. The most prominent examples are the following (Serruya, 1978; LKDB, 1969-1999): 1) Flooding in the winter seasons of 1968-1969 and 1991-1992 when input discharges exceeded 1 x $10^9$ $m^3$ (Rom, 1999); 2) Droughts and input discharges lower than 300 x $10^6$ $m^3$/year (1973; 1986; 1989-1991; 1998-1999); 3) High water levels >209 (1929;1969;1992); 4) Low water levels <212 (1934; 1986; 1989-1991; 1999) (LKDB, 1969-1999); 5) High winter temperatures with a short turnover period (1984) and low winter temperatures and long mixing period (1971; 1991-1992) (LKDB, 1969-1999); 6) High content of phosphorus (P) and nitrogen (N) forming in the water column (1974) (LKDB, 1969-1999); 7) Low biomass (1975; 1996; 1997) and high biomass (1994;1998) of

Pyrrhophyta (Pollingher and Zohary, LKDB, 1969-1999); 8) $N_2$-fixing cyanobacteria summer blooms (1994; 1995) (Berman, 1996; Gophen, 994;Gophen *et al.,* 1999;Hadas *et al.,* 1999;) 9) High densities of small bodied bleaks (1993-1996) (Gophen *et al.,* 1999); 10) massive fish kill during winter (1971-1972; 1991-1992); 11) Heavy blooms and scum during winter of *Microcystis* (1995) (Berman, 1996; Hadas *et al.,* 1999); 12) Long term (1970-1993) decline of zooplankton biomass and an increase during 1993-1999 (Gophen 1992; Gophen *et al.,* 1999); 12) Pelagial invasion by the zooplankter *Eudiaptamus dreischi* (Azoulay, 1999) and the extinction of *Daphnia lumholtzi* (late 1950s-early 1960s) (Gophen, 1979); 13) Intensification of fishery resulted in an increase of fishing pressure and stocking practicies of *Sarotherodon galilaeus* (Gophen *et al.,* 1999).

This is a partial list of man-made and natural modifications in the Kinneret ecosystem (Serruya, 1978). The outcome of these events is a question related to the ecological capacities of the system: Does the Kinneret ecosystem (lake and catchment) represent a stable or unstable structure? Can we identify periods with high or low stability or capacity of ecological bufferization especially with regard to water quality? The answer to these questions might improve long-term prediction of water quality as a result of human management or natural changes. Moreover, population capacities of the biota in the Kinneret ecosystem, like in other aquatic natural structures, have their own amplitude of changes as affected by human intervention and/or natural events. Stability is not always reflected by low amplitude of fluctuations but much more by the time-length existence of the modified population or parameter. Population capacity is flexible and its resiliency reflects stability and not instability (Reynolds, 1997). Instability is reflected by population collapse as a result of long-term overloaded capacity. On the other hand, short-term duration of extreme levels of the population capacity accompanied by the ability to return to the former (pre-modified) level reflect stability. Therefore, modification of ecological parameters does not necessarily reflect instability as do long-term enhancement or collapse of the population. Long-term (55 years) records of rainfall in the Kinneret drainage basin (4 stations: Har-Knaan, Kfar-Blum, Dafna and Ayelet Hshachar) (Figure 1) indicate a normal distribution curve, i.e., no significant long-term trend of change of inflow discharges (Rom, 1999) with >95% of the values within 2 SD's above and lower than the average (average:575 mm/year; SD=181 mm/year;) (Z-score plot, Figure 1), Skewness positive value of 0.572, and Kurtosis positive value of 0.229 (Median=547 mm/year).

In this paper, I will present an analysis of long-term changes of several major components of the Kinneret ecosystem (plankton, nutrients and water level) (LKDB, 1969-1999) aimed at an insight into the system structure: stable or unstable? Resilience or non-resilience? Resiliency means short-term up and/or down fluctuations or forward and backward ecosystem alterations such as from oligotrophic to mesotrophic status and vice versa whilst instability is unidirectional changmostly towards eutrophication (water quality deterioration) and/or maximum population (food-web organisms) capacity (Reynolds, 1997). The water level (WL) parameter was chosen as an example of human controlled factor while plankton and nutrients were chosen as major components of the system which strongly affect water quality.

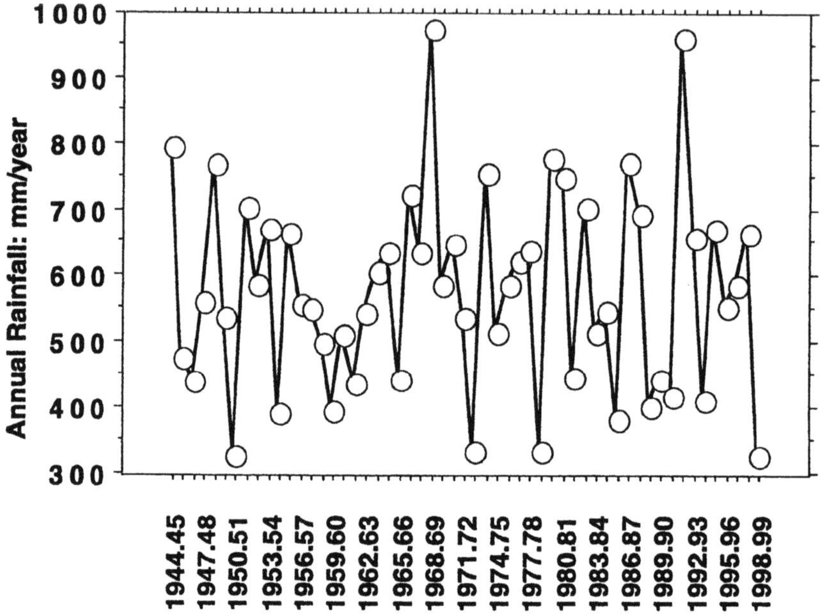


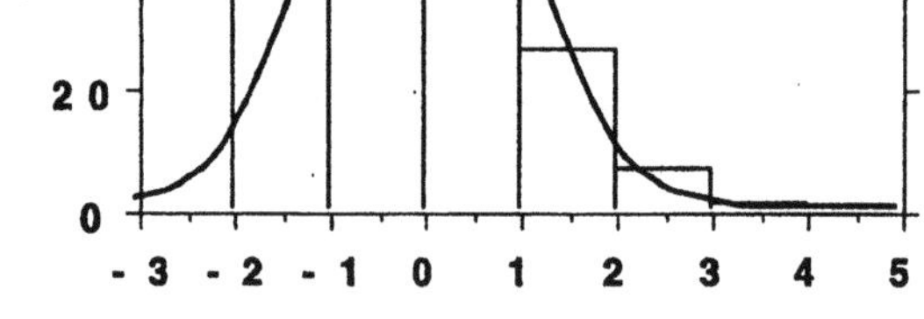


Fig. 1. Average (1944-1999) of annual rainfall (mm/year) in 4 stations in the Kinneret catchment: Kfar Blum, Dafna, Ayelet-Hashachar, and Har-Knaan (upper panel). Z–score histogram plot of the data is presented (lower panel).

## 2. Water Level Effect

Until 1932 the water level (WL) of Lake Kinneret has fluctuated at an amplitude of 1-1.5 m every year (Serruya 1978). The Jordan outlet threshold altitude is ca 211.00 m- bsl or 212.00 m-bsl (Niv 1978). WL increase of 1.0-1.5 m during 2-3 months was the outcome of winter inflow discharges. After the construction of Degania Dam (1932) and deepening of the outflow threshold by 3-4 m, the lake water level became man-controlled. Therefore WL altitude varied between 208.66 m bsl (1969) and 213.30 m bsl (1999) (Figure 2) (LKDB, 1969-1999). Regression analyses of biological parameters as dependent variables vs. water level altitude (independent variable) are presented in Table I. Results in Figure 2 indicate high amplitude of water level fluctuations (maximal amplitude during 1969-1999 was 4.80 m), but biotic communities and nutrient responses in the lake were mostly unaffected by WL variations (Table I). Therefore, it can be taken into account that the ecological dependence of nutrients (P, N) and biotic components on WL is low, probably because of internal regulatory processes in the lake (Serruya *et al.,* 1980). On the other hand, nutrient composition changes (slight increase of TP and slight decline of TN) (LKDB, 1969-1999) initiated conditions favored by $N_2$ fixing cyanobacteria bloom, which will be discussed later.

The low values of $R^2$ presented in Table I indicate a dependence of less than 10% of all parameters on the WL (except TN/TP ratio – 12%). Nevertheless it must be clarified whether the slight changes of nutrient composition in the lake are dependent on internal (lake) or external fluxes. Data published by Geifman *et al.* (1969-1999) and Rom (1999) were analysed and results are given in Table II. Rom (1999) reported on significant decline of total suspended solids (TSS), particulate P (PP), total phosphorus (TP), organic-N, and a significant increase of dissolved phosphorus (DP). Available P (soluble) inputs increased and available N declined, and this resulted in a long-term decreases of N-organic/DP mass ratio of external loads (Table II). The lake response to these changes was probably only a slight increase of chlorophyte biomass during the1980s and onwards and the appearance of $N_2$-fixing cyanobacteria bloom. Nevertheless, none of these changes can be classified as maximum population capacity (Reynolds, 1997) and/or long-term consistent water quality deterioration trend. The enhanced biomass of green algae was compensated by intensification of grazing pressure by zooplankton (Serruya *et al.,* 1980; Gophen *et al.,* 1999) and cyanophyte blooms significantly declined prior to summer 1995 as a result of improvement of nutrient composition in the lake (LKDB, 1969-1999).

TABLE I

Linear regression parameters ($R^2$, p; * = not significant) between plankton (zooplankton and phytoplankton) ($g/m^2$) and nutrient (tonnes/lake) parameters (monthly means) (dependent variables) and water level altitude (monthly parameters) (dependent variable) during 1969-1999. Trend: Increase (I) or Decrease (D) of dependent variable vs. independent variables (Phytoplankton data source: Pollingher and Zohary, LKDB,1969 1999; Nutrients: A. Nishri, LKDB, 1969-1999).

| Independent parameter | $R^2$ | p | Trend |
|---|---|---|---|
| TN (ton./lake) | 0.097 | < 0.0001 | D/D |
| TP (ton./lake) | 0.022 | 0.0051 | I/D |
| TDP (ton./lake) | 0.003 | 0.3671 | I/D* |
| TN/TP (mass ratio) | 0.116 | < 0.0001 | D/D |
| TDN/TDP (mass ratio) | 0.087 | < 0.0001 | D/D |
| *Copepoda* ($g/m^2$) | 0.009 | 0.0725 | D/D |
| *Cladocera* ($g/m^2$) | 0.023 | 0.0045 | D/D |
| *Chlorophyta* ($g/m^2$) | 0.013 | 0.0336 | I/D |
| *Pyrrhophyta* ($g/m^2$) | 0.104 | < 0.0001 | D/D |
| Diatoms ($g/m^2$) | 0.019 | 0.0075 | I/D |
| *Cyanophyta* ($g/m^2$) | 0.008 | 0.0838 | I/D |

TABLE II

Regression ($R^2$, p; * = insignificant) parameters between nutrient (N-organic, TDP, TIN/DP) loads measured at Huri Bridge (annual loads, tons) and time (years 1970-1999). Trends: D (decline) or I (increase) with time from 1969 to 1998 (Data source: Geifman *et al.*, 1969-1998; Rom, 1999)

| Parameter | $R^2$ | p | Trend |
|---|---|---|---|
| N-organic | 0.363 | 0.0003 | D |
| TIN ($NH_3 + NO_3$) | 0.001 | 0.8735* | no change |
| DP | 0.112 | 0.0658* | I |
| TIN/DP | 0.251 | 0.0041 | D |
| (TIN + N org)/DP | 0.463 | <0.0001 | D |

## 3. Long Term (1969-1999) Measures of Distribution

Two types of distribution analysis of the monthly averages of biotic and nutrient parameters (1969-1999) were carried out: Z-scores and Kurtosis-Skewness (StatView-SAS 1995). The results of Z-score tests indicated normal or very close to normal distribution (68% of the values fall within 1 SD interval of the mean, 95% within 2 SD interval of the mean and >99% within 3 SD interval of the mean (Figures 3-5). The results of Skewness-Kurtosis tests are presented in Table III. All skewness values calculated for the distribution of zooplankton, phytoplankton and nutrient monthly averages (1969-1999) are positively

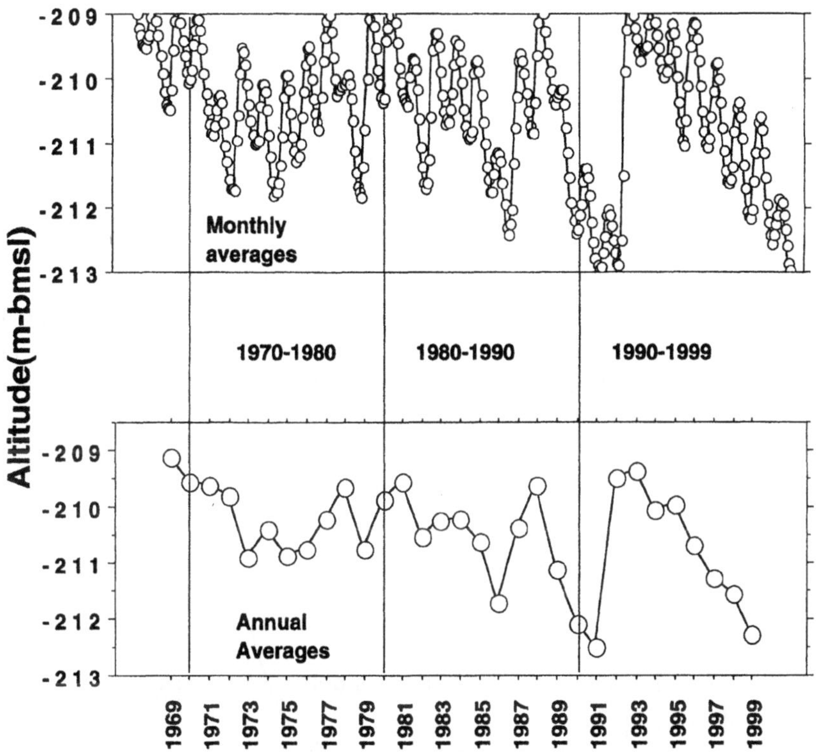


Fig. 2. Monthly (upper panel) and annual (lower panel) averages of water level in Lake Kinneret during 1969-1999

TABLE III

Skewness, Kurtosis, Median and Mean (see also Table I) measures of distribution values (monthly means, 1969-1999) of phytoplankton and zooplankton ($g/m^2$) and nutrients (tonnes/lake) in Lake Kinneret. Counts: Phytoplankton – 369; Zooplankton – 366; Nutrients – 368 (TDP – 308) (Data source: Phytoplankton-Pollingher and Zohary; Nutrients-A. Nishri; LKDB, 1969-1999).

| Parameter | Mean | Medium | Skewness | Kurtosis |
|---|---|---|---|---|
| *Copepoda* ($g/m^2$) | 9.3 | 8.3 | 1.783 | 5.609 |
| *Cladocera* ($g/m^2$) | 17.2 | 14.8 | 1.493 | 3.418 |
| *Rotifera* ($g/m^2$) | 2.2 | 1.1 | 3.119 | 15.363 |
| *Cyanophyta* ($g/m^2$) | 3.3 | 1.2 | 5.156 | 36.998 |
| Diatoms ($g/m^2$) | 7.7 | 1.8 | 5.929 | 41.278 |
| *Chlorophyta* ($g/m^2$) | 10.2 | 8.3 | 2.621 | 10.929 |
| *Pyrrhophyta* ($g/m^2$) | 54.3 | 16.3 | 1.951 | 3.466 |
| TP (t/lake) | 76.7 | 75.7 | 0.524 | 1.323 |
| TDP (t/lake) | 40.3 | 38.9 | 0.663 | 1.173 |
| TN (t/lake) | 2769 | 2580 | 1.182 | 1.689 |
| TIN (t/lake) | 983 | 905 | 1.898 | 6.871 |

skewed, i.e., number of smaller than mean values is greater than those which are larger than the mean, and the mean is greater than the median. Consequently, it can be suggested that there is no trend of zooplankton and phytoplankton populations which consistently approach the "maximum population capacity" (Reynolds, 1997) which by itself is not defined here.

The Kurtosis test is a measure of the amount of values (monthly averages during 1969-1999) not in the central part of the distribution but at the tail end. All Kurtosis values calculated for zooplankton, phytoplankton and nutrient values (Figure 3) were found to be positive, indicating that most of the values are pressed (in the distribution curve) in the middle part and the tails are slim with few extremes (Leptokurtic curve) (StatView, 1995). I suggest that these results make clear that the Kinneret ecosystem dynamics is characterized as no-trend towards "maximum population capacity".

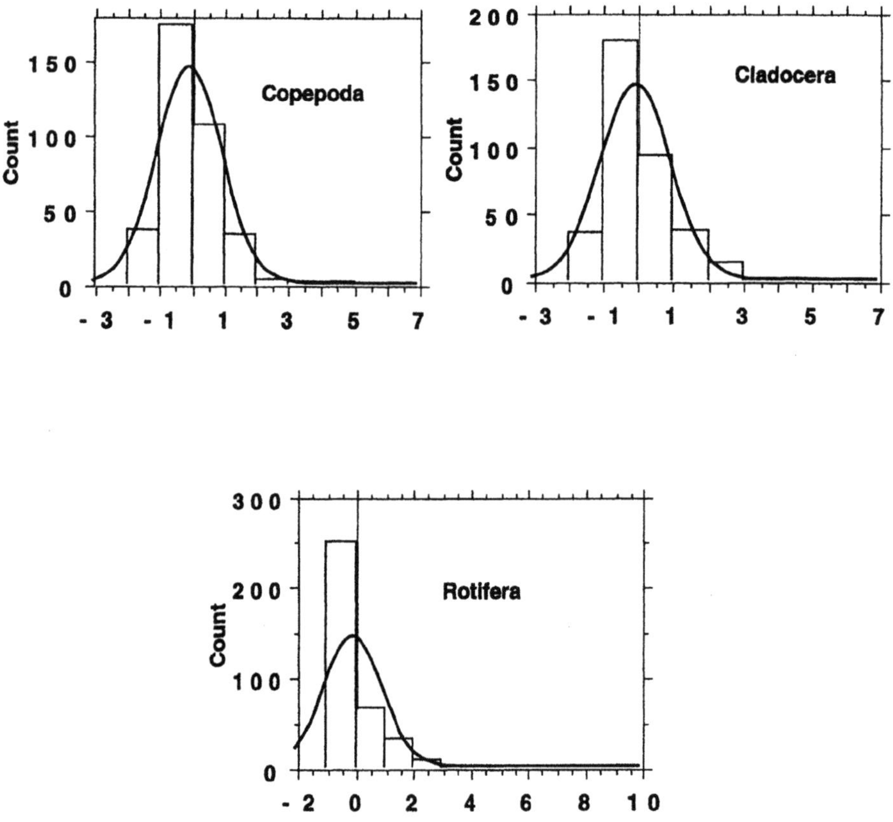


Fig. 3. Z-score histogram plot of zooplankton (*Copepoda, Cadocera, Rotifera*) ($g/m^2$) biomass (monthly averages; 369 counts) in Lake Kinneret during 1969-1999.

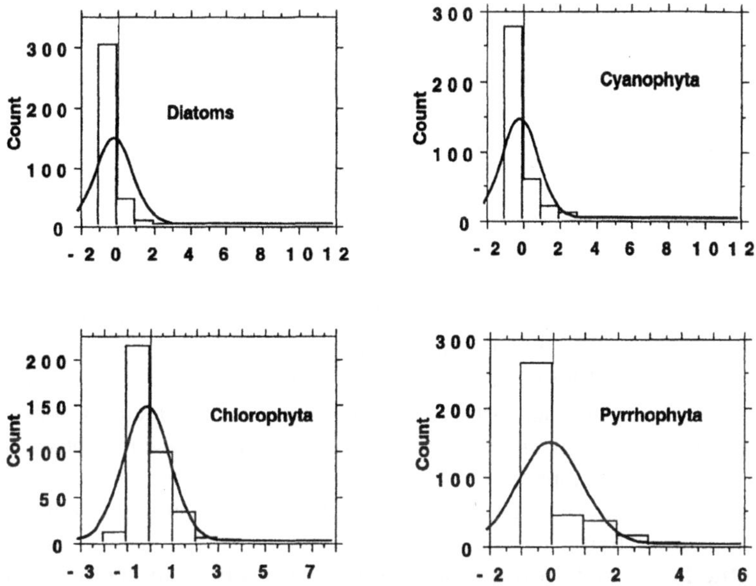


Fig. 4. Z-score histogram plot of phtoplankton (*Cynophyta,* diatoms, *chlorophyta, pyrrhophta*) (g/m$^2$) biomass (monthly averages; 369 counts) in Lake Kinneret during 1969-1999 (Source: Pollingher and Zohary, LKDB, 1969-1999)

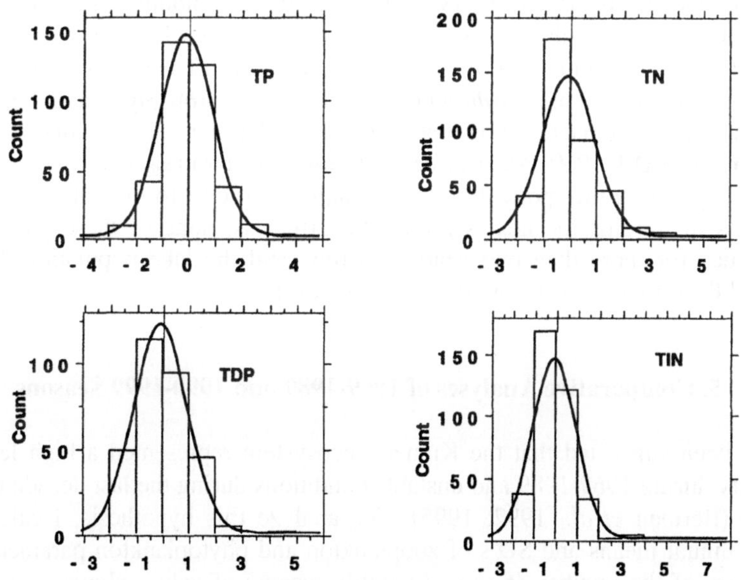


Fig. 5. Z-score histogram plot of nutrient inventory (TP, TN, TDP, TIN) (tonnes/lake) in Lake Kinneret (monthly averages; 368 counts, TDP – 308) during 1969-1999) (Source: A. Nishri, LKDB, 1969-1999)

## 4. The *Peridinium* Case

*Peridinium* is a dinoflagellate which predominantly occurs in Lake Kinneret (Berman *et al.*,1992;1995). It is a bloom forming algae with an annual average (1969-1999) biomass of 54.3 g/m$^2$ (SD 75.6). The bloom normally appears in late February and crashes in late May (Breman *et al.,* 1992; 1995). During the bloom (February-June) season the multiannual average is 111 g/m$^2$ (SD 88.5). Monthly means of *Peridinium* biomass above 200 g/m$^2$ occurred 25 times (months) in the years 1970, 1972, 1979, 1980-1982, 1985, 1986, 1987, 1990-1992, 1994, 1995, 1998, 1999. Monthly means of *Peridinium* biomass lower than 100 g/m$^2$ during February-June were measured 81 times during the years 1969-1999 (except 1998) but only six times in four month seasons in 1974, 1975, 1982, 1988, 1996 and 1997. *Peridinium* can be considered as one of the most adaptive organisms in the Kinneret ecosystem: this algae is exposed to low grazing pressures (Gophen *et al.,* 1999), its ability to migrate supports appropriate timing of light exposure as well as nutrient uptake and photosynthetic activity; the sporulation process and the existence of dormant non-motile heavy cysts which absorb nutrients (mostly P) in bottom sediments enables the motile vegetative cells, after excitation, to maintain a high rate of multiplication under low ambient concentration of P (Berman *et al.,* 1992; 1995; Pollingher, 1978). *Peridinium* is probably limited by two factor combinations: a) nutrients (mostly P and N) availability during the bloom onset (Pollingher, 1978); b) optimal conditions for the sporulation process. The two combinations are not necessarily linked. If these two combinations affected solely or together the ups and downs of *Peridinium* blooms, it was probably not a long-term (several years) impact. It is not impossible that nitrogen decline in Lake Kinneret (LKDB 1969-1999) affected *Peridinium* biomass decrease and/or P deficiency during the sporulation season suppressed cyst formation causing low biomass of blooms the next season. Nevertheless, none of these parameters continued for more than two years in a row, and the lake population did not exceed the maximum capacity or extinction level.

## 5. Comparative Analyses of 1969-1989 and 1990-1999 Seasons

It has been suggested that the Kinneret ecosystem represented a high level of stability during 1969-1989 and unstable conditions during the last decade (1990-1999) (Berman *et al.,* 1992; 1995). To analyze this hypothesis, I calculated multi-annual means and SD's of zooplankton and phytoplankton parameters. A summary of the number of cases (monthly means) of values above mean-plus-SD during the two periods (1969-1989, 21 years and 1990-1999, 10 years) and the number of such cases per year are presented in Table IV. It is prominent that for zooplankton (*Copepoda* and *Cladocera*), there was an increase of high value

TABLE IV

Number of monthly average cases higher than nultiannual mean (X) plus SD, during two periods: a) 1969-1989 (21 years); b) 1990-1999 through October (10 years) of the biomass ($g_{w.w.}/m^2$) of Copepoda, Cladocera, Rotifera, Cyanobacteria, Diatoms, *Chlorophyta* and *Pyrrhophyta* in Lake Kinneret: n = 366 (zooplankton), 369 (phytoplankton) (Source: phytoplankton-Pollingher and Zohary, LKDB 1969-1999).

| Taxon | x (SD) ($g/m^2$) | Case number of >(X+SD) | Number of years with cases of > (X+SD) | | >(X+SD/year) | |
|---|---|---|---|---|---|---|
| | | | 1969-1989 | 1990-1999 | 1969-1989 | 1990-99 |
| *Copepoda* | 9.3 (5.9) | 46 | 13 | 2 | 5.0 | 15.4 |
| *Cladocera* | 17.2 (10.7) | 55 | 18 | 5 | 2.8 | 4.8 |
| *Rotifera* | 2.2 (3.0) | 50 | 18 | 8 | 2.2 | 2.4 |
| *Cyanophyta* | 3.3 (6.2) | 35 | 6 | 5 | 11.1 | 6.4 |
| Diatoms | 7.7 (21.2) | 22 | 6 | 6 | 6.4 | 3.1 |
| *Chlorophyta* | 10.2 (8.7) | 45 | 12 | 7 | 4.1 | 3.4 |
| *Pyrrhophyta* | 54.3 (75.6) | 59 | 19 | 8 | 2.4 | 2.7 |

cases/year during 1990-1999 (not indicating instability) and a decline for phytoplankton (*Cyanophyta,* diatoms and *Chlorophyta*) high value cases/year (also not significant to unstable conditions), and not a great change for *Peridinium* and *Rotifera.* The increase of zooplankton biomass cases (monthly means per year) was the outcome of implementation of a policy to subsidise removal of *Lavnun* (Bleak, *Acanthobrama*) fishes (Gophen *et al.,* 1999). Phytoplankton increase started as early as the 1980s (Berman *et al.,* 1992; 1995).

## 6. The *Aphanizomenon* Case

During the period of 1969-1999, both algal biomass and rates of primary production exhibited consistent repeatable seasonal patterns of development within each year of records, although interannual oscillations were also evident in these parameters (Berman *et al.,* 1992, 1995).

Substantial changes have occurred in the Lake Kinneret watershed, which have altered both inflowing water volume and its chemistry (Gophen, 1990;Gophen *et al.,* 1999; Rom, 1999). A decline in the influent TIN/TIP ratio has occurred. Not surprisingly, significant decline also has occurred in the TN/TP and PN/PP ratios in the epilimnion of the lake itself. Because $N_2$–fixing cyanobacteria are considered to be toxic (Carmichael, 1992; Hadas *et al.,* 1999) and competitively favored over other algae under conditions of N-deficiency and P-sufficiency (NDPS) (i.e., low N/P ratio) (Smith, 1983; Schindler, 1977), it has been predicted (Gophen *et al.,* 1990) that these undesirable blue-greens may bloom as has already happened in the summer of 1994 and, to a lesser extent, in the summer of 1995 (Gophen *et al.*, 1999; Hadas *et al.*, 1999). NDPS conditions primarily enhanced the growth of the $N_2$-fixing cyanobacterium *Aphanizomenon ovalisporum* (Berman, 1995; 1996; Gophen, 1994; Hadas *et al.*, 1999). After the summer of 1994, a trend towards increasing N-sufficiency and P-deficiency (the opposite of $N_2$-fixing cyanobacteria favored trend) has occurred: increase of TDN/TDP ratio. These changes may account for the observed decline in the population of *A. ovalisporum* after 1994 (Gophen *et al.*, 1999). It is suggested that even such an event, like *A. ovalisporum* bloom in Lake Kinneret, can be classified as reversible alteration of phytoplankton community structure with significant effect on water quality and, to a lesser extent, as a symptom of ecosystem instability.

## 7. Conclusion

Instability measures in natural ecosystems are not only presented by number and amplitude of fluctuations. The time length of ups and downs have to be considered in order to analyze whether populations reached the level of maximum capacity followed by long-term alterations and in aquatic ecosystems, mostly by water quality deterioration.

Based on 31 years of limnological records in Lake Kinneret (LKDB, 1969-1999), I suggest that this ecosystem represents a high level of stability as reflected by the distribution of biotic and nutrient values and the frequency and time length of exceptions.

## References


Azoulay, B.: 1999, "The occurrence of the diaptomid *Eudiaptomus cf. Drieschi*" (Poppe & Marazek, 1985) in Lake Kinneret, Israel, Abstract,7th International Conference on *Copepoda.* Curitiba, Brazil, 25-31 July 1999, pp. 53.

Berman, T.: 1995, "Cyanobacteria in the Sea of Galilee (Lake Kinneret)", *Bull. Amer. Soc. Limnol. Oceanogr* **4**,13-14.

Berman, T.(ed): 1996, "Water Quality in Lake Kinneret during 1994-95: exceptional quantitative phenomena, and algal activity and species composition", IOLR-KLL Special Report No. **T5/96**, pp. 20.

Berman, T., Jacobi, Y. Z. and Pollingherm, U.: 1992, "Lake Kinneret phytoplankton: Stability and variability during 20 years (1970-1989)", *Aquat. Sci.* **52**,104-127.

Berman, T., Stone, L., Yacobi, Y. Z., Kaplan, B., Schlichter, M., Nishri, A. and Pollingher, U.: 1995, "Primary production and phytoplankton in Lake Kinneret: A long-term record (1972-1993)", *Limnol. Oceanogr.* **40**, 1064-1076.

Carmichael, W. W.: 1992. "Cyanobacteria secondary metabolites – the cyanotoxins", *J. Appl. Microbiol.* **72**, 445-459.

Geifman, Y., Dexter H. and Shaw, M.: 1970-1998, *Nutrient fluxes into Lake Kinneret, Annual Reports,* Mekorot, Water Supply Co. Nazareth, Jordan District, Catchment area monitoring Uniy. (in Hebrew).

Gophen, M.: 1979, "Extinction of *Daphnia lumholtzi* (Sars) in Lake Kinneret (Israel)," *Aquaculture* **16**(1), 67-71.

Gophen, M.: 1992, "Long-term changes of plankton communities in Lake Kinneret, Israel", *Asian Fish. Sci.* **5**, 291-302.

Gophen, M.: 1994, "Blue-green algae and Lake Kinneret", *Bull. Amer. Soc. Limnol. Oceanogr.* **3,** 5.

Gophen, M., Serruya, S. and Threlkeld, S.: 1990, "Long-term patterns in nutrients, phytoplankton, and zooplankton of Lake Kinneret and future predictions for ecosystem structure", *Arch. Hydrobiol.* **118**, 449-460.

Gophen, M., Smith, V. H., Nishri, A. and Threlkeld, S. T.: 1999, "Nitrogen deficiency, phosphorus sufficiency, and the invasion of Lake Kinneret Israel, by the N2-fixing cyanobacterium *Aphanizomenon ovalisporum*," *Aquat. Sci.* **61**, 1-14.

Hadas, O., Pinkas, R., Delphine, E., Vardi, A., Kaplan, A. and Sukenik, A.: 1999, "Limnological and ecophysiological aspects of *Aphanizomenon ovalisporum* bloom in Lake Kinneret, Israel,"*Jour. Plan. Res.* **21**, 1439-1453.

LKDB (Lake Kinneret Data-Base) (M. Schleichter-manager): 1969-1999, Nutrients: Lake – Geifman, Y., Serruya, C. and Nishri, A., Catchment –Geifman, Y., Rom, M., Shaw, M. and Dexter, H., Phytoplankton – Pollingher, U. and Zohary, T., Zooplankton – Gophen M.

Niv, D.: 1978, "The Geology of Lake Kinneret", in: *Lake Kinneret: the lake and watershed,* Kinneret Authority, 15-26. (in Hebrew).

Pollingher, U.: 1978, "Periinium cinctum fa westii: Life cycle", in: *Lake Kinneret,* Monographiae, Biologicae, Junk Publishers, 271-274.

Reynolds, C. S.: 1997. "Vegetation processes in the pelagic: A model for the ecosystem theory", in O. Kinne (ed), *Excellence in ecology,* Ecology Institute, Olendorf/Luhe, Germany, pp. 371.

Rom, M.: 1999, *The Kinneret water catchment: trends and changes in nutrient contribution,* Mekorot National Water supply Co., Water catchment unit, Jordan District, Special Report, pp. 63 (in Hebrew).

Schindler, D. W.: 1977, "Evolution of phosphorus limitation in lakes", *Science* **195,** 260-262.

Serruya, C. (ed): 1978, Lake Kinneret (Monographiae Biologicae),Vol. **32**, Junk, Amsterdam

Serruya, C., Gophen, M. and Pollingher, U.: 1980, "Lake Kinneret: Carbon flow patterns and ecosystem management", *Arch. Hydrobiol.* **88**, 265-302.

Shaham, G.: 1992, The Hula Project: Dynamics of human intervention in nature. ISEEQS Pub. Jerusalem, Israel (Y. Steiberger ed.) Vol. **vi A/B**, 648-651.

Smith, V. H.: 1983, "Low nitrogen to phosphorus ratios favor dominance of cyanobacteria in lake phytoplankton", *Science* **221**, 669-671.

StatView-SAS: 1995, "Measures of distribution characteristrics", in *StatView Reference*, SAS Institute Inc., 6-7.

# EFFECT OF THE ENVIRONMENT ON THE BACTERIAL BLEACHING OF CORALS

E. BANIN[1], Y. BEN-HAIM[1], T. ISRAELY[1], Y. LOYA[2] and E. ROSENBERG[1]

[1] *Department of Molecular Microbiology and Biotechnology,* [2] *Department of Zoology and the Porter Super Center for Ecological and Environmental Studies, Tel Aviv University, Ramat Aviv, Tel Aviv 69978, Israel*

**Abstract.** Bleaching in stony-corals is the result of disruption of symbiosis between the coral hosts and photosynthetic microalgal endosymbionts (zooxanthellae). Coral bleaching events of unprecedented frequency and global extent have been reported during the last two decades. Recently, we demonstrated that bleaching of the coral *Oculina patagonica* in the Mediterranean Sea is caused by the bacterium *Vibrio shiloi*, when seawater temperature rises and allows the bacterium to become virulent. The first step in the infection process is host-specific adhesion of *V. shiloi* to *O. patagonica* via a β-galactoside receptor on the coral surface. The bacterium then penetrates into the coral tissue and produces extracellular materials which rapidly inhibit photosynthesis of zooxanthellae and bleach and lyse the algae. The inhibition of photosynthesis is due to a low molecular weight, heat stable toxin and ammonia. Bleaching and lysis are due to a heat-labile, high molecular weight materials, probably lytic enzymes. Elevated temperature induces different virulence factors within the infectious agent of the disease, *V. shiloi*. Adhesion was found to be temperature-regulated. When the bacteria were grown at 16°C there was no adhesion to corals maintained at either 25°C or 16°. However, when the bacteria were grown at 25°C they adhered avidly to corals maintained at 16°C and 25°C. In addition, the production of lytic enzymes and the photosynthesis inhibitor was also found to be temperature dependent. Production of the latter toxin was ten times greater at 29°C than at 16°C, and extracellular protease was 5-fold higher in cultures grown at 29°C than at 16°C. The data presented here suggest an explanation for the correlation between elevated seawater temperatures and seasonal coral bleaching.

**Keywords:** bleaching, bacterial infection, coral, *Oculina,* temperature, toxin, zooxanthellae

## 1. Introduction

Coral reefs are an essential part of marine ecosystems. They are highly reproductive and apart from the tropical rain forest, there is no natural environment so rich in species diversity as the marine coral reef. In many countries, coral reefs are used as a main resource for fishing and tourism. Thus, the integrity of these ecosystems is important to humans both commercially and for the long range health of this planet.

During the past 15 years, coral bleaching events of unprecedented frequency and global extent have been reported (Glynn, 1991b; Goreau, 1990 and 1994; Hoegh-Guldberg and Salvat, 1995). Coral bleaching has been described as the dissociation of the symbiotic relationship between the animal hosts and their

*Water, Air, and Soil Pollution* **123:** 337–352, 2000.

photosynthetic endosymbiotic algae, the zooxanthellae (Iglesias-Periato, 1997; Brown, 1997). These algae are primarily responsible for the photosynthetic productivity that makes possible the growth of coral reefs in notoriously nutrient-poor waters, and thus enable many kinds of fish, squid and other organisms to survive and reproduce in this environment. As a result of the loss of the algae and/or their pigments, the stony corals lose their color and turn pale or white, due to increased visibility of their white calcareous skeleton (Fitt and Warner, 1995). An example of a bleached and healthy coral (*Oculina patagonica*) is showed in Figure 1. The loss of zooxanthellae greatly affects the coral host because these photosynthetic symbionts supply up to 63% of the coral's nutrients. Histological analysis of bleached corals following a natural bleaching event showed that a number of cellular mechanisms could result in reduced algal density, the most important mechanism appeared to be the degradation of zooxanthellae *in-situ* (apoptosis and necrosis) (Brown *et al.*, 1995).

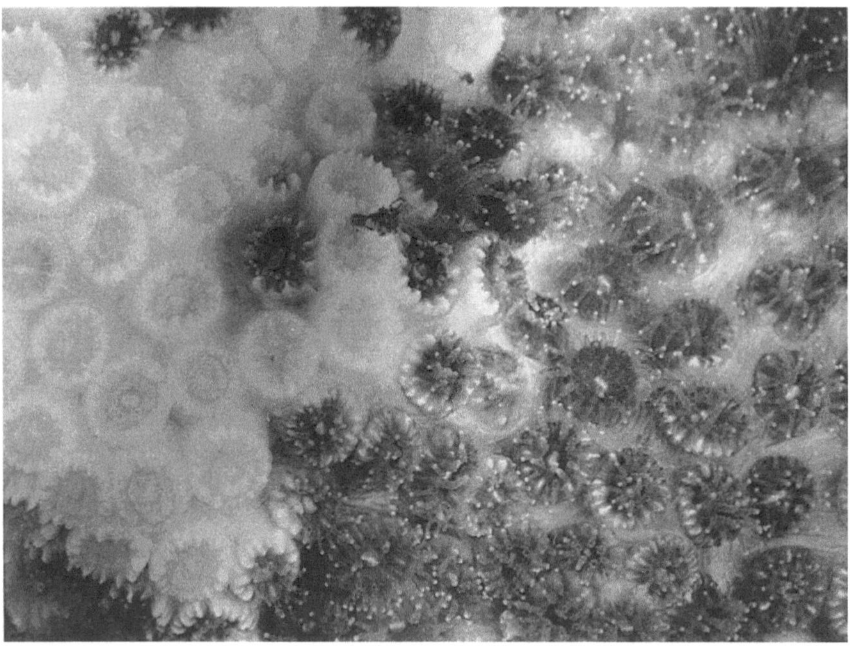

Fig. 1. Photograph of *O. patagonica* showing a bleached area (left) and a healthy area (right). Magnification X4.

Coral bleaching is generally considered to be a disease caused by environmental stress, such as increased sea water temperature (Brown *et al.*, 1996; Glynn, 1993; Hoegh-Guldberg and Smith, 1989; Lesser *et al.*, 1990; Davis *et al.*, 1997; Warner *et al.*, 1996), decreased sea water temperature (Coles and Fadlallah, 1991; Gates *et al.*, 1992; Kobluk and Lysenko, 1994), increased solar radiation, (Fisk and Done, 1985; Gleason and Wellington, 1993; Shick *et al.*, 1995), pollution (Mitchell and Chet, 1975), reduced salinity (Fang *et al.*, 1995), and combination of these stresses (Brown *et al.*, 1995; Lesser *et al.*, 1990).

The evidence supporting stress as the cause of coral bleaching is based both on field studies and laboratory experiments. The field studies involve correlation between environmental parameters and frequency of bleaching. The most common factor believed to be responsible for extensive coral bleaching is elevated sea water temperature. This is particularly important because of the possible link between coral bleaching and global warming (Atwood *et al.*, 1992; Glynn, 1993 and 1991). During the last two years bleaching events of unprecedented frequency have occurred throughout the world. Analysis of satellite-derived sea temperature data has shown that elevated sea temperatures coincide with both onset and duration of major bleaching events in the Caribbean Sea and Pacific and Indian Oceans.

In principle, there are three possible targets of the environmental stress, resulting in coral bleaching: (1) the coral host (2) the endosymbiotic algae and (3) potential coral pathogenic microorganisms. There are several reasons to assume that the coral animal is not the prime target. If the increased temperature affected the coral physiology, one would have expected that genetically identical coral species, exposed to the same temperature stress would all bleach. Frequently a few colonies bleach in the summer, while other closely located colonies do not bleach. Moreover, several authors have reported on the patchy spatial distribution and spreading nature of coral bleaching (Fisk and Done, 1985; Jokiel and Coles, 1990; Lang *et al.*, 1992; Oliver, 1985). The spreading pattern is usually characteristic of bacterial and viral infectious diseases. It has been argued that the random mosaic patterns of bleaching are difficult to attribute solely to the effect of temperature stress, since neighboring regions of the colony must be exposed to the same extrinsic conditions (Hayes and Bush, 1990). Furthermore, the correlation between coral beaching and seawater temperature is not always evident. For example, Oliver (1985) and Fisk and Done (1985) showed that extensive bleaching in the Great Barrier Reef during the summer of 1982 was not associated with any major sea surface temperature increase.

The possibility that the endosymbiotic algae are the target of environmental stress, resulting in coral bleaching, was implied in the ‘adaptive bleaching hypothesis’, suggested by Buddemeir and Fautin (1993). According to their hypothesis, coral bleaching is a normal regulatory process by which genetic variation among the zooxanthellae, populating the coral host, is allowed. Accordingly, increased seawater temperature would lead to the loss of resident algae, allowing other algae that are more heat-resistant to form stable symbioses with the coral. Moreover, the model of Ware *et al.* (1996) showed how the adaptive bleaching hypothesis could explain some features of bleaching events that are difficult to reconcile with mechanisms based on invariant temperature tolerances of the two symbiotic partners. Recently it have been shown that corals can host multispecies communities of symbiotic zooxanthellae (Rowan *et al.*, 1997). The composition of these communities followed a gradient of environmental parameters, and analysis of the symbionts before, during and after a bleaching event suggested that some corals were protected from bleaching by hosting algal populations that contained more tolerant symbionts of the same species to the stress conditions. The fact that different algae may make the corals more resistant to bleaching does not prove that the algae are the primary target of the stress conditions. As discussed below, it is likely that pathogenic microorganisms become more virulent due to the stress conditions, and the different algae show different sensitivities to the pathogen.

The third possible target for an environmental stress condition leading to coral bleaching is potential pathogenic microorganisms. It is known that stress conditions, especially temperature, can cause certain bacteria to become virulent, by 'turning on' virulence genes (Colwell, 1996; Patz *et al.*, 1996; Ware *et al.*, 1996). The surface of living corals is covered by a mucoid material. This mucopolysacchride layer provides a matrix for bacterial colonization, allowing establishment of a 'normal bacterial community' that may be characteristic for a particular coral species (Ducklow and Mitchell, 1979; Ritchie *et al.*, 1994; Sorokin, 1973). Thus, the coral animal lives in association with both endosymbiotic algae and a dense heterogeneous flora of surface bacteria. The role of these bacteria in coral biology is not clear. Based on analogies with other marine animals, the normal bacterial flora may produce antimicrobial compounds that help the coral avoid infection by pathogens (Jensen and Fencial, 1994). It should be noted that recently, the importance of microorganisms as potential pathogens in coral diseases has become more apparent (Santavy and Peters, 1997).

The data summarized in this work indicate that in the well-studied coral *O. patagonica* in the Mediterranean Sea, the causative agent of bleaching is a bacterium, *Vibrio shiloi* AK-1, and that increased seawater temperature is required for the pathogen to cause the disease.

## 2. Materials and Methods

### 2.1. BACTERIAL STRAIN AND GROWTH CONDITIONS

*Vibrio shiloi* AK1, isolated from bleached coral as previously described (Kushmaro *et al.*, 1996; Rosenberg *et al.*, 1999), was used in this study. The strain was maintained on MB agar (1.8% Marine Broth [Difco] plus 0.9% NaCl solidified with 1.8% agar). After streaking, the plates were incubated at 30°C for two days and then allowed to stand for 1 week. For experiments described here, the bacteria were grown either in MB medium (1.8% Marine broth and 0.9% NaCl) for 24 h or MBT medium (1.8% Nutrient broth, 0.9% NaCl and 0.5% Tryptone) for 43 h or 72 h at 29°C with shaking.

### 2.2. LABORATORY AQUARIA BLEACHING EXPERIMENTS

Ten microliters containing either 1.2 x $10^6$ or 1.2 x $10^2$ cells of *V. shiloi* were placed directly on each of six healthy corals and the corals then put back in separate two-liter aerated aquaria maintained at 16°C, 20°C, 25°C and 29°C. For controls, six corals were inoculated with 10 µl of sterile medium, and placed in separate aquaria at 16°C, 23°C, 25°C and 29°C. Prior to the experiment, the corals were acclimated for 10 days to the different temperatures. Percentage of bleaching was determined qualitatively by visual observation.

### 2.3. ADHESION OF BACTERIA TO *O. PATAGONICA*

Ten milliliters of a 24 h culture of *V. shiloi*, grown at 25°C in MBK medium, were centrifuged at 10,000 x g for 10 min at 25°C. The pellet was suspended in 10 ml of filter-sterilized seawater, and 0.1 ml samples were inoculated into 125 ml flasks containing 20 ml sterile seawater, plus a fragment of the regenerated (removed from the aquarium and rinsed in seawater) *O. patagonica* coral (ca. 1 $cm^2$ surface area). The flasks were incubated at 25°C with gentle shaking, 40 rpm on a "Belly Dancer" (Stovall Life Sciences Inc., Greensboro, NC). Samples of the seawater were removed at timed intervals, diluted in seawater and plated on MB agar. Two sets of control experiments were performed as described above: (a) No added *V. shiloi* controls: the number of bacteria that eluted from the nonsterile fragments of coral and formed colonies on MB agar was always <1% of the experimental values. (b) No coral controls: there was a small but significant decrease in viability, even in the absence of coral. In order to calculate net adhesion, the value for the adhesion of the no coral control was subtracted from the experimental values.

## 2.4. PREPARATION OF ZOOXANTHELLAE FROM CORALS

Intact colonies of the coral *O. patagonica* were collected during April 1997 and July 1998 from a depth of 1 to 3 meters along the Mediterranean Coast of Israel. Within 2 h of collection, each colony was split into several pieces and placed into 2-liter aerated aquaria containing filtered (0.45 µm pore size) seawater that were maintained at 25$^0$C. The aquaria were illuminated with a fluorescent lamp in 12 h light:12 h darkness cycles. To obtain zooxanthellae, a healthy coral fragment (~ 1cm$^2$ surface area) was removed from the aquarium, rinsed gently with filter-sterilized seawater and then the tissue was disrupted by water piking with ca. 50 ml sterile seawater. The suspension was centrifuged for 30 min at 2,000 x g. The pellet, resuspended in 1 ml seawater, was then centrifuged in an Eppendorf centrifuge 5402 for 4 min at 10,000 rpm. The pellet, resuspended in 1 ml seawater, was then centrifuged for 21 min at 1,200 rpm. The final pellet was resuspended in seawater to ca. 5 x 10$^6$ algae ml$^{-1}$ (based on hemacytometer counts). Fresh zooxanthellae preparations were used in all experiments.

## 2.5. PREPARATION OF THE EXTRACELLULAR CELL-FREE SUPERNATANT OF *V. SHILOI* (AK1-S)

Cultures of *V. shiloi*, grown in MBT medium for 43 h at 29$^0$C, were centrifuged at 12,000 x g for 10 min at 4$^0$C. The supernatant fluid was then passed through a 0.2 ml Millipore membrane filter. The resulting filtrate, was stored at -18$^0$C in Eppendorf tubes.

## 2.6. MEASUREMENT OF PHOTOSYNTHETIC QUANTUM YIELD OF ZOOXANTHELLAE

The portable underwater Mini Pulse-Amplitude-Modulation (PAM) Fluorometer (Walz Gmbh, Germany) was used to measure the quantum yield of zooxanthellae. This instrument allows for the direct non-invasive measurement of effective quantum yield (Y) of photosystem II under ambient light (Schreiber *et al.*, 1986, 1996 and 1997; Krause and Weis, 1991). Good correlations between measurements of quantum yield and photosynthetic rates (determined by $O_2$ evolution and $CO_2$ uptake) have been reported for plants (Gentry *et al.*, 1990) and cyanobacterial symbionts of lichens (Sundberg *et al.*, 1997).

In the experimental procedure used here, the quantum yields of 0.1 ml zooxanthellae in seawater (5 x 10$^6$ algae ml$^{-1}$) were measured in 10 ml sterile tubes (Yo). Measurements were performed in the presence of a fluorescent lamp (light intensity: 16 µmol photons m$^{-2}$s$^{-1}$) while shaking the tubes in a 29$^0$C water bath. Then, 0.1 ml of sterile seawater or growth media (controls) or experimental sample were added to the algae, and kinetics of quantum yield (Yt) were measured with the Mini-PAM from 1 to 60 min. The percent quantum yield at the different times was Yt/Yo x 100.

### 2.7. MEASUREMENT OF BLEACHING AND LYSIS OF ZOOXANTHELLAE

Zooxanthellae preparations (0.1 ml, 5 x $10^6$ cells $ml^{-1}$) were incubated at $29^0$C with 0.1 ml seawater, sterile media or the experimental sample in Eppendorf tubes. The tubes were shaken in a $29^0$C water bath and illuminated with a fluorescent lamp. At each time point, at least 3 samples were removed and examined microscopically (in a hemacytometer) for total algal cells and bleached algae. The total number of algae counted in each sample was at least 1000. The percent bleaching represents the fraction of the total algae that are bleached compared to time zero. The percent recovery of algae are the concentration of algae determined at the indicated time divided by the concentration at time zero times 100.

## 3. Results

### 3.1. SEASONAL BLEACHING OF *O. PATAGONICA* IN THE MEDITERRANEAN SEA

As described in the Introduction, during the last two decades many studies have indicated a correlation between increased seawater temperatures and coral bleaching. A similar correlation (Figure 2) was observed for the coral *O. patagonica* in the Mediterranean Sea off the coast of Israel. Surveys showed that the number of bleached colonies and the extent of bleaching increased rapidly in the summer following rising seawater temperatures, reaching a maximum of 80% bleaching in August when water temperatures were $29^0$C. In the winter, very few (<3%) of the colonies exhibited bleaching, when the water temperature was $16^0$C.

### 3.2. *VIBRIO SHILOI* IS THE CAUSATIVE AGENT OF BLEACHING OF *O. PATAGONICA*

Koch's postulates were applied to demonstrate that *V. shiloi* was the causative agent of the coral bleaching disease of *O. patagonica* (Kushmaro *et al.,* 1996 and 1997). (1) *V. shiloi* was found in 28/28 different bleached colonies, while it was absent from 24/24 healthy ones.(2) *V. shiloi* was isolated from a bleached coral and maintained in pure culture. (3) Inoculation of as few as 120 *V. shiloi* bacteria caused 83% of the healthy corals to bleach in 20 days at $29^0$C in controlled aquaria experiments, whereas control experiments under the same conditions without adding bacteria or with the bacteria but adding antibiotics showed no bleaching. (4) *V. shiloi* was reisolated from the infected corals.

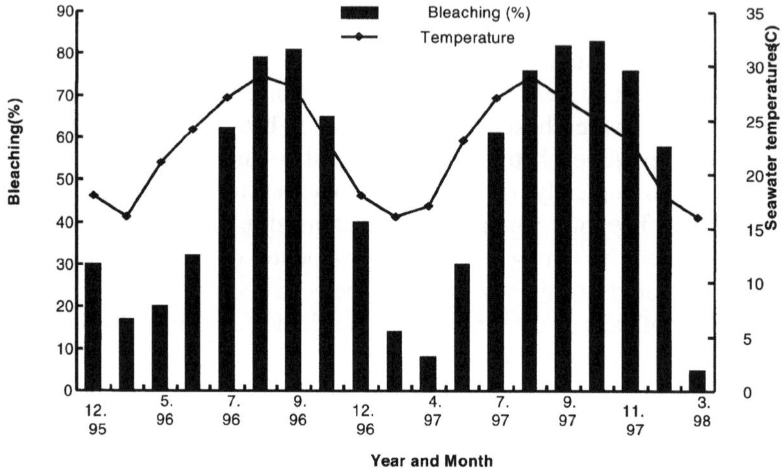


Fig. 2. Bleaching of the coral *O. patagonica* in the Mediterranean Sea from Dec. 1995 to March 1998 as a function of sea water temperature (●). Adapted from Kushmaro *et al.*, (1998).

## 3.3. THE PROCESS OF INFECTION

Infection of *O. patagonica* with *V. shiloi* proceeds in a series of sequential steps: (1) adhesion onto the coral surface, (2) penetration into the coral tissue, (3) multiplication of the bacteria within the tissue, and (4) toxin production. Each of these steps can, in principle, be regulated by different environmental factors, such as temperature as will be discussed subsequently.

### *3.3.1. Adhesion*

Toren *et al.* (1998) demonstrated that *V. shiloi* adheres to *O. patagonica* through a β-D-galactoside receptor on the coral surface. Methyl-β-D-galactopyrnoside (50 μM) completely inhibited adhesion , whereas several other sugars that where tested had no effect on the adhesion. Kinetic experiments indicated that 80% of the *V. shiloi* cells adhered to the coral by six hours after addition of the bacteria to the aquarium.

### *3.3.2. Penetration and multiplication*

Two types of experiments indicated that *V. shiloi* penetrated into the coral tissue from 8-24 h after infection. First, thin sections of coral tissue stained with fluorescent antibodies specific to *V. shiloi* demonstrated that the bacteria are inside the coral cells, beginning at 8 h and increasing for at least 36 h . Second,

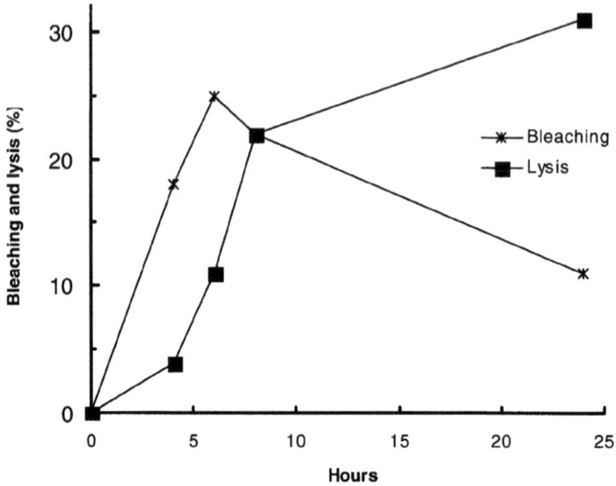


Fig. 4. Kinetics of bleaching and lysis of zooxanthellae by *V. shiloi* supernatant. The experiment was performed as described in Table I, except that the algae were examined for bleaching and lysis at 0, 4, 6, 8 and 24 h.

TABLE I

Bleaching and lysis of zooxanthellae by *V. shiloi* supernatant[a]

| Samples | Algae recovery (%) | Algae bleaching(%) | Lysis and bleaching[b](%) |
|---|---|---|---|
| Sea water control | 95 | 2.7 | 7.7 |
| Medium control | 94 | 2.4 | 8.4 |
| *V. shiloi* sup.[c] | 53 | 8.6 | 55.6 |
| Heated *V. shiloi* sup.[d] | 94 | 1.4 | 7.4 |

[a] Zooxanthellae ($3x10^{-6}$) were incubated for 24 h at pH 8 at 29°C with each of the samples.
[b] Bleaching and recovery of algae were the median differences in values obtained microscopically after 0 and 24 h of incubation.
[c] *V. shiloi* supernatant was obtained after growth on Marine Broth for 72 h. The sample was filtered to remove all bacteria.
[d] The *V. shiloi* supernatant was preheated for 5 min at 95°C.

the photosynthesis inhibition enhancer toxin which is heat stable. The bleaching and lysis could be caused by extracellular enzymes. Preliminary data indicate the presence of proteolytic as well as celluloliytic activities in the extracellular fluid.

### 3.4. THE EFFECT OF TEMPERATURE ON BLEACHING AND VIRULENCE OF *V. SHILOI*

Kushmaro *et al.* (1998) performed infection experiments in aquaria at different temperatures (Figure 5). At 25 and 29$^0$C, coral bleaching was rapid and extensive, reaching 80-100% after 45 days. At 20$^0$C, bleaching was slow and reached a maximum of 32% after 45 days. At 16$^0$C, as well as in the controls with no added bacteria, no bleaching was observed. The failure of *V. shiloi* to infect corals and cause bleaching at temperatures below 20$^0$C is not due to poor growth of the bacterium at low temperatures. At 16°C, the growth rate constants of *V. shiloi* were 0.47 and 0.50 generations $h^{-1}$ in media which contained coral mucus and tissue, respectively, as the sole nutrient.

#### *3.4.1. Effect of temperature on adhesion*

An adhesion experiment summarized in Table II, demonstrated temperature regulated adherence of *V. shiloi* to *O. patagonica*. When the bacteria were grown at 16$^0$C there was no adhesion to corals maintained at either 25$^0$C or 16$^0$C. However, when the bacteria were grown at 25$^0$C they adhered avidly to corals maintained at 16$^0$C or 25$^0$C. The temperatures at which the experiments were carried out (16$^0$C or 25$^0$C) did not significantly affect the outcome. This experiment proves that the temperature of bacterial growth plays a critical role on its ability to adhere to corals. The fact that at 16$^0$C there was no adhesion, provides at least one explanation for the failure of the bacterium to bleach at that temperature. The first step of infection is blocked and the infection can not proceed.

#### *3.4.2. Effect of temperature on toxin and enzyme production*

Production of the photosynthesis toxin was found to be temperature dependent. *V. shiloi* was grown on Marine Broth at 29$^0$C and 16$^0$C for 48 h, at which time both cultures reached the same final turbidity ($A_{600}$= 1.5). After removal of the cells by centrifugation , the inhibitory effect of the extracellular fluid was tested on photosynthesis of fresh zooxanthellae (Table III). At 29$^0$C, toxin production was ten fold higher (10 units compared to 0.9 units) than when the bacteria were grown a 16$^0$C. In addition, proteolytic activity (on azocasien) was 5 fold higher in the cultures grown at 29$^0$C compared to 16$^0$C (activity was measured at 30$^0$C).

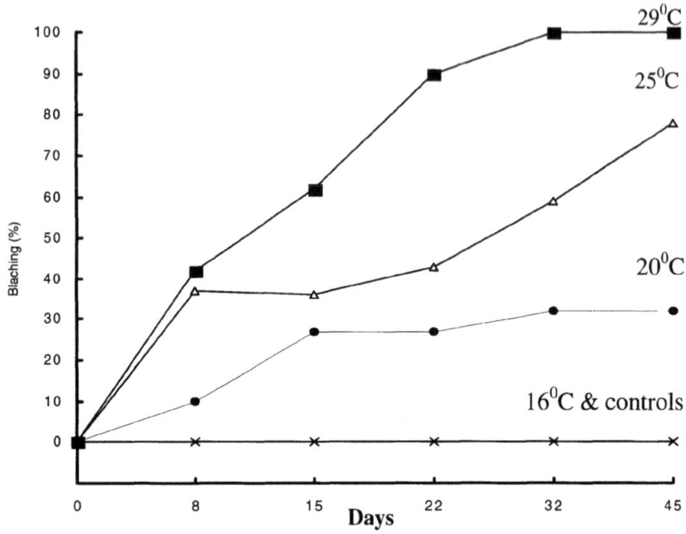


Fig. 5. Bacterial bleaching of *O. patagonica* in experimental aquaria at different temperatures. *V. shiloi* was inoculated into the water of 2 liter aerated aquaria, each containing 5 or 6 coral fragments. The final bacterial cell density was $10^6$ cells/ml. For each temperature, 3 separate aquaria were used. Control aquaria at each of the four temperatures were treated in exactly the same matter but with sterile medium in place of bacteria. Adapted from Kushmaro *et al.*, (1998).

TABLE II

Adhesion of *Vibrio shiloi* to *O. patagonica* as a function of temperature[a]

| Temp of process (⁰C) | | | |
|---|---|---|---|
| Bacterial Growth | Coral Growth | Adhesion expt. | Net Adhesion (%)[b] |
| 16 | 16 | 16 | 7 |
| 16 | 25 | 16 | 0 |
| 16 | 25 | 25 | 0 |
| 25 | 25 | 25 | 79 |
| 25 | 16 | 25 | 47 |
| 25 | 16 | 16 | 53 |

[a] Bacteria were grown overnight at either 16 or 25⁰C, the corals were maintained in an aquarium at 16 or 25⁰C for at least one week, and the incubation temperature during the 6 h adhesion test was performed at 16 and 25⁰C.

[b] The experiment was performed with an initial inoculom of $6.0*10^7$ cells per 20 ml . Adhesion was measured by determining the number of viable *V. shiloi* cells remaining in the sea water

TABLE III

Toxin and proteolytic activty of *V. shiloi* extracellular supernatant after growth at 16$^0$C and 29$^0$C

| Sample | Photosynthesis inhibitor toxin[a] (u/ml) | Proteolytic activity[b] ($A_{450}$/ml*min) |
|---|---|---|
| *V. shiloi* grown at 29$^0$C | 10 | 0.6 |
| *V. shiloi* grown at 16$^0$C | 0.9 | 0.11 |

[a] Extracellular supernatants of *V. shiloi* grown for 48 h at 16$^0$C and 29$^0$C were dialyzed, and the inhibition of photosynthesis was measured as described in Fig. 4, with 15 mM $NH_4Cl$. One unit of activity was defined as a 10% drop in quantum yield after 10 min of incubation. The activity was calculated after subtracting the activity of the 15 mM $NH_4Cl$ seawater control.
[b] Proteolytic activities of the same samples were measured on azocasein.

## 4. Discussion and Conclusions

The most common current speculation for the direct cause of coral bleaching is rising sea water temperatures due to global warming or El Niño events. As discussed in the Introduction, this environmental stress hypothesis fails to explain certain features of bleaching events, such as the patchiness and spreading nature of the disease. Furthermore, the correlation between coral bleaching and seawater temperatures is not always evident (Oliver, 1985; Fisk and Done, 1985). At least to us, observations on the patterns of coral bleaching suggest an infectious disease model. In fact, we have experimentally demonstrated that a new isolated strain, *V. shiloi*, is the causative agent of bleaching of the coral *O. patagonica.* The sequence of the process of infection mechanism is as follows: (1) Specific adhesion of the pathogen to the coral surface via β-D-galactoside receptors on the coral surface, (2) Penetration into the coral cells shortly after adhering to the surface, (3) Multiplication of the pathogen inside the host tissue, and (4) Production of extracellular toxins and lytic enzymes that affect the endosymbiontic algae.

Bacterial infections are a well studied field, especially in warm blooded animals, where the homeostatic conditions (temperature) of the host plays a critical role on the virulence of the pathogen. Corals, on the other hand, are cold blooded animals, and therefore are greatly affected by external temperature changes. Recently, there has been considerable attention paid to the fact that climate factors, including seasonal weather changes and large-scale disturbances such as El-Niño, influence the occurrence and transmission of infectious diseases through multiple direct and indirect effects upon pathogenic microorganisms, vectors, reservoirs and hosts. Many infectious diseases occur cyclically, and the patterns of occurrence often suggest, sometimes quite strongly, that environmentally dependent factors play a role in outbreaks and epidemics. Infectious diseases that are responsive to environmental changes include those diseases, for which there are clearly documented links between incidence and climate factors. These diseases are primarily composed of vector-borne diseases (Colwell and Patz, 1997).

In the case of bleaching of the coral *O. patagonica* in the Mediterranean Sea, a clear correlation was found between seasonal climatic changes and the incidence of the disease. An important question is: how does seawater temperature influence bacterial bleaching of corals? The data presented here demonstrate that elevated temperature induces different virulence factors within the causative agent of the disease , *V. shiloi.* Elevated sea water temperatures induce the pathogen to produce the adhesions that allow it to adhere to the coral surface. Bacterial adhesion is often the first step in the colonization of host tissue and the subsequent establishment of infection in pathogenic systems (Finlay and Falklow, 1997). Adhesion can be considered a primary virulence factor, but other virulence factors must be affected as well, since in the winter some of the corals recover, although the bacteria are already present in the coral (after bleaching in the summer). The production of the toxin and lytic enzymes, which cause bleaching and lysis of zooxanthellae were also found to be temperature regulated. Production of the heat-resistant photosynthesis inhibition toxin was ten times greater at $29^0$C than at $16^0$C. Furthermore, the toxin was more active at $29^0$C than at $16^0$C (data not shown). The photosynthesis inhibitor toxin blocks algal photosynthesis by enhancing the transport of ammonia into the zooxanthellae (Ben-Haim *et al.*, 1999). The equilibrium between ammonia and ammonium ion is also temperature dependent; at higher temperatures there is more free ammonia in the microenvironment of the symbiotic algae, a fact which increases its toxic affect (Azov and Goldman, 1982). Bleaching and lysis of zooxanthellae is probably the result of protease and cellulase activities, which were detected in the extracellular fluid of the pathogen. Extracellular protease activity was 5-fold higher in cultures grown at $29^0$C compared to $16^0$C.

The discovery that coral bleaching of *O. patagonica* is indeed an infectious disease, which is temperature regulated, raises several interesting questions: (1) How general is the phenomenon? Are bacteria the causative agents of coral bleaching in other parts of the world? And if so, which bacteria are involved? (2) How is the disease transmitted? We have been unable to isolate *V. shiloi* from seawater surrounding bleached corals, suggesting that a vector may be involved in transmitting the disease from one coral to another. (3) What other virulence factors are affected by increased temperatures? (4) What is the exact mechanism of bleaching? Finding answers to these questions will help to further understand the disease and possibly find a practical cure.

## Acknowledgements

This work was supported by the United States-Israel Binational Science Foundation grant 95-00177, the Pasha Gol Chair for Applied Microbiology and the Israel Center for the Study of Emerging Diseases.

## References

Atwood, D. K., Hendee, J. C. and Mendes, A.: 1992, *Bull. Mar. Sci.* **51**,118-130.
Avron, M. and Shavit, N.: 1965, *Biochim. Biophys. Acta.* **109,** 317-331.
Azov, Y., and Goldman, C.: 1982, *Appl Environ Microbiol.* **43**, 735-739.
Ben-Haim, E., Banin, E., Kushmaro, A., Loya, Y. and Rosenberg, E. 1999, *Environ. Microb.* **1**,223-229.
Brown, B. E., Dunne, R. P. and Chansang, H.: 1996. *Coral Reefs.* **15**,151-152.
Brown, B. E.: 1997, *Coral Reefs.* **16**,S129-S138.
Brown, B. E., Le Tissier, M. D. A. and Bythell, J. C.: 1995, *Mar. Biol.* **122**,655-663.
Buddemeier, R. W. and Fautin, D. G.: 1993, *BioScience.* **43**,320-325.
Coles, S. L. and Fadlallah, Y. H.: 1991, *Coral Reef.* **9**,231-237.
Colwell, R. R.: 1996, *Science.* **274**,2025-2031.
Colwell, R. R. and Patz, J. A.: 1998, *American Academy of Microbiology coll.report*, June 1997, Montego Bay, Jamaica.
Crofts, A. R.: 1966, *Biochem. Biophys. Res. Commun.* **22**,725-731.
Davies, J. M., Dunne, R. P. and Brown, B. E.: 1997, *Mar. Freshwater Res.* **48**,513-516.
Ducklow, H. W. and Mitchell, R.: 1979, *Limnol. Oceanogr.* **24**,715-725.
Fang, L. S., Lino, C. W. and Liu, M. C.: 1995, *Zool. Stud.* **34**,10-17.
Fisk, D. A. and Done, T. J.: 1985, *Proc. 5th Int. Coral Reef Symp.* **6**,149-154.
Fitt, W. K. and Warner, M. E.: 1995, *Biol. Bull.* (Woods Hole). **189**,298-307.
Gates, R. D., Baghdasarian, G. and Mustatine, L.: 1992, *Biol. Bull.* **182**,324-332.
Genty, B., Briantais, J. M., and Baker, N. R.: 1990, *Biotech Biophys Acta.* **990,** 87-92.
Gleason, D. F. and Wellington, G. M.: 1993, *Nature.* **365**,836-838.
Glynn, P. W.: 1993, *Coral reefs* **12**,1-17.
Glynn, P. W.: 1991, *Trends Ecol. Evol.* **6**,175-179.
Glynn, P. W.: 1991b, *Science.* **253**,69-71.
Goreau, T. J.: 1990, *Nature.* **343**,417
Goreau, T. J.: 1994, *Ambio.* **23**,176-180.
Hayes, R. L. and Bush, P. G.: 1990, *Coral Reefs.* **8**,203-209.
Hoegh-Goldberg, O. and Salvant, B.: 1995,. *Mar. Ecol. Prog. Ser.* **121**,181-190.
Hoegh-Goldberg, O. and Smith, G. J.: 1989, *J. Exp. Mar. Ecol.* **129**,279-303.
Iglesias-Perito, R.: 1997, *Proc. 8th Int.Coral Reef Symp.* **2**,1313-1318.
Jensen, P. R. and Fenical, W.: 1994, *Annu. Rev. Microbiol.* **48**,559-584.
Jokiel, P. L. and Coles, S. L.: 1990, Coral Reefs. **8**,155-162.
Kobluk, D. R. and Lysenko, M. A.: 1994, *Bull. Mar. Sci.* **54**,142-150.
Krause, G. H., and Weis, E.: 1991 *Ann. Rev. Plant Physiol. Plant Mol. Biol.* **42,** 313-349.
Kushmaro, A., Rosenberg, E., Fine M., Ben-Haim Y. and Loya, Y.: 1998, *Mar. Ecol. Prog. Ser.* **171**,131-137
Kushmaro, A., Rosenberg, E., Fine, M. and Loya, Y.: 1997, *Mar. Ecol. Prog. Ser.* **147**,159-165.
Kushmaro, A., Loya, Y., Fine, M. and Rosenberg, E.: 1996, *Nature.* **380**,396
Lang, J. C., Lasker, H. R, Gladfelter, E. H, Hallock, P., Jaap, W. C., Losada, F. J. and Muller, R.G.: 1990, *Am. Zool.* **32**,696-706.

Lesser, M. P., Stochaj, W. R., Tapley, D. W. and Shick, J. M.: 1990, *Coral Reefs*. **8**,225-232.
Mitchell, R. and Chet, I.: 1975,. *Microb. Ecol.* **2**,227-233.
Oliver, J.: 1985, *Proc. 5th Int. Coral Reef Symp.*, Tahity. **4**,201-206.
Patz, J. A., Epstein, A., Thomas, A. B. and Balbus, J. M.: 1996, *JAMA*. **274**,217-223.
Ritchie, K. B., Dennis, J. H., McGarth, T. and Smith, G. W.: 1994, in L. Kass (ed), Bacteria associated with bleached and nonbleached areas of *Montrastrea annularis*. Bahamian field station. *Proc. 5th Symp. Nat. Hist. Bahamas, San Salvador, Bahamas*. pp. 75-79.
Rosenberg, E., Ben-Haim, Y., Toren, A., Banin, U., Kushmaro, A., Fine, M. and Loya, Y.: 1999, The effect of temperature on bacterial bleaching of corals, in *Microbial ecology and infectious disease*, ASM Press, Washington DC.
Rowan, R., Kowlton, N., Baker, A. and Jara, J.: 1997, *Nature*. **338**, 265-269.
Santavy, L. D. and Peters, C. E.: 1997, *Proc. 8th Int. Coral reef Symp.* **1**, 607-612.
Schreiber, U., Gademann, R., Ralph, P. J. and Larkum, A. W. D.: 1997, *Plant Cell Physiol* **38,** 945-951.
Schreiber, U., Kuhl, M., Klimant, I. and Reising, H.: 1996,. *Photosy Res* **47,** 103-109.
Schreiber, U., Schliwa, U. and Bilger, W.: 1986, *Photosy Res* **10**, 51-62.
Shick, J. M., Lesser, M. P., Dunlap, W. C., Stochaj, W. R., Chalker, B. E. and Wu Won, J.: 1995, *Mar. Biol.* **122**,41-51.
Sorokin, Y. I.: 1973, *Nature*. **245**,415-417.
Sundberg, B., Campbell, D. and Palmquist, K.: 1997, *Planta* **201,** 138-145.
Toren, A., Landau, L., Kushmaro, A., Loya, Y. and Rosenberg, E.: 1998, *Appl. and Environ. Microbiol.* **64**,1379-1384.
Ware, R. J., Fautin, G. D. and Buddemeier, W. R.: 1996, *Ecol. Model.* **84**,199-214.
Warner, M. E., Fitt, W. K. and Schmidt, G. W.: 1996, *Plan Cell and Environ.* **19**,291-299.

# BETTER LAWS = BETTER ENVIRONMENT THE ROLE OF THE ENVIRONMENTAL LAWYER IN THE RECONSTRUCTION AND MODERNISATION OF THE WATER SECTOR IN PRIVATE SECTOR PARTICIPATION

I.C. SINCLAIR

*Senior Partner of Sinclair Consultancy Services, Consultant on water law to Severn Trent Water International Ltd., International Environmental and Water Law Adviser*
*18 White House Way, Solihull, W.Mids, B91 1 SE, England, GB*

**Abstract.** In order to secure a better environment, it is essential to ensure that the right and appropriate laws exist and are enforced. Before any major change is made to the physical environment, careful consideration should be given to the existence of the appropriate legal and organisational structure to ensure that the desired changes and improvements can be satisfactorily accomplished. Private sector participation is often seen as a catalyst for change and improvement in the economic and environmental situation, but, of itself, that will not create the desired cure. As water is a natural monopoly, it is essential to secure and maintain an effective, efficient and independent regulator. Environmental law should be kept under constant review to ensure that it is relevant and that it meets the challenges of the future, as well as securing the victories of the past.

**Keywords**: law, private sector participation, reform, regulation, water

## 1. Introduction

"Water for Life"
"No water, no life"
"Water is essential for all existence on earth"

The above are three ways of expressing the most fundamental truth about life on this earth, namely, that water is the most precious and vital substance and, accordingly, everyone has a responsibility for its stewardship.

The role of the environmental lawyer in this context is all the most important. The environmental lawyer's response to the "environmental challenges for the next millenium" must be to assist in the creation of the correct and appropriate laws in regard to the environment and, in particular, in relation to water. However, it is not sufficient merely to have good laws; they will not be effective unless such laws are correctly administered and enforced.

To set the scene in an appropriate world context, it is relevant to quote from the Foreword to Volume 1 "Selecting an option for private sector participation" of the three volume *Toolkits for private participation in water and sanitation*, published by the World Bank:

*Water, Air, and Soil Pollution* **123:** 353–360, 2000.

> Millions of urban dwellers, especially the poor, lack adequate access to safe drinking water and sanitation. Improving services will, in most cases, require more efficient operation of water utilities and investments in rehabilitating and extending supply systems. Many central and local governments are turning to the private sector to help them address these needs. But private sector participation is no simple panacea.
> Its success depends on how well the chosen private sector arrangement fits local circumstances, on whether the regulatory environment is suitable, and on how well the reforms respond to the concerns of those affected (World Bank, 1997).

This paper will review the functions and responsibilities of both the private sector and of environmental lawyers in securing a better environment and major improvements in the water sector.

## 2. The Importance of Water

World water resources are scarce and the availability of useable fresh water is remarkably small and is diminishing. Only about 2.5% of the world's water resources are fresh water, and of that it is estimated that about 70% is locked up in ice-caps and another 29% is not available for economic use for a variety of reasons. Thus only about 1% is available for human use. Overall, the demand for fresh water is growing, especially due to the considerably growth in the world population, but also the availability of clean fresh water is declining at an alarming rate.

Not surprisingly, the phrase "water stress" is increasingly being used.

Such growth in demand arises from the following areas of demand:

1. Human – for safe drinking water and sanitation;
2. Agriculture – for expanding production to meet human population growth;
3. Environment – to protect endangered species, biodiversity and areas of special interest and importance;
4. Industrial – to provide more goods and services for the growing population.

It is estimated that 25% of the world's population is facing water shortages. This is expected to rise to 66% by 2025, only a quarter of a century away. This is against a six-fold rise in global water consumption this century. The World Health Organisation estimates that, at present, some 1.3. billion people, more than 20% of the world's population, do not have access to safe drinking water. Moreover, about half (50%) of the world's population is without access to adequate sanitation facilities. The urban population of the less developed regions is expected to grow from 2.0 billion in 2000 to 4.0 billion by 2005.

According to the Global Environmental Fund, between US $600-800 billion of capital expenditure is required in order to provide a basic level of water and sanitation services in the present decade of 1995 to 2005.

## 3. Worldwide Water Issues

What are some of the major common issues worldwide in relation to water? These include:

1. Under-investment in the infrastructure in all aspects of water provision;
2. Ever-increasing economic and urbanisation pressures;
3. Lack of adequate finance to provide and maintain the infrastructure;
4. Past spending on unsustainable capital solutions;
5. Poor maintenance of existing water assets;
6. Inadequate income arising from a reluctance to pay the economic cost.

In the context of this paper, two additional important aspects must be tackled:

7. Inadequate provision of the necessary institutional and organisational structures to reform and administer the capital assets;
8. Lack of an adequate enforcement and regulation system.

Urbanisation is occurring at an alarming rate with changes in the agricultural economy resulting in the concentration of the population in most countries into smaller geographical areas, with increasing stresses on infrastructures for "public services" (where they exist) and, in particular, for the movement of large quantities of water, both for use as clean water and for the subsequent disposal of foul water. Shifting patterns of population suggest that the number of cities with more than one million inhabitants will more than double within the next three decades, and so the proportion of people living in urban areas will increase from one-third to two-thirds. In addition, it is being increasingly recognised that the provision of a safe water supply and the proper removal of sewage waste is the single most important step to improving the health and well-being for the population of the world.

Thus, adequate water and sanitation services are the essential key to economic development. Traditionally, the delivery of these services has been seen as a role for the public sector, namely by Government. However, in many countries, it is being increasingly recognised that a more practical alternative is to involve the private sector to take the responsibility for service delivery within a regulated framework as this will provide a different and more efficient commercial and performance driven approach.

## 4. Provision of Water Services

Any service provider, whether public or private, must act in an efficient, economic and sustainable manner to carry out the following responsibilities:

1. Find and secure adequate water resources;
2. Treat water to adequate minimum standards for the required use;
3. Distribute the treated water to customers;
4. Collect and remove the used foul water;
5. Treat the used water to an acceptable standard;
6. Dispose of the treated water back into the aquatic environment.

In order to carry out these aspects of dealing with the water cycle, the provider must deal with the following matters while discharging its responsibilities and running its operations:

1. Politics;
2. Legal framework, constraints and regulations;
3. Regulatory framework and regime;
4. Technical and operational aspects;
5. Commercial and financial sustainability.

Before a new operator will be prepared to undertake the responsibility of providing and managing a viable water supply and wastewater disposal service, the operator needs to have confidence in the following:

1. the law – that it is sufficient and appropriate for delivery of the required service;
2. the regulator – that a fair and honest regime will be upheld;
3. the system – that there is no political or external interference or corruption;
4. the finance – that it is sufficient and secure to provide the required improvements;
5. the contract – that it is fair and not subject to unilateral variation;
6. the people – that they are sufficiently capable of carrying out what is required.

## 5. Factors Affecting Change

Changes should not be made to the system of providing water services without considering the effect of such change on the environmental system of a country as a whole. With any major change in the system of provision of water services, there should be a fundamental review of the laws that effect the environment and the water sector as a whole. And so it is in this context that the international water lawyer can make a great contribution.

The order of change is important – start at the beginning and then progress by stages as follows:

1. Amend the substantive law of the relevant sector;
2. Create the appropriate institutions to carry out the revised responsibilities;
3. Write the detailed rules and regulations.

What are some of the guiding principles and principles for the reform of the water sector so as to create the right situation for effective private sector participation?

1. Determine what are the priorities and what are the major, fundamental problems that have to be solved;
2. Ensure that there is a comprehensive review and reform of the sector – a whole-scale change , in the long run, is far more effective and efficient, than "tinkering with the system";
3. Ensure that there is logical division of responsibilities , and that there are not conflicting or over-lapping duties in the same body (the classical division between the legislature, the executive and the judicature).

## 6. Regulation of the Water Sector

A matter that is of particular significance that is often overlooked is that of proper and effective regulation. By their very nature, water utilities operate in a natural monopoly – there is no substitute for water. Regulation in dealing with water has to be both environmental regulation (to protect the natural environment as well as the health and well-being of those affected) and economic regulation (to protect the economic and financial interests of those to whom the service is to be provided).

Economic regulation is required as there is no real economic "market-place" competition. It is particularly importance to ensure that, with the involvement of a private sector operator, the operator does not abuse his monopolistic position – hence, the need for an effective and independent regulator.

Some of the most important issues that the economic regulator must deal with are:

1. Regulation of the prices and tariffs that may be charged for the service provided;
2. Regulation of the levels of service of quantity and quality to be provided;
3. Mechanisms for dealing with failure to comply with the conditions to provide the service, including service standards;
4. Resolution of disputes and conflicts between the various parties concerned with the provision of the services provided (usually called "stakeholders").

The regulator has to provide a mechanism for protecting the legitimate interest of both the provider of the service (so that he can obtain a proper return of income for his services including a reasonable profit) as well as the customers (to ensure that the charges are reasonable having regard to the service provided and are affordable).

## 7. Private Sector Participation

The necessary requirements before there can be effective private sector participation include:

1. Political will for change and the ability to carry it out effectively;
2. Economic, financial and commercial needs; (the project must be "bankable" (i.e., economically viable in such terms as to attract international investment) and be financed by affordable customer tariffs);
3. Organisational and institutional change;
4. Social and welfare benefits for the population and receiving customers;
5. Adequate and appropriate finance to make the necessary changes and to carry out the consequential improvements to the water infrastructure;
6. The terms (including the risk) offered to the investor must be sufficiently secure and attractive as to attract the required injection of capital.

Private sector participation usually involves a fundamental renewal and development of the financial structure underlying the provision and development of the water services. This finance must be sufficient to deal with the immediate requirements in the short term, but also to secure the necessary income stream to support long-term improvements and maintenance of the improved and expanding service.

Private sector participation can take many different forms and so it is essential to ensure that, before undertaking any substantial involvement, the necessary legal and organisational frameworks are in place.

## 8. Creating Better Laws for a Better Environment

Change has too often been brought about in the past in the physical structures and environment without ensuring that the necessary provisions of the substantive law and the appropriate institutions are in place and are able to operate effectively before such physical changes come into operation.

Sadly, my experience in many parts of the world has shown that there is a natural tendency to "do something" (usually to build major capital intensive works) without ensuring that the correct legal and organisational structure is in place first, and then considering the likely effect of the proposed changes to the physical environment on the environment as a whole, or ensuring that questions of maintenance and operation are tackled on a long-term basis.

Thus, to ensure that, indeed, better laws will equal a better environment, the international environmental lawyer can make a major contribution, especially if he is brought in at an early enough stage, and that he is involved in the thinking and planning of the proposed changes. This will not be effective if the lawyer is merely involved as "the scribe" or "wordsmith", merely putting into legal language the ideas of others.

Points to be emphasised include the following:

1. Major changes in the planning, management and regulation of the environment must not be undertaken unless there is a review of the situation in the country (and considering the international implications) and involving all relevant disciplines and skills – engineers, environmentalist, financial experts, lawyers – drawn from all parts of the water cycle.
2. Consideration should be given to major and fundamental issues. For example, in relation to water-related matters, such as the integrated management of the water cycle, the organisation of water on a basin /catchment basis, the interaction between surface water and ground water, the effects of (and interaction between) abstraction of raw water and the discharge of used water, the division of responsibilities between provision of resources and distribution of water and wastewater services, and the role and responsibilities of Government.
3. Maintenance and improvement in the environment depends on the existence of the right and appropriate laws in sufficient detail to make them workable.
4. However good and appropriate the laws are, they are only effective if they are observed, controlled and enforced, requiring adequate and appropriate precautionary remedies, detection of offences, prosecution of offenders and penalties for breach.
5. Enforcement depends upon independent regulation, by a regulator with sufficient powers and resources to carry out its functions.
6. To be effective, environmental laws require adequate finance and the support of the community for whose benefit they are imposed and enforced.

## 9. Responsibilities of the Environmental Lawyer

The environmental lawyer has a major contribution to make in advising on the creation and amendment of laws in the environmental sector, and thus seeking to achieve a real improvement in the environment. These include:

1. Drawing on experience gained in other jurisdictions and advising on what is applicable in the particular circumstances of the particular country;
2. Knowledge of what "works" and what does not, including appropriate precedents in documents, laws and practical examples, as well as appropriate reference material (in the case of private sector participation, this will include the World Bank's *Toolkits for private participation in water and sanitation*);
3. Breadth and width of vision to see what has been "left out" of the instructions, so as to ensure the solution proposed takes the long-term view and does not merely deal with the short-term problems;

4. Courage to challenge the concepts that are presented and to enter into constructive debate with the employer/client.

## 10. Summary

In order to sum up how to achieve better laws in order to secure a better environment, it will probably suffice to quote from three different sources:

1. From a well known source (probably "Anon"): "In order that evil may succeed, all that is necessary is for good men to do nothing". The foundation of a good and secure society is a just body of law which is properly and impartially enforced. Laws have to be changed and adapted to the needs of the society at the time, and so environmental laws have to be kept under constant review, to see if they are achieving the desired purposes and to anticipate problems.
2. From the start of the article "Enforcement strategies in Israel" in the Special Enforcement Issue of *Israel Environment Bulletin*: "How best to change human behaviour so that environmental requirements are complied with ? Environmental agencies worldwide have been grappling with the problems for years. The solution may well lie in just the right blend of education, motivation, administrative enforcement, judicial enforcement and prosecution" (Ministry of the Environment, 1996).
3. From the article "The future of sustainable development in Israel" in *Israel Environment Bulletin*: "The development and implementation of a sustainable development policy is a long process. However, in the past two years, Israel has begun planting the seeds which will, hopefully, provide for a better quality of the environment, not for the present generation, but for future generations as well.
   To achieve the vision of a better environment, Israel [and other countries as well – comment of the Author of this Paper] will have to move from an environmental paradigm built on control and clean-up to one based on efficient use of limited resources and avoidance of environmental harm.
   The challenge of to-day is to transform these concepts into practical policies which are relevant to Israel's [and other countries'] physical, social and economic reality.
   It is the moral obligation of the present generation toward those to come to leave them the possibility of realising their own future" (Ministry of the Environment, 1998).

## References

Ministry of the Environment: 1996 Winter, *Israel Environment Bulletin* **19,** no.1, Jerusalem.
Ministry of the Environment: 1998 Summer, *Israel Environment Bulletin* **21**, No.3, Jerusalem.
World Bank, 1997: *Toolkits for Private Participation in Water and Sanitation.*

# EUROPEAN ENVIRONMENTAL TAX LAW AND POLICY: GREENSPEAK !

K. DEKETELAERE
*Director, Institute for Environmental and Energy Law, University of Leuven*

**Abstract**. This article examines the use of fiscal and non-fiscal policy instruments in European environmental policy. Although a shift in the choice of environmental policy instruments has been noted for a number of years, the use of market-based instruments in general and fiscal instruments in particular remains problematic. Notwithstanding a large number of policy declarations and proposals, the development of a European environmental tax policy and legislation remains, until now, greenspeak !

**Keywords:** environmental policy instruments, European environmental law, fiscal instruments, tax law

## 1. Introduction

Once a government has defined its environmental goals, it can execute them by means of different policy instruments. The following environmental policy instruments can be distinguished (Dente, 1995; Golub, 1998; OECD, 1989): (i) instruments of social regulation (OECD, 1989), such as transfer of information (environmental education, environmental labels) (Jadot, 1993), environmental impact reports (Elni, 1997; CEDRE, 1991), self-regulation (environmental policy agreements) (Aalders, 1992; Bocken, 1991; Klok, 1989), self-control and environmental care systems (Platteau, 1998); (ii) instruments of financial aid (OECD, 1989; OECD, 1996a), such as subsidies, soft loans, and fiscal incentives (Jenkins, 1992, I; Lockhart, 1998; IISD, 1994b) (investment deduction, tax reduction and tax exemption); (iii) instruments of planning (Deketelaere, 1997a), such as macro-planning and micro-planning, binding planning and non-binding planning, sectoral planning and non-sectoral planning; (iv) instruments of direct regulation (OECD, 1989; OECD, 1997c), such as permits, prohibitions and restrictions, and different sorts of requirements (quality demands, product demands, emission demands, design demands, construction demands and production demands); (v) instruments of market regulation (Baumol, 1988; Helm, 1991; Guppes, 1992; Jenkins, 1992b; OECD, 1989; OECD, 1985; OECD, 1991; NCM, 1996; OECD, 1993a; OECD, 1997a), such as liability rules (Wetterstein, 1997), insurance obligations (Wetterstein, 1997), marketable emission rights, deposit and refund system (OECD, 1989), acceptance obligations and take-back obligations (OECD, 1989), enforcement incentives and environmental levies (Gaines, 1991;

*Water, Air, and Soil Pollution* **123:** 361–377, 2000.

Hamilton, 1998; O'Riordan, 1997; Paulus, 1995; Rowan-Robinson, 1998; Schlegelmilch, 1998; Smith, 1995; Smith, 1997; Tindale, 1996; Van Humbeek, 1997; Vermeend, 1998; OECD, 1992, II; OECD, 1995a; OECD 1996b; OECD, 1994; OECD, 1995b; OECD, 1997b; IISD, 1994a; IISD, 1994b; EFILWC, 1996; EC, 1995; EC, 1997; EEA, 1996; OECD, 1993c).

In this contribution I will first examine (cf. 2) which instruments have already been used within European environmental policy. In the second part (cf. 3), closer attention will be paid to the use of fiscal environmental policy instruments in the European environmental policy.

## 2. The Use of Non-Fiscal Environmental Policy Instruments

### 2.1. INTRODUCTION

The question of how the responsibility for the development and the introduction of environmental policy instruments should be divided between the European Community and its Member States is not easy to answer. The answer will differ for each environmental policy instrument (Deketelaere, 1993a; Thieffry, 1992). Hereafter (cf. 3), this paper will specifically relate to the use of fiscal environmental policy instruments, which are the most frequently used new environmental policy instruments in most Member States today. We will, however, start with a general survey of the use of other important, non-fiscal, environmental policy instruments

### 2.2. SURVEY

For many of the important non-fiscal environmental policy instruments, European initiatives have already been taken or are being considered. Following is a survey of these initiatives:

- Environmental impact and safety reports (Elni, 1997): a satisfying European framework is established by Council Directive (EEC) 85/337 concerning the assessment of the effects of certain public and private projects on the environment, and Council Directive (EC) 96/82 on the control of major accident hazards involving dangerous substances;
- Environmental labels (Jadot, 1993): a satisfying European framework is established by Council Regulation (EEC) 880/92 concerning a Community system for the award of environmental labels;
- Environmental policy agreements (Deketelaere, 1998, I): a satisfying European framework is established by the Communication of the Commission of 27 November 1996 concerning environmental agreements and the Recommendation of the Commission of 9 December 1996 concerning environmental agreements implementing Community Directives;

- Environmental care systems (Platteau, 1998): a satisfying European framework is established by Council Regulation (EEC) 1836/93 concerning the voluntary participation by companies from the industrial sector in a communitarian environmental management and audit system;
- Environmental subsidies (Van Calster, 1998): a satisfying European framework is established by the Community guidelines on state aid for environmental protection and specific (environmental) financial aid instruments (LIFE, Cohesion-Fund);
- Environmental planning (Deketelaere, 1997): different forms of national, regional and sectoral environmental planning have a European basis in different Directives and/or Regulations. However, no general European framework has been created until now for national, regional and sectoral environmental planning in general, unless the five European environmental action programs (EAPs) can be (and should be) considered as such a framework;
- Direct regulation (EC, 1996): many of the concrete national and regional instruments of direct regulation (prohibitions, restrictions, permit systems and quality requirements) have a basis in the hundreds of European environmental Directives and Regulations. This is a category of policy instruments of which the European Community in the past has made too much use, through too many and too detailed Directives. Preferable future European action concerning these instruments is deregulation: less and less detailed Directives, based on a multi-media approach (cf. Council Directive (EC) 96/61 concerning integrated pollution prevention and control);
- Market regulation (Golub, 1998; Dente, 1995): In the fifth EAP, there is a strong plea for a shift in environmental policy instruments: more market regulation, less direct regulation. Different aspects of this new approach can already be found in the earlier European environmental policy: (i) appeal to the Member States to use market-based instruments (e.g., Directive (EEC) 91/157 concerning batteries and accus containing dangerous products); (ii) introduction in the legislation of the principle of cost allocation (e.g., Council Directive (EEC) 75/439 concerning the removal of lubricated oil); (iii) introduction in legislation of maintenance incentives (e.g., Regulation (EEC) 259/93 concerning the transfrontier movement of waste); (iv) introduction in legislation of liability rules (e.g., the Green Paper of the Commission on repair of environmental damage and the modified proposal of the Directive concerning the legal liability for damage and environmental deterioration caused by waste); (v) introduction in legislation of deposit-refund systems (e.g., Directive (EC) 94/62 concerning packaging and packaging waste); (vi) introduction in legislation of tradeable permits (e.g., Regulation (EC) 3093/94 concerning products breaking down the ozone layer). As to the future, the Commission is planning a greater use of these instruments of market regulation. However, one must say that the purely

market-oriented initiatives of the Commission have not been (so) successful yet; the direct impact of these instruments on the internal market is probably the main reason for this, and will certainly have to be (more) considered in future action.

## 3. The Use of Fiscal Environmental Policy Instruments

### 3.1. INTRODUCTION

To what extent does the European Community intend to make use of environmental taxes and ecological fiscal incentives in its future environmental policy, in order to realize its environmental policy goals more quickly, more effectively and more efficiently? Depending on a negative or positive answer to this question, different other subquestions can be asked. If a negative answer is given, the evident question to be asked is: why do Community institutions in their policy declarations again and again stress the necessity to develop a European environmental tax policy? Taxes are always said to be the environmental policy instrument of the future. If positively answered, however, a logical question to be asked is: what has the Community already done, what has been planned and what could and/or should it do? In the following, all these questions will be discussed. The starting-point in these matters will be the consideration that in the developing environmental tax policy, the protection of the environment should be the main concern.

### 3.2. GREENSPEAK

This ecological starting-point is not without importance: more than a score years after the actual initiation of the European environmental policy, the quality of the European environment is worse than ever. This remark could cause some surprise, as this policy was characterized by ambitious environmental action programs (EAP's) and the use of 'expensive' environmental policy principles like 'polluter pays', 'sustainable development', 'best available technology' during the past decades. This apparent contradiction illustrates the typical 'Greenspeak' of the European Community in environmental issues: countless minimal norms are set and several environmental projects are initiated to meet public opinion and to give the impression that protection of the environment is a priority of the Community. In reality, however, environmental protection is still subordinate to the aims and priorities of other European policy areas like industry, agriculture and transport. This powerlessness of the European environmental policy is illustrated in particular by the European environmental policy instruments which have been used until now: namely, 'direct regulation Directives'. These are Directives which impose a considerable amount of permit systems, prohibitions, limitations and quality requirements upon the Member States. Though this may give the impression that the Community is tackling environmental issues, it is precisely these instruments of direct regulation that cause the slowness, the inefficiency and the ineffectiveness of European environmental policy. Often protection is only offered to a small extent, Directives are carried out improperly and instructions are not fulfilled as they should. The Community has been aware of these problems for a long time: the yearly reports on the supervision of the application of Community environmental law leave practically no doubt about that. But the Community has just not yet succeeded in raising the efficiency of the environmental policy instruments, neither through stronger direct regulation – more supervision, more reports, providing more information – nor through weakened direct regulation or an increased market-conform regulation – environmental taxes, marketable emission rights, deposit systems – which is often even more problematic due to the possible consequences for the internal market. (Gee, 1997; Guscuolo, 1998; Rosembuy, 1998).

## 3.3. POLICY REPORTS

In spite of this disenchanting reality, the Community institutions have reported ever since the start of European environmental policy, echoing the OECD (OECD, 1985; OECD, 1989), that they are seriously considering using other environmental policy instruments. Only these instruments can lead to a fairer implementation of the 'polluter pays'-principle. The many policy reports on this matter, however, are characterized by differing and even contradictory views. What seems clear from these declarations is that until the beginning of 1993 (cf. the fifth EAP) the protection of the environment was considered the central purpose of the would-be European environmental tax policy: the proceeds of the environmental tax measures and the possible ways of spending were only marginally mentioned. Afterwards (cf. 'the White Paper' on growth, competitiveness and employment), the Community took a totally different course: today, the proceeds of the environmental tax measures are the main issue and environmental protection has degraded to an expected side-effect of this policy, which concentrates on job creation and other social policy targets (Deketelaere, 1995b).

## 3.4. ADOPTED FISCAL ENVIRONMENTAL POLICY INSTRUMENTS

Though there are many policy reports, the number of adopted European fiscal policy instruments is very small. As of 1999, there were not more than five. Firstly, there is the use of fiscal incentives recommended by the Community in the sector of air pollution to promote use of 'clean' vehicles (Council Directives (EEC) 70/220 and 88/77), the European differentiation of minimum excise duty rates for leaded and unleaded petrol (Council Directive (EEC) 92/82 and the European harmonization of vehicle taxes and tolls and charges raised for the use of several infrastructures (former Council Directive (EEC) 93/89, now annulled but maintained as regards contents). Further, there is the recommended use of fiscal incentives by the Member States in the sector 'noise nuisance' to encourage the use of so-called 'silent' vehicles (Council Directive (EEC) 70/157). In the sector 'waste', finally, we have the use of national taxes, allowed by the Community, on waste oils and dangerous and toxic waste (former Council Directive (EEC) 78/319).

### 3.5. PROPOSED/SUGGESTED FISCAL ENVIRONMENTAL POLICY INSTRUMENTS

Apart from these realizations, there are, on the one hand, two specific proposals and, on the other, countless suggestions for possible European ecological fiscal incentives and environmental taxes. The specific (modified) proposals (which will be discussed in detail in the final version of this paper) regard the differentiation of maximum excise duty rates for fuels from agricultural sources and other fuels and the tax on carbon dioxide emissions and the use of energy, the so-called $CO_2$ – and energy tax. (These proposals will be discussed hereafter).

Both proposals are already pending for a long time. Their chances of approval are considered to be very low.

### 3.6. TOWARDS A EUROPEAN ENVIRONMENTAL TAX LAW

The above analysis of adopted and proposed European fiscal environmental policy instruments shows that the Community institutions, in general, and the European Commission, in particular, do not have a clear view of the development of a European environmental tax law. In the absence of a general debate on this topic, there is no elaborated policy line and, as a consequence, all adopted and proposed European fiscal environmental policy instruments are 'single-shots', which do not fit into a general framework. Therefore, in this discussion about the development of a European environmental tax law, the absolute priority must be the creation of a concrete European policy vision and the elaboration of a general European policy framework, on the basis of which European fiscal environmental policy instruments are to be developed, proposed and adopted. Evidently, the Community will only develop, propose and adopt fiscal environmental policy instruments if it is convinced, after an economic analysis, that such fiscal environmental policy instruments are the most appropriate instruments for the environmental problems which have to be dealt with.

Before elaborating the three fundamental pillars of a European environmental tax legislation, it may be useful to indicate some arguments in favor of the development of such European environmental tax legislation: (i) the necessity of a European 'coordination and/or harmonization' of the different national environmental tax systems, in order to avoid trade barriers and distortions of competition; (ii) the necessity of a European organized 'ecological clean-up' of the different national tax systems, in order to eliminate environmentally-unfriendly national tax measures; (iii) the increase of the effectiveness and efficiency of the European environmental policy, since the instruments of direct regulation have proven their ineffectiveness and inefficiency; (iv) the necessity of elaborated European fiscal environmental policy instruments to deal with worldwide environmental problems, taking into account that national fiscal environmental policy instruments will mostly be insufficient to deal with these problems; (v) the necessity of European environmental tax initiatives because of the fact that Member States are not always prepared to introduce national environmental tax measures, out of fear for their own competition position. (Deketelaere, 1993a; Deketelaere, 1995a; Deketelaere, 1996; Deketelaere, 1997b; Deketelaere, 1998b; OECD, 1989).

## 3.7. POLICY VISION FOR A EUROPEAN ENVIRONMENTAL TAX LAW

A possible and desirable policy vision, fulfilling the above mentioned necessities, is one by which the Community, via ambitious ecological goals and regulating fiscal environmental policy instruments, resolutely chooses in favor of an environmental tax policy, in which the protection of the environment is the central goal, and not the creation of new jobs or financial means. Jobs and money can be interesting side-effects of such a policy, but they are not necessary and they should certainly not influence the contents of the environmental tax policy. Due to the bad condition of our environment, the fundamental aim of a more effective and more efficient environmental protection and policy is a matter which is far too important, too large-scale and too delicate (Deketelaere, 1993a; Deketelaere, 1995a; Deketelaere, 1996; Deketelaere, 1997b; Deketelaere, 1998b; OECD, 1989).

## 3.8. POLICY FRAMEWORK FOR A EUROPEAN ENVIRONMENTAL TAX LAW

A possible and desirable legal framework, corresponding to this policy vision, is one by which the Community chooses the following three pillars, based on the principle of subsidiarity and the principle of integration: pillar one, a European conducted 'clean-up' of the present environment-unfriendly fiscal regulations in regional and national (fiscal) legislation; pillar two, a European-conducted use of ecological fiscal incentives; pillar three, a European-conducted use of environmental taxes.

As regards the first pillar, it should be clear that it is no use starting up a European environmental tax policy when the present environment-unfriendly (fiscal) regulations are not abolished (e.g., the deductibility of professional transportation costs). Hence, organizing a European-coordinated and/or led 'ecological clean-up' of such regulations should be the first task of the Community. This way, modernization, rationalization and ecologization of the existing fiscal systems could be started. Depending on their contents, such 'clean-up' measures can be based on Article 175 (2) EC-Treaty (ecological measures which are primarily of a fiscal nature), Article 175 (1) EC-Treaty (ecological measures which are not primarily of a fiscal nature), Article 94 EC-Treaty (harmonization of direct taxes) or Article 93 EC-Treaty (harmonization of VAT and excise duties). However, it must be noted that in case of use of Article 175 (2) EC-Treaty, Article 94 EC-Treaty or Article 93 EC-Treaty, unanimity in the Council is still needed (even with the Treaty of Amsterdam), so that in fact every Member State can veto fiscal policy proposals.

As to the second pillar, a tripartite system of ecological fiscal incentives should be developed by the European Community, taking into account the application of the subsidiarity principle and the integration principle. This system could be designed as follows.

The first level consists of European-controlled national and/or regional ecological fiscal incentives for national environmental problems. The European control on the conformity of the involved incentives with the primary and secondary Community legislation, is described in the Communication of the Commission on "Environmental Taxes and Charges in the Single Market". The relevance of this Communication for national and/or regional ecological fiscal incentives will be examined under the third pillar, together with its relevance for environmental taxes.

A second, following, level consists of national and/or regional ecological fiscal incentives, recommended and/or harmonized by the European Community, for supra-national environmental problems. Two scripts are possible here: if the relevant incentives are already being used by a majority of the Member States, harmonization can be immediately undertaken. If the incentives are used only in a minority of Member States, harmonization can be preceded by a period of (European) recommended use. It should be quite clear that the recommended incentives, via sideways imposed conditions, should serve as a model for the harmonized incentives.

Finally, a third level consists of ecological fiscal incentives, constructed by the European Community, for international environmental problems. Incentives of this latter kind could be designed via the European harmonized excise-duty legislation. From an environmental tax viewpoint, however, Council Directive (EEC) 92/12 contains only one relevant Community excise duty product: mineral oils. As a consequence, a preliminary ecologically inspired broadening of these product categories will be necessary. Once this has happened, ecologically differentiated excise duty rates for these products can be established by the Community. On the other hand, these incentives could also be constructed via the VAT legislation. In this matter as well, the provisional European harmonization is still an obstacle: since the VAT-Council Directive (EEC) 77/388 contains only a standard rate of minimum 15% and one or two reduced rates of minimum 5% for the goods and services which are mentioned in appendix H, a conceptual alteration of the VAT-harmonization seems to be unavoidable. This could be achieved by a revision of appendix H of Council Directive (EEC) 77/388, based on the distinction between the delivery of environment-friendly and environment-unfriendly products and services, through which ecologically differentiated VAT-rates could be fixed by the Community.

The recent proposal for a Council Directive restructuring the Community framework for the taxation of energy products (Rose, 1998) can be an illustration of this third level approach. In view of the global environmental problem of greenhouse gas emissions (Adebavale, 1998; OECD, 1991b; OECD, 1992a; OECD, 1993b) and the envisaged reduction of these emissions (as set forward in the Kyoto Protocol), the Commission is proposing an overall tax system for the taxation of energy products. With this overall tax system, three objectives are envisaged: (i) improving the functioning of the internal market; (ii) encouraging behavior conclusive to protection of the environment; (iii) promoting the greater use of the factor labor. The contents of the proposal may be summarized by the following elements:

(1) Member States are to impose taxation on energy products in accordance with the proposal, and endeavor to avoid any increase in their overall tax burden. In order to achieve this, it is suggested that they reduce the statutory charges on labor.

(2) The proposal for a Directive lists the energy products which it covers and prescribes minimum levels of taxation for them. Member States may not apply a level of taxation which is lower than the minimum levels prescribed.

(3) The minimum levels of taxation applicable to motor fuels are the following: for petrol: £ 417 per 1000 liters as of 1 January 1998, £ 450 on 1 January 2000 and £ 500 on 1 January 2002; for gas oil and kerosene: £ 310 per 1000 liters as of 1 January 1998, £ 343 on 1 January 2000 and £ 393 on 1 January 2002; for liquid petroleum gas: £ 141 per 1000 kg as of 1 January 1998, £ 174 on 1 January 2000 and £ 224 on 1 January 2002; for natural gas: £ 2.9 per gigajoule as of 1 January 1998, £ 3.5 on 1 January 2000 and £ 4.5 on 1 January 2002. The minimum levels of taxation are modified where these motor fuels are used for certain industrial or commercial purposes. The proposal refers to: agricultural, horticultural or piscicultural works, and forestry; stationary motors; plant and machinery used in construction, civil engineering and public works; vehicles intended for use off the public roadway; passenger transport and captive fleets which provide services to public bodies.

(4) The minimum level of taxation applicable to heating fuels is fixed as follows: for gas oil: £ 21 per 1000 liters on 1 January 1998, £ 23 on 1 January 2000 and £ 26 on 1 January 2002; for heavy fuel oil, depending on

its category: £ 18 or £ 22 per 1000 kg on 1 January 1998, £ 23 or £ 28 on 1 January 2000, £ 28 or £ 34 on 1 January 2002; for kerosene: £ 7 per 1000 liters on 1 January 1998, £ 16 on 1 January 2000 and £ 25 on 1 January 2002; for liquid petroleum gas: £ 10 per 1000 kg on 1 January 1998, £ 22 on 1 January 2000 and £ 34 on 1 January 2002; for natural gas and solid energy products: £ 0.2 per gigajoule on 1 January 1998, £ 0.45 on 1 January 2000 and £ 0.7 on 1 January 2002.

(5) The minimum level of taxation on electricity is fixed at £ 1 per megawatt hour as of 1 January 1998, £ 2 on 1 January 2000 and £ 3 on 1 January 2002.

(6) Provided that they comply with Community law and with the minimum levels of taxation set out in the proposal, Member States may apply differential rates of taxation according to the use or quality of a product.

(7) The following are exempt from taxation: energy products used for purposes other than as motor fuels or as heating fuels; energy products used to produce electricity and heat generated during its production. However, Member States may, for reasons of environmental policy, subject these products to taxation; energy products supplied for use as fuels for the purpose of air navigation other than private pleasure flying for as long as such products are obliged to be exempted under international obligations; energy products supplied for use as fuel for the purposes of navigation within Community waters (including fishing), other than private pleasure craft. Member States may limit the scope of these exemptions and in particular tax national air or sea travel or even, where bilateral agreement exists, travel within the Community.

(8) Member States may apply total or partial exemptions or reductions in the level of taxation to: energy products used under fiscal control in the field of pilot projects for the technological development of more environmentally-friendly products or in relation to fuels from renewable sources; biofuels; forms of energy which are of solar, wind, tidal or geothermal origin, or from biomass or waste; forms of energy which are of hydraulic origin produced in hydroelectric installations with a capacity of less than 10 MW; heat generated during the production of electricity; energy products used for the carriage of goods and passengers by rail; energy products used for navigation on inland waterways other than in private pleasure craft; natural gas where the gas market is in the process of development. These exemptions or reductions in taxation are granted in the form of a refund of all or part of the amount of taxation paid. Since such measures may constitute state aid, they must if necessary be notified to the Commission.

(9) Member States may refund all or part of: the tax paid in relation to investment expenditure incurred by a firm in improving the efficient use of energy; the amount of tax paid by a firm on any part of its non-transport-related energy costs which exceeds 10% of its total production costs. When the part of a firm's non-transport related energy costs exceeds 20% of its total

production costs, Member States must refund the whole of the tax paid by the firm on the part of its non-transport related energy costs which exceeds 10% of its total production costs. Where renewable energy sources are used, the tax is paid by the consumer on electricity and heat generated during its production and, in this case, it is the producer who receives the refund.

(10) For specific policy reasons and for a specified period, Member States may apply levels of taxation below the minimum levels specified in the proposal. If they wish to do so, they must make a request for a derogation. The proposal describes the procedure applicable to such authorizations.

(11) Changes in one or more rates of taxation may be passed on to stocks of energy products already released for consumption.

(12) Energy products released for consumption in a Member State and used by motor vehicles as such or by refrigeration systems, oxygenation systems, thermal installation systems or other systems fitted in containers transported by those same vehicles may not be subject to taxation in any other Member State.

(13) Member States are to inform the Commission of the: levels of taxation which they apply to the energy products covered by the proposal for a directive; measures they have taken to comply with the objective of tax neutrality as defined by the proposal.

As regards the third pillar, here as well a tripartite system of environmental taxes should be developed by the European Community, taking into consideration the application of the subsidiarity principle and the integration principle. This system could be worked out as follows.

The first level might consist of European controlled national and/or regional environmental taxes for national environmental problems. Just as in the case of ecological fiscal incentives, the Communication of the Commission on environmental taxes and charges in the single market indicates what those points of control must be (For a detailed analysis, see de Wit, 1997):

(i) checking with the Articles 23-25 EC-Treaty, concerning forbidden custom duties and taxes of equal effect;

(ii) checking with Articles 28-30 EC-Treaty, concerning forbidden quantitative restrictions and measures of equal effect;

(iii) checking with Article 87 EC-Treaty, concerning forbidden state aids;

(iv) checking with Article 90 EC-Treaty, concerning forbidden tax discrimination;

(v) checking with Article 93 EC-Treaty, concerning indirect taxation;

(vi) checking with notification requirements in the field of state aid, in the field of (former) Council Directive (EEC) 83/189 and secondary Community environmental law and concerning national measures transposing Community Directives.

The second level consists of national and/or regional environmental taxes, recommended and/or harmonized by the European Community, for supra-national environmental problems. Just as with ecological fiscal incentives, here too two scripts are possible: if the environmental taxes are already being used by a majority of Member States, harmonization can immediately be worked out. If the environmental taxes are being used by only a minority of Member States, harmonization can be preceded by a period of recommended use. Evidently, the recommended environmental taxes, should, via sideways imposed conditions, serve as a model for the harmonized ones. This harmonization aims at least at streamlining the tax bases and the tax rates. The Member States should still have enough room to be able to take into consideration (for the determination of the tax base and the tax rate) local circumstances like the degree of pollution and kind of pollution. Because avoiding trade barriers and distortions of competition here will be the main aim, the most appropriate bases for this harmonization, are the Articles 93 and 94 EC-Treaty.

Finally, a third level consists of environmental taxes, constructed by the European Community, for international environmental problems. As measures aiming at ecological purposes are at stake, which are moreover of a primarily fiscal nature, Article 175, second part, first line EC-Treaty seems to be the most advisable basis for these European environmental taxes.

For European environmental taxes, three policy options can be formulated:

- Starting from the developed policy vision, regulating European environmental taxes (which are interdisciplinary in design and announced a long time beforehand), should be resolutely chosen;
- The demand for fiscal neutrality is not relevant to European environmental taxes, considering the Community's incompetence as to compensation, on the one hand, and the third policy option, on the other;
- The proceeds of regulating European environmental taxes should be reserved for ecological purposes. It seems, moreover, advisable that a compartmentalized European Environmental Fund should be founded, into which the proceeds of the regulating European environmental taxes are deposited, together with the European environmental budget share, in order to ensure the continuity of this Fund. Accounting for the ecological purposes of the created financial measures would mean that they can not and will not be used to compensate the lowering of the fiscal pressure on labor and capital. The frequent demand to lower charges and to modernize and rationalize tax law is desirable and necessary, but preferably not by using the environment as milkcow. Authorities, companies and individuals will already have to pay a lot for a preventive and curative environmental policy in the coming years; it would be rather hypocritical to use the little (compared to the cost price of the environmentally passive) money, won through the environmental taxes, for the struggle against unemployment. Other methods and means seem better suited to handle this problem.

An example of such a European environmental tax, would have been the first proposal of a carbon dioxide ($CO_2$) and energy tax. The objective of this proposal was the introduction of an additional harmonized tax on $CO_2$ emissions and energy content in order to limit the emission of greenhouse gases and to promote the efficient use of energy. The contents of the proposal may be summarized by the following elements (Pearson, 1991; Shome, 1998; OECD, 1993d):

(1) The tax was to be levied on the following products used as motor or heating fuels, whether in crude or processed form: coal, lignite, peat and their derivatives; natural gas; mineral oils; ethyl and methyl alcohol obtained by distillation from fossil fuels; electricity and heat generated in hydroelectric installations with a capacity of over 10 MW or in nuclear power stations.

(2) The tax did not apply to raw materials or renewable sources of energy.

(3) The features of the tax were: it was to be introduced at Community level but levied by the Member States; the chargeable event was to be extraction or production on the territory of the Community, or importation; the tax was to be chargeable when the taxable products are released for consumption or when shortages are recorded; it was to borrow from the arrangements for excise duties the rules on product movement and monitoring with respect to mineral oils, with special provisions for the other products; 50% of the taxable amount was to be based on the products' carbon content and 50% on their energy content; the tax was to be additional to existing excise duties harmonized at Community level; the rate was to be common to all Member States and was to vary by product on the basis of £17.70 per TOE; Member States were to be free to apply higher rates.

(4) The Directive would take into account: the situation of energy-intensive firms by including additional provisions (reductions and exemptions) in order to safeguard the competitiveness of industry; the need to save energy and reduce carbon gas emissions by allowing Member States to introduce, subject to Community competition rules, tax incentives for new investment in this field.

(5) The Member States might be authorized by the Council, acting unanimously on a proposal from the Commission, to suspend temporarily the application of the tax in response to economic problems or progress made in stabilizing carbon dioxide emissions.

(6) The entry into force of the tax was to be conditional on a similar tax or measures having an equivalent financial impact being brought in by other OECD countries and was to be decided on by the Council, acting by qualified majority.

(7) The carbon dioxide tax was to be based on the principle of tax neutrality and be offset in full, as soon as it is introduced, by tax incentives or reductions in direct taxes and other statutory contributions.

(8) A Committee on the Carbon Dioxide/Energy Tax was to be set up which would take part in preparing the measures for the application of the Directive.

(9) After three years the Commission was to present to the Council with a report on the operation of the tax system.

(10) The arrangements for granting tax exemptions, reductions and refunds and those for tax incentives were to be reviewed every three years.

(Deketelaere 1993a; Deketelaere, 1995a; Deketelaere, 1996; Deketelaere, 1997b; Deketelaere, 1998b; OECD, 1989).

*3.8.1. Greenspeak*

By way of conclusion on this point, the hope can be expressed that the European Community takes into full account the above-mentioned vision and pillars. The development of a European environmental tax policy has been considerably threatened by the typical disease of the general European environmental policy, viz. the above-mentioned Greenspeak: 'Much theoretical green ado, about nothing practical.' By stressing so often the fact that the European Community will use fiscal environmental policy instruments, but not realizing very much concretely, the Community loses a lot of its credibility. We therefore should hope that we will not have to add the concept 'European environmental tax law' to the list of hollow and grandiloquent environmental policy terms of the European Community.

## 4. Conclusion

The general conclusions of this paper can be summarized as follows:

(i) For each environmental problem, the most adapted policy instrument(s) must be applied;

(ii) For the general environmental policy, a well-balanced mix of policy instruments must be pursued;

(iii) The introduction of new policy instruments must be announced well before the moment of application;

(iv) The introduction of new policy instruments must be done in consultation with all related and concerned governments (local, regional, federal, European);

(v) The new policy instruments must be analysed, before their introduction, from a legal, economic and ecological perspective.

## References

Aalders, MVC and Van Acht, R.J.J. (eds): 1992, *Afspraken in het milieurecht,* W.E.J. Tjeenk Willink, Zwolle.

Adebavale, M.: 1998, "Climate change and taxation policy in Britain – an overview", *Environmental Taxation and Accounting*, **1**, 62-68.

Baumol, W.J. and Oates, W.: 1988, *The theory of environmental policy,* Cambridge University Press, New York.

Bocken, H. and Traest, I. (eds): 1991, *Milieubeleidsovereenkomsten*, E. Story-Scientia, Brussels.

CEDRE: 1991, *L'évaluation des incidences sur l'environnement: un progrès juridique?* CEDRE, Brussels.

CEDRE: 1994, *L'introduction des écotaxes en droit belge*, CEDRE, Brussels.

De Clercq, M: 1996, "The implementation of green taxes: the Belgian experience", in *Environmental taxes and charges – National experiences and plans*, European Foundation for the improvement of living and working conditions, Dublin, 45-66.

de Wit, W.: 1997, *Nationale milieubelastingen en het EG-Verdrag*, Kluwer, Deventer.

Deketelaere, K.: 1993a, "The use of economic instruments in the European environmental policy", *Elsa Law Review*, 45-67

Deketelaere, K.:1993b, "The use of economic instruments in the European environmental policy", *Elsa Law Review*, 45-67

Deketelaere, K.:1995a, "Towards a European environmental tax law: Greenspeak?", in F. Abraham, K. Deketelaere and J. Stuyck, J (eds), *Recent economic and legal developments in European environmental policy*, Leuven University Press. Leuven.

Deketelaere, K.:1995b, *Naar een Europese milieufiscaliteit – Het gebruik van fiscale instrumenten in het Europees milieubeleid (doctoraal proefschrift)*, Institute for Environmental and Energy Law, Leuven (Towards a European environmental tax law – The use of fiscal instruments in the European environmental policy (hereafter: doctoral thesis).

Deketelaere, K.: 1996, "European environmental tax policy: proposal for a policy vision and a legal framework", *European Environmental Law Review*, 9-16.

Deketelaere, K.: 1997a, "European environmental and spatial planning, *European Environmental Law Review*, 278-286 and 307-317.

Deketelaere, K.:1997b, "The European Union and fiscal environmental policy instruments - Policy tracks for the future European environmental policy", *Environmental Taxation and Accounting* **3**, 8-20.

Deketelaere, K.:1998a, "Voluntary environmental agreements: European Community viewpoints", in X, *European Union Environmental Law*, Finnish Society for Environmental Law, Helsinki, 81-125.

Deketelaere, K.:1998b, "New environmental policy instruments in Belgium", in J. Golub (ed), *New instruments for environmental protection in the EU*, Routledge, London.

Dente, B. (ed): 1995, *Environmental policy in search of new instruments*, Kluwer Academic Publishers, Dordrecht/Boston/London.

EC: 1990a, *The environmental dimension – Task force report on the environment and the internal market , EC*, Bonn.

EC: 1990b, *Report of the working group of experts from the Member States on the use of economic and fiscal instruments in EC environmental policy*, EC, Brussels.

EC: 1990c, *The environmental imperative – declaration by the European Council*, EC, Dublin.

EC: 1993, *White paper on growth, competitiveness and employment*, EC, Luxembourg.

EC: 1995, *Waste-water charge schemes in the European Union*, ECSC-EC-EAEC, Brussels/Luxembourg.

EC: 1996, *European Community environment legislation* – volumes 1-7, EC, Luxembourg.

EC: 1997, *Tax provisions with a potential impact on environmental protection*, EC, Luxembourg.

EEA: 1996, *Environmental taxes – Implementation and environmental effectiveness*, European Environment Agency, Copenhagen.

EFILWC: 1996, *Environmental taxes and charges – National experiences and plans*, European Foundation for the improvement of living and working conditions, Dublin.

Elni (ed): 1997, *International environmental impact assessment – European and comparative – Law and practical experience* , Cameron May, London.

Gaines, S.E. and Westin, R.A.:1991, *Taxation for environmental protection – a multinational legal study*, Quorum books, Connecticut.

Gee, D.,:1997, "Eco-nomic tax reform in Europe", *Environmental Taxation and Accounting* **1**, 21-26.

Golub, J. (ed): 1998, *New instruments for environmental policy in the EU,* Routledge, London/New York.

Guppes, G., Van der Vogt, E., Van der Naald, W., Manson, P. and Vonkeman, G.: 1992, *New market-oriented instruments for environmental policies,* Graham and Trotman, , London.

Guscuolo, A.: 1998, "Environmental policy and fiscal policy in the Community: the challenge of national eco-taxes", *Environmental Taxation and Accounting* **3**, 52-70 and **4**, 10-30.

Hamilton, C.: 1998, "Ecological tax reform in Australia", *Environmental Taxation and Accounting* **2**, 9-42 and **3**, 13-37.

Helm, D. (ed): 1991, *Economic policy towards the environment,* Blackwell Publishers, Oxford.

IISD: 1994a, *Making budgets green,* IISD, Winnipeg.

IISD: 1994b, *The greening of government taxes and subsidies,* IISD, Winnipeg.

Jadot, B. and De Sadeleer, N. (eds): 1993, *Le label écologique et le droit,* E. Story-Scientia, Brussels.

Jenkins, G.P. and Lamech, R.: 1992a, "Fiscal policies to control pollution: international experience", *Bulletin*, 483-502.

Jenkins, G.P. and Lamech, R.: 1992b, "Market-based incentive instruments for pollution control", *Bulletin*, 527-536.

Klok, P.I.: 1989, *Convenanten als instrument van milieubeleid*, Enschede.

Lockhart, J.A.: 1998, "Environmental tax subsidies in the United States", *Environmental Taxation and Accounting* **2**, 43-54.

NCM: 1996, *The use of economic instruments in Nordic environmental policy.* Nordic Council of Ministers, Copenhagen.

OECD: 1985, *Environment and economics,* OECD, Paris.

OECD: 1989, *Economic instruments for environmental protection,* OECD, Paris.

OECD: 1991a, *Environmental policy: how to apply economic instruments,* OECD, Paris.

OECD: 1991b, *Responding to climate change: selected economic issues,* OECD, Paris.

OECD: 1992a, *Convention on climate change – Economic aspects of negotiation,* OECD, Paris.

OECD: 1992b, *Umweltschutz durch Abgaben und Steuern,* R. v. Decker's Verlag, G. Schenk, Heidelberg.

OECD: 1993a, *Economic instruments for environmental management in developing countries,* OECD, Paris.

OECD: 1993b, *International economic instruments and climate change,* OECD, Paris.

OECD: 1993c, *Taxation and the environment – Complementary policies*, OECS, Paris.

OECD, 1993d, *Taxing energy – why and how* , OECD, Paris.

OECD: 1994, *Environment and taxation: the cases of the Netherlands, Sweden and the United States,* OECD, Paris.

OECD: 1995a, *Environmental taxes and charges,* Kluwer Law International,The Hague/London/Boston.

OECD: 1995b, *Environmental taxes in OECD countries,* OECD, Paris.

OECD: 1996, *Subsidies and environment – Exploring the linkages,* OECD, Paris.

OECD, 1996b, *Implementation strategies for environmental taxes,* OECD, Paris.

OECD: 1997a, *Applying market-based instruments to environmental policies in China and OECD-countries,* OECD, Paris.

OECD: 1997b, *Environmental taxes and green tax reform,* OECD, Paris.

OECD: 1997c, *Reforming environmental regulation in OECD countries,* OECD, Paris.

O'Riordan, T (ed): 1997, *Ecotaxation,* Earthscan Publications Limited, London.

Paulus, A.: 1995, *The feasibility of ecological taxation,* Maklu, Antwerpen/Apeldoorn.

Pearson, M. and Smith, S.: 1991, *The European carbon tax: an assessment of the European Commission's proposal,* The Institute for Fiscal Studies, London.

Pittevils, I.: 1996, "Ecotaxes on products in Belgium: the need for a proper point of imposition", in *Environmental taxes and charges – National experiences and plans*, European Foundation for the improvement of living and working conditions, Dublin, 201-210.

Platteau, H. and Declerck, B.: 1998, *Milieuaudit en Accounting* , Kluwer Editorial, Brussels.

Rose, I.: 1998, "EU initiatives to control air pollution: taxation of energy products", *International Business Lawyer*, 436-440.

Rosembuy, T.: 1998, "Issues on European environmental taxation", *Environmental Taxation and Accounting* **3**, 71-83.

Rowan-Robinson, J.: 1998, "The use of taxes to achieve environmental goals in Britain", *Environmental Taxation and Accounting* **2**, 61-66.

Schlegelmilch, K (ed): 1998, *Green budget reform in Europe,* Springer, Berlin/Heidelberg/New York.

Shome, P.: 1998, "A 21st century global carbon tax", *Environmental Taxation and Accounting* **1**, 11-29.

Smith, S.: 1997, "'Green tax reforms and road transport in Britain and Germany", *Environmental Taxation and Accounting* **3**, 21-37.

Smith, S., 1995, *"Green" taxes and charges: policy and practice in Britain and German,* The Institute for Fiscal Studies, London.

Thieffry, P.: 1992, "Les nouveaux instruments juridiques de la politique communautaire de l'environnement", *Revue Trimestriel de droit Européen*, 669-685.

Tindale, S. and Holtham, G.: 1996, *Green tax reform,* Institute for Public Policy Research, London.

Van Calster, G.: 1998, "State aid for environmental protection: has the EC shut the door?", *Environmental Taxation & Accounting* **3**, 38-51.

Van Humbeek, P.: 1997, "Environmental taxation in Flanders", *Environmental Taxation and Accounting* **4**, 52-64.

Vermeend, W. and Van der Vaart, J.: 1998, *Greening taxes – the Dutch model,* Kluwer, Deventer.

Wetterstein, P. (ed): 1997: *Harm to the environment: the right to compensation and the assessment of damages,* Clarendon Press, 1997.

Zonnekeyn, G.: 1997, "Eco-taxes in Belgium and Europe", *Environmental Taxation and Accounting* **1**, 9-20.

# THE USE OF ECONOMIC INSTRUMENTS AND GREEN TAXES TO COMPLEMENT AN ENVIRONMENTAL REGULATORY REGIME

R. HAWKINS
*Eden Hall, Kelso, TD5 7QD*
*e-mail: richardhawkins.edenhall@btinternet.com*

**Abstract.** In 1999, some countries are developing and effectively applying economic instruments for environmental protection and natural resource management, whilst others are relying on command and control regulatory procedures under-enforced by sometimes inadequately trained and motivated enforcement officers. This paper considers the current and future role of economic instruments as policy instruments for use by governments. In many developed countries, past over-regulation allied to a serious shortfall of experienced environmental enforcers required regulatory regimes to be supplemented by well targeted economic instruments and green taxes. Their application to countries which do not have developed environmental control systems is more questionable. The purported threats of the longer term effects of global warming, damage to the ozone layer and an apparent loss of biodiversity have led environmentalists to adopt the so-called precautionary principle. Sustainable development has added to the pressures for further national and transfrontier legislation. The challenge facing policymakers, therefore, is to design policies to enable market forces to operate in the environmental sphere, for example through a system of pollution charges, principally intended to promote greater environmental efficiency. These charging systems can be of many kinds but their main defining feature is their reliance on markets and the price mechanism to internalise environmental externalities, thereby attempting to make polluters pay through facing the full social costs of their activities. Some of the applications of these charging systems, financial and fiscal instruments and tradable emission systems are explained and illustrated.

**Keywords**: economic instruments complementary to command and control regimes, green taxes

## 1. Introduction

### 1.1. 1999 ENVIRONMENTAL TRUTHS EXAMINED

During the last decade, the environmentalists' movement has renewed its attack on the desirability of economic growth, and has re-doubled its call for drastic measures to ward off environmental catastrophe. A leading, if mistaken, environmentalist, Jonathan Porritt, at the beginning of this decade wrote that, "in the current state of the planet, the idea that economic growth is necessarily and automatically a good idea is no longer tenable" (Porritt, 1990).

Those who have been practising environmental law (Hawkins, 1998) in the field for the last quarter of a century can well remember in 1972 the Club of Rome publishing a highly influential, if specious, report entitled, "Limits to Growth." In it was stated that the total global oil reserves amounted to some 550 billion barrels, which were anticipated to be used up by 1990. By 1990, the

*Water, Air, and Soil Pollution* **123:** 379–394, 2000.

world had used 600 billion barrels of oil but, also by that year, unexploited reserves amounted to at least 900 billion barrels. The Club had used the word "reserves" with a rather restrictive interpretation. It also made similarly wrong predictions about aluminium, copper, lead, natural gas, silver, tin, uranium and zinc.

Environmentalist pressures eased off in the middle 1970s when the sudden rise in the oil price led to deflationary policies so that the public was more concerned with the lack of economic growth than with a fear that economic growth was harmful.

In 1980, Global 2000 was a report to the President of the United States by a committee of the great and the good which predicted that population would increase faster than world food production so that food prices would rise by between 35 and 115% by 2000. Yet, with six months to go, the World Food Commodity Index has fallen by 50%. No doubt there will be some form of explanation, some time.

In the early 1970s, there were assumptive statements that, whilst meteorologists disagreed about the cause and extent of the cooling trend (the new Ice Age), they were almost unanimous in the view that that trend would reduce agricultural productivity for the rest of the century. Contrast this with what Vice President Al Gore said in 1992, "Scientists conclude almost unanimously that global warming is real and the time to act is now."

From the 1860s to the 1880s, Britain lived through a prolonged cold snap. It is this period which is taken as the base line to measure the global warming increase of 0.5°C to the present day. So, are we as certain about the effects of global warming as all the politicians seem to think? Can it be one of those mantras which should be subject to greater scrutiny than heretofore?

The unanimity of scientists has undergone a sea of change in the last 20 years, with greenhouse gases apparently accumulating like nemesis waiting to punish us for the fatal hubris of achieving real economic growth. Mr. Gore, incidentally, also said in 1992 that 20% of the Amazon had been deforested and that deforestation continued at the rate of 80 million hectares a year. The real figures are agreed to be 9% of deforestation in the Amazon Basin and are now falling to about 10 million hectares a year.

## 1.2.PRECAUTIONARY PRINCIPLE AND SUSTAINABLE DEVELOPMENT

The alleged threats of the longer-term effects of global warming, damage to the ozone layer and the apparent loss of biodiversity have led environmentalists to adopt the so-called precautionary principle. In addition, the old Club of Rome myth that supplies of so-called finite resources were indeed finite has given rise to the pursuit of only sustainable development.

These two new catch phrases are repeated parrot fashion by environmental policymakers, commissions and committees, national and international, set up to supervise and report on the adoption of policies to promote sustainable development and to implement the precautionary principle.

Politicians, particularly in America, Germany and England, who fail to pay due lip service to this principle do so at their peril. Many politicians and public figures, particularly in the fields of academe, are eager to demonstrate their sense of social responsibility. The precautionary principle has, of course, been around since time immemorial. Plaques on the facades of houses in the City of London in the 17th century denoted that the occupier had subscribed to a particular local commercial fire brigade on whose extinguisher help he could rely at the time the first wisp of smoke appeared. Now there are lightning conductors, insurance policies, car deadlocks and smoke alarms.

Indeed, insurance itself is one of the best working examples of an economic instrument. It involves the quantification of environmental risk and an explicit financial discouragement to high risk activities. Pollution insurance can be seen as part of a broader universe of economic instruments which act as direct financial incentives away from polluting processes.

Sustainable development for its part cannot be interpreted, as it sometimes is, as implying that all other components of welfare are to be sacrificed in the interests of preserving the environment exactly in the form it happens to be in today. It is in fact no different from the accepted goal of maximising society's welfare.

It is misleading to introduce the concept that it constitutes some important new insight to guide environmental policy. This merely creates confusion, although it has a high feel good factor rating and is an excellent moveable step ladder in order to occupy the moral high ground.

There have been attempts to focus more closely on whether sustainable development has any helpful meaning in environmental management apart from the received truth that it depends on only development that meets the needs of the present without compromising the ability of future generations to meet their own needs.

A strong form of sustainability in which the current generation leaves future generations with an unchanged stock of natural resources is not only unrealistic but is an unnecessarily restrictive requirement. A weaker requirement would be that future generations should receive at least as much total capital, aggregating natural resources and physical and intellectual capital as the present generation inherited (Markandya, 1989).

The challenge facing policymakers, therefore, is to design policies to enable market forces to operate in the environmental sphere, e.g., by a system of pollution charges where property rights cannot be applied. It is on this area that policymakers should concentrate their efforts rather than flying in expensive seats to vast international jamborees to discuss the latest apocalyptic predictions (Beckerman, 1995).

The worst environmental conditions are those found in lower income countries, e.g., a lack of acceptable sanitation and clean drinking water which, when available, is often carried for over half a mile in an earthenware pot on a head. A billion people in developing countries have no access to safe drinking water, and at least two billion have no access to proper sanitation (UNDP, 1990). There is also the urban squalor of almost all of the main cities in developing countries with their intolerable air pollution as a result of dung fires. The worst threat to the environment arises largely from poverty in the Third World.

Governments should not be distracted from attempts to solve these problems by attention paid to spurious disaster scenarios. All too often, however, politicians are deluded by the environmentalist movements' predictions that we are on the verge of environmental catastrophe and that governments must be pushed into taking far more drastic action than they appear to be taking. Alarm bordering on hysteria is no guide to balanced policy.

In times of tightening science budgets, and in an age where scientists feel misunderstood and unloved, some of them are being seduced by money and popularity to encourage, or at least acquiesce in the latest well-funded, well-publicised "green" scares. Predicting catastrophe is a good way of obtaining money to continue your research and showing what a compassionate scientist you are (Kenny,1994). People will more readily send a donation to an organisation that promises to save the planet from imminent catastrophe, than to one which merely wants to spend more money on drains for over-populated shanty towns outside Durban or Djakarta.

Edward Goldsmith, as if bent on balancing the commercial successes of his late brother, claims that what we must aim for is not growth but negative growth or economic and demographic contraction (Goldsmith, 1988). The present South African Government is unlikely to endorse that view.

## 1.3. OVER-REGULATION AND UNDER-ENFORCEMENT

So pressures to take drastic action to save the world from imminent environmental disaster will only push governments further in the direction of regulations and controls, like those used in the Soviet economies with the disastrous effects that are now well known. Command and control policies suit the bureaucratic instinct to regulate matters rather than help markets operate more efficiently within an appropriate, even-handedly, transparently enforced legal framework.

It is harmful to pile law upon law without regard to the cumulative effect of measures which individually may be laudable. Priorities become confused, resources misapplied, the public bewildered as to basic environmental truths and what standards to expect.

The result is that we become over-regulated and under-enforced, misleadingly educated and erratically resourced. Such over-regulation through its surfeit of laws devalues public respect for command and control regimes and, as important, their enforcers.

However, with this over-regulation, enforcement will become more difficult, the interpretation of those laws complex and there will be an inadequately trained cadre of well-motivated, experienced, reasonable enforcers with a natural discretion when to use the carrot or the stick. The UK Environment Agency in the last year had suffered not unexpected senior personnel losses for reasons not unrelated to that state of affairs.

## 1.4. COMPLEMENTARY AIDS TO A COMMAND AND CONTROL REGIME

Are there, therefore, other complementary systems which will make society not so dependent upon regulatory regimes? These regimes need trained enforcement officers who will probably not attain the necessary experience either to determine which offences are serious enough to be prosecutable or to command the respect of local and regional industry, without at least 5 years' experience.

Economic instruments as policy instruments are seen generally as an evolution in policy making from the conventional reliance on laws and regulations which polluters must adopt under penalty of fines or other sanctions. This command and control approach has been increasingly criticised by economists on grounds of inefficiency in that:

- It requires compliance with the same standards by all pollution sources, irrespective of marginal compliance costs;
- It provides little incentive to technical improvement once compliance has been achieved.

Economic instruments address both these issues directly, allowing polluters with relatively low abatement costs to treat their wastes, whilst allowing those with relatively high abatement costs to buy permits and thus avoid abatement costs.

Systems in which economic forces operate effectively are ones in which technological change occurs. Such change is significantly retarded where effective pricing signals are absent. Pollution control technologies have perhaps been slow to develop for this very reason. Other advantages to be gained through the use of economic instruments are:

- It is on the basis of prices and charges that society makes its decisions with respect to resource use. Market prices and charges are the most direct way to identify and quantify the many variables involved in making a decision. Command and control mechanisms can only approximate this information function and then only at much greater cost.
- Industries can react to pollution control in a wide variety of ways, including process change, technology development and product modification. This flexibility in response can be initiated through signals from the pricing system tending to produce least cost, that is efficient solutions.

- Likewise, actions oriented to direct impacts on the profit picture of a firm tend to initiate very rapid responses. The adoption of energy efficient programmes following the oil price shocks of the 1970s is an excellent example of market-based response speed.

Command and control will of course still be preferred if those cases where the consequences of non-compliance are especially serious. For instance, for minimising exposure to a highly toxic substance, regulatory control, including perhaps an outright ban on that substance, will always be preferred over a policy which would discourage use through a seemingly prohibitive product tax.

The use of economic instruments requires the existence of a strong institutional framework whereby the economic signals created can travel efficiently. For developing countries with weak and transforming institutions, the introduction of economic instruments generally should only follow a strengthening of these market structures. Likewise in Central and Eastern Europe, economic instruments will only be able to play a significant role after the systems for monitoring and enforcing existing environmental legislation are strengthened. This necessity for a strong institutional pollution law enforcement framework would preclude, for example, Italy from qualifying for the acceptance of economic instruments for environmental control.

## 2. Some Economic Instruments Illustrated

Fourteen Members of the Organisation for Economic Co-operation and Development (OECD) use between one and 21 economic instruments for environmental protection. Germany, Sweden and the Netherlands are the most progressive in their use. Half of the economic instruments were charges, one-third financial subsidies (UNEP Annual Report, 1996).

### 2.1. CHARGE SYSTEMS

Charge systems have been typically applied for the protection of resources from waste emissions and discharges. In Malaysia, for instance, effluent charges have been in operation for 20 years to protect water quality from effluents arising from the palm oil and rubber industries.

In Singapore, charging drivers for using roads in the city centre during peak hours resulted in a 73% reduction in traffic in the restricted zone and carbon monoxide levels.

At least 10 cities in the UK are set to sign up to a low emission vehicle programme designed to exclude all motor vehicles from town centres except those with low or near zero emissions. The idea is that random exhaust spot checks at the roadside will be conducted and that if offenders have entered the Low Emission Zone, they will be given a pollution charge, namely a fixed financial penalty notice to be paid within 14 days, probably about $50 or $100 if paid after these 14 days.

Effective monitoring and enforcement remains a strong pre-condition on the application of charge systems which can lessen their comparative efficiency in relation to straight command and control.

Table I

Economic Instruments For Environmental Protection
And Natural Resources Management

| | |
|---|---|
| CHARGE SYSTEMS | MARKET CREATION |
| Road Tolls | Tradable Emission Permits |
| Access Fees | Tradable Catch Quotas |
| Pollution Charges | Tradable Development Quotas |
| User Charges | Tradable Water Shares |
| Betterment Charges | Tradable Resource Shares |
| Impact Fees | Tradable Land Permits |
| Administrative Charges | Tradable Offsets/Credits |
| FINANCIAL INSTRUMENTS | BONDS AND DEPOSIT REFUND SYSTEMS |
| Eco Funds/Environmental Funds | Environmental Accident Bonds (e.g., Oil Spills) |
| Financial Subsidies | Environmental Performance Bonds (e.g., Forest Management) |
| Soft Loans | Land Reclamation Bonds (e.g., Mining) |
| Grants | Waste Delivery Bonds |
| Location/Relocation Incentives | Deposit Refund System |
| Subsidised Interest | Deposit Refund Shares |
| Hard Currency at below Equilibrium Exchange Rate | |
| Revolving Funds | |
| Sectoral Funds | |
| FISCAL INSTRUMENTS | LIABILITY SYSTEMS |
| Pollution Taxes (on Emissions or Effluents | Legal Liability |
| Product Taxes | *Non-Compliance Charges |
| Input Taxes | *Joint and Several Liabilities |
| Export Taxes | Natural Resource Damage Liability |
| Import Tariffs | Liability Insurance |
| Tax Differentiation | Enforcement Incentives |
| Royalties and Resource Taxes | |
| Land Use Taxes | |
| Investment Tax Credits | |
| Accelerated Depreciation Subsidies | PROPERTY RIGHTS |
| | Ownership Rights (Land, Water, Mining) |
| | User Rights |

Source: UNEP and Impax Ltd.

## 2.2. FINANCIAL INSTRUMENTS

These involve the direct use of subsidies or investments to accelerate the development of environmentally benign technologies. Sometimes they can be seen as negating the polluter pays principle.

## 2.3. FISCAL INSTRUMENTS

These relate specifically to a government tax or fiscal policy. Taxes on the landfilling of waste and the differential tax on leaded and unleaded fuel are obvious examples.

## 2.4. MARKET CREATION

These instruments involve the creation of a national or international market which is defined by the total permissible pollution or discharge allowed within that market. An artificial currency is created in the form of credits or permits which are traded amongst the players in the market, allowing those who are efficient in the control of emissions to transfer or sell the costs of overall compliance to those which are less efficient.

Economists are usually keen on markets and, with Tradable Emission Permits, they see possibilities in creating a market where none exists. There is a precedent in America, where a law allowing power companies to trade their right to emit sulphur dioxide has proved highly successful.

The Government determines what are the allowable emissions from each power plant. Those plants that can clean up cheaply, and thus emit less than the allowed amount, are then free to sell their unused rights to those for whom pollution control would be costly. Overall, this has cut sulphur emissions faster and more cheaply than anyone predicted (Economist, 1997).

Emissions trading can be transfrontier based. If a German coal fired power station finds that meeting its allocation of emissions is unexpectedly expensive, it might contract to buy the unused emissions of a Russian chemical plant working far below capacity.

This may well leave a difficult accounting computation. German emissions will rise, which might not please the Green Party, but those in Russia will fall. What matters is that global emissions are being limited in a cost effective way with the cuts being made where they are cheapest.

A variation is that of joint implementation in which one country introduces a practice that reduces carbon dioxide levels in another country. Two examples could be replanting a logged out forest or modernising a smoke belching smelter. Part of that reduction in the neighbouring country is applied against its own commitments.

Many such projects would possibly involve poorer countries because they have more opportunities for emissions reductions. Determining which activities should get the credit would not be easy. Nor would an independent audit of integrity. Corruption and venality could well throw out of kilter the most carefully laid schemes.

But if the details can be worked out, and the standards performed to, enormous benefits will beckon. If poorer nations accept the principle of an international permit trading system, which would require overall limits on emissions, they could possibly receive more monies than they now do from aid programmes, although into which pockets those monies may proceed will always be, for some countries, a matter of international concern.

A strong objection is, however, that it could allow rich countries to avoid taking domestic action to curb greenhouse gases. Yet global emissions will still be reduced. Another concern is that emissions trading does nothing to address emissions from homes or vehicles. But those will have to be the subject of other measurement and monitoring systems.

After Kyoto, the European Union accepted the principle of variable emission targets for different countries and its inter-European institutions may make it easier to trade emissions rights among Members.

## 2.5. BONDS AND DEPOSIT REFUND SYSTEMS

These are both based on the principle of paying up front for the externality imposed and being able to recoup the cost, if certain conditions or measures are met. Such performance bonds are increasingly used in mineral extraction, forestry and waste management as a form of insurance against long term environmental damage.

## 2.6. LIABILITY SYSTEMS

These involve the assignment of whole or part liability of the requirement to cover liability through insurance. Pollution insurance can be seen as part of a broader universe of economic instruments which act as direct financial incentives away from polluting processes. The European Commission may well adopt a European Union wide system of strict liability for environmental damage. Whether this will result in a common system after a number of years' application through case law within different EU countries is another matter (EC, 1998).

## 2.7. PROPERTY RIGHTS

These are seen mostly in the developing world and are intended primarily to slow the rate of natural resource depletion. Privately held land is converted to communal ownership so allowing local communities to share in the economic benefits of maintaining the asset.

In Papua New Guinea, where more than 90% of the land remains communal, only 13% of the forest land has been converted to other uses. In India, private water rights provide incentives for the efficient management of an increasingly scarce resource (Burnett, 1997).

## 3. Principal Environmental Areas for Application of Economic Instruments with Some Examples

### 3.1. WASTE

Environmental issues concerning waste are largely concerned with the consequences of different methods of waste disposal. These are mainly the pollution and amenity costs of landfill disposal, incineration, licensed marine disposal, etc. Other forms of waste management may also involve pollution and amenity costs, materials reclamation, packaging re-use, glass recycling etc. All can impose amenity costs, for example noise from a bottle bank impacting on local residents.

One important opportunity for the imposition of a Green Tax has been avoided by politicians, certainly in the UK and Germany. It is unacceptable that arrangements for the collection and disposal of household waste currently provide households with little or no individual financial incentive to reduce the amount of waste requiring disposal, thus discouraging their purchase of superfluous packaging.

In Britain, household refuse collection and disposal is provided free of charge by local authorities although, since the economically sensible introduction of compulsory competitive tendering for this service, a high proportion of household refuse collection and disposal services is supplied by private sector firms under contract to the local authority to provide a defined standard of service. Whilst the cost of household waste disposal will affect the level of council tax to be levied, there is no direct link between waste disposal costs and the incentive for individual householders to reduce the amount of waste that requires disposal.

In Germany, however, rather than financing the service through general local taxation, explicit charges are levied for household waste collection and disposal but, in practice, they provide, so far at least, a very weak incentive to reduce the amount of household waste.

#### *3.1.1.Packaging Taxes*

Even with consumption related charging for waste collection and disposal services, it is unlikely that clear enough signals will be received by manufacturers and packagers to modify product design and packaging to reduce disposal costs. Packaging taxes, e.g., for beverage containers, will establish a more direct incentive for substitution away from packaging which has high disposal costs.

Until there are sensible charges on individual household waste volumes, it is unlikely that environmental packaging taxes will change consumers' purchasing decisions. Thus the incentive for producers to modify their products or packaging would appear to be of no great account in the coming years. The city of Kassel, however, introduced a packaging tax on disposable plates, cutlery and packaging for take-away food and drink in 1992. This was subject to legal proceedings to challenge the power of the municipality to levy a tax in this form, but the Courts decided that packaging taxes may be levied at the municipal level, so long as similar taxes are not levied by the Federal Government (Smith, 1995).

*3.1.2. Working Example of a Contemporary Green Tax: The Landfill Tax In England, Wales and Scotland*

Since 1990, the UK Government has made clear that it intends to complement traditional forms of environmental control with economic instruments. Waste has proved a fruitful paradigm with the introduction of two economic instruments.

The first is the waste recycling credit scheme, which operates between waste collection authorities and waste disposal authorities so as to pass onto the former savings in waste disposal costs derived from waste collection authorities' own recycling initiatives. The UK's November 1994 Budget announced that the Government intended to introduce a new tax on waste disposed in landfill sites (UK Department of the Environment, 1990).

The additional tax burden that this would impose on business would be off-set by a corresponding reduction in the level of employer National Insurance contributions to be made when the tax entered into force. Thus the burden of taxation on business would remain constant, but the base of the tax would switch away from labour costs to waste disposal costs.

The Landfill Tax came into operation in October 1996 (HM Customs, 1996) in England, Wales and Scotland with the UK Government's announced intention of:

- ensuring that the full cost of the environmental implications of disposal is taken into account in the price; and
- seeking to move the emphasis from disposal towards re-use, re-processing, reclamation and re-cycling.

In addition to the disposal gate fee which waste owners have to pay to the 1,400 registered landfill operators in order to deposit their waste (taxable disposals) at a particular licensed site, the tax now sets a charge of:

- £2 per tonne for inert or inactive wastes (which really ought to be hand sorted to conform to that description); and
- £10 per tonne for active wastes which, through degradation, could potentially release pollutants into the immediate environment.

The landfill operator is charged with collecting the tax for the waste owner and passing it on to the Treasury. As an incentive to environmental improvement, the landfill operator is allowed to pay up to 10% of the tax into a special Environmental Trust Fund and, in return, will receive a proportionate rebate on National Insurance contributions for employees. The Trusts are non-profit-making private sector bodies engaged in the restoration of landfill sites or research into waste management and are generally required to provide services of public value without direct benefit to their contributors.

Landfill operators can, therefore, choose whether to pay the landfill levy to Government or to an Environmental Trust, except that the rebate against the landfill levy allowed for contributions to the Environmental Trusts are limited to 90% of their value in order to ensure that the contributors have an incentive to ensure that resources provided to the Trusts are spent efficiently.

There are certain exemptions as follows:

- material produced by dredging in fresh or sea water;
- by-products of mining or quarrying; and
- pet cemeteries (being England).

Difficulties have been posed for the management and engineering of wastes. The waste industry has told a House of Commons Environment Committee (House of Commons, 1998) that much of the waste which has been diverted from landfill is construction waste such as soil and aggregate, which the industry uses for constructing landfill cells and restoring sites after use.

It is interesting to contemplate what could be the possible exemptions in Israel. Certain contaminated land reclamation excavations for building work as an incentive?

*3.1.3. The Environment Agency and the Landfill Tax*

The possibility that the Landfill Tax would encourage fly tipping was a major concern for environmental groups, local government, the public and the enforcing authorities. The evidence, however, is still mainly anecdotal. One example is that of the National Farmers' Union which gave evidence to the House of Commons Committee. The National Farmers' Union carried out its own survey of members which revealed a clear increase of fly tipping onto agricultural land since the introduction of the tax.

The Environment Agency's survey of fly tipping with the Tidy Britain Group stated that: "There may have been an apparent increase in incidents of fly tipping but it does not actually prove a link to the Landfill Tax."

3.2. WATER

Incentives through economic incentives can be used to discourage water pollution at the same time as a price mechanism can be used to charge for water abstraction and water consumption. In the UK, a water pollution charge is set to recover costs of operating the pollution monitoring and control system. The system could form the basis for an evolution to a more comprehensive incentive charge for water pollution.

In the UK, a gradual rise in rates and a move to structure the charge to reflect pollution damage more closely could transform the existing administrative charges into an incentive system. The tax structure could be more directly related to the actual emissions performance which would strengthen their environmental impact.

In the UK, however, households which are a key category of water users, do not, in the main, face individual incentives for water conservation. Although the proportion of water consumers who are metered has risen in recent years, well over 90% of households still pay for water according to rateable value, or on some other basis unrelated to the amount of water consumed. This contrasts sharply with the situation in Germany, where volumetric charging of household consumers, based on water metering, is the rule.

### 3.3. ENERGY

Environmental taxes on energy should reflect the wide range of energy-related environmental problems. A carbon tax, such as the tax proposed by the European Commission, has considerable attractions as an environmental incentive mechanism. Indeed, the £5.2 billion windfall levy on the excess profits of the energy utilities brought in by the Blair Government has been used to fund Labour's "New Deal" for the young unemployed. A recent report into the merits of an industrial energy tax recommended the recycling of carbon tax revenues into environmental protection.

The European Commission proposal has, however, raised concerns about the impact of the tax on industrial competitiveness and of the energy intensive industry in particular, especially if countries outside the EU do not take similar measures. Britain has so far resisted the imposition of such a carbon tax on the general principle that it cannot accept further tax impositions from the EC – *a fortiori* after the 15 Commissioners' resignations in March 1999. Some would say not before time.

### 3.4. TRANSPORT

Road transport provides considerable opportunities for enhancing the environmental orientation of the UK tax system. Road transport is already heavily taxed, e.g., vehicle purchase, initial registration, annual charges on vehicle use, and taxes on motor fuels. There is thus scope for introducing environmental incentives by restructuring these existing taxes rather than establishing wholly new green taxes or charges.

Yet nothing is so simple. The range of social costs involved, the complex interactions between road transport and other transport modes, issues of spatial development and the general countryside aesthetic, all demand an integrated transport policy not easy to deliver.

Congestion costs and accident costs, global and local air pollution, noise and aesthetic losses, and the uncharged costs of consumption of the publicly provided road infrastructure, all should arguably be reflected in the costs of road use faced by individual road users (Royal Commission, 1994).

Ministers have now confirmed that they are considering allowing fines on polluters to be used to fund environmental schemes. However, it is worth recall that in the early $20^{th}$ century both the Road Fund and the National Insurance Fund started life as hypothecated (i.e., dedicated) taxes. Now they are firmly in the Treasury's grip. The National Lottery was the subject of solemn ministerial pledges at the outset that none of its receipts would be used for general public spending. But now £1.4 billion, 13% of the proceeds, is being diverted into health, environmental and education spending (Economist, 1999).

## 4. Conclusion

It may well be still too early to assess how effective economic instruments can be both as policy instruments and as supplements to command and control regulatory regimes. In both Britain and Germany, for example, actual policy measures to introduce environmental taxes have been slow in forthcoming. To date, the number of specific tax measures implemented with a primary rationale in terms of their environmental effects, is very limited in all EC countries.

In the UK, the only explicit environmentally motivated tax reforms have been to the excise duty differentials between leaded and unleaded petrol, and between petrol and diesel, and the landfill levy.

Does this very slow progress that has been made in the implementation of environmental taxes in the EC countries give any reason to question the feasibility or the merits of green taxation? What is certain is that all countries' constitutions and fiscal configurations are by no means the same. For example, in order to finance re-unification, the overall tax burden in Germany has had to rise sharply. In the UK, an integrated transport policy may well produce opportunities for the deployment of economic instruments through:

- higher road fuel and use duties;
- office parking tax;
- a transportation quarrying tax;
- higher waste disposal taxes.

Whilst political will always remains the key stumbling block, and there is a learning curve to be climbed in terms of proper planning and structuring of economic instruments, they should reinforce the drive to attaching considered value to the environment. That will be no mean achievement in itself.

## References


Beckerman, W.: 1995, *Small is Stupid.*

Burnett, S.: 1997, March, *Economic Incentives to Achieve Environmental Control,* IMPAX Capital Corporation Limited Address to Conference on "Insuring Pollution Risk", London.

EC: 1998, October, *EC White Paper on Strict Liability for Environmental Damage.*

Goldsmith, E.: 1998, *The Great U-Turn, De-Industrialising Society,* Green Books, Devon.

Hawkins, R. G. P.: 1998, March, *Some 1998 Potential Environmental Concerns for the Insured in UK Industry* (Address to the United Kingdom Environmental Law Association Annual Conference at the University of Sussex).

HM Customs & Excise: 1996, *HM Customs & Excise Landfill Tax Regulations,* Statutory Instrument No 527.

House of Commons All Party Select Committee on Environment, Transport and Regional Affairs: 1998, 17 June, *Sustainable Waste Management,* Sixth Report.

Kenny, A.: 1994, 12 March, *The Earth is Fine, the Problem is the Greens,* (Spectator Magazine).

Markandya, Pearce D.W. and Barbier, E.B.: 1989,*Blueprint for a Green Economy,* London Earthscan.

Porritt, J.; 1990, *Friends of the Earth Handbook*, 22.

Royal Commission on Environmental Pollution: 1994, *Transport and the Environment,* 18th Report, 1994.

Smith, S.: 1995, *Green Taxes and Charges: Policy and Practice in Britain and Germany* (Institute for Fiscal Studies), 67.

The Economist: 1997, December.

The Economist: 1999, 9 January.

UK Department of the Environment:1990, *This Common Inheritance, Command Paper 2068*, 35.

UNEP Annual Report: 1996.

UNDP: 1990, *Global Consultation on Safe Water and Sanitation for the 1990s* (a Background Paper for the 1990 New Delhi Conference of the same name).

# THE LAND AND ENVIRONMENT COURT OF NEW SOUTH WALES A MODEL FOR ENVIRONMENTAL PROTECTION

M. L. PEARLMAN
*Chief Judge of the Land and Environment Court of New South Wales, Level 4, 225 Macquarie Street, Sydney, New South Wales, Australia, 2000*

**Abstract**. The Land and Environment Court of New South Wales was established in 1979 as a superior court of record with exclusive jurisdiction in environmental matters. It is a specialist court and deals with both civil and criminal cases. It may make declarations and injunctions and impose criminal sanctions. It is unique in Australia and comparisons are made with courts and tribunals in other states of Australia. This article considers the effect of the court since its inception in developing environmental jurisprudence in Australia. Examples are given in key areas of environmental prosecution; public participation in environmental protection and decisions-making; transboundary pollution; and ecologically sustainable development principles.

**Keywords**: Australian specialist environmental courts, environmental jurisprudence, environmental prosecutions

## 1. Introduction

The Land and Environment Court of New South Wales celebrates its twentieth anniversary this year. The creation of a superior court, specialising in planning and environmental litigation, was something of a bold experiment in 1979. The aim of this paper is to describe the Land and Environment Court, and to consider how that bold experiment has come to fruition in the Court's contribution to the development of environmental jurisprudence.

Upon introducing into the Legislative Assembly the five statutes which created a new environmental planning system in New South Wales, the Minister for Planning and the Environment, the Hon D. P. Landa, stated the legislature's objectives and aspirations for the role of the Court;

> "The government's decision to create the new court attempts to rationalise the present diversified jurisdiction of a number of courts or tribunals, all pertaining to the use and development of land, land values and taxes, and the enforcement of those laws. This rationalisation was in my view essential as the planning legislation in many respects travels well beyond the boundaries of existing town planning legislation ... A specialist court, such as that proposed by the Land and Environment Court Bill, will have a vital role to play in the task of judicial interpretation of the new legislation and its operation ..." (Landa, 1979).

*Water, Air, and Soil Pollution* **123:** 395–407, 2000.

In many ways, the establishment of the Land and Environment Court was part of the process of the early evolution of environmental law in New South Wales. The growing amount of comprehensive and specialised environmental legislation in the 1970s in the areas of planning and environmental pollution created the necessity for a specialist court with specialist knowledge and expertise. What is interesting is that, in this evolutionary journey, environmental law in Australia, which had traditionally arisen out of administrative law inherited from the British system (particularly in the areas of planning and local government law) has in more recent times been influenced by the American model in terms of environmental impact assessment. That is not to say that the New South Wales model is simply a hybrid of both the British and American models. Rather, it is a unique system which has been inspired by external sources, but ultimately it has been engineered to meet the specifications of its domestic jurisdiction.

The state of New South Wales was undoubtedly the leader in developing a comprehensive system for addressing all aspects of environmental law in a "one-stop shop". The Court was established by the Land and Environment Court Act 1979, replacing the Land and Valuation Court and the Local Government Appeals Tribunal. Its genesis reflects its status as a superior court of record able to perform all of the acts and functions of the New South Wales Supreme Court, because Supreme Court judges with dual commissions formerly presided over the old Land and Valuation Court. Therefore, when the Court was established, the standard had in many ways been set as to the status of the Court and it has rank and status equivalent to the Supreme Court in the hierarchy of courts in New South Wales.

The Land and Environment Court is comprised of five judges and nine non-legal technical commissioners. It enjoys exclusive jurisdiction arising out of a myriad of environmental and planning statutes. Its jurisdiction is divided into classes which cover merit appeals in planning and building matters heard mainly by commissioners (classes 1 and 2), Aboriginal land rights, valuation and compensation for resumption cases (class 3), civil enforcement, including judicial review (class 4), criminal environmental prosecutions (class 5), and appeals from environmental offences heard by magistrates in the local courts (class 6).

Judges sitting alone hear all proceedings in classes 4, 5, and 6 of the Court's jurisdiction, and some applications in class 3. When hearing appeals in classes 1 and 2, judges may sit alone, or may, if time and resources permit, sit with a commissioner. Appeals from the decisions of judges of the Court are heard, in civil matters, by the Court of Appeal, and in criminal matters, by the Court of Criminal Appeal.

## 2. State Comparison

When the Land and Environment Court was established, it was unique in Australia. The states of Queensland and South Australia have now followed by establishing specialist environmental courts, but what sets the Land and Environment Court apart from its state equivalents is its exclusive jurisdiction. In New South Wales, the Land and Environment Court had and still has exclusive jurisdiction in all matters of environmental and planning law, and it is the only court in this State (except the appellate court) which may administer all forms of legal redress in those fields, including judicial review, civil enforcement, equitable orders and remedies, and summary criminal prosecutions.

In all states of Australia, except New South Wales, Queensland and South Australia, environmental proceedings are somewhat fragmented. Most states have planning tribunals, such as the Resource Management and Planning Tribunal set up in Tasmania in 1993. Tribunals such as these are largely concerned with appeals involving a review on the merits of decisions of local authorities. Judicial review, civil enforcement generally and criminal proceedings for environmental offences are heard in other courts.

In Queensland, the Local Government Court was renamed the Planning and Environment Court in 1990 under the Local Government (Planning and Environment) Act 1990. It is empowered by various statutes to make declarations and orders, and more recently, it has acquired limited power to grant equitable relief in the form of restraining orders and the like. Jurisdiction arising under the various statutes is exclusive, saving the right to appeal to an appellate court on an error of law or a question of the court's jurisdiction.

The Queensland Court is an intermediate court in the state hierarchy. Furthermore, the Queensland Land Court has not been amalgamated into the new court, resulting in a system of environmental law that is still fragmented. However, the Planning and Environment Court in Queensland is developing along the lines of the Land and Environment Court, particularly now that it has exclusive power to impose maximum penalties for serious offences including penalties of up to five years imprisonment.

Under the Environment, Resources and Development Court Act 1993, South Australia established an environment court which commenced operation in 1994. It is a specialist court dealing exclusively in building, environmental and planning disputes quite separate from other courts. Criminal and civil enforcement proceedings may also be heard. However, the court does not have exclusive criminal jurisdiction, but its jurisdiction is concurrent with other courts. Nevertheless, there have been a number of prosecutions in the court for environmental offences, and it is likely that the number will increase. The court, of intermediate status, is presided over by legal and non-legal personnel including two judges, two commissioners, a magistrate/master and 24 part-time commissioners.

## 3. The Development of Environmental Law

An important catalyst for the development of environmental law in New South Wales has been the right of an individual or a body of persons to initiate environmental litigation in the public interest. Since the early 1980s, third parties have been able to commence proceedings in order to restrain a breach of an Act which may cause environmental harm. Whereas traditionally a direct right or interest was required to invoke standing (particularly in a superior court), wider awareness of the public interest nature of protection of the environment has resulted in legislative reform acknowledging and accommodating public interest litigation. Indeed these New South Wales provisions, that were innovative and revolutionary in Australia at the time, have provided a model, which has been replicated in many states. The consequence of open standing provisions is that they provide the opportunity for proceedings to be taken in the Land and Environment Court which analyse and test relevant legislation and such proceedings provide the impetus for the development of environmental jurisprudence.

One of the benefits of a specialist court has been that the judges acquire specialist expertise and a large body of case law has been developed in a number of key areas including polluter pays, custodial sentencing for serious environmental offences, open standing provisions for public interest litigation, transboundary pollution and principles of ecologically sustainable development including the precautionary principle. The development of legal principles by the Court is crucial to the evolution of environmental jurisprudence. In many instances, a line of authority has developed, and the judges have managed to push the boundaries, instigating change over time. In this paper, I consider some of that developing environmental jurisprudence.

## 4. Environmental Prosecutions

In 1997, Justice Lloyd of the Land and Environment Court in the case of the Environment Protection Authority v Gardner (No 50072 of 1996 and 50074 of 1996), imposed the first custodial sentence in New South Wales for an environmental offence. Mr. Gardner was convicted under s 5(1) of the Environmental Offences and Penalties Act 1989 for the offence of willfully disposing of waste in a manner which was likely to harm the environment. He was sentenced to 12 months imprisonment and fined $250,000, the maximum penalty available for an individual.

An explanation of the facts demonstrates just why a sentence of such magnitude was imposed. Mr. Gardner owned and operated a caravan park in northern New South Wales on the edge of the Karuah River. The caravan park was not connected to the main sewerage system. Instead, sewage had to be held in septic tanks which required regular emptying about eight times per week via a tanker. In order to reduce the necessity for emptying to only three times per week and therefore to make a considerable financial saving, Mr. Gardner devised a system of concealed underground pipes to pump raw sewage from the septic tanks directly into the wetlands of the Karuah River. Mr. Gardner's actions did not come to light until after he sold the caravan park to a new owner, who discovered and reported the underground system. It was estimated that approximately 130,000 litres of untreated effluent per week had been pumped out via this illegal system over a period of 128 weeks. Based on evidence of viral contamination and significant degradation near the outlet pipe, and the serious health risk to oyster farms in the vicinity, the judge had little difficulty in establishing the presence of environmental harm.

In sentencing the defendant, Justice Lloyd. said :

> "This case contains a number of aggravating features. Your actions were not an isolated or single act of pollution, as are most cases that come before the Court. It was a deliberate act repeated a number of times a week for the 128 weeks of the offence period. ... It was ... done for the motive of financial gain. It had the most serious consequences of environmental harm and likely environmental harm imaginable. Moreover, harm to the environment in this instance affects not one or two people but the community as a whole. You were aware that it would cause harm to the environment. You were aware that what you were doing was illegal. You went to a great deal of trouble to conceal what you were doing. I cannot imagine a worse case than this." (EPA v Gardner, 1997, pp 4-5).

This case was a landmark decision because it indicated that the Land and Environment Court is willing to impose a maximum penalty where extensive environmental pollution is perpetrated in a deliberate and dishonest manner. It also reflects public perception, which is the seriousness of environmental crime and that, rightfully, the "polluter pays". This decision attracted widespread interest and media coverage and it is likely to have significant future deterrence value for individuals and industry alike.

Important though the decision in Environment Protection Authority v Gardner is, it is only one of a large number of environmental prosecutions which come before the Court. The principal prosecutions are those for pollution of water (under the Clean Waters Act 1970), for air pollution (under the Clean Air Act 1961), for willful or negligent causing a substance to leak, spill or escape in a manner which harms or is likely to harm the environment (under the

Environmental Offences and Penalties Act 1989) for pollution from spills of polluting material into State waters, including Sydney Harbour and the sea (under the Marine Pollution Act 1987) and for noise pollution (under the Noise Control Act 1975).

The maximum penalty which the Court may impose (which the legislature has prescribed in the case of some offences which are the most grave) is $AUD 1,000,000 in the case of a corporation, and $AUD 250,000 or seven years imprisonment, or both, in the case of an individual. The Court may also make orders for remediation or mitigation of the harm caused to the environment by the commission of any offence.

For less grave offences the Court may impose a maximum penalty of $AUD 125,000 in the case of a corporation, and $AUD 60,000 in the case of an individual. In such cases, the Court has, in cases over the last few years, imposed penalties ranging from 4% to 80% of the maximum. An example of this kind of offence is the case of Environment Protection Authority v Sydney Water Corporation (No 50019 of 1996) heard by Justice Talbot. in the Land and Environment Court.

The defendant operated sewerage treatment works in Sydney, in respect of which it had been granted a licence by the Environment Protection Authority. It was charged with a contravention of a condition of its licence which required it to maintain its plant in an efficient condition.

Justice Talbot. found that the defendant failed to comply with its obligation to maintain certain equipment in an efficient manner. The consequence was that untreated sewage discharged over a period of 56 hours into the ocean in the vicinity of sensitive water recreation areas.

Justice Talbot. said:

> "The ultimate result is that the defendant has been found guilty of an offence which led to environmental consequences of the most disgusting and abhorrent kind. The remedy was in its own hands. It is only through its own careless omission to properly check the efficiency of an essential integral part of the plant that the consequences occurred." (EPA v Sydney Water Corporation, 1997, p. 16).

The maximum penalty for the offence was $AUD 125,000. Justice Talbot. fined the defendant $AUD 100,000.

## 5. Open Standing

Section 25 of the Environmental Offences and Penalties Act 1989 permits any person to make an application to the Land and Environment Court to restrain a breach of an Act or statutory rule if such a breach is causing or is likely to cause harm to the environment. Such application may only be brought with the

Court's leave which is dependent on the Court being satisfied of certain matters such as the *bona fides* of the proceedings, their likelihood of success and that they are genuinely brought in the public interest (ss(3)).

The Land and Environment Court has construed s 25 liberally to give effect to its unequivocal purpose of removing barriers to the commencement of *bona fide* proceedings by third parties which are brought in the public interest. Brown v Environment Protection Authority (1992a), is the leading example.

Mr. Brown commenced proceedings against the Environment Protection Authority and North Broken Hill Limited (trading as Associated Pulp and Paper Mills) ("APM"). APM had operated a paper mill on the banks of the Shoalhaven River for many years. It had constructed, for the discharge of effluent, waste disposal outfalls or drains which protruded into the river. Mr. Brown sought a declaration of the invalidity of APM's licence granted by the EPA which permitted it to lawfully pollute via by release of a prescribed maximum amount of effluent into the river. His challenge was grounded on the allegation that pollution levels permitted by the licence exceeded target limits and contravened a ministerial directive as to licensing conditions.

In granting leave to Mr. Brown to bring the proceedings (1992b), Justice Stein. interpreted the statutory criteria for leave in a liberal fashion, requiring no more than the Court's satisfaction on the balance of probabilities that the case sought to be brought is within its jurisdiction. At the subsequent trial, I held that once leave is granted, the door is unlocked, and the trial should proceed without any question as the satisfaction of the statutory criteria for the grant of that leave. Furthermore, I held that the relevant statutory provision incorporated a wide definition of "environmental harm" and the Court could make whatever order it considered appropriate to ensure that whatever was causing or likely to cause environmental harm was discontinued.

## 6. Public Interest Litigation

Oshlack v Richmond River Council and Iron Gates Developments Pty Ltd., the 1994 decision of Justice Stein. (when he was a judge of the Land and Environment Court), was recently upheld by the High Court of Australia (the final court of appeal)(1998). Justice Stein's decision encapsulated the culmination of a long line of authority that had been developed over the years by judges of the Land and Environment Court. The Court's approach to exercising its discretion under s 69(2) of the Land and Environment Court Act 1979, when determining an application for costs by the successful party, favoured the consideration of the public interest nature of the litigation as a relevant factor to be taken into account.

The applicant, Mr. Oshlack, had commenced proceedings under the open standing provisions of the Environmental Planning and Assessment Act 1979, s 123 which provides the following:

> "Any person may bring proceedings in the Court for an order to remedy or restrain a breach of this Act, whether or not any right of that person has been or may be infringed by as a consequence of that breach."

Ultimately, the applicant was unsuccessful in his challenge to the validity of a development consent granted by the Richmond River Council ("the council") to Iron Gates Developments Pty Ltd., to subdivide and clear environmentally sensitive land.

The general rule in the Court (except in planning and building appeals) is that costs follow the event, that is the successful party is to be compensated by an award of costs in his or her favour. Justice Stein. held that, in weighing up all of the competing relevant facts and circumstances of the proceedings, including the public interest nature of the litigation, there should be no order as to costs. This had the effect of causing each party to bear their own costs of the litigation.

The decision was immediately challenged in the New South Wales Court of Appeal and the appeal was upheld. That led to an ultimate challenge in the High Court.

The High Court held that the fact that the proceedings had been brought in the public interest was a relevant factor to take into account in the exercise of a discretion to award costs. The judges of the High Court were at pains to point out that it is not a determinative factor, but one among others which may be considered.

The High Court decision is important because it recognises that the financial implications for public interest litigants bringing proceedings, which are attended by the risk of failure and the real possibility of a costs order against them, could act as a disincentive to utilising the open standing provisions for public interest litigation. This view of the futility of open standing provisions alone was articulated some 10 years ago by Justice Toohey. (former Justice of the High Court), in an address to an International Conference on Environmental Law:

> "Relaxing the traditional requirements for standing may be of little significance unless other procedural reforms are made. Particularly this is so in the area of funding environmental litigation and the awarding of costs. There is little point in opening the doors to the courts if litigants cannot afford to come in."

## 7. Transboundary Pollution

The concept of transboundary pollution between states was first recognised in the Land and Environment Court by Justice Cripps in Brownlie v State Pollution Control Commission. On appeal to the New South Wales Court of Appeal (1992), the central issue was whether a resident of the northern adjoining state of Queensland could be liable for engaging in conduct on his property which had the consequence of polluting the waters of New South Wales, contrary to the Clean Waters Act 1970.

The conduct concerned the spraying of a cotton crop with the insecticide endosulfan, on a property in Queensland where the boundary between New South Wales and Queensland was in the middle of a river. Following spraying, there was heavy rain and soon afterwards thousands of dead fish were found in the river.

Chief Justice Gleeson dismissed the appeal, upholding Justice Cripps' finding that transboundary interstate pollution was culpable under New South Wales law because of the nature of the Clean Waters Act which created "result offences". By characterising the offence in this way, the Court of Appeal held that the Clean Waters Act could apply to acts or omissions outside New South Wales, so long as those acts were likely to have adverse environmental consequences in New South Wales. Chief Justice Gleeson said the following:

> "... offences sometimes relate to conduct which may take place across territorial boundaries. Upon the basis of the so-called "terminatory theory" of criminal jurisdiction, there is no doubt about the jurisdiction of a New South Wales court to try a person who is charged with a "result offence" where the result is one that occurs, or, is likely to occur, in New South Wales, even though the acts bringing about that result took place outside New South Wales ...
>
> Where a certain result is an essential part of conduct constituting a given offence, then that conduct may be relevantly regarded as local if the result in question is one occurring within the territory in question." (Brownlie v State Pollution Control Commission,1992, p. 424).

In the judgment (at p. 425) reference is made to the established international environmental law principle of transboundary pollution, affecting relations between neighbouring sovereign states. Although cases have been successfully prosecuted at the international level, the difficulty which arises is whether a sovereign state should punish persons for conduct that has caused transboundary pollution where their conduct has had no harmful consequences which are culpable under that state's law. However, Chief Justice Gleeson drew the distinction between sovereign states and bordering territories within the federation of Australia, and focused on statutory interpretation to determine the legislative intent of the Clean Waters Act, which he found to be the purpose of preventing the occurrence of pollution in New South Wales.

## 8. Ecologically Sustainable Development Principles

The objective of ecologically sustainable development principles (ESD) has been incorporated into much environmental legislation in New South Wales. But this has been a recent development. Prior to those principles having statutory force, however, the Land and Environment Court had cause to consider their application, at least in regard to the precautionary principle.

That principle was recognised as having persuasive merit in Leatch v Director General National Parks and Wildlife Service and Anor (1993) and Greenpeace Australia Ltd v Redbank Power Pty Ltd and Anor (1994).

Leatch v Director General National Parks and Wildlife concerned the proposal of a local council in New South Wales to site a new road in an area which provided habitat to endangered fauna, including the giant burrowing frog. The council submitted a fauna impact statement to the National Parks and Wildlife Service ("NPWS") as required by the Endangered Fauna (Interim Protection) Act 1991. The NPWS then issued a licence to "take or kill" the endangered species, enabling the council to proceed with construction of the road.

The applicant, Ms. Leatch, appealed against the granting of the licence. The task of the Court, under the legislation as it then stood, was to consider the application for the licence on its merits, and to grant or refuse the licence. The Court refused the licence. Justice Stein. noted that there was almost no evidence of the population, habitat and behavioural patterns of the giant burrowing frog, and that accordingly it could not be concluded with any certainty that a licence to "take or kill" should have been granted. In reaching his decision, Justice Stein gave due consideration to the relevance of the precautionary principle as a factor for decision-makers to take into account under endangered fauna legislation such as the National Parks and Wildlife Act.

The precautionary principle was subsequently considered in Greenpeace Australia Ltd v Redbank Power Company Pty Ltd and Anor. Greenpeace Australia Ltd appealed against the grant of a development consent by a local council to the Redbank Power Company for the construction of a power station in the Hunter Valley in rural New South Wales. Greenpeace contended that the impact of air emissions from the project, particularly carbon dioxide ($CO_2$), would unacceptably exacerbate the "greenhouse effect" in the earth's atmosphere, while Redbank contended that the project would have an environmentally beneficial effect in utilising tailings, a sludge-like by-product of coal washing.

Although I found that the relevant statutes did not at that stage expressly incorporate ESD principles, I applied the precautionary principle in the following way:

> "Greenpeace's contention was that scientific uncertainty should not be used as a reason for ignoring the environmental impact of $CO_2$ emission. In other words, the Court should take into account the 'precautionary principle' ...
>
> There are, however, instances of scientific uncertainty on both sides of the issues in this case. For example, Redbank has contended that tailing dams pose environmental problems, whilst Greenpeace has denied that there are serious environmental problems surrounding current methods of tailing disposal. On the other hand, Greenpeace has asserted that $CO_2$ emission from the project will have serious environmental consequences, whilst Redbank has asserted that there is considerable uncertainty about its consequences ... The application of the precautionary principle dictates that a cautious approach should be adopted in evaluating the various relevant factors in determining whether or not to grant consent; it does not require that the greenhouse issue should outweigh all other issues". (Greenpeace Australia v Redbank Power, 1994, p 154).

Subsequent to these two decisions, statutory amendments and a raft of new legislation in New South Wales expressly require consent authorities to consider ESD principles. From this perspective, it would appear that the approach taken by the Court in the development of the common law demonstrated foresight which is likely to have been influential in a number of ways, firstly, upon consent authorities such as councils (in the case of Greenpeace) and upon the National Parks and Wildlife Service (in the case of Leatch) and of course further afield, but, secondly, and importantly, upon the wider community and the legislature.

Another example of the development of relevant environmental principles by the Land and Environment Court is its approach to the contents of an environmental impact statement (EIS) which is required in connection with environmental impact assessment. Such assessment is mandated, for example, by s 111 of the Environmental Planning and Assessment Act 1979, which provides that, in connection with certain activities, the consent authority must "... examine and take into account to the fullest extent possible all matters affecting or likely to affect the environment by reason of that activity", and by s 112, which calls for the production of an EIS.

A long line of authorities has laid down a number of relevant principles. It has been held that, in the context of the statutory provisions, the purpose of an EIS is to alert the approval authority and the public to the inherent problems of the proposed development, to encourage public participation in the decision making process by public exhibition of the EIS, and to ensure that the approval authority takes a hard look at what is proposed. The EIS must therefore be comprehensive and objective in its approach.

The Court's approach reflects an understanding that environmental impact assessment is an important feature of environmental protection, and that environmental impact statements are key documents that are subject to the scrutiny of the Court.

## 9. Conclusion

These examples demonstrate the value of specialist judges within a specialist court whose knowledge and expertise continue to play a key role in the evolution of case law in environmental jurisprudence. Initially, the Land and Environment Court had its detractors and from time to time there have been arguments that environmental law should be integrated rather than segregated. This argument was often raised over concern that environmental issues might somehow be marginalised. However, the opposite is true and in the 20 years that the Court has been in existence, awareness of environmental law issues has been raised significantly amongst government, industry and the community in general. Undoubtedly, the open standing provisions and the public interest considerations of the Court have been an influential factor.

Additionally, it can be said that the substantial criminal penalties imposed by the Land and Environment Court for serious environmental offences have been responsible for fostering the public perception that environmental harm is a crime. The Court has consistently imposed penalties in line with precedent and has not departed from its duty to impose penalties with regard to their deterrence value.

On a procedural level, the underlying philosophy of the Land and Environment Court, (which is reflected in the Land and Environment Court Act 1979 and particularly in its rules), is to settle disputes quickly and efficiently with as little unnecessary formality as possible. There has been a deliberate attempt by its creators to dispense with cumbersome rules of legal practice where they are unnecessary. For example, the rules of evidence do not apply to merits review proceedings (in classes 1 and 2 of its jurisdiction). Similarly, the Court has endeavoured to curtail litigation by ambush by promulgating rules of the Court which set out the procedures for defining the issues and for the filing and serving of evidence. In addition the Court also offers a comprehensive mediation service where parties can, alternatively, elect to mediate their disputes with a trained mediator.

Overall, the Land and Environment Court has been an overwhelming success and it has become a model for other States to follow not least in terms of costs, efficiency and justice, but spectacularly in terms of the evolution of environmental jurisprudence.

## Acknowledgements

I record with thanks the assistance I have received in the preparation of this paper from my research assistant, Ms. Katherine Gardner LL.B (Hons). I also record with thanks the information furnished to me about their respective courts by Judge Charles Brabazon QC of the Queensland Planning and Environment Court and by Judge Christine Trenorden, of the South Australian Environment, Resources and Development Court.


## References


Brown v Environment Protection Authority and Anor: 1992a, 78 LGRA, 119.

Brown v Environment Protection Authority and Anor: 1992b, 75 LGRA, 397.

Brownlie v State Pollution Control Commission: 1992, LGRA, 419.

Environment Protection Authority v Gardner: 1997, 7 November, Justice Lloyd, NSWLEC (unreported).

Environment Protection Authority v Sydney Water Corporation: 1998, 11 December, Justice Talbot. NSWLEC (unreported).

Greenpeace Australia Ltd. v Redbank Power Pty. Ltd. and Anor: 1994, 86 LGERA, 143.

Landa, D.P.:1979, 17 April, *Hansard*, NSW Legislative Assembly, 3349-3350.

Leatch v Director General National Parks and Wildlife Service and Anor: 1993, 81 LGERA, 270.

Oshlack v Richmond River Council and Anor: 1994, 82 LGERA 236.

Oshlack v Richmond River Council: 1998, 193 CLR, 72.

Richmond River Council v Oshlack and Ors: 1996, 39 NSWLR, 622.

State Pollution Control Commission v Brownlie:1990, 2 August, Justice Cripps, NSWLEC (unreported).

Toohey, J.L.:1994, in 82 LGERA 236, 238.

# PRACTICE, POLICY, AND PEDAGOGY IN A MANDATORY ENVIRONMENTAL LAW COURSE

M. A. WOLF[1] and J. B. EISEN[2]
[1] *Professor of Law and History,* [2] *Associate Professor of Law*
*University of Richmond, School of Law, Richmond, VA 23173, USA*
*(email: mwolf@richmond.edu, jeisen@richmond.edu)*

**Abstract.** All students at the University of Richmond School of Law are required to complete a first-year course in Environmental Law, as an introduction to the strategies and technicalities of statutory and regulatory regimes. This requirement poses special pedagogical challenges for the instructors. How, for example, is the class to address pollution questions regarding diverse media? Which of the many federal and nonfederal statutory provisions deserve the most (and least attention)? How does the instructor hold the attention of, and gain the most effective work product from, students whose major interests lie outside the area of Environmental Law? What is the proper mix of statutory, regulatory, and judge-made (pre- and post-regulatory) law?

Our solution has been to craft a unique combination of traditional lectures and readings regarding specific statutes, regulations, and case law; weekly discussion questions that immerse the students in the intricacies of statutory and regulatory provisions; a series of small-group simulations designed to highlight substantive legal points and to place the students in a wide range of legal and extralegal settings (client interviews, appellate argument, advocacy before administrative bodies, negotiation, lobbying public officials, legislative drafting); and periodic written assignments that focus the students' attention on technical legal and public policy issues.

**Keywords:** environmental law, legal education, pedagogy, regulations, simulations, statutes

## 1. Introduction

A decade ago, Joe Sax, one of the most prominent American environmental law professors and scholars, surveyed his colleagues regarding their teaching experiences. The news from the front lines was not encouraging:

> The subject seems to have overwhelmed us. Virtually every law teacher–however broad his or her overlook–wants to introduce students to the specific materials in the field, and to provide some experience and familiarity with it. Yet, every such attempt is an encounter with statutes of numbing complexity and detail. . . .
>
> Complexity as such does not seem to be the problem. Lawyers enjoy puzzles. What discourages law teachers is rather a sense of being drawn into a system in which enormous energy must be expended on something that is ultimately vacuous. . . . (Sax, 1989).

In 1992, despite this discouraging appraisal, faculty and students at the University of Richmond embarked on an experiment in learning environmental law, as a required, first-year course.

*Water, Air, and Soil Pollution* **123:** 409–418, 2000.

This essay is intended to share some of the lessons learned from this joint enterprise with others active and invested in the future of environmental law. The text reviews the faculty rationales for requiring a first-year environmental law course, the pedagogical decisions the instructors made in fashioning the course, and the special challenges posed by this endeavor.

## 2. Curricular Shift: Rationales for a Regulatory, First-Year Course

Environmental law joined the legal education canon at the University of Richmond as part of a comprehensive curricular revision enacted during the 1990-1991 academic year, after the requisite retreat, committee compromise, and drawn-out gatherings of the full faculty. The Curriculum Committee's successful recommendation to the faculty included the following in its straightforward, if not simple, charge:

> While the Committee felt confident that traditional legal analysis was being well taught in the first year, there is concern that the first year curriculum reflects or expresses a view of the legal world that no longer exists in that it is primarily a common law curriculum. To the extent statutes are considered, the statutes are designed to regulate behavior among individuals as opposed to addressing the impact of statutory and regulatory schemes which create public rights and responsibilities. . . .
>
> The consensus of the Committee was that environmental law was both *suitable for pedagogical reasons* and *important enough for substantive reasons* to merit addition to the first year. Environmental law allows the systematic study of a statutory scheme allowing *development of statutory interpretation skills* and allow[ing] rigorous study of *how such schemes develop and evolve* (a process quite analogous in the statutory field, as, say, the development of product liability in torts, landlord tenant in property, or personal jurisdiction in civil procedure). The course would also *introduce students to the regulatory aspect* with an introduction to *administrative enforcement*. (Curriculum Committee, 1990, emphasis added).

The Committee also included a suggested course description, which drew heavily on the description of Columbia law school's then-new, first-year offering titled "Foundations of the Regulatory State." The following phrases peppered the description: "justification for regulation," "market failure and externalities," "command and control regulation vs. market-based incentives," and "regulation and redistribution" (Curriculum Committee, 1990).

In essence, environmental law was selected as the new substantive, first-year course because of its pervasiveness and its suitability as the ideal vehicle for introducing students to the nuances of the regulatory state. The relevance and timeliness of the topic also played a significant part in achieving majority approval of the faculty. The two instructors who designed the course aimed to meet goals articulated by the committee and endorsed by the faculty. Drawing directly on the faculty's charge, the course was crafted to meet student needs in three areas: (1) understanding statutory and regulatory implementation, interpretation, modification, enforcement; (2) appreciating the pervasiveness of environmental regulation in the American legal system; and (3) considering alternative strategies for balancing private rights and public needs.

### 3. Eight Pedagogical Decisions

The instructors, with the assistance of Professor Sax (whose visits were funded by a local environmental foundation), confronted a number of preliminary, pedagogical issues. First, as this was an introductory, first-year course, providing the students with substantive depth was not as important a goal as it is in upper-level, specialty courses. Instead, the theme of the course would be to introduce various devices in the regulatory "toolbox," such as technology controls, ambient standards, incentives and pollution trading, environmental impact statements, outright bans, pollution taxes, and criminal sanctions. For example, instead of exploring state and federal clean air regulations in great detail, the notion was to select one representative tool from the regulatory regime, such as the ambient clean air standards contained in Title I of the federal Clean Air Act and accompanying regulations.

The second challenge was to decide upon the proper mix of environmental topics. The instructors' desire was to have the students confront a meaningful mix of topics regarding media (air, water, waste), procedure and jurisdiction (Administrative Procedure Act, standing), constitutional and common-law (regulatory takings, nuisance), and intergovernmental relations (local and state land-use and environmental controls). As is often the case in introductory courses, there was a constant tension between the aspiration to present the students with a representational landscape of the area of knowledge and the need to furnish meaningful depth.

Third, while the faculty's undeniable mandate was to create a course with a heavy emphasis on statutes and regulations, the instructors still felt compelled to provide a background for the students in the kind of judge-made law that has preceded, succeeded, and run parallel to the significant amount of legislative and administrative activity at all levels of American governance. It would be left to instructors in other courses to make sure that students mastered the art of interpreting judicial opinions. The focus in environmental law would be to provide understandings of the common-law analogues for modern environmental regulation and the ways in which judges, on the one hand, contributed substantive provisions in the event of (often purposeful) statutory ambiguity and silence and, on the other hand, erected significant barriers (chiefly demanding standing requirements) to persons and organizations seeking judicial relief.

The fourth challenge was selection of required texts that suited the instructor's goals and matched the level of student knowledge. Nearly all of the commercially available, law school texts were designed for second- and third-year students and took an "in-depth" approach, typically devoting hundreds of pages to providing details regarding the substantive provisions, operation, interpretation, and enforcement of different federal acts regulating air and water pollution and waste treatment and disposal. Given this mismatch, the initial plan was to compile an original set of course readings–statutes, regulations, cases, testimony before legislative bodies, excerpts from scientific and popular journal articles – that met the substantive and pedagogical goals discussed previously.

Fortunately, a new environmental law text appeared about a year before the class was scheduled to begin. That volume, *Environmental Law and Policy: A Coursebook on Nature, Law, and Society*, took a new approach to the topic:

> In the face of the numbing mass and complexity of modern environmental law, this coursebook uses the structure of the legal system as its organizing principle, selecting the best examples of how the process works, including a sampling of classic environmental cases, without particular regard for the type of pollution or policy involved (Plater *et al.*, 1992).

A chief component of the text is its "Taxonomy of Environmental Statutes":

> No one book can cover all environmental statutes – the Clean Air Act, Clean Water Act, or Superfund could each more than fill an entire semester of readings. The compromise chosen here is to study the selected statutes, analyzing their structures, process, and enforcement mechanisms, their successes and failures, and then to generalize from them. Each statutory type emphasizes a particular strategy for implementing environmental quality, each presenting tactical advantages and disadvantages. (Plater *et al.*, 1992)

This representative approach, along with the fact that the coursebook originated in materials developed for undergraduate and graduate students (that is, non-law students), made *Environmental Law and Policy* a highly suitable basic text for Richmond's new, first-year course.

To accompany the coursebook, the instructors assigned a volume of selected federal environmental statutes and regulations and prepared a set of additional readings, consisting chiefly of additional cases, regulations, and supplemental readings. Two years ago, primarily because the published coursebook failed to keep up with rapidly changing environmental law developments, the instructors substituted a new text. This book provides short summaries of several environmental law topics, followed by detailed examples and explanations. The instructors also developed a set of specially edited cases (roughly two per week) that were distributed on the noticeboard and web site.

The fifth decision regarding the design of the new course involved classroom format. Would the socratic method, typically found in first-year classes in American law schools, be appropriate for a regulatory, as opposed to common-law, course? In other first-year classes, the focus of class discussion is typically on the logic, rationale, and structure of appellate judicial decisions. Moreover, it is not unusual for casebooks to feature dissenting as well as majority opinions, sending the not-too-subtle message to students that often there is no one, "right" answer. The instructors anticipated that, while some class discussion would focus on judicial interpretation, students in this environmental course would more often be asked to provide citations to specific statutory or regulatory provisions. Indeed, the chief skill that the instructors hoped to emphasize was the ability to maneuver through and interpret an intricate body of legislative and administrative provisions that were often closely interrelated and just as often apparently contradictory.

These significant differences in course content and goals led to the sixth decision–to devote a significant amount of class time to a weekly set of discussion questions (Exhibit I) that would be distributed via electronic noticeboards on the law school computer network. This means of communication was made feasible by the University of Richmond's decision to be the first law school in the nation to require each student to purchase and use laptop computers and by the fact that each entering student was provided with an individual work space in the law school (typically a large carrel in the law library) with electrical power and an internet connection. In addition, students and faculty were given remote, call-in access to the law school network. In recent years, the instructors have also distributed discussion questions, along with other course materials, on a web site carried on the law school's intranet.

Some discussion questions posed hypothetical sets of facts involving pollution and government controls; others introduced real-world factual situations gleaned from recent newspaper articles. Although some questions asked students to consider the political and social implications of governmental and private-sector decision, the inquiries were designed primarily to immerse the students in the statutory supplement in search of specific provisions that were relevant to the facts at hand (real or imaginary).

The seventh and most innovative element of the new course was the requirement that students participate in "Environmental Law Labs" (Exhibit II) in addition to their three hours of weekly class meetings. Each late Wednesday afternoon, nearly every available classroom, large and small, was set aside for four-or five-person groups of first year students. The students would participate in fifteen-minute simulations designed (1) to review substantive concepts addressed in assigned readings and discussion questions, and (2) to introduce the wide range of professional settings in which lawyers and their clients face environmental issues (interviewing and counseling clients, lobbying government officials, arguing before regulatory agencies and appellate courts, engaging in negotiating and collective bargaining sessions, advising corporate directors). These simulations – distributed over the network noticeboards (and later on the intranet web site) – would take place nearly every week, with the instructors often providing secret facts designed to stimulate on-the spot problem-solving and creative advocacy.

The instructors decided to provide feedback to students on their performance in completing the simulations in a number of ways. First, the instructors would observe and evaluate students completing the problem for that week. Second, the instructors or their research assistants would videotape selected groups completing their problem in the Moot Court Room, Jury Room, or Judge's Chambers. A schedule would be set in advance that ensured that each student would be videotaped doing a problem one time during the semester. In later years, students were asked to provide two videotapes to their instructors, from the first and second halves of the course.

Exhibit I
Sample Discussion Questions

---

The following article appeared in the January 29, 1998, *Richmond Times-Dispatch*:

**4 STATES SEEK STUDY OF RISKS FROM PFIESTERIA**
**CDC MAY FUND 5-YEAR EXAMINATION**
**BY LAWRENCE LATAN III**

Virginia and three other coastal states are planning a major study of the possible human health effects of a fish-killing microbe that appeared in the Chesapeake Bay last year. The organism, Pfiesteria piscicida, was linked to a host of physical and cognitive problems in several people. But because states like Virginia and Maryland reviewed their cases differently, little is known about Pfiesteria's risk to humans. This year's proposed study would ally Virginia, Maryland, North Carolina and Delaware in a coordinated effort funded through the Centers for Disease Control. . . .

Pfiesteria burst upon the scene last August when it killed thousands of fish in the Eastern Shore's Pocomoke River that straddles the Virginia-Maryland line. Soon after, Maryland medical researchers documented skin rashes and memory problems in two dozen people who had been on the river. . . .

The Center for Disease Control and Prevention has the authority to award the states up to $450,000 each for the study this year. Lawrence Posey of the health studies branch at the National Center for Environmental Health said the study is expected to take five years. The center, he said, hopes to approve money for the study by early March. . . .

**A.** Apparently, the Department of Health and Human Services (HHS) did not prepare an EA or EIS for the CDC program. Imagine that a group of concerned citizens called ESPN! (Everybody Stop Pfiesteria NOW!) comes to your law office, asserting that HHS has failed to comply with NEPA. Given what you now know about the program, pfiesteria, and NEPA (from your assigned readings for the weeks of January 26 and February 2), be prepared to show how specific provisions of NEPA, the CEQ regulations, and NEPA case law regarding each of the following concepts apply to the potential challenge ESPN! wishes to bring: Ripeness, Proposal, PEIS, Small Handle

**B.** Now is your chance to show your creativity. Make up additional facts that would make each of the following NEPA concepts relevant to ESPN!'s challenge. Here's an example: Assume that the concept is "connected action." The additional facts could be that the pfiesteria program is part of a new cost-cutting approach at HHS to "farm out" epidemiological research projects involving environmental risks to state agencies, agencies that traditionally do not have the technological expertise of CDC employees. Here are the other concepts: Cumulative Actions, Similar Actions, Tiering, Segmentation, Statutory exemption, Statutory conflict

---

The eighth and final issue confronted by the instructors concerned evaluating student performance, specifically providing graded exercises in addition to the traditional, cumulative final examination. Eventually, the instructors settled upon an additional feedback mechanism. Most of the simulations would be revised to require one or two group members to prepare a written document of some kind (for example, a memorandum of understanding, a draft of a statute or regulation, or a letter to a client). Each student would submit (by e-mail) a few of these short papers during the term, and the instructors would review and grade these documents.

## 4. Special Challenges

It should be evident that this new course would be particularly labor-intensive for students and instructors. Students would be asked to prepare for class by reading a wide array of judicial, legislative, administrative, scholarly, and extra-legal materials, to participate in weekly labs, and to prepare written answers to graded exercises throughout the semester. For the instructor, the tasks of drafting the problems and discussion questions, selecting and training upper-level students to provide assistance in the course, coordinating the administrative details required to administer the problems and discussion questions, providing feedback to students,

Exhibit II
Sample Simulation

---

**SETTING WETLANDS POLICY**
You are all members of a working group convened by U.S. Army Corps of Engineers (Corps) and charged with the task of reviewing a redraft of portions of the federal regulation defining "waters of the United States" (located at 33 C.F.R. § 328.3), a regulation that was held to exceed the congressional authority of the agency in *United States v. Wilson*, 133 F.3d 251 (4th Cir. 1997). In the *Wilson* case, the Corps of Engineers had used this regulation to regulate dredging activities in wetlands isolated from freestanding bodies of water.

**The Lab Exercise**
Person 4 in your group is the Corps representative. The Corps representative will go first and will run the meeting. The Corps representative is instructed to take no more than three minutes to introduce the proposed redraft and explain the nature of the changes and why they were made. Person 4 will then hear comments on the proposed redraft from each other member of the group. **There are secret facts for Person 4.** The website for this organization is http://www.usace.army.mil/.

Person 3 in your group is a member of The National Wetlands Coalition (http://www.thenwc.org/). At the meeting you are to review the Corps' draft, discuss the Coalition's position on the proposed changes, suggest new or amended language, and contest suggestions offered by others that you deem unwise.

Person 1 in your group is a representative of the federal Environmental Protection Agency (EPA) (http://www.epa.gov/owow/wetlands/). At the meeting you are to review the Corps' draft, discuss

EPA's position on the proposed changes, suggest new or amended language, and contest suggestions offered by others that you deem unwise.

Person 2 in your group is a member of the National Audubon Society (http://www.audubon.org/campaign/wetland/index.html). At the meeting you are to review the Corps' draft, discuss the Society's position on the proposed changes, suggest new or amended language, and contest suggestions offered by others that you deem unwise.

Here are excerpts from the actual regulation:

§ 328.3 Definitions.
For the purpose of this regulation these terms are defined as follows:

(a) The term *waters of the United States* means

(1) All waters which are currently used, or were used in the past, or may be susceptible to use in interstate or foreign commerce, including all waters which are subject to the ebb and flow of the tide;

(2) All interstate waters including interstate wetlands;

(3) All other waters such as intrastate lakes, rivers, streams (including intermittent streams), mudflats, sandflats, wetlands, sloughs, prairie potholes, wet meadows, playa lakes, or natural ponds, the use, degradation or destruction of which could affect interstate or foreign commerce including any such waters:

(i) Which are or could be used by interstate or foreign travelers for recreational or other purposes; or

(ii) From which fish or shellfish are or could be taken and sold in interstate or foreign commerce; or

(iii) Which are used or could be used for industrial purpose by industries in interstate commerce;

(4) All impoundments of waters otherwise defined as waters of the United States under the definition;

(5) Tributaries of waters identified in paragraphs (a)(1)-(4) of this section;

(6) The territorial seas;

(7) Wetlands adjacent to waters (other than waters that are themselves wetlands) identified in paragraphs (a) (1)-(6) of this section. . . .

(b) The term *wetlands* means those areas that are inundated or saturated by surface or ground water at a frequency and duration sufficient to support, and that under normal circumstances do support, a prevalence of vegetation typically adapted for life in saturated soil conditions. Wetlands generally include swamps, marshes, bogs, and similar areas.

(c) The term *adjacent* means bordering, contiguous, or neighboring. Wetlands separated from other waters of the United States by man-made dikes or barriers, natural river berms, beach dunes and the like are "adjacent wetlands." . . .

**Secret Facts for Corps Representative (Person 4)**
Here is the change to the definitions found in 33 CFR Part 328 that the Corps is proposing (at the meeting on April 1, 1998):
33 C.F.R. § 328.3(a)(3) should now read:

All other waters such as intrastate lakes, rivers, streams (including intermittent streams), mud flats, sand flats, wetlands, sloughs, prairie potholes, wet meadows, playa lakes, or natural ponds *which are hydrologically connected to or within 100 miles of interstate waters that can actually be navigated,* including any such waters: . . . [the remainder of the subsection would remain intact]

**Written Problem for This Week**

After the working group meeting, **Person 2 and Person 4** will submit his or her own redraft of 33 C.F.R. § 328.3(a)(3) and (6). Your redraft must be different from the language submitted by the Corps at the Wednesday session and from the actual regulation. **Your redraft cannot exceed 100 words.** Then, **in 400 words or less**, you are to explain why you made the changes you did, and how your draft fits in with relevant provisions of the Clean Water Act and with the holding in *United States v. Wilson*.

---

and meeting the diverse goals set for the course by the law school faculty, have required an unusual amount of preparation. Moreover, as the instructors sought to make the course as topical as possible, it was often impossible to anticipate before the course began some of the specific topics that would be addressed in discussion questions and simulations.

For the most part, despite these additional demands, the course has proved to be quite stimulating for the five faculty members who have taught the course during its seven-year existence. This is true in large part because of the opportunity the course has provided for cooperative teaching and learning. The preparation of discussion questions, simulations, and examination questions has consistently been a joint effort, with various instructors bringing different pedagogical, experiential, and scholarly strengths to the table. For example, the first year of the course teamed a professor with nearly two decades of experience teaching and writing about environmental law topics with a junior colleague with teaching and writing interests in land-use planning, nuisance law, and urban redevelopment. The third tenure-track instructor brought to the course his significant experience as counsel on congressional committees with extensive environmental responsibilities and a scholarly interest in brownfields and alternative dispute resolution. One visiting instructor contributed insights gleaned for several years of environmental law practice in the private sector, while the other brought special skills in designing simulation exercises.

The first-year, mandatory environmental law course at the University of Richmond is very much a work in progress. The course has posed and continues to pose special – even unique – challenges to the instructors. First, upper-level, elective courses in environmental law are typically populated by students interested in environmental law and environmentalism. Unfortunately, a significant minority of the students in the mandatory course resent the environmentalist bias of the assigned readings and of their instructors. Throughout the course, the instructors need to remain mindful of this undercurrent, and to remind the students that, no matter what their feelings are regarding pollution, overconsumption, and development of raw land and natural resources, environmental law is a pervasive area of inquiry and practice that promises to be with American lawyers and politicians for decades (if we are lucky, centuries) to come.

Ironically, the first-year environmental law course at the University of Richmond is probably responsible for stimulating more paper usage that any other course in the entire university. While the instructors continually emphasize that many of the readings, particularly background readings for the simulations, are to be read selectively (and, in most cases, in electronic form), this advice is often ignored by students who are apprehensive about "missing out" on important information. The result has been a widespread perception of document overload. As the semester moves into the second month, students become less and less reliant on the printed page, as they learn the important skill of analyzing a large body of diverse materials and selecting out those items that are most relevant.

This mandatory course provokes a special reaction not only from those students who are "cool" to environmentalism, but also those who identify themselves as "green." At the beginning of the course, like their classmates on the opposite end of the enviro-political spectrum, green students often identify environmental law with environmentalism. They are then disappointed to discover that the practice of environmental law is much more tedious, intellectually demanding, and indirect than the kind of environmental activism practiced by mainstream and more radical, earth-first nongovernmental organizations. Environmental law battles are often won by delay tactics facilitated by manipulation of regulatory procedures, not by civil disobedience or mass demonstrations. Instructors who themselves are committed to environmentalist causes thus find it a special challenge to neutralize the disenchantment of their formerly enthusiastic students, to demonstrate through their scholarship and through their work with local and national organizations, that personal commitment to conservation, preservation, and anti-pollution causes is possible despite the morass of American environmental law.

The final challenge faced by instructors of Richmond's mandatory, first-year environmental law course – indeed by all environmental law teachers – has been to keep current with changes in judge-made, statutory, and administrative law, to anticipate the "hot topics" that will garner the attention of lawyers, activists, and politicians who are engaged in the field. The environment is an area that continues to receive a high degree of public and scholarly attention, to generate deep emotional responses,, and to stimulate a growing body of law. As the teaching and learning of environmental law continues in the new millennium, there is every indication that this is one challenge that all legal instructors will continue to confront, and, it is hoped, to meet more responsively, responsibly, and creatively.

## References

Curriculum Committee, University of Richmond School of Law: 1990, *1990 Curriculum Report.*

Plater, Z.J.B., Abrams, R.M., Goldfarb, W.: 1990, *Environmental Law and Policy: A Coursebook on Nature, Law, and Society*, West Publishing.

Sax, J.: 1989, *Environmental Law Reporter* **19**, 10251.

# THE GREAT SYDNEY WATER CRISIS OF 1998

P. L. STEIN

*Judge of New South Wales Court of Appeal, Supreme Court, GPO Box 3, Sydney, Australia, 2001*

**Abstract.** This paper traces the history of the contamination of Sydney's water supply between July and September 1998, its impact and consequences. Routine testing found persistently high readings of *Cryptosporidium* and *Giardia* in the water supply. After initial official inactivity, health warnings were issued leading to comprehensive boil water alerts. The paper examines the issues of scientific uncertainty and the response of the Government to the crisis. In particular, the paper examines the establishment and course of the Sydney Water Inquiry, chaired by Peter McClellan, QC, which delivered its final report in December 1998. The recommendations of the Inquiry are assessed, in particular the establishment of an independent catchment management authority, augmentation of treatment processes, and upgraded monitoring and research. Implementation of the recommendations by the Government by legislative and executive act is also addressed. Finally, the lessons to be learned are discussed, especially care of the catchment and restrictions on certain types of development within it. *Postcript*: Amazingly, no one got sick!

**Keywords:** *Cryptosporidium* contamination, *Giardia* contamination, recommendations arising out of inquiry, resource sustainability, Sydney Water Crisis, upgraded monitoring and research of water quality, water contamination, water quality

## 1. Introduction - Sydney and its Water Supply System

The provision of drinking water has been a key problem for communities for over 3,000 years, including, for example Greek, Roman and Mesopotamian civilisations. Late 20th century advances in science and technology have spurred government regulators (health and environmental) and societal concern with the quality of drinking water.

Sydney is a large modern city by any standard. It sprawls to the north, west and south and its population is close to 4 million. When foreigners think of Sydney they invariably conjure up images of its picturesque harbour, its beaches and the Olympic Games to be held in September 2000. Sydney Water (a state owned corporation) supplies 1600 million litres of water each day to 1.5 million properties in Sydney and its outlying areas. The city has a large and complex catchment with nine major dams and several storage reservoirs. About 21,000 km of water main, almost 200 pumping stations and many tunnels deliver water from four main river systems. Sydney Water is also responsible for maintenance of 15,000 km of private water services between the main and the meter. The water is filtered through eleven treatment plants. Seven are owned by Sydney Water and four are privately owned. These plants provide 90% of Sydney's drinking water and one plant, Prospect, provides up to 80%.

The purpose of this paper is to present the response of Government to Sydney's water crisis in 1998, and to increase the awareness and responsibility of all factors, decision makers and general public alike, to the vital issues revolving around water quality and sustainable development.

## 2. Australian Drinking Water Standards and Contamination Dangers

Drinking water standards in Australia are subject to the Australian Drinking Water Guidelines 1996, developed by the National Health and Medical Research Council (the NHMRC) and the Agriculture and Resource Management Council of Australia and New Zealand. Also relevant are the World Health Organisation's 1993 guidelines. The 1996 guidelines state that drinking water should be safe to use and aesthetically pleasing. However, water travels a long way before it reaches our taps. A number of problems can occur along the way. For example, water in storage reservoirs, rivers or bores may contain microbiological or chemical pollutants. Water treatment plants seek to rid these potential pollutants and upgrade the quality of the water. However, the chemicals used in this process may have the potential to cause their own problems. Treatments may fail and water pipes may be contaminated; even home plumbing may be a problem. One should also mention that there are policy objectives other than water quality in relation to the regulation of potable water. For example, its sustainability as a resource (National Audubon Society v Superior Court of Alpine County, 1983), its availability to consumers and its affordability.

The main dangers for water contamination are bacteria, blue-green algae, protozoa and viruses. The bacteria principally come from water contaminated by human and animal faeces. Coliforms are therefore routinely monitored by water supply authorities. Disinfection, mainly by chlorine, usually kills all bacteria. Blue-green algae (cyanobacteria) have become common in Australia. By way of illustration, in 1991-1992, more than 1,000 kms of the Murray-Darling River system were affected by algae bloom. Toxins may be produced which can damage the liver and nervous system. Removing the toxins requires special treatment and boiling the water will not help. Viruses can also be found in water. They are usually killed by disinfecting the water. The extent of this problem is unknown as the source of a viral infection is difficult to trace. Lastly, protozoa, a group of micro-organisms which include *Cryptosporidium* (relevantly *C. parvum*) and *Giardia*, can cause severe illness if above certain levels. They are often resistant to disinfection but must be filtered out of the water. There is no general agreement on the efficiency of water filtration to remove *Cryptosporidium* and *Giardia*.

## 3. Discovery of *Cryptosporidium* and *Giardia* in Routine Testing of Sydney's Water Supply

*Cryptosporidium* and *Giardia* were first detected in the water supply on 21 July 1998 (the first event). The levels did not then raise health concerns. Low levels of *Cryptosporidium* and *Giardia* are common in water supplies around the world. Testing continued at various points in the distribution system and, by 26 July, high and extremely high readings were reported. On 27 July 1998 the first precautionary boil water alert was issued for the eastern Central Business District (CBD). The incident was treated as localised. However, by late on 29 July 1998 high readings were found in samples at the Prospect Filtration Plant, in a reservoir and at a location further down the system. A new boil water alert was announced for the south of Sydney Harbour. On 30 July 1998 high readings were obtained from samples at Palm Beach, well to the north of the harbour. A Sydney wide alert was then declared. Virtually the whole population of Sydney was eventually required to boil water before drinking, even for cleaning teeth. An expert panel was formed to advise the Government on water safety and the crisis was placed under direct ministerial control.

On 4 August 1998 the water supply was declared safe. However, high readings were again found on 13 August, 1998 (the second event), although it was believed that most organisms were likely to be dead. More positive readings were found on 14 August 1998, although lower. Further contamination was identified on 24 August and an extended boil water alert was declared. This was progressively lifted suburb by suburb until further contamination was reported on 5 September 1998 (the third event). A two-week alert was then instituted. The alert was finally lifted on 19 September 1998.

The crisis caused a mixture of fear, cynicism and anger in the community. Consumer surveys of customer response to the events will be examined later in the paper. Claims for compensation have been brought against Sydney Water by in excess of 12,000 businesses and individuals. Schools, hospitals and other institutions were also severely affected. The cost of the crisis to Sydney Water is estimated at A$33 million which includes $20 million paid in rebates to customers, $13 million in lost revenue, water testing and staff costs and at least $2.5 million for damages claims. These costs do not include capital expenditures on improvements to the system and infrastructure referred to later in the paper.

The crisis, once over, has led to important changes to the structure of water delivery to consumers, to catchment management and for continuing research. The amazing fact was that there were no reported illnesses!

### 3.1. THE CAUSE OF THE CONTAMINATION AND THE SCIENTIFIC UNCERTAINTIES

An independent inquiry was ordered by the State Government. It was chaired by a prominent environmental lawyer, Mr. Peter McClellan, QC. The Inquiry published a number of reports between August and December 1998 making significant recommendations (91 in total) the majority of which have been accepted by the State Government and many already implemented. The Inquiry was well resourced, received more than 200 submissions, and interviewed in excess of 130 witnesses. It drew on the expertise of experts (local and international) on all relevant issues.

The McClellan Inquiry concluded that it was 'apparent that the catchment waters for much of Sydney's water supply contain significant sources of *Cryptosporidium* and *Giardia*. Heavy rains ... which followed a period of significant drought carried the organisms into the stored waters of [the] dams'. The Inquiry found that operational difficulties at the treatment plant allowed pathogens to pass in greater than usual numbers.

The Commissioner drew attention to scientific uncertainties which made it impossible to conclusively determine the actual levels of contamination. The second and third contamination events (out of three events) were caused by pathogens being washed into the catchment waters and passed through the treatment plant at a time when it was required to treat highly turbid water. It appeared that contaminated water had accumulated at a specific level of water density (a thermocline) in the dam and water was intermittently drawn into the offtake. One theory is that the level of organisms in the raw water was so great that the plant allowed high numbers to pass.

Mr. McClellan drew attention to the many scientific and medical uncertainties relating to *Cryptosporidium*, *Giardia* and water treatment. Notwithstanding that it was unlikely that any person suffered illness from ingesting *Cryptosporidium* and *Giardia* during the events, their potentially fatal affects demanded a conservative response. Given the present level of scientific knowledge, further research was mandated.

## 4. Management of the Contamination Events

Sydney Water and the Government came under heavy attack, particularly in the media, for the way in which the first event was handled. On 25 and 26 July 1998, the first extremely high readings of *Cryptosporidium* and *Giardia* were detected. The initial response of the Chief Executive Officer of Sydney Water (Mr Pollett) was to query the test results. The Health Department was notified of the readings. The inquiry found that as of late on 26 July 1998 the Health

Department failed to give consideration to the appropriate public health response.

The Commissioner stated in his Final Report (December 1998):

> I understand that officers of Sydney Water and NSW Health were faced with an unusual situation, one that they had never previously encountered. However, the extremely high level of independently confirmed results demanded a more urgent response. On any view, these levels justified consideration of a boil water alert and its rejection only after consideration at the highest level. These levels should have caused Sydney Water to respond by questioning the integrity of all components, including the filtration system. Appropriate sampling from the commencement of the distribution system at Prospect should have been instituted without delay. In addition, raw water and backwash water should have been sampled. (p 56).

On the morning of 27 July 1998 the CEO and the Chair of the Board of Sydney Water (Mr David Hill) met with the Minister for Planning for a regular briefing. The CEO did not advise Mr Hill or the Minister for Planning of the extremely high readings of the previous two days. By contrast, the Minister for Health and the State Premier had already been informed via the Health Department. The Commissioner was critical of Mr Pollett for failing to inform his Minister. A teleconference took place later on 27 July 1998 between Sydney Water and the Health Department. It was agreed that a precautionary boil water alert should be issued for the eastern CBD. Mr Pollett and Mr Hill were advised of this decision as well as the Minister. There then ensued a delay of around five hours caused by infighting between Sydney Water and the Health Department about the content of the media release. The Commissioner was critical of the delay, stating that it could have had serious health consequences. On 28 July 1998 the Health Department issued a Press Release on the Water Crisis. Amongst other things the release said:

> No relationship has been established between finding *Cryptosporidium* in drinking water at any level (in Australia or elsewhere) and effects on human health (Final Report, 1998, p. 61).

This paragraph was quoted in newspapers. The statement was not accurate and its publication caused Sydney Water to limit its media release on 29 July 1998 to *Giardia* only. By 29 July 1998 it was clear that the relationship between the media units of Sydney Water and the Health Department had totally broken down. At 5.30 pm high levels of contamination were found at the Prospect plant. The Health Department was not informed for four hours and the Commissioner was also critical of this. It was agreed that a new and more extensive boil water alert would be issued as a precautionary public health measure. However, once again an unacceptable delay occurred. The delay occurred because of highly charged arguments about the content of the warning. Mr Hill became involved doubting the wisdom of issuing a boil water alert without harder evidence. Whilst the argument was raging, the Health Department issued the draft Sydney Water press release. Pandemonium ensued with Mr Hill berating his own and Health Department officers. His instructions were to 'kill' the media release. The CEO and the Chair then set about redrafting the release. This included omitting references to the most potentially dangerous organism, *Cryptosporidium*. It would be facetious to say that the warning was 'watered down'.

Not surprisingly the media highlighted the apparent confusion within Government, describing the handling of the crisis as a 'shambles'. Following this unfavourable publicity, the Government announced the Sydney Water Inquiry headed by Mr McClellan QC, and the Minister assumed control. Unfortunately, this did not stop further misleading press releases, one suggesting that the Prospect Filtration Plant had been 'completely bypass[ed]' when it had not and never has been. Indeed, Mr Pollett reiterated this in an interview with ABC radio. The Commissioner found that the failure to provide prompt and accurate advice to the Minister was a serious breach of trust and the public were entitled to better advice from Sydney Water.

Mr. McClellan was highly critical of the management of the first event by the agencies. He said:

> Public confidence in the water quality will only be achieved:
>
> - if information about water quality is published; and
> - a transparent process for the issue of a health alert is determined.

Some of the problems indicate a lack of effective application of the incident management systems in place at the time. Others may have arisen from the Corporation's structure which required it to give equal consideration to its business objectives, protection of the environment and the protection of public health (Final Report, 1998, p. 89).

The Commissioner found that the decision to limit the alert was one whereby concerns for the reputation of the water supply corporation were favoured over public health interests. The blame for the wrong decision was placed at the feet of the CEO and the chair of Sydney Water. Subsequently, these officers (along with some other senior officers) resigned their positions.

## 5. Sources of the Contamination

The most significant cause of the events was contamination from sources within the catchment. The Inquiry found a number of sources of contamination in the Inner Catchment including cattle, residential development, native and feral animals. Given the role of the inner catchment as a filtration system, the continuation of human activity was of concern. The Outer Catchment constituted a long-term source of contamination. Catchment conditions, especially in the Outer Catchment, required immediate attention to control future development and protect water quality as a priority. While parasitic contamination was of concern to human health, also important was pollution of waters by pesticides, nitrates and phosphorous, industrial pollutants and oil and grease. Mining was also a potential threat to catchment waters, in particular coal mining (including many unworked and unrehabilitated mines) as well as base metal extraction and sand mining.

No definitive conclusions could be reached by the Inquiry on the exact sources of the *Cryptosporidium* and *Giardia*, however sewage treatment plants (STPs) constituted the highest risk. Of the nine plants within the catchment, many were performing badly. Discharges of poorly treated or untreated sewage were found to have occurred. In addition, some smaller package STP's were performing poorly and unsewered areas were experiencing septic tank seepage into creeks within the catchment. Sewer overflows were another problem. There are around 21,500 on-site sewage systems in the catchment, including two villages.

Agricultural activities posed another significant problem. These activities included sheep and cattle grazing, poultry farms and chicken hatcheries, piggeries, dairies, saleyards and abattoirs. Heavy rainfall caused run-off into rivers and creeks. The problem had been exacerbated by pasture and soil loss from a long drought and removal of riparian vegetation. The protection of this vegetation was crucial to water quality. Another deleterious by-product of agriculture was chemical contaminants.

A further issue was the levels of *Giardia* and *Cryptosporidium* in biosolids from sewage treatment. These provided a reservoir of inactive pathogens which could be released by rain and washed into watercourses. Stormwater run-off flowing directly into waterways was also a problem. Management of stormwater run-off was fragmented and standards were below Best Practice.

The Commissioner concluded that the state of the catchment posed serious risks for the safety of Sydney's drinking water. The highest risk sources were the STPs and unsewered areas. However, all of the sources constituted ongoing threats to water quality and required urgent attention. The catchment was compromised and there were multiple sources of *Cryptosporidium* and *Giardia* within it. The lengthy drought followed by significant rainfall mobilised pathogens in the catchment and transported them to the water storage dams.

## 6. Efficiency of the Treatment Plants and Public Health Impacts

It is clear that no treatment plant can guarantee removal of all *Cryptosporidium* and *Giardia* and that some of these organisms passed through the plants. Notwithstanding, a number of the treatment plants had significant problems which required immediate attention. One of the difficulties was the 'picture of uncertainty' which the Inquiry saw concerning the ability of analysts to accurately identify *Cryptosporidium* and *Giardia*. There was no method of testing which allowed for constant monitoring for parasites. Also, there were many problems with sampling techniques and testing. While there was dispute between experts and laboratories as to the accuracy and consistency of readings giving rise to the alerts, the Commissioner was satisfied that both oocysts and cysts were present in drinking water at levels of public health concern.

While there was no general agreement about the quality of the data generated by tests and uncertainty regarding their reliability, there were, without any doubt, *Cryptosporidium* and *Giardia* present in raw and treated waters. The public health response throughout the contamination incidents was found to be appropriate.

The public health impacts of *Cryptosporidium* and *Giardia* in drinking water can cause gastroenteritis which, in the case of immuno-compromised people, can cause intractable diarrhoea and even death. *Cryptosporidium* and *Giardia* in drinking water have been widespread throughout the world with massive outbreaks reported overseas. Following are some extracts from the final report regarding overseas outbreaks related to drinking water:

> Massive outbreaks of both cryptosporidiosis and Giardiasis have been linked to contaminated drinking water systems in recent years in both Europe and North America.
>
> Reported recent outbreaks of cryptosporidiosis due to contaminated municipal drinking water supplies include:
>
> - 13,000 people in Carrollton, Georgia, USA in 1987, due to suboptimal filtration of drinking water;
> - 15,000 cases in Jackson County, Oregon, USA in 1992 due to water treatment failures;
> - 403,000 cases among the 1.6 million persons in Milwaukee, Wisconsin, USA, in 1993 due to failure of effective filtration;
> - deaths of at least 20 persons with HIV infection in Las Vegas, USA, in 1994 (served by a state-of-the-art water treatment plant).
>
> Reported outbreaks of Giardiasis due to contaminated municipal drinking water supplies include:
>
> - 5,300 cases in Rome, New York in 1975;
> - 3,800 cases in Pittsfield, Massachusetts, population 50,000, in 1985-86 in chlorinated but unfiltered water supply;
> - from 1965-1984, 90 outbreaks of 23,776 cases of *Giardia*sis were reported in the United States, of which 69% of the outbreaks and 74% of the cases were linked to contaminated public water supplies, mostly from surface water supplies with inadequate filtration or chlorination;

- in 1993-94, five outbreaks of *Giardia*sis were reported in the USA, linked to community drinking water, and affecting 385 persons. (Final Report, 1998, p. 172).

The Commissioner concluded that the contamination of Sydney's water supply was unlikely to have caused any increase in infection with *Cryptosporidium* and *Giardia* but it was not possible to determine the reason for this. Nonetheless, the precautionary boil water alerts were found to be an appropriate response given the significant health risks to the population of Sydney.

## 7. The Regulatory Framework

Until 1995 Sydney Water was a statutory authority directly under ministerial control. It was both regulator and operator. Pursuing the mantra of microeconomic reform, it was seen as imperative to change its structure to introduce market pricing, based on user-pays principles, through a commercial framework and to separate the operator/regulator roles. Political compromise lead to Sydney Water being corporatised as a State Owned Corporation with many innovations brought about by government negotiation with lobby groups. For example, the governing statute established explicit environmental and public health objectives that had equal standing with the Corporation's commercial objectives. It was a novel deviation from the classic model of corporatisation and arguably, according to McClellan, left external regulators unprepared as to how to deal with this new 'creature'.

The Water Board (Corporatisation) Act 1994 provided for an Operating Licence to be issued setting forth the operating and customer standards to be met by Sydney Water. The Minister responsible for issuing the Licence was the Minister for Urban Affairs and Planning. That Minister was also responsible for the administration of the Corporatisation Act, as well as the planning laws. Mr McClellan was critical of this system and its implications for standards and catchment quality. The Commissioner saw the Operating Licence as an ineffective regulatory mechanism with insufficient accountability. Sydney Water essentially wrote its own licence with no requirement for public consultation. The Operating Licence was also clearly out of date by nominating the old 1980 NHMRC guidelines. Sydney Water has since agreed to endeavour to meet the 1996 NHMRC guidelines. The Licence also contained no requirements for *Cryptosporidium.*

Mr. McClellan believed that the Operating Licence should be replaced by a new licence with clearly defined roles and outcomes for Sydney Water and for the new Sydney Catchment Authority. He stressed the need for transparency and consultation in developing the new regime. The Commissioner also recommended amending the structure of Sydney Water to increase Ministerial direction and accountability.

## 8. Water Quality Standards

In the case of *Cryptosporidium* and *Giardia* it was difficult for the Inquiry to specify what standards should apply for operational levels. This was because of the state of scientific knowledge. Nonetheless, the Inquiry was of the opinion that water suppliers should be required to endeavour to produce drinking water free of viable *Cryptosporidium* and *Giardia*. The experience in the United Kingdom was useful to draw upon. The Commissioner recommended that the process of assessing and imposing operational water quality performance standards begin immediately and be informed by the current and continuing work of the NHMRC. The expressed hope was to impose a standard by the end of 1999. However, the acknowledged deficiencies in scientific knowledge have meant that this is not a feasible proposition. In this regard, it is of interest to note that the UK Expert Group on *Cryptosporidium* in Water Supplies recently concluded that it was not possible to recommend a health-related standard ($3^{rd}$ Report to the UK Government).

## 9. Monitoring for *Cryptosporidium* and *Giardia*

So far as my researches have been able to ascertain, there is no monitoring programme for *Cryptosporidium* and *Giardia* in the Northern Territory, Western Australia and parts of Victoria. Western Australia argues that its catchment practices negate the need for monitoring! Limited monitoring is undertaken in Tasmania and the Australian Capital Territory. More comprehensive monitoring is carried out in Brisbane, Melbourne and South Australia. Monitoring by water suppliers in New South Wales varies. The Inquiry recommended upgrading of Sydney Water's monitoring programme and to include an Event Monitoring Programme. There was also seen to be a need to develop a plan for the management of drinking water quality incidents. It was recommended that Best Practice Models be adopted.

## 10. Improvements to the System

The most effective approach to keeping *Cryptosporidium* and *Giardia* from a water supply was a multi-barrier one. This meant minimising its entering watercourses; managing storage to retain water as long as possible to allow for settlement and die-off; using a number of treatment processes and maintaining distribution systems. A number of improvements were recommended.

The principal improvement necessary was in catchment management. The present catchment was compromised and likely to further deteriorate unless action was taken to address sources of contamination. While it was acknowledged that the catchment would never be pristine, proper management would diminish the risk of contamination by *all* pollutants. The establishment of the Sydney Catchment Authority was an important step taken by the Government to implement an interim report recommendation on the crisis. But an accurate picture of the condition of the catchment was necessary, as well as research to understand and rank the risks posed by the diffuse sources of *Cryptosporidium* and *Giardia* and other pollutants. While it was essential to develop remedial and preventative strategies, there were other specific actions required. These included:

- enhancing the capacities of STPs in the catchment;
- accelerating the sewering of urbanised areas close to the main storage dam at Warragamba;
- addressing pollution from intensive agriculture;
- minimising cattle faeces entering watercourses; and
- investigating the effects of biosolid application in the catchment.

Many other improvements were seen as necessary. Better management of the storages was required. Improvements were also required to optimise performance of the Prospect filtration plant and the distribution system. A comprehensive package was essential.

## 11. Recommendations of the Inquiry and Implementation

### 11.1. ADMINISTRATIVE RECOMMENDATIONS

The Inquiry recommendations stressed the need to overcome poor management of the incident by the authorities. Sydney Water is developing a comprehensive Incident Management Plan and staff training. Mr McClellan also recommended that NSW Health have clear authority to make public health alerts in relation to drinking water incidents and develop the expertise to make those decisions. By way of response the Government requested the NHMRC to accelerate its revision of the 1996 water guidelines focussing on *Cryptosporidium* and *Giardia*. In the meantime, NSW Health issued an Interim Health Protocol. The Public Health Act 1991 was amended to give the Chief Health Officer the exclusive responsibility to issue boil water alerts. The amendments have strengthened the regulatory power of NSW Health regarding drinking water. A revised Memorandum of Understanding between Sydney Water and Health has also been completed.

## 11.2. COMMUNITY EDUCATION

An extensive public education programme and campaign on *Cryptosporidium* and *Giardia* was recommended. Greater public transparency was required to restore public confidence. A public education programme has been initiated. A brochure on 'Drinking Water and Public Health' and Fact Sheets on *Cryptosporidium* and *Giardia* are available in hard copy and on the Internet. Discussions are being pursued between Sydney Water and NSW Health to develop a public education programme. NSW Health is considering a campaign focusing on advice to the immuno-compromised, and to medical practitioners. Also recommended was a ministerial advisory committee to advise the Minister for Health on public information issues associated with *Cryptosporidium* and *Giardia*. This has been established.

## 11.3. WATER QUALITY DATA

The Inquiry made a series of recommendations for greater co-ordination within government and sharing of data on water quality. Greater transparency in reporting of water quality data was necessary to restore public confidence. Action taken includes legislation to require Sydney Water to publish quarterly consumer confidence reports on drinking water quality. Summaries are to be provided to each customer with their water bill. The first of these was published on 30 April 1999. Also, the Chief Health Officer is now required to publish an annual report on drinking water quality. Sydney Water is publishing all treated water test results on its Website: (http:www.sydneywater.com.au).

## 11.4. MINISTERIAL CONTROL

To implement the Inquiry recommendations for greater ministerial control over Sydney Water, legislation was passed changing its structure to give the Minister the power to obtain information and give directions in matters of public interest.

## 11.5. CATCHMENT PROTECTION

The Third Interim Report of the Inquiry made 32 detailed recommendations to protect the catchment and minimise sources of contamination. These included developing, as an interim measure, a State Environmental Planning Policy (SEPP) to provide parameters for acceptable development.

SEPP 58 - Protecting Sydney's Water Supply (Department of Urban Affairs and Planning, 1998), was promulgated on 24 December 1998 and came into effect on 1 February 1999. The effect of the State Policy is that almost all development within the catchment (certainly development of any significance) has to be assessed for its effects on water quality. The principal aim of the policy is to ensure that development does not have a detrimental impact on water quality. Specific considerations are to be given to any development proposed within the catchment which is within the SEPP. These are:

(a) whether the development or activity will have a neutral or beneficial effect on the water quality of rivers, streams or groundwater in the hydrological catchment, including during periods of wet weather;
(b) whether the water quality management practices proposed to be carried out as part of the development or activity are sustainable over the long term; and
(c) whether the development or activity is compatible with relevant environmental objectives and water quality standards for the hydrological catchment when these objectives and standards are established by the Government.

In most circumstances approvals require the concurrence of the Director-General of the Department of Urban Affairs and Planning.

The SEPP is to be followed by a Regional Environmental Plan (REP) for the catchment. The REP, which is being developed by the Department of Urban Affairs and Planning, is to be a prescriptive instrument and is to include water quality objectives. The estimated time scale is 12 to 18 months. Central to the Inquiry's strategy was the establishment of a Catchment Management Authority responsible for providing water quality of a prescribed standard and controlling infrastructure. The catchment authority should conduct a full assessment of the state of the catchment. The authority should also be responsible for and conduct water quality monitoring. The Environment Protection Agency (EPA) should be the primary regulator in the catchment and its role strengthened. The Inquiry also recommended an independent Water Auditor to critically review the performance of all parties.

The Government has also responded to many of these key recommendations. The Sydney Water Catchment Management Act 1998 has been passed to establish the Sydney Catchment Authority and vest in it the responsibility for management of Sydney's drinking water catchment. It assumes Sydney Water's responsibilities in the catchment and gives the new authority concurrence power over development in both the Inner and Outer Catchments. It also has an inspection and enforcement role. The Act also mandates the making of an REP to set water quality objectives and requires that consent can only be given to developments with a neutral or beneficial effect on water quality.

## 11.6. THE OPERATING LICENCE AND LICENCE REGULATOR

As discussed earlier, the Inquiry was of the view that the current Operating Licence be replaced with a new licence which clearly outlines the obligations of Sydney Water and the new Catchment Authority. The new Operating Licence should be developed through a transparent process at arm's length from the two authorities. The Licence Regulator should be renamed the Water Auditor and its role strengthened and better defined. It is to independently audit the operations of Sydney Water and the Sydney Catchment Authority, including water filtration plants. The Water Auditor should report to Parliament.

These proposals are being implemented with operating licences being developed by the Independent Pricing and Regulatory Tribunal (IPART) through a transparent consultative process. A water taskforce is considering the role and powers of the Water Auditor.

## 11.7. MANDATORY OPERATIONAL WATER QUALITY STANDARDS FOR *CRYPTOSPORIDIUM* AND *GIARDIA*

A series of recommendations were made with the aim of imposing operational water quality standards. However, the Inquiry considered it inappropriate to have mandatory health related standards for *Cryptosporidium* and *Giardia* given the current state of scientific knowledge. Nonetheless, the operation of all water filtration plants is being reviewed and a comprehensive sampling programme has been implemented to monitor their performance.

## 11.8. IMPROVEMENTS TO FILTRATION PLANTS

Again, a number of recommendations were made by Mr McClellan, including the maintenance and utilisation of the prototype plant developed for the Inquiry. This was notwithstanding that it could not entirely replicate all of the conditions of the outbreak. The plant is being maintained in operational readiness while further research and development projects are being pursued with existing treatment processes.

## 11.9. RESEARCH

The Commissioner placed much stress on the need for research into *Cryptosporidium* and *Giardia* and their health impacts. His recommendations included research on the viability of pathogens and the specific detection of *C. parvum*. NSW Health is working with the NHMRC and industry to develop a research programme. Sydney Water is addressing research holistically through the development of the Draft Drinking Water Quality Management Plan. This includes water quality research projects on the health impacts of pathogens and chemicals in drinking water, upgrading infrastructure for water delivery and improving utility management and system reliability.

## 12. Reaction of Consumers

The most notable reaction of consumers to the events of July to September 1998 was a loss of trust, in particular customer trust in Sydney Water. Research by Sydney Water into consumer attitudes to the boil water alerts is illuminating. The majority of customers prefer to be immediately notified when any contamination is detected. Sydney Water also ranked very low in the public's esteem as to a reliable source of information. Customers surveyed also thought that the incidents revealed a management problem and that profit had been put before water quality. People believed that the incident did in fact pose a serious health risk and a large majority kept strictly to the instructions.

Public reaction to the events changed as they unfolded. People were fairly forgiving of the first event and just wanted the problem fixed. The second incident, however, eroded the public's trust in the quality of the water and in the authorities. Anger was the prevailing response. After the third incident people thought that there was something seriously wrong with the system and that the incidents posed a serious risk to health. 72% of respondents thought profit had been put before water quality. Two thirds saw the incidents as a management problem. Use of bottled and filtered water rose from 23% before the incidents to 30% one month after the incidents. By February 1999 (five months later) this had risen to 32%. Satisfaction with tap water was high before the incidents, 85% being very or quite satisfied. By October 1998 this had reduced to 56% with 42% being not very or not at all satisfied.

By February 1999 public satisfaction with water quality had recovered somewhat but was still well down on pre-June 1998. In terms of trust of Sydney Water, the figure below is indicative of a sharp fall between pre-incident trust and during the incidents (September). This trust has slowly started to return. Sydney Water has been working hard to rebuild consumer confidence in its ability to provide safe drinking water (Figure 1).

One message, however, which comes through loud and clear from the market research is that the vast majority of customers were not enthusiastic about the $15 rebate which the Government required to be credited to customer accounts. Almost universally people believed that the $20 million could have been better spent on the catchment or direct water quality issues.

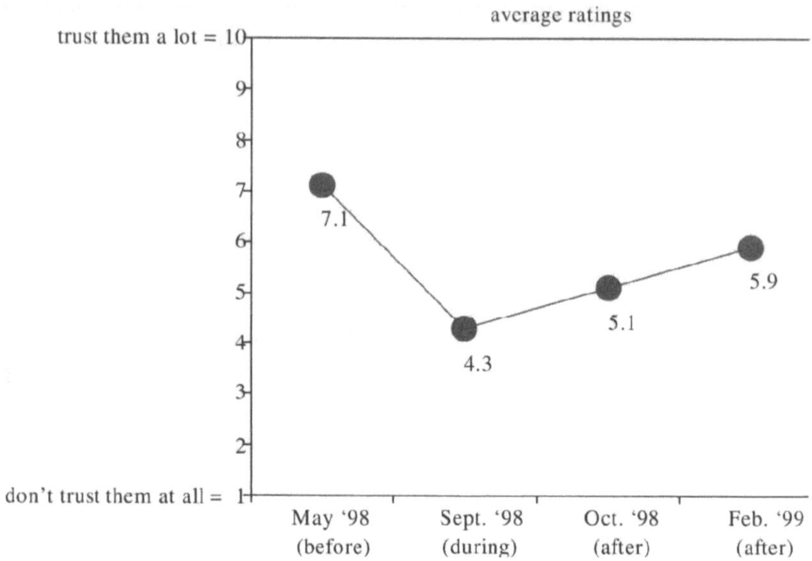


Fig. 1. "To what extent do you trust Sydney Water as an organisation?"
Sydney Water March 1999

## 13. Reaction of Sydney Water

The impact of the crisis on Sydney Water has been dramatic. Indeed one commentary has described the events and the Inquiry as having a profound impact on the Australian Water Industry as a whole (Health Stream, 1999). Sydney Water has lost top executive staff, been restructured, placed under ministerial control and partly dismembered. Needless to say, staff morale has been badly affected. Sydney Water acknowledges that it has learned a lot from the 1998 events. It accepts that there is considerable work to be done. For example, it is aware that it must develop a fast and accurate method of identifying *Cryptosporidium* and *Giardia* and assess any risks. It needs to be able to determine whether the organisms are alive or dead and whether they are infectious.

Sydney Water has established a five-year management plan for drinking water quality in the Sydney region. Part of the objective of the plan is to rebuild consumer confidence in the ability of Sydney Water to provide safe drinking water. The development of the plan has highlighted the need for performance improvement. The plan also recognises that there are emerging water quality issues which need to be evaluated. These include disinfection by products such as trihalomethanes which may have adverse health impacts - increased incidence of miscarriage and some cancers. Blue green algae is another example because they may produce toxins which can cause illness or promote tumours. Another issue is whether some chemicals may cause endocrine disruption in humans and wildlife. There is considerable scientific uncertainty as to which chemicals are involved, mechanisms of exposure and effects on humans and wildlife. All of these issues are to be the subject of research. Other aspects of the Plan include stricter catchment management, improvements in operational efficiency of water filtration and improvements in water quality monitoring programmes for raw and treated water. This may include revision of water quality objectives and criteria. The installation of particle counters in all plants is part of the plan. A new Incident Management Plan is being developed, coupled with training for crisis preparedness.

There have been, as already mentioned, financial implications for Sydney Water. Additional capital expenditure of $39 million has been scheduled. This is to sewer two villages within the catchment and to improve water treatment. This expenditure is in addition to Sydney Water's pre-existing capital works water programme, which includes water main renewals and remedial works to one of the largest reservoirs. Operating expenditure is also being increased, particularly for additional monitoring of drinking water quality. For example, $10,050,000 is to be spent on monitoring for each of the next 5 years. In addition, around $3 million per annum is to be spent on additional operating expenditure for the water treatment plants.

## 14. Conclusion

It could be argued that what occurred during the Sydney drinking water crisis involved an application of the 'precautionary principle' by the Government. The decision to issue the precautionary boil water alerts accepted the grave public health risks involved and the high degree of scientific uncertainty. It demonstrated a precautionary approach to decision-making which acknowledged these uncertainties in the context of possible poisoning a vast number of water consumers. It has served to focus the agencies involved, private industry and the Government, away from profit to the protection of the public. The recommendations will continue to have a crucial influence on future developments in the water industry. The reforms which have and are taking place are clearly much needed and, although they come at a cost, at least there has been no cost in human health.

David Getches (1997) has stressed the need for principles to guide us in resolving modern sectoral conflicts over water use and demands. He emphasises efficiency, equity, ecological sustainability and balance. In dealing with equity, he draws attention to the Mono Lake decision in the USA where the court reminded us that a state's water is effectively held on a 'public trust'.

Getches' thesis is that the goal of sustainability reflects the ideal of balance because developing water resources for present needs has to preserve sufficient resources for the use of future generations. The core principle of inter-generational equity, accepted into international law and reflected in many national laws, requires this balanced view of resource sustainability.

The Sydney water crisis has driven home to the public many issues surrounding water quality. No longer do they take drinking water quality for granted. While consumers may not realise it, many drinking water issues revolve around core principles of ecologically sustainable development, e.g., the precautionary principle, the pollutor pays principle, inter-generational equity, public participation, adequate environmental impact assessment procedures and the conservation of biological diversity and ecological integrity. Allied to this has been a largely unstated demand for a fundamental citizen's right to safe drinking water, hitherto accepted as the norm in Australia. Such a right does not find a place in the Australian Constitution, although specifically included in some new Constitutions of national states.

Disruptive as the events of July to September 1998 were, the Australian public has undoubtedly been the winner.

## Addendum

The author wishes to emphasise that this paper is written from the perspective of a lawyer and not a scientist.

## References


Cooperative Research Centre for Water Quality and Treatment: 1999, March, *Health Stream* **13**, Melbourne, Australia.

Department of Urban Affairs and Planning,: 1998, December 24, *State Environmental Planning Policy No 58 - Protecting Sydney's Water Supply*, Government Gazette No **178**, 24.

Getches, D.H.:1997, *Sectoral Conflicts over Water: Resolving tensions among Agricultural, Municipal and Industrial and Ecological Demands*, Paper presented to Rosenberg International Forum on Water Policy, San Francisco.

McClellan, P.: 1998, December, Sydney Water Inquiry - *Final Report* (2 volumes).

National Audubon Society v Superior Court of Alpine County: 1983, 189 Cal Rptr 346 at 356

Sydney Water Catchment Management Act: 1998 (NSW).

# CATCHMENT BASIN MANAGEMENT OF WATER

R.E. LASTER
*Law Offices of Laster and Gouldman, 48 Azza Road, Jerusalem, Israel 92384*
*e-mail: laster@netvision.net.il*

**Abstract.** With the declaration of the State of Israel in 1948, Israel inherited the Ottoman law for water and not the English Rule of Riparian Rights. The Ottoman law allowed the State to control the water sources and, in fact, did not allow private ownership of water sources. In its early years, Israel promulgated some of the most forward looking legislation in the world concerning protection of water sources. But as time went on, Israel lost its preeminence in the water protection field, while other countries revised their water laws in light of the environmental revolution. It is now time for Israel to redesign its water system along catchment basin lines.

**Keywords:** catchment basin management, comparison with England and France, management of water resources, Israel, water legislation, water pollution

## 1. Introduction

For the last forty years Israel has moved in the direction opposite to that of Britain and France in the management of its water resources. Israel has also achieved the opposite results: poor water quality and a bloated, inefficient water management system. It didn't have to be that way. In the 1950s Israel had passed four of the most forward looking water laws in the western world. But, as the State developed, vested interests, on the one hand, and a lack of political will and foresight, on the other, gradually dismantled the water management system developed in the 1950s.

Today Israel can no longer serve as a beacon for other countries in its water management system. As this article will show, Israel should learn from other countries before it is too late.

## 2. France

The French water management system underwent a major overhaul during the last thirty years (Sironneau, 1993; 1996). This overhaul can only be described as revolutionary in scope. It required the passage of major legislative acts and the restructuring of some of France's executive bodies.

Beginning in 1958, France began the slow process of integrating its water management system. Prior to 1958, the Minister of Agriculture was responsible for rural water supplies, the Minister of Public Works for shipping and flood protection, the Minister of Industry for industrial power requests, the Minister of Interior for supervision of the communes – the local authorities that are

*Water, Air, and Soil Pollution* **123:** 437–446, 2000.

responsible for water distribution to their residents, the Minister of Health for aspects relating to hygiene, and the Treasury for finance. In 1959 a Water Commission was created as a think tank to develop the principles for reorganization of the system. The commission held that 1) the interdependence of water uses in the hydrographic basin requires solidarity to encourage all users to minimize pollutant outfalls; 2) national water management is only possible if one recognizes water's economic value and water pollution's negative value; and 3) water resources of a given basin must be considered a single resource so that long term water planning must consider the needs of all users in the context of the basin as a whole.

The commission report was followed by the creation of a Permanent Secretary for the Study of Water Problems. At the same time, France instituted political and administrative changes in its executive branch with its creation of regions. These changes encouraged the Secretariat to draft the Water Act of 1964. The Act divided France into six hydrographic units governed by basin committees and served by Water Boards. It introduced water quality objectives and it increased the supervisory power of the state to minimize pollution. The government then created the Ministry of Environment in 1971 and one of its immediate goals was to create uniformity and add a measure of professional responsibility in the field of water management. Among the new ministry's responsibilities was supervision over the Water Boards along with its responsibility for water policy, water quality regulations and planning.

The Draft Act of 1964 was passed into law in 1968 and it led to improvements in every facet of water management. With its implementation, the Act highlighted further inadequacies in water management. For example, nonpoint sources of pollution, the treatment of wastewater and storm water. These inadequacies, together with the new demands of the European water directives for water quality led France to revise the 1968 law by the French Water Act of January 1992. The new law increased the public's involvement in water management, recognized water as part of France's heritage, upgraded the powers of the local authorities (the communes) with respect to sewerage services and introduced a regional system of planning.

Probably the two most exciting aspects of France's water management system are the basin committee and its planning apparatus, the Water Resource Development and Management Plan (SDAGE). Each of the six hydrographic units has a basin committee which acts as a water parliament. These "parliaments" are composed of users, local representatives and professionals (40-45% of the seats); representatives of the region, departments and the communes (38% of the seats) and representatives of the state (19-23% of the seats). The total number of members in each of the six basin committees ranges between 61 and 114, depending on the size of the basin. Here debates among users, polluters and government representatives lead to the approval of the water board's annual budget and the SDAGE.

The SDAGE defines the general criteria for the use, development and qualitative and quantitative objectives of surface and ground water in the basin. The SDAGE and the budget allocations approved by the basin committee are then implemented by the Water Boards, presently called the Water Agencies (Agence de l'Eau). The Water Agencies collect fees from abstractors and polluters and this money is used to purify water and aid in its conservation. The agencies also have the duty of studying the basin, monitoring water quality and quantity and carrying on research projects and assistance in training programs. Their main role, however, is providing financial incentives to improve and preserve France's water resources.

## 3. England

England has chosen a different system than France, but still using the hydrographic unit (Sinclair, 1985; 1996) (Howarth, 1990;1992). Similar to France, Parliament created a central advisory water committee, which recommended a major reorganization of the structure of water management. This led to the Water Act of 1973 which legislated integrated, water management in a catchment basin scheme. Prior to the 1960s, supply of water was the responsibility of the local authorities, as was sewage disposal. Drainage authorities handled drainage and flood control; fishing was the responsibility of fisheries; and hardly any authority handled amenity uses and conservation.

In a series of laws beginning in 1963, the English overhauled their water management system, culminating in its full integration in the law of 1973. The 1973 Act divided England into 10 River Authorities, along hydrographic basins. The River Authorities, under the 1973 Act, obtained responsibility for all aspects of water use and reuse in the basin. This includes water abstraction, supply and sewage disposal, amenity uses, fisheries and some control over drainage and navigation. The River Authorities were so successful in their endeavor that the quality of drinking water and sewage water increased tremendously, pollution was reduced and rivers and streams began to breathe again. This so impressed the Thatcher government that following in the ideological footsteps of many of Mrs. Thatcher's moves, the conservative government decided to privatize two aspects of the River Authorities, water supply and sewage purification (Sinclair, 1999). In the law of 1989, the government privatized the end uses of the river authorities by setting up statutorily created public companies. These PLCs, as they were called, were given the entire water and sewage infrastructure and then allowed to go public on the London Exchange. The result has been a mixed blessing (Schofield and Shaoul, 1997).

On the one hand, service has improved and water quality in England now meets the European standards without the government having to charge its taxpayers for these improvements. On the other hand, water costs have increased while salaries at the water companies have increased as well (to the dismay of most of the residents), and some water companies failed miserably to meet standards from the drought of 1995.

Once parliament legislated out the supply and sewage aspects of the river authorities from the integrated system, there arose a need for supervision over water supply and water quality. This required the creation of a National River Authority responsible for quality and quantity of water in the ten catchment basins. At the same time, an Office of Water was created to supervise pricing and consumer complaints against the ten water companies. In addition, a Drinking Water Inspectorate was created to monitor drinking water quality. This has led to some friction between government agencies, which was never part of the previous integrated system of river authorities.

What is significant about both the French and English systems, however, is that these two "old democracies" had the courage to revamp an outmoded system. What is interesting about both countries is that even though their legal systems are different, France is a civil law country while England uses the common law approach, both decided to revise their water management systems along catchment basin units. Israel, a young democracy, needs to learn from her older sisters about political courage in the face of vested interest.

## 4. Israel

When Israel won its independence in 1948, it inherited the Ottoman water law (the Majelle) and not the British Rule of Riparian Rights (Palestine Order and Council, 1922) (Hooper, 1933). This was a blessing in disguise. Under the Majelle, flowing water was incapable of private ownership. The Majelle was Roman law unadulterated by European and later American theories of property rights (Wiel, 1908). Thus Israel was never burdened with the Riparian Rights doctrine nor the Theory of Prior Appropriation (Teclaff, 1985). This enabled the Israeli Government a free hand to legislate the management of its water resources and the government wasted no time in doing so. In just ten years after the founding of the state, Israel had passed four water laws, each more comprehensive than its predecessor.

The first two laws passed in 1955, the Water Drilling Control Law and the Water Metering Law enacted into law two basic tenets for proper water management. First, no one could supply water from any source without first measuring it; second, no one could drill a well nor abstract water from a water free source without a government permit.

Following these two basic laws, the Knesset passed a comprehensive drainage law - The Drainage and Flood Prevention Control Law of 1957. To an outsider it might seem strange that such an arid country as Israel would need a major flood control law. Yet, in the early part of this century, the land cover in Israel had suffered from overgrazing and it was inadequate to prevent flooding. The English tried to reduce flooding during the Mandate by proper planning but not until the drainage law was passed in 1957 did Israel take the subject seriously in hand.

The Drainage Law of 1957 was an early attempt at catchment basin management. It empowered the Minister of Agriculture to divide the country into drainage districts, with each district to be composed of representatives of the local authorities in the district. The drainage boards themselves were independent bodies; they appointed their own chair and hired their own staff, while funding was to come from a land assessed drainage fee. Each drainage authority was empowered to implement a drainage plan for the protection of its region after approval of the plan by the national drainage council and the district planning commission.

In short, the drainage law contained the seeds of catchment basin management: statutorily created authorities, independently funded with national supervision over their planning policies. Yet the system failed, and, on the eve of its reorganization in 1996, the drainage boards were bankrupt. They had failed to raise funds for drainage projects and were totally dependent on government handouts. Without funds, they were inadequately staffed, they failed to produce comprehensive drainage plans to prevent construction in flood plains, to reduce the amount of excess drainage, to coordinate drainage policies with local authorities and to prevent flooding. And after the floods of 1991-92, the drainage boards were left with responsibility for damages but without the tools to cope. At the present time, there are suits that have been filed against Israel's Drainage Boards in a sum exceeding one billion Israeli shekels for damages.

The system failed for the following reasons. First of all, the major failure was the Minister of Agriculture's refusal to create drainage boards along catchment basin or hydrological boundaries. He gerrymandered the original drainage boards, creating 26 in all, to make sure each board was controlled by the agricultural local authorities. The result was that only this sector benefited from drainage activities and the boards lost support from cities and towns. Second, due to plain inefficiency, the boards never assessed land for a drainage fee and they were left with government handouts when available and support from the rural local authorities benefiting from their services. Finally, the law itself ruled out the possibility of the drainage boards also acting to purify sewage. They, therefore, had limited control of sewage quality, further reducing their effectiveness to manage water uses in the basin. But all this need not have happened had the drainage law of 1957 been properly implemented.

## 5. The Water Law

In 1959 the Knesset passed The Water Law, a brilliant legislative code to protect all aspects of Israel's water and the recipe for its proper management. The law opens with the following declaration: "The water resources in the State are public property; they are subject to the control of the State and are intended for the use of its inhabitants and for the development of the country." It then defines water sources to include every type of water used by man and animal; surface, runoff, ground water, floods, storm water, sewage, even stream beds with intermittent flows (wadis). After the water sources were defined and entrusted to the care of the State, the law then goes on to create Israel's water institutions. A national water company is created to be the bulk water supplier for the country. The position of Water Commissioner is created "to manage Israel's water system." A National Water Advisory Board is created to advise the Water Commissioner and the Minister of Agriculture on water policy. A Water Court is created to resolve water disputes. The Ministry of Agriculture is appointed the ministerial authority over the law, and he is given the right to divide Israel into rationing areas to protect the precious little water there is.

The Water Law was brilliantly and beautifully crafted. But it was flawed. To expect the Water Commissioner to "manage the water needs of the state" was too demanding a challenge. No one person can manage all aspects of a country's water sources, and the Water Commissioner proved no exception to the rule. He moved with the flow of pressure, but not the water pressure, the political pressure. And as he moved inexorably to protect his constituency, he found himself capturing the headwaters of Israel's streams and rivers, authorizing engineering enterprises to bring water from the North to the South, ignoring sewage flow into what was left of Israel's streams and wadis. And, he was always attentive to agricultural interests who were using 90% of Israel's water production even as late as the 1980s.

This could not go on for long, but the Water Commissioner's failure was misinterpreted by the legislative branch. In their infinite wisdom, but finite knowledge, the Knesset members acted to dismember the Ministry of Agriculture's power over all water uses. Instead of developing an overall strategy for water use, instead of looking outside of Israel's boundaries for inspiration, the Knesset simply began distributing powers granted to the Ministry of Agriculture to other authorities without replacing it with a ministry at the center of a holistic system.

In 1962 the local authorities were given control over sewage flow and treatment. This, added to the fact that they were already supplying drinking water to their residents, prevented area wide management of piped water. Then the Ministry of Health was given control over drinking water and sewage water quality (The Public Health Ordinance, 5730/1970) and the Minister of Agriculture had to consult him in the proper management of these two areas.

The Ministry of Environment was created in 1988 and took from the Ministry of Agriculture its supervision over the quality of natural waters. This led to the creation of Israel's first River Board, The Yarkon River Authority, created in 1989. Finally, the Ministry of Infrastructure was created in 1996 and took the Water Commissioner from the Ministry of Agriculture and placed it under its supervision. To make matters even more complicated, the Ministry of Finance took control over water pricing. Consequently, today the management of the hydrological cycle has become so convoluted that the following chart only shows a part of the picture (Table I). As matters have now reached catastrophic proportions, there is hope for a positive reaction. The buds of that reaction are popping out in different forms, and with a little germination, Israel can catch up to where it started 40 years ago.

## 6. About Buds, Germination and Killer Bees

In 1996, the Ministry of Agriculture set up a team of experts to revive Israel's drainage board system. The team, in less than three years, revised the structure of the boards and reduced their number from 26 to 11 along catchment basin lines.

At the same time a new drainage law was drafted to give drainage boards powers comparable to England's River Boards, no longer just drainage and flood control, but control over sewage and amenity uses of streams and lakes as well. Portions of this proposed law were copied from a law passed as early as 1965, but not implemented until 1988, the Streams and Springs Authorities Law of 1965. Under this law, the Yarkon River Authority previously mentioned, operated.

The Yarkon River Authority was another small "bud" germinating towards catchment basin management. Composed of representatives of the government, local authorities and landowners bordering on the Yarkon River, it has served as a model for river and stream reclamation. The Ministry of Environment, which supervises river authorities, also took action to create the Rivers Administration, a nonprofit organization with drainage boards, local authorities, the Israel Lands Authority and the Jewish National Fund, all participating in its activation. The Rivers Administration has been pressing for years to create catchment basin boards and has suggested that the power of the drainage boards be transferred from the Ministry of Agriculture to the Ministry of the Environment.

With the reorganization of Israel's drainage boards, two River Authorities now in operation and the River Administration investing in stream reclamation, it is right to lead to the next stage of germination, combining river authorities and drainage boards under one catchment basin format. But there are clouds on the horizon, and not rain clouds, but hoards of killer bees, each with the potential to undermine reorganization of Israel's water system.

The major challenge comes from the drones of the Ministry of Finance. The ministry has been impressed by Margaret Thatcher's theories of privatization. If the Treasury had its way, the water cycle would be divided up under theories of the free market, with the supply of water and sewage control going to the highest bidder. To that end, the Treasury has drafted a bill for the creation of water companies to replace local authority control over water supply and sewage purification (Proposed Bill for Water and Sewage Companies, 5758/1997). But this bill is too early; Israel has yet to restructure its water system, as the English

Table I
Government Responsibility for Water in Israel - 1999

| Subject | Direct Responsibility | Indirect Responsibility |
|---|---|---|
| A. Natural water | | |
| 1. Surface water | | |
| (a) in nature | | |
| (1) streams & springs | River Authority/Drainage Authority | Use: Water Commissioner |
| (2) wadis | River Authority/Drainage Authority | Quality: M. of Environment |
| (3) lakes | Kinneret Administration Drainage Authority | |
| (b) Supply | Mekorot (National Water Company) | Responsibility of Water Commissioner, M. of Infrastructure, Except to agriculture: M. of Agriculture Recharging: Water Commissioner |
| 2. Ground water | | |
| (a) in nature | | |
| (b) supply | Mekorot, water associations, local authorities | Drilling-Water Commissioner Pollution prevention: M. of Environment |
| 3. Floods | | |
| (a) Drainage | Drainage Authority | |
| (b) City Drainage | Drainage Authority, local authorities | M. of Agriculture M. of Interior |
| B. Piped water | | |
| 1. Potable | | |
| (a) water supply | Mekorot | Licensing: Water Commissioner |
| (b) to the consumer | local authorities, water associations | Quality: M. of Health |
| 2. Sewage water | | |
| (a) Untreated | local authorities, companies | M. of Interior |
| (b) Treated | local authorities, companies | M.of Environment M. of Health Water Commissioner M. of Infrastructure |
| 3. Price | | |
| (a) Drinking water | Water Commissioner | M. of Infrastructure |
| (b) Sewage | local authorities | M. of Interior M. of Infrastructure |
| (c) Treated sewage | local authorities | M. of Interior M. of Infrastructure |
| (d) Charging fee | M. of Infrastructure with agreement of the Treasury and approval of Knesset | |

did before they sold off the water and sewage infrastructure. If private companies take over water supply under today's management system, what incentive do they have to save water in a parched country like Israel and what incentive to purify sewage to a quality meeting the ecological needs of the receiving streams? To these questions, the Treasury has a ready answer. Supply of fresh water is only a question of purchasing power. When there is need, we will desalinate. As for sewage quality, if the Ministry of Environment or the Yarkon River Authority want better quality effluent, let them pay for it. This, however, is too facile an approach to a complicated subject.

## 7. Conclusion

Israel needs catchment basin management for the following reasons: First, from a hydrological standpoint, it is the only logical way to manage the water cycle. One cannot make intelligent decisions about water use and reuse, without controlling the entire aspect of the system. Second of all, catchment basin boards bring all interests to the bargaining table, so whether the final decision is intelligent or not, at least the stakeholders involved have had their say. Third, the system works. Look around and see that water management is a give and take proposition, affecting all interests with high stakes. Fourth, there are serious moral and ethical issues at stake before practical economic interests are decided. How clean do we want our water? How much water do we want to use? How much flood protection? What sectors get most favored status? Finally, Israel is not alone in a single catchment basin (Twite and Isaac, 1994). The country straddles the Jordan Basin, the Yarmook and the Litani, all flowing on or outside of the borders of Israel. How can we negotiate river use without an overall catchment basin approach that takes into account other countries' needs? Water can be a source of sharing and a catalyst for common good or common greed. Just as all of us want a holistic approach to life, we need a holistic approach to our water environment.

## References

Hooper, C.: 1933, *The Civil Law of Palestine and Transjordan*, vol. **1**.

Howarth, W. (ed.): 1990, "The Law of National Rivers Authority", in *The Law of National Water Authorities.*

Howarth, W. (ed): 1992, *Wisdom's Law of Water Courses*, 5th ed.

Palestine Order and Council: 1922, Article 46, III *Dreighton's Laws of Palestine*, 2569-2580.

Schofield, R. and Shaoul, J.: 1997, Regulating the Water Industry: Swimming Against the Tide or Going Through the Motions? Public Interest Report, August 1996; in *The Ecologist* **27**, January-February 1997.

Sinclair, I.: 1985, International Legal Aspects of Water Management, *International Seminar on Institutional and Legal Aspects of Water Management*, Madrid.

Sinclair, I.: 1996, Management of an Inland Water System, Vol. VIa, *Proceedings of the 6th International Conference of the Israel Society for Ecology and Environmental Quality Sciences*, Jerusalem.
Sinclair, I.: 1999, Better Laws Equal Better Environment: The Role of the Environmental Lawyer in the Reconstruction and Modernization of the Water Sector and Private Sector Participation, in the *7th International Conference of the Israel Society for Ecology and Environmental Quality Sciences*, Jerusalem.
Sironneau, J.: 1993, *Institutional Framework for Water Management in France.*
Sironneau, J.: 1996, Water Management in France. French Concepts and Mechanisms Existing in the Field of Water Basin Management, vol VIa, *Proceedings of the 6th International Conference of the Israel Society for Ecology and Environmental Quality Sciences*, Jerusalem.
Teclaff, L.A.: 1985, Water; in Teclaff, L.A. (ed.): Water Law In Historical Perspective.
Twite and Isaac (eds.): 1994, *Our Shared Environment: Israelis and Palestinians Thinking Together about the Environment of the Region in Which They Live.*
Wiel, S.C.: 1908, "Running Water", *Harvard Law Review*, 190.

# SUSTAINABILITY LAW FOR THE NEW MILLENNIUM AND THE ROLE OF ENVIRONMENTAL LEGAL EDUCATION

B. BOER
*Australian Centre for Environmental Law, Faculty of Law, University of Sydney and Commission on Environmental Law, IUCN-The World Conservation Union*

**Abstract.** The first half of this paper discusses the development of sustainability law. It takes as its basis that environmental law plays a central role in the achievement of sustainable development and that environmental lawyers and those from associated disciplines must come to terms with the imperatives of the internationally accepted concept of sustainable development. The second half of the paper deals with the implications of sustainability law for the teaching of environmental law, including the need to further liberate environmental law from the confines of law schools and lawyers, and to make it part of the common discourse of government, business and communities at large. A number of current initiatives in environmental legal education and training are canvassed, focussing on the Asia Pacific region.

**Keywords**: environmental conventions, environmental law, environmental law education, globalisation, international, sustainable development

## 1. Introduction

If we are to achieve sustainability of the earth's environment, a holistic and integrated approach to environmental management should be taken in all sectors of economic activity and covering all environmental media: atmospheric, marine, riverine and terrestrial. Without a holistic and integrated approach, the continuation of business as usual is for the most part expected, with national governments often going in different directions from each other, and the private sector going about its usual business.

At a global level, states, in close collaboration with institutions such as the United Nations Environment Programme (UNEP) and the United Nations Development Programme (UNDP), are beginning to more closely cooperate to achieve the implementation of conventions and strategies. These efforts are often backed by financial institutions such as the various components of the World Bank, and with inputs from individual countries through overseas development assistance programmes. Environmental conventions now exist to cover a myriad of environmental problems, including biodiversity conservation, the establishment and maintenance of protected areas, climate change, desertification, air pollution, marine pollution and freshwater pollution.

At a regional level, in order to address these issues more efficiently and in ways which are not as seemingly overwhelming, a trend is emerging towards the generation of regional strategies and conventions, often promoted through

*Water, Air, and Soil Pollution* **123:** 447–465, 2000.

regional offices of global environmental bodies. These include the Regional Office for Asia and the Pacific of UNEP situated in Bangkok, and the regional offices of UNDP, as well as regionally-focussed intergovernmental organizations. These include the South Pacific Regional Environment Programme, based in Samoa, covering some 20 Pacific Island developing countries and dependencies, the ASEAN Environment Programme, covering 9 South East Asian countries, and the South Asia Cooperative Environment Programme, catering to 9 South Asian countries. They also include regional banks such as the Asian Development Bank and the African Development Bank.

In recent years, the concept of sustainable development has begun to be recognized in international law, though not unambiguously. It is clearly the basis of the Rio Declaration on Environment and Development and is the fundamental concept behind Agenda 21. With these two documents, generated through the process of the 1992 United Nations Conference on Environment and Development, we already have a set of basic principles, and the plans and strategies to carry out those principles (Boer, 1995). Clearly it is now time to operationalise these principles in all sectors of economic activity, through legal and policy mechanisms at the national level.

We also see the concept of sustainable development endorsed through inclusion in various environmental conventions. Variations of the term and associated principles are also found in a number of recent instruments, particularly in preambular statements, such as in the 1992 Convention on Biological Diversity, the Framework Convention on Climate Change and the Desertification Convention. Further, the preparation of *the* IUCN Draft Covenant on Environment and Development of 1995 and the Earth Charter (1999 draft) promise to facilitate the incorporation of the notion of sustainability into national policy and legislation. Some of these documents will be briefly explored below.

This paper uses the term "sustainable development" for the most part. On occasion, the term "sustainability" is used, especially when referring to the broader dimensions of environmental management, in an attempt to remind readers not to assume that development is always necessary, or that it is always acceptable as long as it is somehow sustainable.

This paper focuses on the central role that environmental law plays in the achievement of sustainable development and relates to the implications of sustainability law for the teaching of environmental law. As an illustration, a number of current initiatives in environmental legal education and training are presented, focussing on the Asia Pacific region.

## 2. The Evolution of Sustainable Development Law

### 2.1. THE LEGAL AND POLICY PRESCRIPTIONS OF AGENDA 21

Agenda 21 is aimed particularly at the achievement of sustainable development across every sector of human activity. It represents a comprehensive framework for the cooperative generation of strategies for sustainable development and environmental management at a global level. It lays the basis for a new way of thinking in relation to environmental regulation based on the concept of sustainability. From the point of view of environmental law, Chapter 8 is perhaps the most significant in this context, with an explicit recognition and promotion of the role of environmental law and policy.

Chapter 8 deals with the integration of environmental and developmental issues in decision-making at policy, planning and management levels, with the overall objective of improving or restructuring the decision-making process in order that socio-economic and environmental issues are fully integrated. It urges the adoption of a national strategy for sustainable development, to build on and harmonise sectoral economic, social and economic policies in each country. The overall objective, 'in the light of country-specific conditions', is the promotion of the integration of environment and development policies through appropriate legal and regulatory policies, instruments and enforcement mechanisms at every level of government.

The more pertinent paragraphs are:

8.13. Laws and regulations suited to country-specific conditions are among the most important instruments for transforming environment and development policies into action, not only through "command and control" methods, but also as a normative framework for economic planning and market instruments. Yet, although the volume of legal texts in this field is steadily increasing, much of the law-making in many countries seems to be *ad hoc* and piecemeal, or has not been endowed with the necessary institutional machinery and authority for enforcement and timely adjustment.

8.14. While there is continuous need for law improvement in all countries, many developing countries have been affected by shortcomings of laws and regulations. To effectively integrate environment and development in the policies and practices of each country, it is essential to develop and implement integrated, enforceable and effective laws and regulations that are based upon sound social, ecological, economic and scientific principles. It is equally critical to develop workable programmes to review and enforce compliance with the laws, regulations and standards that are adopted. Technical support may be needed for many countries to accomplish these goals. Technical cooperation requirements in this field include legal information, advisory services and specialised training and institutional capacity-building.

8.15 The enactment and enforcement of laws and regulations (at the regional, national, state/provincial or local/municipal level) are also essential for the implementation of most international agreements in the field of environment and development, as illustrated by the frequent treaty obligation to report on legislative measures....

In many developing countries, there are major barriers to the implementation of environmental law and the resolution of environmental and natural resource disputes. These barriers are of an economic, political and cultural character. It is the task of environmental lawyers, working with experts from other disciplines, to develop the legal and policy mechanisms to ensure that those barriers are overcome, and in ways appropriate to the economic, political and cultural contexts in which these mechanisms are intended to work.

This and subsequent sections of chapter 8 reflect and further promote a trend towards homogenisation of approaches, policies and principles. In examining developments in national environmental policy, it is clear that the provisions of chapter 8 are beginning to form the basis of decision making on environment and development matters for governments, intergovernmental organisations, non-government organisations and the private sector.

Agenda 21 thus sets out the framework for the cooperative generation of strategies for sustainable development and environmental management at a global level. The combination of the political commitments made at Rio, and the momentum for the promotion of its programmes through the Commission on Sustainable Development and by international intergovernmental organisations and NGOs, is ensuring that its major suggested programmes will be carried into effect at national level.

Whilst Agenda 21 is not legally binding in international law, the major programmes and activities that it promotes have had and will continue to have a very important influence on developments at a national level, assisted and encouraged by the United Nations Commission on Sustainable Development.

In the past seven years, many of the provisions of Agenda 21 have begun to form the basis of decision making on environment and development matters for governments, intergovernmental and non-governmental organisations and the private sector. In particular, with the establishment of the United Nations Commission on Sustainable Development in 1993, many nations are reporting regularly on their efforts in achieving sustainable development. These formal indications of implementation must, however, be closely examined, to ascertain whether there is real progress being made, or whether the statements are little more than rhetoric.

A United Nations General Assembly Special Session "Rio + Five" or the "Second Earth Summit" took place in June 1997. The session demonstrated the difficulties of attempting to achieve sustainability on a global basis over a short period of time. It also highlighted the inherent political and economic difficulties faced by individual countries in trying to achieve sustainability in various economic sectors. It is apparent that there is as yet no common agreement on an international or regional level about the approaches and concepts on the achievement of sustainable development. Whilst an enormous amount of intellectual effort has been focused on the concept, a great deal more work needs to be done, particularly at the individual economic sector level, to develop practical approaches for its achievement.

## 2.2. IUCN DRAFT COVENANT ON ENVIRONMENT AND DEVELOPMENT

The IUCN Draft Covenant on Environment and Development has its origins in the work of the IUCN Commission on Environmental Law on the preparation of the instrument which became the World Charter for Nature in 1992. It gained further impetus from the call of the UN Conference on Environment and Development (UNCED) and its action plan, Agenda 21, to identify ways in which to integrate environment and development. The role of environmental law was recognized as one of the essential tools to achieve this integration at national, sub-regional, regional and international levels, including:

(a) elaborating the balance between environmental and developmental concerns;
(b) clarifying the relationships between the various existing treaties; and
(c) ensuring national participation in both developing and implementing these legal measures, with particular focus on developing countries (Agenda 21, 1992, par 39.1)

The Commission on Environmental Law (CEL) of IUCN, the World Conservation Union, responded to UNCED's call by preparing a Draft International Covenant on Environment and Development. A special working group of the Commission on Environmental Law of the World Conservation Union was established in 1992 and produced the Draft Covenant in 1995.

The reasons for preparing the Covenant are set out in the introduction to the document. They include:

- to provide the legal framework to support the further integration of the various aspects of environment and development;
- to create an agreed single set of fundamental principles like a "code of conduct", as used in many civil law, socialist, and theocratic traditions, which may guide States, intergovernmental organizations, and individuals;
- to consolidate into a single juridical framework the vast body of widely accepted, but disparate principles, of "soft law" on environment and development (many of which are now declaratory of customary international law);
- to facilitate institutional and other linkages to be made between existing treaties and their implementation;

- to reinforce the consensus on basic legal norms, both internationally, where not all States are party to all environmental treaties, even though the principles embodied in them are universally subscribed to, and nationally, where administrative jurisdiction is often fragmented among diverse agencies and the legislation still has gaps;
- to fill in gaps in international law, by placing in a global context principles which only appear in certain places and by adding matters which are of fundamental importance but which are not in any universal treaty;
- to help level the playing field for international trade by minimizing the likelihood of non-tariff barriers based on vastly differing environmental and developmental policies;
- to save on scarce resources and diplomatic time by consolidating in one single instrument norms, which thereafter can be incorporated by reference into future agreements, thereby eliminating unnecessary reformulation and repetition, unless such reformulation is considered necessary; and
- to lay out a common basis upon which future lawmaking efforts might be developed.

It is this last objective which is perhaps the most vital. By applying the common principles being developed through the Covenant at both international and national level, the possibility exists that a common approach will be followed by all countries that may become party to the Covenant. At this stage however, the Covenant has no international legal status. It is presently undergoing further negotiation and revision, which may include integration, or at least a close linkage, with the Earth Charter.

## 2.3. THE EARTH CHARTER

The Earth Charter originated from the call for a new charter by the World Commission on Environment and Development (the Brundtland Commission) in its report *Our Common Future* (1987) "to consolidate and extend relevant legal principles to guide State behavior in the transition to sustainable development". The Commission had commissioned a separate report on the legal aspects of global environmental issues from a group of environmental law experts. In their report they included a set of legal principles which were to set a framework for future negotiation. It had been hoped that the new charter would be an outcome of the United Nations Conference on Environment and Development, and would form the ethical basis for Agenda 21 and the Rio Declaration on Environment and Development. The Charter itself did not emerge, and the Rio Declaration itself was not seen to meet the expectations of a Charter. The present Benchmark Draft of April 1999 adopts sustainable development as its basis. After a broad preamble setting out its ethical approach, it states:

> Having reflected on these considerations, we recognize the urgent need for a shared vision of basic values that will provide an ethical foundation for the

emerging world community. We, therefore, affirm the following principles for sustainable development. We commit ourselves as individuals, organizations, business enterprises, communities, and nations to implement these interrelated principles and to create a global partnership in support of their fulfillment.

The Earth Charter's principles are elaborated under the following headings:

I. GENERAL PRINCIPLES

1. Respect Earth and all life.
2. Care for the community of life in all its diversity.
3. Strive to build free, just, participatory, sustainable, and peaceful societies.
4. Secure Earth's abundance and beauty for present and future generations.

II. ECOLOGICAL INTEGRITY

5. Protect and restore the integrity of Earth's ecological systems, with special concern for biological diversity and the natural processes that sustain and renew life.
6. Prevent harm to the environment as the best method of ecological protection and, when knowledge is limited, take the path of caution.
7. Treat all living beings with compassion, and protect them from cruelty and wanton destruction.

III. A JUST AND SUSTAINABLE ECONOMIC ORDER

8. Adopt patterns of consumption, production, and reproduction that respect and safeguard Earth's regenerative capacities, human rights, and community well-being.
9. Ensure that economic activities support and promote human development in an equitable and sustainable manner.
10. Eradicate poverty, as an ethical, social, economic, and ecological imperative.
11. Honor and defend the right of all persons, without discrimination, to an environment supportive of their dignity, bodily health, and spiritual well-being.
12. Advance worldwide the cooperative study of ecological systems, the dissemination and application of knowledge, and the development, adoption, and transfer of clean technologies.

IV. DEMOCRACY AND PEACE

13. Establish access to information, inclusive participation in decision making, and transparency, truthfulness, and accountability in governance.
14. Affirm and promote gender equality as a prerequisite to sustainable development.
15. Make the knowledge, values, and skills needed to build just and sustainable communities an integral part of formal education and lifelong learning for all.
16. Create a culture of peace and cooperation.

Whilst most of the provisions of the Charter are no more than statements of ethical aspiration, they nevertheless form an important basis for the elaboration of legal duties and obligations.

In its final paragraphs, under the heading "A New Beginning", the draft Charter includes reference to the IUCN Covenant:

> In order to build a sustainable global community, the nations of the world must renew their commitment to the United Nations and develop and implement the Earth Charter principles by negotiating for adoption a binding agreement based on the IUCN Draft International Covenant on Environment and Development. Adoption of the Covenant will provide an integrated legal framework for environmental and sustainable development law and policy.

Thus the Earth Charter and the Covenant appropriately linked, if not integrated, may, when further negotiated, provide a coherent legal framework in the form of an international instrument through which common obligations might be carried out and promoted at a national level. One of the many questions that then arises is just how such an instrument will be implemented and enforced, given the relatively weak institutional framework which the Commission on Sustainable Development, UNEP and UNDP represent. It may well be time to look more closely at a Global Environmental Agency, as suggested by some commentators, or to use the United Nations Trusteeship Council for these purposes, as suggested by Maurice Strong, the Secretary General of the 1992 United Nations Conference, and now Chair of the rather grandiosely titled Earth Council.

## 2.4. THE AARHUS CONVENTION

The recently concluded Convention on Access to Information, Public Participation in Decision Making and Access to Justice in Environmental Matters (known as the Aarhus Convention) (1998) also acknowledges the central role to be played by sustainable development, and the principles derived from it. Its preamble includes "the need to protect, preserve and improve the state of the environment and to ensure sustainable and environmentally sound development". Importantly, in the context of environmental legal education, the preamble also states that the Convention desires to "promote environmental education to further the understanding of the environment and sustainable development and to encourage widespread public awareness of, and public participation in, decisions affecting the environment and sustainable development." The point needs to be made here that environmental education should these days automatically include environmental *legal* education.

The objective, of the Convention is found in Article One:

> In order to contribute to the protection of every person of present and future generations to live in an environment adequate to his or her health and well-being, each Party shall guarantee the rights of access to information, public

> participation in decision-making, and access to justice in environmental matters in accordance with the provisions of this Convention.

Whilst this Convention leaves something to be desired in terms of its outlook and enforceability, its adoption by a large number of European states to date indicates that there is a good deal of commitment to its provisions. It is clearly desirable that it be exported to other regions for adoption, either in its present form (non-European members of the United Nations are able to become members upon approval by the Meeting of the Parties (article 19.2), or with suitable modifications to address regional differences.

## 2.5. THE INTERNATIONAL COURT OF JUSTICE: THE DANUBE DAM CASE

Another strand in the evolution of sustainable development at international level is found in the work of the International Court of Justice. In response to the international concern regarding environmental matters, the Court decided in 1993 to set up a special Environmental Chamber under the statute of the International Court of Justice. A recent landmark case concerned the building of a system of locks at Gabcikovo in what is now Slovakia, and at Nagymaros, in Hungary. The intention of the project was to produce hydroelectricity, improve the navigation of the river, and protect that part of Danube River against flooding. Hungary and the former Czechoslovakia had concluded a treaty to jointly finance, build and operate the system of locks. The case is now commonly referred to as the Danube Dam case (Case Concerning the Gabcikovo-Nagymaros Project, 1998). In this case, the International Court of Justice had before it a splendid opportunity to make a series of major statements on issues of international environmental law, but, for reasons that need not concern us here, the majority of the Court did not avail itself of that chance to do so (A-Khavari and Rothwell, 1998). The majority judgment did, however, pay some regard to the question of sustainable development in a backhanded statement, endorsing the concept in the following way:

> Throughout the ages, mankind has, for economic and other reasons, constantly interfered with nature. In the past, this was often done without consideration of the effects upon the environment. Owing to new scientific insights and to a growing awareness of the risks for mankind — for present and future generations — of pursuit of such interventions at an unconsidered and unabated pace, new norms and standards have been developed, set forth in a great number of instruments during the last two decades. Such new norms have to be taken into consideration, and such new standards given proper weight, not only when States contemplate new activities but also when continuing with activities begun in the past. This need to reconcile economic development with protection of the environment is aptly expressed in the concept of sustainable development (paragraph 140 of the main judgment).

Judge Weeramantry, in a separate opinion, did take the opportunity to try to develop international environmental law in the case, making a forceful statement indicating that sustainable development is now a part of customary international law:

> The Court must hold the balance even between the environmental considerations and the developmental considerations raised by the respective Parties. The principle that enables the Court to do so is the principle of sustainable development.
>
> The Court has referred to it as a concept in paragraph 140 of its Judgment. However, I consider it to be more than a mere concept, but as a principle with normative value which is crucial to the determination of this case. Without the benefits of its insights, the issues involved in this case would have been difficult to resolve.
>
> Since sustainable development is a principle fundamental to the determination of the competing considerations in this case, and since, although it has attracted attention only recently in the literature of international law, it is likely to play a major role in determining important environmental disputes of the future, it calls for consideration in some detail. Moreover, this is the first occasion on which it has received attention in the jurisprudence of this Court.
>
> When a major scheme, such as that under consideration in the present case, is planned and implemented, there is always the need to weigh considerations of development against environmental considerations, as their underlying juristic bases — the right to development and the right to environmental protection — are important principles of current international law.

A-Khavari and Rothwell comment on the main judgment and the Weeramantry judgment as follows:

> The Danube Dam Case was much anticipated by many commentators because it was seen as presenting the ICJ with a real opportunity to decide a case fully on the merits of the environmental issues raised before it. Unfortunately, this proved to be a false hope. Ultimately, international environmental law ran a poor third to treaty law and international watercourse law as the basis of the court's judgment........Despite this assessment, the Danube Dam Case does break new ground in the development of international environmental law because, for all its faults, it probably remains the most important decision in which environmental law principles and concepts have genuinely influenced the final orders. In this respect, the reference by the court to sustainable development is perhaps the most significant, especially given the status the concept has obtained in international environmental law. While in some respects the court's discussion of sustainable development raises more questions than it answers, the apparent acceptance by the court that this concept plays in balancing environment and development issues may well prove to be a turning point in its legitimacy in international law (A-Khavari and Rothwell, 1998, 534-535).

## 2.6. THE ROLE OF INTERNATIONAL ORGANISATIONS

Over the past few years, global and regional environmental organisations, both intergovernmental and non-governmental, have played an increasingly important role in the development of international environmental law, in contributing to and sponsoring the preparation of environmental conventions and other instruments, as well as in fostering the concept of sustainable development. Among the most important organizations in this respect are UNEP, UNDP, and the World Conservation Union and its Environmental Law Programme. The Environmental Law Programme is constituted by the Environmental Law Centre, based in Bonn, and the Commission on Environmental Law. The Commission includes some 600 members expert in environmental law and policy from most of the world's regions. It has been active for the past 30 years in a wide range of areas. For example, in recent years it has been effective in promoting legal mechanisms at an international and national level in areas as diverse as biodiversity, protection of the environment in armed conflict, strengthening the role of the judiciary, encouraging the implementation of the World Heritage Convention, participating in the 1998 Aarhus Convention on Public Participation and Access to Environmental Information, and developing further mechanisms in relation to the management and sustainable development of forests. Currently it is also becoming more closely involved in the implementation of the 1972 World Heritage Convention, and in the investigation of an international agreement on a sustainable soils instrument and associated national legislative mechanisms.

A feature of the work of these and related organisations has been an increasingly collaborative approach, in terms of the preparation of international environmental instruments, the implementation of Agenda 21 and the Rio Declaration on Environment and Development and assistance to individual states with the development of their environmental law. For example, UNEP's Environmental Law and Institutions Programme provides legal technical assistance to developing countries in enhancing their legislation and relevant institutions. The collaborative approach is well illustrated by the UNEP/UNDP Joint Project on Environmental Law and Institutions in Africa. In that project, UNEP cooperates with UNDP, FAO, World Bank, WHO and IUCN to provide systematic technical assistance to a number of African countries. Through its regional offices, (in Asia and the Pacific, Europe, Latin America and the Caribbean and West Asia) it provides similar assistance in relation to environmental legislation and institutions as well as in relation to the implementation of international environmental conventions.

## 2.7. THE LEGAL TOOLS FOR ACHIEVING SUSTAINABILITY AT NATIONAL LEVEL

In order for there to be a proper basis for sustainability, an essential task is to draft and enact adequate legal frameworks to allow principles of sustainability to be implemented and enforced. Whilst a number of countries have begun to include sustainable development within their environmental legislation, it is normally mentioned in preambles and object clauses rather than in substantive provisions which oblige decision makers to use the imperative of sustainability as the fundamental guiding force of whether or not any particular development proposal or activity should be allowed to go ahead.

### *2.7.1. An Australian example*

Australia provides some excellent examples of where sustainable development has been incorporated into policy documents and a wide range of legislation at both national and state level. Unfortunately however, the development of jurisprudence exploring the implementation of the concept in most Australian jurisdictions has been slow and a little uncertain to date. At national level, there are some 22 pieces of legislation which include reference to the Australian-coined term "ecologically sustainable development" (ESD). At state level, there are over 100 statutes that contain some reference to the concept (Stein, 2000, pp. 22-23).

One of the first Australian State enactments to incorporate ESD was the Protection of the Environment Administration Act 1991. This statute established the New South Wales Environment Protection Authority, and contains explicit ESD provisions which have been used as the baseline for many other pieces of legislation in that State. Section 6(1) of the Act reads:

The objects of this Act are as follows:

(i) to protect, restore and enhance the quality of the environment in New South Wales, having regard to the need to maintain ecologically sustainable development

The Act states in s 6 (2) that "ecologically sustainable development" "requires the effective integration of economic and environmental considerations in decision making processes, and that this can be achieved by the following principles and programmes:

- the precautionary principle;
- intergenerational equity;
- the conservation of biological diversity;
- the improved valuation and pricing of environmental resources.

The ambiguity in these provisions centers on the use of the word "maintain". This ambiguity is seen at two levels. First, the phrase can be read as implying that there is a need to maintain *development*, as long as it is ecologically sustainable, when clearly in a range of situations, development of any kind is inappropriate; second, there is the implication that ecologically sustainable development is already being achieved, and that there is thus only a need to maintain the status quo. Clearly, at the time the legislation was enacted, ESD was by no means being achieved on any indicator and, seven years later, regulatory authorities at national, state and local level are still struggling with the policies, guidelines and strategies that are meant to achieve it. If a developed country such as Australia cannot put the basic legally enforceable framework in place to make sustainability an absolute imperative in all environmental decision making, what chance do developing countries have to do so? Imitating the Australian legislative rhetoric on ESD is not going to achieve it. The fundamental problem, however, is the fact that the provisions appear in objectives, rather than in the substantive provisions of the legislation. As Stein and Mahony point out:

> A shortcoming of "objective-led ESD" is that the substantive activities and decisions made under legislation are not circumscribed by the principles of ESD. Accordingly, the EPA, for example, may be entitled to make decisions without reference to sustainability. Nowhere does the EPA's enabling legislation *require* it to treat sustainability as a mandatory objective (Stein and Mahony, 1999).

One answer to inadequate drafting of legislation will be found in the innovative approaches taken by lawyers and judges in the courts. In Australia, this has already begun, although so far, only a small number of cases have tested the concept.

In various Asian countries, notably India and the Philippines, we are beginning to see the emergence of jurisprudence which is extending constitutional interpretation and common law understandings of environmental obligations, in circumstances where the environmental and natural resources legislation is not adequate or non-existent (see Boer, 1999, pp. 1516-1517 and 1534-1537). As more legislation includes sustainability as a basic principle, and with an increasingly vigorous public interest environmental law movement in various countries, the provisions of legislation will continue to be challenged.

## 3. Sustainability and Environmental Legal Education

Agenda 21 (1992) provides an explicit charter for the improvement of environmental legal education. Chapter 8.20 states:

> Establishing a cooperative training network for sustainable development law: Competent international and academic institutions could, within agreed frameworks, cooperate to provide, especially for trainees from developing countries, postgraduate programmes and in-service training facilities in environment and development law. Such training should address both the effective application and the progressive improvement of applicable laws, the

related skills of negotiating, drafting and mediation, and the training of trainers. Intergovernmental and non-governmental organizations already active in this field could cooperate with related university programmes to harmonize curriculum planning and to offer an optimal range of options to interested Governments and potential sponsors.

This provision has been a starting point for a range of programmes organised and/or funded by UNEP, the World Bank, the World Conservation Union and by overseas development assistance programmes of various kinds, several of which are mentioned below.

## 3.1. ENVIRONMENTAL LEGAL EDUCATION IN THE ASIA PACIFIC

Universities and other institutions in many areas of the world, both in developed and developing countries, have begun to upgrade their offerings in environmental law. This reflects the role that environmental law itself has begun to play in the development of environmental management mechanisms at an international, regional, national and local level. UNEP has been particularly active in organizing training workshops on environmental law and in implementing specific international environmental conventions, at the regional level, through global training programmes in Environmental Law and Policy, subject to the availability of resources. Comprehensive training programmes have been held regularly since 1993. UNDP and the United Nations Institute for Training and Research, as well as the Asian Development Bank and the World Bank have also begun to rise to the challenge that environmental law represents, in becoming involved in the training of governmental officials and university lecturers. In addition, the World Conservation Union, particularly through its Commission on Environmental Law, has been very active in supporting environmental legal education (see further below).

In the Asia Pacific region, in particular, environmental legal education has been undergoing something of a revolution in recent years. The World Conservation Union and UNEP, with the assistance of the Asian Development Bank and the World Bank, have been involved in various comprehensive environmental law-training programmes that have been established in the last several years. These initiatives are canvassed below.

### *3.1.1. National University of Singapore: The Asia Pacific Centre for Environmental Law (APCEL)*

The Asia-Pacific Centre for Environmental Law (APCEL) was formally launched at the Faculty of Law, National University of Singapore, in July 1996. Initiated by the World Conservation Union's Commission on Environmental Law, APCEL is operating in collaboration with UNEP and with support from the Asian Development Bank and the National University of Singapore. APCEL was established in response to the need for capacity building in environmental legal education and the need for enhancing awareness of environmental issues.

APCEL is to serve as a regional training center for the teaching of environmental law and as both a regional and global clearinghouse for environmental law information in conjunction with the IUCN/UNEP Joint Environmental Law Information System (JELIS). APCEL projects include the development of tertiary level educational curricula for regional instructors in environmental law; the creation of an Internet accessible database containing both international and regional instruments; and various environmental fora and workshops.

One of the most significant of these workshops has been the "Training the Trainers" programme, conducted in collaboration with a wide range of resource persons from universities and organizations around the world. This Asian Development Bank-funded programme attracted over thirty participants from over twenty countries in the Asian and Pacific Region. The programme lasted for some four weeks in 1997 and 1998, drawing resource persons from both international organizations and a range of universities around the world.

*3.1.2. National Law School of India University: Centre for Environmental Advocacy, Education, and Research (CEERA)*

The National Law School of India University is a one faculty university established through the Bar Council of India in the late 1980s. The Centre for Environmental Advocacy, Education, and Research (CEERA) is one of a number of specialist centers within the National Law School. It is pertinent to note that the Bar Council of India has made environmental law a compulsory subject in all Indian law schools. In 1997, CEERA attracted a major World Bank grant for the training of environmental law lecturers and government officials in the area of environmental law. This programme was inspired by the Singaporean programme. It is planned to be conducted over the next five years, in order to train several hundred environmental law lecturers, and eventually, around 5000 government officials in the field of environmental law. A number of intensive programmes have already taken place in various parts of India coordinated by the National Law School of India University. As with the Singaporean programme, outside resource persons are being used together with Indian experts. Comparative study of environmental law is an important part of some of the programmes. A significant feature of the Indian programme is an emphasis on the indigenous customary practices, with lectures on ancient as well as modern environmental management assistance.

*3.1.3.The Research Institute for Environmental Law, Wuhan University*

The Research Institute for Environmental Law, within the Law School at Wuhan University in the People's Republic of China, has been established for over ten years. It is an official research body for the Chinese State Environmental Protection Administration. The Institute runs Master's level programmes and carries out high-level research projects for the Chinese government.

In recent years, the Institute and the Australian Centre for Environmental Law-Sydney have established official links, and have conducted several postgraduate coursework programmes entitled *Sustainable Development Law in China and Australia*, as well as an exchange programme for staff between the two universities. This initiative has been funded by several grants from the Australian overseas development programme, AusAID. In late 1999, the Institute held an international conference celebrating the first 20 years of China's Environmental Protection Law. It also established a new programme entitled Promoting Environmental Law in China, in collaboration with a number of Chinese universities and the Commission on Environmental Law of the World Conservation Union.

*3.1.4. Environmental law teaching in Indonesia*

In Indonesia, environmental law has been a compulsory subject for some ten years. With around 200 more school spread over the archipelago, the task of training environmental law lecturers is large. A longstanding programme of environmental law teacher training is found at Airlangga University in Surabaya.

Several comprehensive environmental law training programmes for Indonesian judges, government officials and non-government organisations have also recently commenced, funded by AusAID. Presently there is also an initiative by the IUCN Commission on Environmental Law to establish a capacity-building programme on environmental and natural resources law, which will include a number of Indonesian universities, in partnership with overseas universities.

## 3.2. THE PROPOSED IUCN ACADEMY OF ENVIRONMENTAL LAW

The World Conservation Union's Commission on Environmental Law recognizes the need for establishing a worldwide approach to the on-going scholarly and professional study of environmental law. It is accepted that a transnational approach is required to establish elements of environmental law in countries where present systems are inadequate, to link the levels of international, regional, national and local environmental into an effective continuum, to link the environmental sciences with environmental law, and to integrate the various sectors of law with environmental law. Advanced training courses can be conducted for the teachers of environmental law, and advanced courses of study can be organized for attaining proficiency in environmental law, as well as conducting special conferences and seminars. This training, further education and meetings will, in turn, promote research for law reform in the participants' countries. To advance the systematic study of environmental law internationally, IUCN has endorsed the prospective establishment of the IUCN Academy of Environmental Law as an autonomous unit working in partnership with the IUCN Environmental Law Programme.

The mandate of the Academy, as presently envisaged, is planned to be:

1. To provide for the formation of a core network of IUCN-endorsed Environmental Law Teaching Institutions in selected member countries;
2. To establish the secretariat of the IUCN Academy of Environmental Law at a selected partner institution;
3. To establish the IUCN Academy of Environmental Law Fund and to ensure that the Academy can take independent initiatives as an academic body, whilst working in close collaboration with the IUCN Commission on Environmental Law and the partner institutions;
4. To endorse Master's-level environmental law courses around the world, graduates of which would qualify for entry into the proposed IUCN Postgraduate Diploma in Environmental Law;
5. To conduct Master's level and post-Master's level courses of study in environmental law in association with the partner institutions, as well as independently;
6. To confer an IUCN Postgraduate Diploma in Environmental Law in association with the core partner institutions;
7. To act as a clearinghouse for the conduct of and participation in advanced courses of training for environmental law professors, in collaboration with partner institutions.

If this project, which is still in the early planning stages, obtains the necessary support, there is little doubt that it will make a substantial difference to the quality and extent of environmental law teaching both in developed and developing countries.

### 3.3. LIBERATING ENVIRONMENTAL LAW FROM LAW SCHOOLS

In teaching environmental law, part of the law professor's duty is to ensure that every course includes at least some rudimentary coverage of the history, meaning, and ethical dimensions of sustainable development, and the study of its implications for environmental management and regulation at international, national and local levels.

Further, if we are serious about achieving sustainability across every sector of economic activity and every relevant manifestation of human behaviour, the teaching of environmental law must be liberated from the confines of law schools and must become part of every major related discipline. The role for university law schools and specialist environmental law centres is to ensure that environmental law is taught within relevant faculties and departments, such engineering, architecture and planning, science, economics, etc., and as part of training programmes for government officials and workers in the private sector. To be effective, all relevant stakeholders need to know the elements of environmental law which govern their particular area of activity, be it pollution control, environmental impact assessment, planning, conservation and exploitation of natural resources or the protection of cultural heritage.

### 3.4. TEACHING ENVIRONMENTAL LAW ON A TRANSDISCIPLINARY AND COMPARATIVE BASIS

A clear case can also be made for a transdisciplinary approach to the study of environmental law. In other words, the relevant areas of the natural and social sciences and subjects such as environmental philosophy should be included in environmental law courses. Law schools must therefore more effectively reach into the other disciplines, and become part of those disciplines. Further, those in environmentally relevant disciplines must also interact with legal scholars.

Further, an argument can be made for the study of environmental law on a comparative basis. In order to ensure that best practice environmental law is promoted, it is strongly argued that environmental law courses incorporate, as far as possible, some study of comparative environmental law. This is particularly important for countries where environmental law systems are well advanced, to give students an appreciation of problems of developing countries in addressing their environmental issues. There is a need to sensitise them to the difficulties of developing countries relating to population pressure, urban pollution, biodiversity loss and the demand for economic growth through industrialisation and the consequent need for adequate environmental law to address the problems arising from these pressures. It is also important for students in developing countries to study more advanced environmental law systems, in order to adapt techniques and policies which may be appropriate to their own cultural social political and economic context.

## 4. Conclusion

The development of environmental law on a global and national level over the past two decades has been rapid, but highly variable from one region and country to another. The same is true of the area of environmental legal education in the Asia Pacific region. The programmes discussed here are one way of achieving a more coherent and collaborative approach. Through cooperation and collaboration between environmental law teachers and practitioners in environmental law and related disciplines and the relevant United Nations organizations and the Commission on Environmental Law, a more coherent and effective approach may be achieved. This collaborative approach promises to be one of the defining characteristics of sustainability law.

Finally, one might reflect on the ethical responsibility of law professors. who teach environmental law. Our primary responsibility may be to imbue law students with a sense of ethical duty, not to the client, to whom an ethical duty is owed as part of a lawyer's professional obligations, but more fundamentally, an ethical duty to the environment itself. Where these two duties ostensibly conflict, the environmental lawyer must work to resolve that conflict in a way that integrates the need for development with that of environmental conservation. The same responsibility is reposed in us when we teach environmental law to those from other disciplines. If we can transmit that sense of ethical duty, we may be a few steps closer to achieving the generation of effective sustainability law, the environmental law for this new millenium.

## References


A-Khavari, A. and Rothwell, D. R.: 1998, "The ICJ and the Danube Dam Case: A missed Opportunity for International Environmental Law?" *Melbourne University Law Review* **22,** 507.

Agenda 21: 1992, United Nations.

Boer, B. W.: 1995, "Institutionalising Sustainable Development; The Roles of National, State and Local Governments in Translating Grand Strategy into Action", *Willamette Law Review* **31,** 307.

Boer, B. W., Ramsay R. and Rothwell, D. R.: 1998, *International Environmental Law in the Asia Pacific,* Kluwer.

Boer, B. W.: 1999, "The Rise of Environmental Law in the Asian Region" *University of Richmond Law Review* **32**, 1503.

Case Concerning the Gabcikovo-Nagymaros Project (Hungary and Slovakia) (Judgment): 1998, *International Legal Materials* **37**, 162.

Convention on Access to Information, Public Participation in Decision Making and Access to Justice in Environmental Matters:1998, Economic and Social Commission for Europe, Committee on Environmental Policy, United Nations Economic and Social Council, ECE/CEP/43, Aarhus, Denmark.

Our Common Future:1987, Oxford.

Stein, P. and Mahony, S.: 1999, "Sustainable Development: From Theory to Practice", in P. Leadbeter, B. Boer and N. Gunningham (eds), *Environmental Outlook: l Law and Policy* **No 3,** Federation Press.

Stein, P.: 2000, "Are decision-makers too cautious with the Precautionary Principle? *Environmental and Planning Law Journal* **17(1)**, pp. 3-23.

# A FUTURE FOR SUSTAINABILITY?

B. JICKLING
*Arts and Science Division, Yukon College, P.O. Box 2799, Whitehorse, Yukon, Canada Y1A 5K4*

**Abstract.** Sustainability has become a focal topic and important goal for many people concerned about environmental issues. It is, therefore, important for educators, and others, to talk about sustainability with their students and colleagues—about its meaning, curricular application, and practice. However, I do not think this is sufficient. In this paper I will examine limitations of the language of sustainability and implications for environmental thinking.

Before launching into a critique, I do acknowledge the importance of "sustainability" and the usefulness of this concept. Many ecological processes are not sustained—not kept going continuously. Species are going extinct at an alarming rate and whole ecosystems are at risk. So, sustainability is important. However, we must also pay attention to what sustainability is not. And, we should consider why these limitations matter.

For example, "education for sustainability" has gained rapid acceptance, yet little critical attention has been given to the term. Just as many environmental educators have expressed reservations about "education for sustainable development," I believe there are serious problems associated with allowing our work to be subsumed by the term "education for sustainability." In this presentation I will explore reasons for educators, and others, to be concerned about relying on the language and goals of sustainability. These areas of concern, or limitations, are discussed in terms of determinism, exclusivity, and conceptualization. Some suggestions are provided to help refocus our direction.

**Keywords**: education, environment, environmental education, environmental philosophy, philosophy, sustainability, sustainable development

## 1. Introduction

Over the past decade there has been much talk, and some lively debate, over the terms "sustainable development" and "sustainability." Those seeking to care for the environment and human-environment relationships have sought goals and rallying concepts around which to organize their efforts. Beginning with the report of the World Commission on Environment and Development (1987) and followed by Agenda 21 (1992), signed by 179 nations in Rio, adherents of sustainable development and sustainability have gained much momentum in their efforts to establish environmental guidelines and goal statements.

This momentum can be seen in the field of environmental education, particularly in the programs and documents sponsored by UNESCO. Six months following Rio, a conference dubbed "EcoEd" was convened in Toronto and billed as the first educational follow-up to the Brazilian meeting. More recently, a series of regional meetings were held in preparation for an international conference hosted by UNESCO and the Government of Greece at

Thessaloniki in December 1997. The resulting proceedings (Scoullos, 1998) revealed that the conference had been organized to highlight "the critical role of education and public awareness in achieving sustainability" (Mayor, 1998, p. 7). Moreover, the "Declaration of Thessaloniki," distributed for participant adoption, reaffirmed UNESCO's preferred goals of "education for sustainable development" and "education for sustainability."

In November 1997, I had the opportunity to attend the regional meeting for the global francophone community hosted in Montreal and called "Planet 'ERE." I was impressed by the vigour of the research and the critical stance taken by many participants who were not convinced of the efficacy and appropriateness of the UNESCO goals. Yet I am left to wonder, what became of this discourse? In spite of these, and a variety of other extant analyses of the sustainability agenda, a reader is hard pressed to find citations of such critiques in the 862 page proceedings from Thessaloniki.

Juxtaposed against this observation we have questions posed by Ann Jarnet of Environment Canada (1998). After promising to implement Chapter 36 recommendations devoted to education, public awareness, and training, why do the recommendations remain so many while the implementation is so little? She further ponders the nature and effectiveness of the terms "education for sustainable development" and "education for sustainability." "Are," she asks, "these terms blueprints for a particular type of action which may actually constrain our possibilities? Or are they simply stepping stones in the evolution of our thinking about education?" (p. 214).

Following Jarnet's (1998) questions, the Canadian Journal of Environmental Education, working together with Environment Canada, Université du Québec à Montréal, and Yukon College, hosted an on-line colloquium in October 1998 to discuss environmental education in Canada, and beyond. Many of the posted papers have been published in Volume 4 of this journal. Taken together, these papers provide readers with a broad range of perspectives and an interesting basis for further discussions about the evolution of our field. Interestingly, some authors felt quite comfortable with terms like "education for sustainability" and sought to infuse this term with meaning, or to use it to address issues under-represented by environmental education (Huckle, 1999; González-Gaudiano, 1999; Gough and Scott, 1999). Others were clearly uncomfortable with the continued sustainability focus (Sauvé, 1999; Berryman, 1999). They expressed concerns about the "globalizing" nature of the "education for sustainability" agenda and the need to nurture alternative perspectives. Yet another author, while recognizing limitations to this terminology, sought means to accommodate the global political agenda (Smyth, 1999). As a tentative step in this direction, he spoke about "education consistent with Agenda 21."

I, too, have concerns as I question the future of "sustainability." Of course this is, in many ways, an important term. Many ecological processes are not sustained. Species are going extinct at an alarming rate and whole ecosystems are at risk. However, the degree to which it remains helpful will depend on how well we recognise its shortcomings as an organizing concept. In this paper I will identify and discuss three pitfalls of a sustainability focussed agenda. I argue that we must first recognize the educational limitations of education for sustainable development or sustainability. Second, I believe that we should seek to avoid the intellectual exclusivity that such an agenda brings. And, third, I discuss the conceptual errors inherent in using sustainability as an aim. I argue that it is not an omnibus term, but rather, one with serious limitations. In answer, to Jarnet's (1998) questions, I believe that "sustainability" is a stepping stone in the evolution of our thinking, and I will conclude with a few suggestions about how we might nurture this process.

## 2. Limitations

### 2.1. DETERMINISM

Perhaps the inherent determinism in goal statements like "education for sustainability" can best be seen in a recent quotation taken from the proceedings from the Thessaloniki conference. For its author, and other adherents, education "should be able to cope with determining and implanting these broad guiding principles [of sustainability] at the heart of ESD [education for sustainable development]" (Hopkins, 1998, p. 172). When highlighted in this way, most educators find such statements a staggering misrepresentation of their task. Teachers understand that sustainable development, and even sustainability, are normative concepts representing the views of only segments of our society. And, teachers know that their job is primarily to teach students how to think, not what to think.

When I look more closely at this quotation, I find at least two problems. The first is the assumption that there might be something like a coherent and cohesive set of guiding principles that can define sustainability and infuse meaning into the term sustainable development. Second is the assumption that education is an instrumental endeavour that can be used to achieve pre-determined goals. Regarding the first point, the authors are clearly building their case on a very shaky foundation. According to recent work by British political scientist Andrew Dobson (1996), there are now three hundred available definitions for sustainability and sustainable development. In light of this new evidence it seems that after more than a decade working to bring meaning to these terms there is less coherence and understanding and perhaps more divergence than previously imagined. The likelihood of arriving at some common understanding of these terms is more remote than ever. The more adherents attempt to infuse these terms with meaning, the more available definitions and the more confusion we seem to get. Moreover, the tenor of these definitions will be clearly dictated by the "stripe" each adherent wears. Particularly alarming for many concerned with the global ecological imperative is the fact that most of the work on these concepts has been done in the economic sector (Dobson, 1996). The best we can hope for from the myriad possibilities is a tussle over whose conception of sustainability and/or sustainable development ought to prevail.

However, as implausible as it seems that we might actually arrive at some consensus on suitable conceptualizations of sustainability and sustainable development, there is still a serious problem with the second assumption. Is it educationally justifiable to "implant" a new normative system into the minds of students? Imagine trying to implant the principles of Marxism into the heart of an American educational system! This would be clearly seen as indoctrination. While presented more subtly, this is precisely what is said when educators speak of education "for" sustainability or sustainable development. In spite of the misgivings described in the previous paragraph, adherents are saying that they know best, and that their value system must prevail.

When "educators" speak with such certitude, with such confidence in their ability to determine the best outcomes for students, they do not see sustainability and sustainable development as stepping stones to future visions; rather, the effect is to constrain possibilities. This approach also exposes educators to serious, and sometimes hostile, criticism. (See, for example the debate between Sanera, 1998 and Courtenay-Hall, 1998; Simmons, 1998; Smith, 1998; and Bowers, 1998). Clearly, if there is to be a future for sustainability within education, we must begin to recognize the educational limitations of the deterministic manifestations of the sustainability agenda. As John Smyth (1999) has said, the practice of education labeled "for" something should have been, by now, demolished.

## 2.2. EXCLUSIVITY

A number of years ago I began to explain why I did not want my children to be educated for sustainable development (Jickling, 1992). This critique has since provoked a variety of responses, many supportive, but some perplexed. Amongst the latter group are those who have wondered just what it is that I do want. The easiest, and still the best, reply has been to express my hope that in ten or fifteen years my children would not even consider a value system, a visionary formulation, or a management compromise that I might accept today. Again, this position can be expressed in terms of Jarnet's (1998) questions. I like to think that present ideas, such as sustainability and sustainable development, are just glimpses of an on-going evolution of environmental thought. We are engaged in a process, and this process is threatened when proponents see these ideas as outcomes.

Aldo Leopold (1949) was an early advocate on behalf of an evolving conception of ethics. In his famous essay "The Land Ethic," he optimistically predicts that our ethical systems may one day extend to be inclusive, of respect for, and duties towards, the land. Affirming his belief in this process, he went so far as to suggest that nothing so important as an ethic should be written down. More recently Anthony Weston revisited these ideas (1992). He argues that traditional ethics, concerned with human, and human-society, relationships, developed over hundreds of years. In this relatively well-established field there have evolved ethical systems which have matured and are now incorporated into many aspects of contemporary decision making and governance. By contrast environmental ethics, with its concern for duties and obligations to the more-than-human (following Abram, 1997) world, is relatively new and we just cannot predict in advance where this thinking will lead. To Weston, environmental ethics is in its originary stage and, for this reason it should be a creative, open-ended process. It must resist temptations to reflect particular normative stances, however enlightened or politically fashionable they claim to be. I agree.

If environmental thought and ethics are evolving processes, then our role as educators is to engage students in this process. Moreover, if environmental thinking is to continue evolving, and if my children are to be participants in an environmental discourse unimagined today, then we must resist temptations to exclude a wide suit of emerging ideas in favour of a sustainability or sustainable development agenda. I want my children to be exposed to a diversity of ideas. I want them to know about bioregionalism, deep ecology, ecofeminism, ecozoic thinking, Leopold's land ethic, environmental justice, social ecology and other emergent forms of environmental thought. Education should be about creating possibilities, not defining the future for our students. And, these creative possibilities can best arise when we embrace exploration, evaluation, and critique of emerging ideas. In this way sustainability and sustainable development are best seen as only two of many stepping stones.

## 2.3. CONCEPTUALIZATION

Much has been said about the difficult, if not impossible, conceptual nature of sustainable development (e.g., Disinger, 1990; Jickling, 1992; Livingston, 1994). For many, a more viable option has been to shift focus to "sustainability." Indeed, this term has seemed unassailable. However, it is now important to look closely at problems associated with emphasizing sustainability as an organizing concept. I argue that as important as it is, this term is not sufficient to explain, or direct, our imperatives; it may even mask important distinctions. Once again, understanding the limitations of sustainability will enhance the future usefulness of this term. To illustrate my point I will draw from two scenarios that have emerged within my own community. The first is about the future of mining and the second concerns wildlife management.

It will be no surprise that citizens differ in their opinions about mining activities and the future of this industry. Embedded in their opinions are environmental, economic, and lifestyle issues. A number of years ago a group of school children embarked on a project to examine these issues in the context of mining. The fruits of their labour were revealed in a collection of letters to the editor of the local newspaper. Like anyone engaged in critique of difficult social issues, the children struggled to balance the environmental health of our region against economic interests and their own consumer desires. While some letters seemed naive, they did reveal an earnest desire to mediate tensions between competing interests. The results inevitably led to recommendations designed to moderate, not eliminate, mining.

Not surprisingly the mining industry responded with incredulity. Surely, we were told, the curriculum must be biased, and that steps must be taken to correct the imbalance (Buckley, 1993). The result was an alliance between industry and government to produce a mining curriculum (Burke and Walker). It is not possible to critique their entire program at this time, but for purposes of illustration I will consider images represented on the poster provided as part of the curriculum package to advertise its arrival into schools and to capture interest of the targeted children. The first thing an observer notices is that the image is full of happy people enjoying products derived from mining activities. Such images complement the thematic slogan running across the bottom of the poster in large letters and reading, "What is Mined is Yours." Of particular interest is additional text that states, "at the heart of our modern lifestyle is a diverse and healthy mining industry."

While true that our society consumes many mining derived products, what is left unchallenged is the possibility that present consumptive lifestyles are not sustainable. In fact, the poster's implicit message is that present lifestyles *ought* to be sustained. This point is driven home by another example form my region. Here the "Concerned Atlin Residents for Economic Sustainability" have mobilized to defend mining from the lobbying efforts of ecological activists (Simpson, 1999). Interestingly, what unites the environmental community arguing on behalf of regional ecology, and the mining community on behalf of economic development, is the word sustainability. Their differences are absorbed by use of this single term and the concept has become cliché. Now ecologists and mining promoters can, with public approval, both use the term sustainability to support radically different values.

Unfortunately, the mantra of sustainability has conditioned many to believe that this term carries unconditional or positive values. Yet critical thought depends on transient elements in ordinary language, the words and ideas that reveal assumptions and worldviews, and the tools to mediate differences between contesting value systems. As the example illustrates, sustainability tends, instead, to flatten out contradictions. And worse still, it is leading us in the direction of Orwell's (1989) famously satirical notion "doublethink" whereby ordinary citizens can increasingly accept contradictory meanings for the same term and accept them both. Seen this way sustainability tends to blur the very distinctions required to thoughtfully evaluate an issue.

The second example draws from a government decision to kill wolves in order to enhance caribou populations. This, too, was a controversial issue with a number of public voices. Some of these voices included First Nations representatives who supported the wolf kill arguing, that freedom to hunt caribou is a deeply cultural experience. For these people, ability to live with the land is closely linked to identity and well being. Other willing participants included local "wildlife managers." For many of these folks the central task is to find less intrusive ways of controlling wolves than shooting them from helicopters. Their interest lies in the ever-more-effective utilization of "resources." A third voice arises from members of the local fish and game association. They see themselves as hunters, along with wolves. It also seems that both hunters deserve a share of the caribou, but that the wolves take too big a share and must be managed. Finally, there are opponents to wolf kills. For many in this group, wolves are scapegoats for past excesses and poor management practices. For opponents, wolf kills run contrary to care and respect for wildlife.

Of course, wolf kill programs are very complex and the purpose here is not to settle issues of public value. However, this brief introduction does reveal further pitfalls in the language of sustainability. In spite of differences between each of these voices, they would all be united by a desire for ecologically sustainable ecosystems, which include wolves. What is really at stake are questions about why particular sets of actions, derived from particular sets of values, should be privileged over others. Wolf kills, like other controversial issues, are not fundamentally about sustainability. Rather, they are issues about cultural identities, respect, society-nature relationships and tensions between intrinsic and instrumental values. Again, sustainability talk can mask central issues.

## 3. Possible Directions

As we enter the new millennium pressure for change will grow; this is an unavoidable reality. At the same time tensions between competing interests and divergent value systems will also grow. While educators can play an important role in preparing the next generation of citizens to mediate these tensions and create new possibilities, their actions will be scrutinized like never before. As the old saying goes, if we enable students to think, there is the danger that one day they might actually do it. At the heart of this joke is the realization that education can be threatening to those with vested interests in the status quo. To some a thinking public is troublesome. To survive, and be effective, environmental educators must become evermore thoughtful about their purpose, evermore vigilant about their weaknesses. This is not a time for leaping on "bandwagons" or grasping for new slogans.

I have acknowledged that sustainability is a useful term. It has the capacity to capture important issues and inspire imagination. But, as I have shown, it alone is not sufficient to organize the educational preparation of thoughtful students and citizens, who will need to examine and evaluate complex issues. I have described some of the limitations of the language of sustainability. We ignore these limitations at our peril; our critics will not. Sustainability, and discussions such as this, can however, lead to new possibilities—provide a stepping stone for the evolution of our ideas. In this section, I provide a few modest suggestions that may help to refocus our direction.

First, we must be less deterministic. The kind of determinism discussed here is easily identifiable (e.g., Sanera, 1998a and b) but not so readily defended. We will more justifiably speak about education "and" sustainability, education "concerning" sustainability, or as John Smyth (1999) has suggested education "consistent with" Agenda 21. These are but a few possibilities that are less ideologically bound and thus stand on a more solid educational footing. These simple adjustments, or others like them, can diffuse some criticism and allow students the intellectual space to move beyond sustainability if they judge this necessary.

Second, we can seek more inclusive language. There are many scholars and citizens who do not feel at home in the sustainability, or sustainable development, "club." Others are simply cynical. Most important, however, are the myriad formulations of other possibilities for environmental thought. Environmental ethics, ecofeminism, and social ecology are all fields in their infancy; let's not cut short the possibilities they, and other emergent fields, offer by focusing so heavily on sustainability. For a start, when we describe environmental education programs for the 21st century let us speak more inclusively about the importance of examining society-environment relationships.

Finally, complex environmental issues are about more than sustainability; this concept alone is too limited to capture the essential issues in environmental education. For example a Mexican colleague speaks about her grassroots project which aims to "promote life which is just, equitable, and ecologically sustainable" (Alvarez-Ugena, 1997). There is a clear indication that citizens of this project believe questions of justice and equity are different from those of ecological sustainability. The issues discussed in this paper reveal that not all values can be sustained, that there are deeper and more important philosophical questions than simply talking about sustainability. We need to speak more confidently about assumptions, lifestyles, worldviews, and conceptions of human place and purpose in ecosystems. We need to talk about cultural, spiritual, and aesthetic values, and not try to subsume these ideas beneath inadequate labels and limited conceptualizations. And, we must find space to discuss cultural identities, respect, society-nature relationships, tensions between intrinsic and instrumental values and other ideas that lie beyond sustainability.

## 4. Conclusion

In the end I do argue that sustainability is an important idea, but one with serious limitations. It is a mistake to think of it as omnibus term, an organizing concept, or an aim of education. Education for sustainability, as seductive as the idea is, falls short of environmental education's largely unrealized potential. So, is there a future for sustainability in the next millennium? Cautiously yes, if we return to Jarnet's (1998) questions, and decide that "sustainability" is a stepping stone in the evolution of our thinking and if we recognize the limitations of this term. Much good work has been done by educators in the name of sustainability, work that we can build upon. But, the real challenge for the next century is to go where sustainability cannot.

## Acknowledgment

Special thanks are given to Priscilla Clarkin, editorial assistant for the Canadian Journal of Environmental Education, for her helpful suggestions during the preparation of this paper.

## References

Abram, D.: 1997, *The spell of the sensuous*, Vintage, New York.

Alvarez-Ugena, E.: 1997, Presentation to the 6th International TOUCH Conference Centre for Environmental Education and Ethics Rychory - SEVER, Horni Marsov, Krkonose, Czech Republic, April 26 to May 2, 1997.

Buckley, A.: 1993, June 2, School's mining curriculum "biased," *Yukon News*.

Burke, J. and Walker, E.: undated, *Rock on Yukon*, Yukon Chamber of Mines, Government of Canada, and Yukon Department of Education, Whitehorse.

Disinger, J.F.: 1990, *J. of Environ. Education*, **21**(4), 3-6.

Dobson, A.: 1996, *Environ. Politics*, **5**(3), 401-428.

Berryman, T.: 1999, *Canadian J. of Environ. Education*, **4**, 50-68.

Bowers, C.A.: 1998, *Canadian J. of Environ. Education*, **3**, 57-66.

Courtenay-Hall, P.: 1998, *Canadian J. of Environ. Education*, **3**, 27-40.

González-Gaudiano, É.: 1999, *Canadian J. of Environ. Education*, **4**, 176-192.

Gough, S. and Scott, W.: 1999, *Canadian J. of Environ. Education*, **4**, 193-212.

Hopkins, C.: 1998, *Environment and Society: Education and Public Awareness for Sustainability*, in M. Scoullos (ed), Proceedings of the Thessaloniki International Conference organised by UNESCO and the Government of Greece (8-12 December 1997), 169-172.

Huckle, J.: 1999, *Canadian J. of Environ. Education*, **4**, 36-45.

Jarnet, A.: 1998, *Canadian J. of Environ. Education*, **3**, 213-215.

Jickling, B.: 1992, *J. Environ. Education*, **23**(4), 5-8.

Leopold, A.: 1949, *A Sand County Almanac*, Oxford University Press, Oxford.

Livingston, J. A.: 1994, *Rogue Primate: An Exploration of Human Domestication*. Key Porter Books, Toronto.

Mayor, F.: 1998, *Environment and Society: Education and Public Awareness for Sustainability*, in M. Scoullos (ed), Proceedings of the Thessaloniki International Conference organised by UNESCO and the Government of Greece (8-12 December 1997), 7.

Orwell, G.: 1989, *Nineteen eighty-four*. Penguin Books, London. (First published in 1949).

Sanera, M.: 1998a, *Canadian J. of Environ. Education*, **3**, 9-26.

Sanera, M.: 1998b, *Canadian J. of Environ. Education*, **3**, 67-81.

Simmons, D.: 1998, *Canadian J. of Environ. Education*, **3**, 41-47.

Simpson, S.: 1999, February 19, *The Whitehorse Star*, 11.

Smith, G.: 1998, *Canadian J. of Environ. Education*, **3**, 48-56.

Smyth, J.: 1999, *Canadian J. of Environ. Education*, **4**, 69-82.

United Nations Conference on Environment and Development: 1992, *Agenda 21*, the United Nations Programme of Action from Rio, UN Department of Public Information, New York.

Weston, A.: 1992, *Environ. Ethics*, **14**(4), 321-338.

# TRANSDISCIPLINARITY IN GROUNDWATER MANAGEMENT – TOWARDS MUTUAL LEARNING OF SCIENCE AND SOCIETY

R.W. SCHOLZ[1], H.A. MIEG[2] and J.E. OSWALD[1]

[1] *Department of Environmental Sciences, Natural and Social Science Interface (UNS), ETH Zurich, HAD, Zurich, CH,* [2] *Department of Environmental Sciences, People and Environment Interaction (MUB), ETH Zurich, HAD, Zurich, CH*

**Abstract.** Transdisciplinarity is defined as a specific type of knowledge prodution that has to be distinguished from interdisciplinarity. Transdisciplinarity deals with relevant, complex societal problems and organizes processes of mutual learning between agents from the scientific and the non-scientific world. We introduce principles of transdisciplinarity when presenting projects on groundwater and soil management in the Klettgau trough, a catchment of 156 $km^2$. Currently, the groundwater aquifer shows a critical nitrate load due to intense agriculture.

The projects are the ETH-UNS Case Study *Responsible Soil Use* in the German-Swiss Area and the Interreg II EG/EU project *Developmental Strategy for the Klettgau Trough.* The ETH-UNS study is part of a new type of undergraduate course. The project organization, the methods of knowledge integration, process details and results are introduced as an example of transdisciplinary activities. The project was linked to the INTERREG II program that also used participatory methods in soil and groundwater management.

**Keywords:** groundwater management, interdisciplinarity, knowledge integration, mutual learning, nitrate, transdisciplinarity

## 1. Introduction

We define groundwater management as all planned activities that assess and affect the quality and quantity of phreatic water. Groundwater management is considered a typical object of transdisciplinary research. This is due to the establishing of environmental standards, appropriate solutions and the implementation of technology and problem solving requiring joint problem solving between science, technology, and society. In Switzerland and Germany and other Central European countries there is no shortage of water. However, the water quality is poor in some places. This also holds true for the Klettgau Trough which is a glacial valley crossing the Swiss-German border west of Schaffhausen (see Figure 1). The 30 km long trough was formed by the Rhine glacier during the Riss ice age. Today it is an area of intense agriculture causing a groundwater load infringing the EU standard of 50 mg $NO_3$/l. In this paper, we will introduce the concept of transdisciplinarity and will describe two projects to illustrate this concept.

*Water, Air, and Soil Pollution* **123:** vii, 2000.

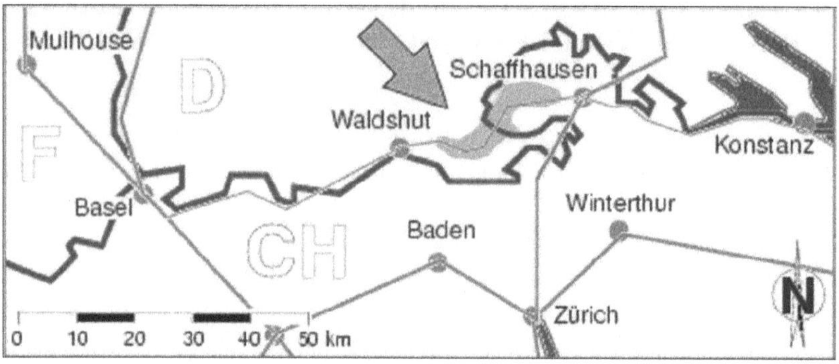


Fig. 1. The Klettgau Trough is a catchment of 156 km$^2$ crossing the Swiss-German boarder west of the city of Schaffhausen. Due to intense agriculture there is an increasing nitrate-load in the main stream of the aquifer below the trough.

## 2. What is Transdisciplinarity?

There have been many attempts to classify scientific knowledge production. Disciplinarity is the label of science production within a single subject area. "Pluri- or multidisciplinarity is characterized by the autonomy of the various disciplines and does not lead to changes in the existing disciplinary and theoretical structures, interdisciplinarity is characterized by the explicit formulation of a uniform, discipline-transcending terminology or a common methodology" (Gibbons *et al.*, 1994, p. 28). Interdisciplinarity thus is featured by crossing boundaries between different sciences and coordinated exchange of ideas striving for a fusion of concepts and methods from different disciplines (Holzhey, 1974; Gibbons *et al.*, 1994).

The term transdisciplinarity is occasionally referred to as a "perfected interdisciplinarity" (Mittelstrass, 1996; Thompson Klein, 1996) or as the transfer of concepts or methods from one discipline to another (Arber, 1993). However, Jantsch (1972) already differentiated interdisciplinarity as a "coordination by higher-level concept" from transdisciplinarity as a "multi-level coordination of entire education/innovation system". Thus, transdisciplinarity goes beyond disciplines and also beyond sciences. Giovannini and Revéret (1998) point out that "transdisciplinarity is often invoked in connection with important societal problems. Mostly it has turned out that a close collaboration of academics with stakeholders from society on a problem involving different disciplines in a complex was one of the methods that has led to a true transdisciplinarity attitude from the academic participants." We will consider transdisciplinarity as a type of scientific activity that:

- supplements the traditional disciplinary and interdisciplinary scientific activities by incorporating processes, methodologies, knowledge and goals of stakeholders from science, industry, and politics;
- deals with relevant, complex societal problems and thus has the potential to contribute to sustainable development;
- organizes processes of mutual learning between science and society so that also persons from non-academia participate in transdisciplinary processes.

## 3. Why Do We Need Transdisciplinarity?

There are many reasons for the need of transdisciplinarity when dealing with environmental systems. We argue that problem definition, problem representation and problem solving should be considered as transdisciplinary activities. From our point of view the residents of an area are the problem owners. Furthermore they may also represent valid knowledge about the environmental quality (Wynne, 1982). Giovannini and Revéret (1998) revealed, that the traditional mode of theory-practice transfer may be denoted a linear/reservoir model of science-society interaction. Society spends money on sciences that generate a reservoir of knowledge. Via applied research or special programs the theory-practice transfer is then organized according to a broadcasting model. This traditional way of problem solving usually leads to a new problem, i.e., the implementation problem. Scientists have mostly developed technical strategies and options for problem solution. However economical, psychological and social issues often provide obstacles of implementing these strategies. We argue that the barrier in the theory-practice relationship can be overcome by transdisciplinarity. Transdisciplinarity requires a new role of the researcher. "To the criterion of intellectual interest and its interaction, further questions are posed, such as 'Will the solution, if found, be competitive in the market?' 'Will it be cost effective?' 'Will it be socially acceptable?'" (Gibbons *et al.*, 1994, p.8). Further the transdisciplinary researcher will encounter criteria of responsibility and accountability.

Though we will not go into detail here, we want to note that the strategy of transdisciplinarity may be highlighted by the *Theory of Probabilistic Functionalism* developed by Egon Brunswik (1943, 1952; see Scholz 1999). This theory suggests a series of basic principles for good (organismic) achievements when coping with complex problem structures. If multiple agents are sampling information in an incomplete, probabilistic manner that also represents different perspectives and fractions of reality, the redundancy and sufficiency principle are decisive for a successful performance. Thus joint problem solving in groundwater management must obey these principles and represent a set of different agents from science and society that allow for sufficiency in problem definition, representation and solving. A critical issue in this process (see Scholz, 1999) is that the different agents must focus on one and the same problem and synthesize the information sampled in an effective manner. A significant point in this process is how the different agents coordinate and adjust their activities. We label this process of coordination as *mutual learning*. It is a decisive part of the transdisciplinary process and encompasses the (evolutionary successful) adaptation process that is linked to the interaction and to the joint problem solving between science, technology and society.

## 4. Transdisciplinarity in Education

The ETH-UNS Case Study on the Klettgau area is part of the five-year master program in Environmental Sciences. The study is a compulsory course for all students in the 8th semester, 14 weeks in duration. 82 students of Environmental Sciences specialized in different fields (i.e., environmental hygiene, terrestrial systems etc.) participated for 18 hours a week. The students were coached by 28 professors, assistants, and scientists from outside universities who were hired for 25 days each. Additionally three persons were responsible for project management. Further scientists from 13 different chairs of the ETH were involved. Most remarkably, about 150 residents, farmers, persons from ministries, business, etc. also participated.

The ETH-UNS case study is a hybrid as it combines training, research, and application:

- Teaching objectives: The students should learn to master ill-defined environmental problems through knowledge integration and cooperation with case agents. Problem definition and "solving" should be arranged in a process of mutual learning between science and society. This means that the case study team and the agents from the case have to initiate a process of cooperative learning. The case study should contribute to developing an encompassing "environmental problem solving ability" (Scholz *et al.*, 1997).
- Research on knowledge integration and environmental problem solving: The elaboration of and research on the methods of knowledge integration are part of the course (see below and Scholz & Tietje, 1999).
- Applied problem solving: The case study should support the problem solving process and foster local sustainable development in the Klettgau Area.

### 4.1. THE KLETTGAU

The Klettgau Trough is a valley of 156 km$^2$, of an average altitude of 450 meter above sea level. It was a former river bed of the Rhine with a current decent of 100 m at a total length of 33 km and an average width of 3 km. The annual average temperature is 9.0°C (January: mean –0.9° C, min. –22° C; July: mean – 18.4° C, max. –36° C). The precipitation yields 979 mm per year. The water balance renders 343 mm p.a. (≈35%) water discharge which provides a total amount of 56.9 Mio. $m^3$ p.a. About 37.3 Mio. $m^3$ of this discharge flows off under ground (Kühnle-Baiker *et al.*, 1992; Prasuhn 1998). The subterranean ground is extraordinarily heterogeneous including fluvial loess, lime, clay, marl, and quartz. Most of the groundwater aquifer flows in an east-west direction, i.e., from the Swiss (60% of the area) to the German part (40% of the area).

There are 24,000 residents living in small villages with anywhere from 300 to 3000 inhabitants. There is intense agriculture and forestry with 35% arable land, 16% permanent grassland, 3.9% vineyards and 45% forest in the Swiss part. The numbers for the German part are 28%, 22%, 0.8% and 49% respectively. Only about 5% of the area in the Swiss part is in a 'close to nature' shape with 14% at the German side (Reineking & Fendt, 1999).

Today, there is a critical nitrate load. In the western German parts of the Klettgau Trough the loads exceed 50 mg/l which is the EU-standard (see Figure 1). The nitrate balance yields a discharge of 444 t N (≈62 %) for the Swiss and 179 t N for the German part. This corresponds to a negative balance of 26.8 kg N/Ha for the Swiss and 25.7 kg N/ha for the German part (Prasuhn, 1998, S. 288). The most important share comes from arable land with 67% followed by the wooded area at 11% and grassland at 9%. The rest is from vineyards, erosion, etc. 74% of the nitrate load is classified as diffuse anthropogene, 11% as punctual anthropogene (e.g., from sewage plants) and only 15% is derived from natural sources.

## 4.2. THE ARCHITECTURE OF KNOWLEDGE INTEGRATION FOR ORGANIZING TRANSDISCIPLINARY PROCESSES

We will distinguish between a vertical and a horizontal structure of the study (Scholz & Tietje, 1999). The vertical structure is comprised of three levels of cognition (see Figure 2). On the first level is the "case", the Klettgau with its history, constraints and dynamics. All participants in a transdisciplinary process have to develop an empathic understanding. The researcher should always and only focus on the Klettgau and not only a particular issue of specific interest. On the second level are the "syntheses". This is the level of the knowledge integration. The methods introduced organize the synthesis of different levels and types of knowledge. Thus they help to develop a more valid case understanding. The methods are both tools of knowledge organization and of managing cooperation in the study team. Thirdly, we distinguish between two types of data. On the one hand there is the data from the case, i.e., the Klettgau. These are observations, measurements, surveys, documents, expertise, etc. On the other, there is the data from the scientific body of knowledge that is consulted in order to analyze and understand the case. The data-level work is often organized in sub-projects. Note that these sub-projects and the data that are generated have to be strongly organized along the requirements of the synthesis level. All types of analytic methods from different sciences may be used on the data level.

**Responsible Soil Use**

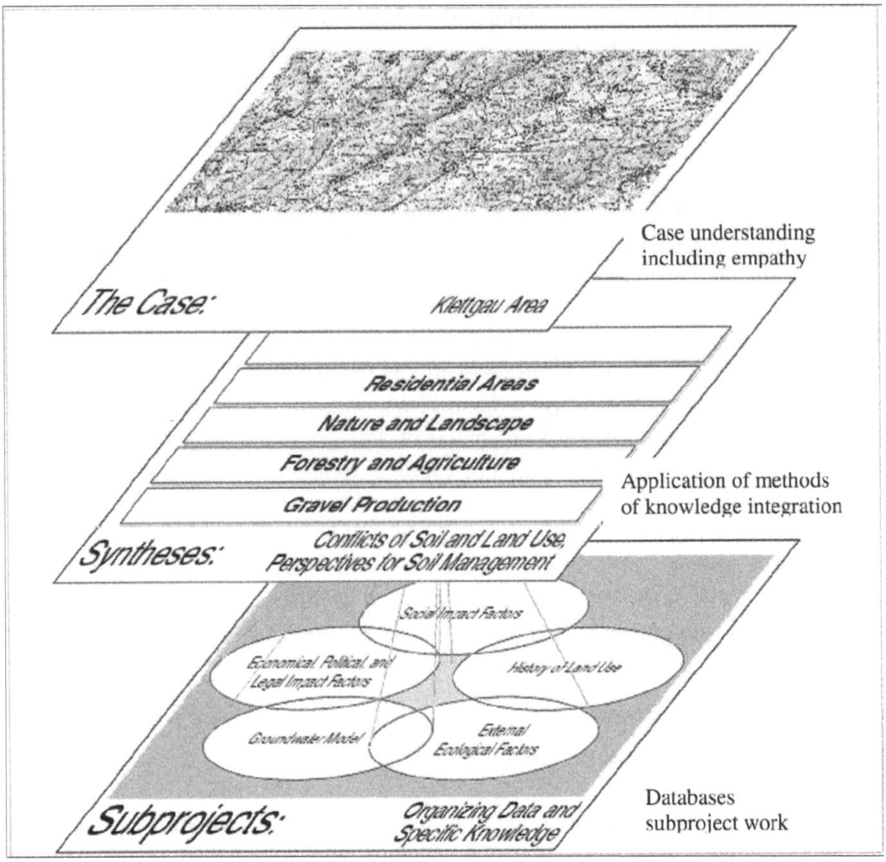


Fig. 2. "Architecture" of an ETH-UNS case study, with the three levels case, syntheses, and subprojects.

The horizontal structure of the study is comprised of the three phases (see Figure 3). Phase 1 was organized by a commission and started with the election of a suitable case. In the ETH Case Study Approach, possible candidates from outside the university and the members of the department are invited to suggest potential cases for a groundbreaking case study. The Klettgau case study was suggested by the Ministerium für Umwelt und Verkehr (Ministry for Environment and Traffic) of the German state Baden-Württemberg. The minister was concerned about ground-water protection, as well as about questions of safety management, sewage cleanup, industrial soil contamination and steep-in water and agricultural land use and land use development.

The first phase also entailed a firm coordination with the Interreg II EG/EU Program: Border-Crossing Cooperation in Groundwater Protection for the

Klettgau Trough (Regli *et al.*, 1998), that entailed studies on the nitrate balance (Prasuhn, 1997) and the groundwater model (Bühl, 1998). A participatory study on the reduction of nitrates by biological agriculture was also financed (Freyer *et al.*, 1998). In a longer discourse with key agents from environmental authorities of both countries (including a written questionnaire) we faceted the case Klettgau. Faceting a case means that the most salient aspects are defined which represent the major issues of the topic *Responsible Soil Use*. The subsequent issues were considered as facets: *Water Management; Gravel Production; Forestry and Agriculture; Nature and Landscape Protection; Residential Areas.*

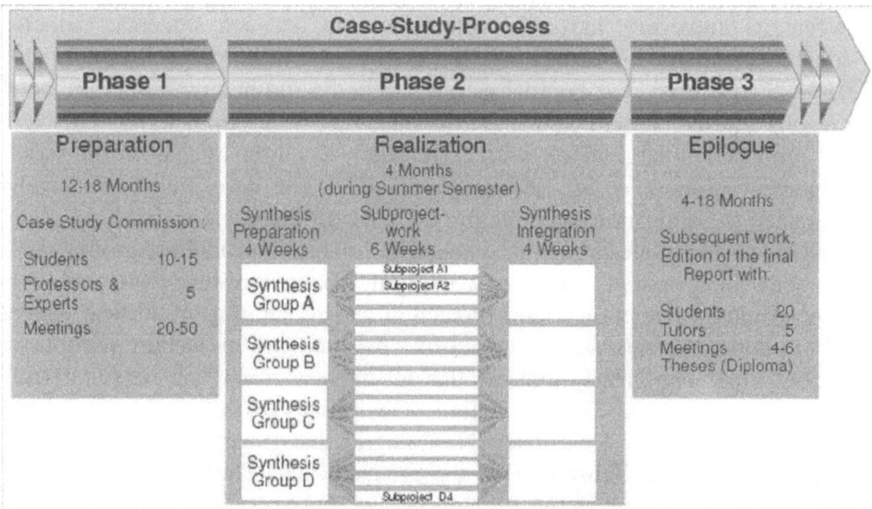


Fig. 3. The three phases of an ETH-UNS Case Study.

Phase 2 is the core of the project. For 14 weeks the whole study team worked on the case. The facets of the case (see Figure 2) defined the topics of the synthesis. Four to five tutors supervised a group of 15–20 students. Phase 2 had three parts. First, the students had to understand the case, to find the relevant issues for the project target. Each of the synthesis groups had to constitute a thematic partner group from the area and received feedback from the advisory board. Furthermore, methods of knowledge integration had to be selected (Scholz & Tietje, 1999). These are methods for representing a case (such as System Dynamics or Formative -Scenario Analysis), for the evaluation of cases (such as Multi-Criteria Utility Models, Integrated Risk Assessment or Life Cycle Assessment), or for case transformation (such as Mediation or Future Workshops). The role of these methods is to allow for a scientific sounding work in a transdisciplinary process.

In Phase 3, the reports of the synthesis groups are finished. There is an extended review process including extended feedback from scientists and from the case. The study ends with an official event. All members of the case and the study team are invited for the presentation of the final results. More than 250 persons attended. Lastly, the case report is submitted to the public (Scholz *et al.*, 1998).

## 5. Transdisciplinary Processes and Outcomes

The ETH-UNS Case Study on the Klettgau was not simply a cooperation between the university and local authorities, but between students, farmers, residents, scientists, authorities and the public (see also Mieg, 1996). In the ETH-UNS case study, the external partners were the "problem owners" and both the results and the process results are the outcomes of a joint activity. The case study produced a large book with 318 pages entailing a chapter on water management (Scholz *et al.*, 1998, pp.165-203). The case study intensively investigated the allocation dilemma provided by different action fields in water management. The preferences of 55 members of different interest groups (i.e., farmers, decision makers, scientists working in local authorities, residents) with respect to nitrate reduction, flood protection and re-naturing of surface water, and hazardous incidents were assessed in a psychometric procedure (Scholz *et al.*, 1998). The data clearly indicate that all interest groups are sensitive with respect to groundwater contamination. Obviously government, population, environmentalists, local enterprises (N=37) agree that gravel should be quarried in a responsible way taking care of the ground water. There exists, however, some knowledge deficits in the population: The risks of groundwater pollution due to inadequate refilling of gravel pits is highly underestimated, the value of ancient gravel pits as biotope is not fully known. The book did not provide specific recommendations, but rather orientations that may be used as decision aids. As a special feature, this book has to be regarded as a product of a transdisciplinary process. This means that it reflects and presents the process of the study in a way that the authorities and residents from the area may “feel and see” that their knowledge and their understanding of the area is incorporated in the analysis and the conclusions. A couple of hundred books were disseminated in the Klettgau.

Clearly the outcomes on the groundwater quality may only be assessed in the long run. However there is already a series of impacts that may be numerated: more than fifty newspaper articles intensively reported about the case study and the results. The students participating in the case study initiated an interest group for the promotion of bioecological projects and provided a handbook that allowed for networking (Fendt & Schaffhauser, 1998). In the field of transportation, the result was a co-operation with the German Motorway Administration Office, which went carefully through the project-book and tried to optimize the planning according to the orientations. Various villages invited the case study students to present results and ideas.

The INTERREG II project provided reports with "hard data" (see Regli *et al.*, 1998). The groundwater use with 600 l/s is considered as sustainable (see above). However, an improvement of the groundwater quality can only be attained if a variety of activities are simultaneously implemented including a reduction of fertilizing and silage corn, altering cornfield to grassland, introducing catch crops, crop rotation, gentle tillage operations etc.. These measures, however, are not free. They will cause a reduction of income of about 960 Fr./ha for a Swiss and of 890 DM/ha for a German farm (Freyer *et al.*, 1998).

Also, the INTERREG II chose a transdisciplinary approach. Thus, for instance, a group of 25 German and 15 Swiss accompanied the project on nitrate reduction from the beginning to its end. Thus already the project goals and the problem solution became a joint activity of science and society.

## 6. Conclusion: Challenges of Transdisciplinarity

Transdisciplinarity is a challenge for students, teachers, and researchers, as well as for society and sciences.

For students, transdisciplinarity is (still) an exceptional situation. It requires a new "attitude" (Givannini & Revéret, 1998). Some of the prerequisites are the willingness to learn, to listen, to cooperate and to accept other interests and values. Furthermore the principle of self-organized learning must be accepted. We found that not all students of a traditional polytechnic sufficiently cope with this challenge. Though a majority of students feels content with transdisciplinarity, about one quarter seems to prefer the traditional lecturing.

Transdisciplinarity requires entirely new teacher aptitudes. The teacher must have not only the ability to educate and convince the students, but also the members of the case. Both may be mastered with the discussion pedagogy. This requires a "Socratic approach", which is another name for pro-active interaction between teachers and students (Ronstadt, 1993). The critical issue for many teachers is that the situation and the problem are not under (complete) control. What is to be dealt with is a yield of discourse and can not be preplanned.

Transdisciplinarity sets sciences on its head. Usually the researcher screens the universe from a narrow theoretical perspective and focuses only on those phenomena that may be somewhat explained by his/her theory. Within a transdisciplinary discourse, the societal relevant issues have to be explained. Thus the researcher must have techniques at his/her disposal to scan the body of scientific knowledge related to a certain problem and to make those parts available which help for that very problem. This is not a simple task and requires (new) methods of knowledge management and knowledge integration.

Sciences are disciplinarily organized. Though environmental programs ask for joint problem solving and often for interdisciplinarity, transdisciplinarity asks for a new type of knowledge integration. One critical issue is that the sciences keep their independence and develop appropriate strategies and standards for transdisciplinarity which extend and complement the traditional criteria.

Quite often society feels uneasy with science. In particular, the theory-practice relationship is unsatisfactorily solved. Quite often the researcher is at a distance to society. In particular, sciences up to now have not yet developed enough to cope with complex, societal relevant problems. Thus, for instance, the fragility of large systems is not well understood and waits for joint problem solving between science and society.

To summarize, transdisciplinarity defines a type of scientific activity and knowledge production that is different from the traditional disciplinary and interdisciplinary scientific activities since it incorporates processes, methodologies, knowledge and goals of stakeholders from science, industry, and politics. Transdisciplinarity is a principle for organizing processes of mutual learning and problem solving between science and society. Thus, transdisciplinarity may contribute to sustainable development.

## References


Arber, W. (ed.): 1993, Inter- und Transdisziplinarität: Warum? – Wie? (Inter- et Transdisciplinarité, Pourquoi? – Comment?), Haupt, Bern.

Brunswik, E.: 1943, Organismic Achievement and Environmental Probability, *Psychological Review* **50**, 255-272.

Brunswik, E.: 1952, The Conceptual Framework of Psychology, *International Encyclopedia of Unified Science* **Vol. 1, No. 10**, University of Chicago Press, Chicago, pp IV, 102.

Bühl, H.: 1998, Interreg II EG/EU: Entwicklungskonzeption Klettgaurinne. Hydrogeologie und Risikovorsorge, in R.W. Scholz, S. Bösch, H.A. Mieg, & J. Stünzi (eds.), *Klettgau – Verantwortungsvoller Umgang mit Boden. ETH-UNS Fallstudie 1997*, Rüegger, Zürich.

Fendt, R., & Schaffhauser, M.: 1998, *Grundlagenordner. Arbeitsmittel für naturrelevante Projekte in der Region Klettgau*, Eidgenössische Technische Hochschule, Zürich, Professur für Umweltnatur- und Umweltsozialwissenschaften.

Freyer, B., Hartnagel, S., Rennenkampff, K., Lüscher, A., & Zeh, H.:1998, Erarbeitung von Massnahmen zur Reduktion der Nitratauswaschung ins Grundwassser durch Anpassungsmassnahmen in der Landwirtschaft im Klettgau, in R.W. Scholz, S. Bösch, H.A. Mieg & J. Stünzi (eds.), *Klettgau – Verantwortungsvoller Umgang mit Boden. ETH-UNS Fallstudie 1997*, Rüegger. Zürich.

Gibbons, M., Limoges, C., Nowotny, H., Schwartzman, S., Scott, P., & Trow, M.:1994, *The New Production of Knowledge*, Sage, Thousand Oaks.

Giovannini, B. & Revéret, J. P.: 1998, (in preparation), *The Practice of Transdisciplinarity in Sustainable Development, Concept, Methodologies and Examples*, University of Geneva: Departément de la matière condensée.

Holzhey, H. (ed.):1974, *Interdisziplinär* (Interdisciplinary), Schwabe, Basel.

Jantsch, E.:1972, *Towards Interdisciplinarity and Transdisciplinarity in Education and Innovation, Interdisciplinarity, Problems of Teaching and Research in Universities*, OECD, Paris.

Kästli, P., Krapf, H., Weber, P., Wüthrich, F. & Weber, O.:1998, Natur und Landschaft, in R.W. Scholz, S. Bösch, H.A. Mieg, & J. Stünzi (eds.), *Klettgau – Verantwortungsvoller Umgang mit Boden. ETH-UNS Fallstudie 1997*, Rüegger, Zürich.

Kuehnle-Baiker, E.K.: 1992, *Erkundung des Grundwasservorkommens in der Klettgaurinne (deutscher Teil)*,Dissertation, Universität Freiburg.

Mieg, H.A.:1996, Managing the interfaces between science, industry, and education, in UNESCO (ed.), *World Congress of Engineering Educators and Industry Leaders* **Vol. I,** UNESCO, Paris, 529-533.

Mittelstrass, J.: 1996, *Stichwort Interdisziplinarität. Mit einem anschliessenden Werkstattgespräch*, Europainstitut an der Universität Basel, Basel.

Palmer, S.E.:1978, Fundamental aspects of cognitive representation, in E. Rosch, & B.B. Lloyd (eds.), *Cognition and Categorization,* Erlbaum, Hillsdale, 259–303.

Prasuhn, V.:1998, Abschätzung der Stickstoffverluste aus diffusen Quellen in die Gewässer und deren Wirkung von Massnahmen in der Landwirtschaft im Klettgau, in R.W. Scholz, S. Bösch, H.A. Mieg, & J. Stünzi (eds.), *Klettgau – Verantwortungsvoller Umgang mit Boden, ETH-UNS, Fallstudie 1997*, Rüegger, Zürich.

Regli, K., Roth, H., Biedermann, R., Pabst, W., & Scholz, R.W.:1998, *Entwicklungskonzeption Klettgaurinne: Interreg II EG/EU: Grenzüberschreitende Zusammenarbeit beim Grundwasserschutz: Schlussbericht.* (Developmental Strategy for the Klettgau Trough: Interreg II EG/EU: Border-Crossing Cooperation in Groundwater Protection), Zweckverband Klettgauwasserversorgung. Klettgau.

Reineking, B. & Fendt, R.: 1999 (in press), Auf dem Weg zu einer integrierten Landschaftsgestaltung – Das Beispiel Klettgau, in R.W. Scholz, S. Bösch, L. Carlucci, & J. Oswald (eds.), *Chancen der Region Klettgau – Nachhaltige Regionalentwicklung. ETH-UNS Fallstudie 1998*, Rüegger, Zürich.

Ronstadt, R.:1993, *The Art of Case Analysis* (3rd edition), Lord Publishing, Pepperdine.

Scholz, R.W.: 1999, *Mutual Learning und Probabilistischer Funktionalismus – Was Hochschule und Gesellschaft voneinander und von Egon Brunswik lernen können*, UNS-Working Paper 21, Eidgenössische Technische Hochschule, Professur für Umweltnatur- und Umweltsozialwissenschaften, Zürich.

Scholz, R.W., Flückiger, B., Schwarzenbach, R.C., Stauffacher, M., Mieg, H.A. & Neuenschwander, M.:1997, Environmental Problem Solving Ability Profiles in Application Documents of Research Assistants, *Journal of Environmental Education* **28** (4), 37–44.

Scholz, R.W., Bösch, S., Mieg, H.A., & Stünzi, J. (eds.):1998, *Klettgau – Verantwortungsvoller Umgang mit Boden. ETH-UNS Fallstudie 1997*, Rüegger, Zürich.

Scholz, R.W., Bösch, S., Carlucci, L., & Oswald, J.: (eds.): 1999 (in press), *Chancen der Region Klettgau – Nachhaltige Regionalentwicklung, ETH-UNS Fallstudie 1998*, Rüegger, Zürich.

Scholz, R.W. & Tietje, O.: 1999 (in press), *Integrating Knowledge with Case Studies. Formative Methods for Better Decisions*, Sage, Thousand Oaks.

Thompson Klein, J.: 1996, *Crossing Boundaries: Knowledge, Disciplinarities, and Interdisciplinarities,* University Press of Virginia, Charlottesville.

Wynne, B.:1982, *Rationality and Ritual: the Windscale Inquiry and Nuclear Decisions in Britain*, British Society for the History of Science, Chalfont St. Giles.

# SUSTAINABLE ENVIRONMENTAL EDUCATION FOR A SUSTAINABLE ENVIRONMENT: LESSONS FROM THAILAND FOR OTHER NATIONS

J. GALLAGHER[1], C. WHEELER[1], M. MCDONOUGH[1] and B. NAMFA[2]
[1] *Michigan State University, East Lansing, Michigan. USA,* [2] *Thailand Ministry of Education, Bangkok, Thailand*

**Abstract.** The Thailand Ministry of Education and Michigan State University collaborated to plan, design, implement, and evaluate this project in which school children and adults in 11 communities worked together to prevent forest degradation and improve forest quality. School children in grades 5 though 8 studied the causes and consequences of deforestation and forest degradation and brought the results of their investigations to local community members. This resulted in identification of local environmental problems that were forest-related and joint efforts in which adults and children worked together to resolve these problems. The results of this intergenerational work were very promising as children, adults, schools, communities, and the environment benefited from collaboration. Important lessons for others embarking this work are reported.

**Keywords**: environmental education, forestry education, improving teaching and learning, intergenerational education, school-community relations

## 1. Introduction

Two issues, arising from different backgrounds, lay at the genesis of this project which had its beginnings in 1991. The first is an educational issue: How to make education more effective. The second is an environmental issue pertaining to the rapid loss of forests, especially in tropical and sub-tropical regions of our earth.

The events behind this project began in 1991 when several members of Thailand's Ministry of Education visited several U. S. universities, including Michigan State University. With USAID funding, the Ministry had embarked on an ambitious project to improve the curriculum in environmental education. The Ministry's team saw this as an opportunity to improve the effectiveness of Thai schools in helping to prepare youth for their future. They sought to partner with a U. S. university to accomplish these goals.

The concern for quality resided in the character of Thai schools at the time. Prior studies indicated that instruction in Thai schools tended to be highly teacher-centered, with much teaching described as "chalk and talk." By the end of primary school, student motivation tended to be low. Approximately 75% of the students did not continue past the sixth grade. The curriculum was described as textbook-centered, with little application of knowledge that was taught to the

*Water, Air, and Soil Pollution* **123:** vii, 2000.

practical world outside of school (Wheeler *et.al,* 1989). As a result, many students were poorly prepared for continued study and for life in the real world outside the classroom.

In making a presentation to the Thai Ministry officials about alternative approaches to schooling, members of the Michigan State University team suggested an approach to environmental education that was viewed as promising. The proposed plan involved school children and adult community members working together to resolve environmental problems at the local level. After some discussion about alternative concerns such as land use and water supply, the group focused on forests and the widespread deforestation that had occurred, and continued to occur, in Thailand. Further, it was recognized that forests and deforestation constituted a rich area for school students to study that involved application of many important concepts that are included in the normal school curriculum relating to study of plants, photosynthesis, soils, water cycle, air quality, and human/environment interactions. Moreover, the issue of deforestation was seen not only as a local problem, but it was viewed as a serious national and international problem as well. Thus an important concern for inclusion as part of a school curriculum was met by viewing deforestation as a problem that could be understood at many levels of complexity, involving concepts from several of the school subject fields. As a result of these discussions, Ministry officials invited Michigan State University to collaborate in developing a pilot project.

The key question underlying this project was "can we formulate a project that will improve both the quality of schooling and the quality of forests in rural north Thailand?" This paper aims to present a response to this question.

## 2. Development of the Project Plan

The Thailand Ministry of Education officials and the Michigan State University colleagues concluded their meeting with the beginnings of a plan that was to take shape over the next eighteen months. The plan led to formulation of a pilot study for a program that involved schools and communities working together to find ways of protecting and renewing forests while also improving the quality of schooling in local communities. An additional factor in our initial success related to the Ford Foundation's interest in forest reclamation and improvement. They had invested in community action projects in Thailand with only limited success. When they heard of our interest in joint school-community projects, they expressed willingness to provide funding to help initiate the project. The details of this program, its development, implementation, evaluation, and beginnings of expansion will be described in the following pages. It is more fully described in Wheeler *et al.* (1997a & b) and in McDonough and Wheeler (1998).

Beginning in February 1992, the authors of this paper, a professor of geography, and several staff members from the Thai Ministry of Education worked together, intermittently, for several months to develop an acceptable plan for this project. The project built on early efforts by the Ministry of Education to improve school effectiveness and by the Royal Forestry Department and the Department of Interior to reduce deforestation and improve forests. However, both of these efforts had met with limited success. Further, research on schooling has shown how difficult it is to bring about significant, lasting changes in teaching. It also appeared that efforts by governmental agencies regarding forests had not been successful because local community members had not been persuaded that the long-range goals of protecting and renewing the forests were of more value than meeting more immediate needs of fuel, building materials, and food. It was hypothesized that through engagement of children *and* adults in collaborative learning and cooperative work related to forest management, the longer-range goal could be accomplished. Moreover, we felt that children could provide information and ideas to adults, and conversely, adults in the community had knowledge that could benefit children. Thus our project had an *intergenerational* education component that appeared to have promise in supporting renewal of forests and reversal of deforestation.

With knowledge about the forests of the region and efforts that had occurred previously as a basis, the planning team formulated a strategy that appeared to have promise for engaging schools and communities in collaborative problem-solving related to local forests. The plan consisted of several steps that involved changes in teaching and in school community interactions:

1. We would identify a small set of schools and communities in provinces near Chiang Mai in northern Thailand to be a pilot group for this study. Schools would be invited to participate. School personnel would then work with community leaders. The choice to participate would be left to local people. Work would begin in grade 5 and continue through grade 8. Teacher commitment would be especially important.
2. The general focus would be on renewal of forests and prevention of further deforestation. The project would foster development of important curricular concepts in science and social studies, but it would teach them in a more effective manner that would engender deeper understanding and ability to apply these concepts in addressing forest related problems in the local community. The plan was limited to forest-related issues including soil erosion and water supply protection because of teachers', students', and community members' need to develop technical knowledge about trees, forests, soil, etc. to work effectively on forest renewal and prevention of deforestation. To allow the focus to be broader in scope would complicate staff development and instructional resource needs beyond what was reasonable for this pilot study.

3. A generic plan of action was developed that would guide staff development and activities at the local level. This plan envisioned teachers, students, and community members working together as follows:
   (a) The first step was to explain this new way of teaching to village elders and parents at a village meeting. This served to reassure villagers as to why students would be seen in the village during school hours and to emphasize that the project had a definite focus on the curriculum. Then, teachers and students would begin the project by studying forests and forest product usage in their local community. Teachers and students would begin by preparing questions for adults about changes that had occurred in local forests over the past several years and about the many uses of forest products in the local community. Students would then refine and rehearse their questions in preparation for interviewing community members, especially older people, using these questions as a guide.
   (b) Students, supervised by teachers, would go to the community and conduct the interviews with adult community members.
   (c) Students and teachers would then return to the classroom and process the information they had obtained. If more information were needed they would create new questions, rehearse them, and return to collect the added information.
   (d) Students would also study local forests to learn about the number, kinds, and condition of trees in them.
   (e) Data from these two types of study would then be organized into a report to community members about the conditions and trends in local forests. Charts and tables would be prepared to aid in their presentations. Presentations would be rehearsed so that information would be presented clearly and effectively.
   (f) Meetings would be arranged so students could present their findings to appropriate community leaders and interested members. At such meetings, or subsequently in smaller meetings with village leaders, students and adults would work together to identify a problem which would serve as a focus for subsequent collaborative activity. Over time, a plan for resolving the problem would be formulated by students working with community leaders.
   (g) Community members and students would work together to implement the plan and monitor its effectiveness.

   It was a premise of the plan that work on this project would be a part of the school curriculum and that it would continue for more than two years. Further, as students moved from grade to grade, the level of sophistication of the work would change to match the students' development and experience.

4. It was recognized that staff development and ongoing support would be an essential part of this project. This was perceived to be a three-stage process involving (a) staff development of a leadership team that would be

responsible for (b) staff development of teachers, and (c) utilization of supervisory personnel to provide continuing support for teachers as they engaged in very different, demanding work that this project entailed.

The leadership team, responsible for staff development for teachers, would draw on people with expertise in the content pedagogical areas that are central to the project. Therefore, the members of the leadership team included Ministry of Education staff members and supervisory staff aided by the Director of the Regional Community Forestry Training Center (RECOFTC) at Kasetsart University in Bangkok. The Leadership Team would work with the Ministry of Education and the Michigan State University advisors in the staff development of teachers for this project. Orientation and staff development for the members of the Leadership Team was also included in the plan.

Leadership Team members, supervisory personnel, and the teachers were to be provided staff development using a *teacher as learner model.* This meant that all members of each group would experience this new project in a manner similar to that expected of students in grades 5-8. That is, teachers, supervisors, and leadership team members would go through the same processes of planning, interview question formulation, data collection, data analysis and presentation, etc, that are outlined in step 2 above. Their work would be in an actual field setting, where tasks expected of students would be performed by the adult project staff-members.

5. A research team would be created to monitor the project development in each of the collaborating schools and communities. This team would study (a) changes in teaching, (b) changes in interactions between teachers, students, and community members during the project, and (c) the effects of the program on forests and forest product usage in each community. Orientation and staff development for research team members was also part of the plan.
6. Materials to help teachers and students with new content and methods would be prepared by the Ministry of Education. This would be in the form of a *Handbook-guide to conducting a case study.* It would contain both procedural guidelines and supporting conceptual information about trees, forests, and related environmental problems including flooding and soil erosion.

## 3. Selecting Schools and Communities

Selection of schools and communities for a pilot study was an important part of the preliminary work for this project. Six rural primary schools (grades K-6) and two lower secondary schools, (grades 7-9) associated with eleven communities were selected for participation in the study. The initial working group of teachers included 24 teachers in eight schools – 9 science and social studies teachers at grades 7 and 8 at the two participating lower secondary schools and 14 elementary teachers at grades 5 and 6.

Eleven communities that were associated with these eight schools were selected. All of the communities in the study were generally typical of rural villages in Thailand in terms of size, occupations and social-economic status. Like all villages in rural Thailand, each village had a formal governmental structure with an elected, salaried Village Headman. In addition, all villages had a Buddhist Temple with an Abbott in charge of it.

## 4. Project Implementation

Over a five semester period, beginning in 1993, the project plan was implemented in eight schools and eleven communities in rural northern Thailand. Project work included staff development for the research team, the leadership team, supervisors, and teachers. All staff development was based on the *teacher as learner* model. Support from Ministry personnel, Michigan State University faculty, and supervisory staff continued throughout the project, largely as planned. However, there were some bureaucratic complications that are not uncommon when large agencies join forces. Some of these will be highlighted in a later section of this report.

## 5. Outcomes

### 5.1. VILLAGE HISTORY

During initial staff development sessions, teachers in grades five and six decided to focus activities during the first semester largely on studying the history of each of the local villages. The rationale for this decision arose from teachers' perception that all students needed to learn basic skills related to data collection and interpretation, and that village history would be a useful context for such learning. This inquiry focus would not be laden with highly technical content, whereas study of forests required a knowledge base that students and most teachers lacked. Teachers and Leadership Team members felt that in subsequent years, grade five would continue to work on village history, with forests as a central theme, as a means of developing investigative skills, and begin some study of forest-related concepts. Students in grades six through eight would work at more advanced levels studying forest-related issues and techniques.

Some villagers who were interviewed were more than eighty years old. As a result, their recollections went back to the 1920s. Many of these elderly people also related stories from their parents and grandparents providing some information about forests that extended into the mid-nineteenth century.

Students learned about the decline of forests and wildlife in their part of Thailand over the past decades.

## 5.2. CHANGES IN TEACHING

During the first semester, at least half of the teachers made significant changes in teaching and all but four made some changes. Nineteen teachers worked with students and community members to organize some form of data collection activities related to village history, including forest related investigations. The work involved preparing questions to ask of villagers to elicit needed information about the history and present conditions of forests in and near the community and the reasons for changes that had occurred. Twelve of the teachers worked with students to refine the questions and rehearse questioning techniques. These same teachers arranged for adult villagers to be available for student interviews. In some cases, the arrangements were made with students' help. Moreover, these teachers provided supervision and support for students as they conducted interviews with adults.

During all five semesters, after data collection in the village and forests, teachers and students worked together in their classrooms to analyze and interpret the new information. The results of this part of the project work often led to formulation of new questions to gather more complete information, a new round of interviews and forest study, and further analysis. Students and teachers worked together to prepare for, and deliver, a presentation of their information to village members. In the first semester, action projects to resolve problems identified by students were initiated in two elementary schools and one lower secondary school. These projects involved tree planting and community-wide actions to care for forests. Eventually students and teachers worked with community members to develop a range of projects relating to forest protection and/or improvement.

In subsequent semesters, following increased staff development and better support from supervisors and each other, more teachers became more effective in implementing the program in their schools and communities. As all teachers learned from each other about techniques for preparing students for data collection, as they learned how to support students in data analysis and interpretation, and as they learned how to prepare students for effective interaction with adults during presentations to villagers, project activity continued to improve. As a result, by the fourth and fifth semesters of the pilot study, 21 of the 24 teachers in the project had made changes in their teaching.

## 5.3. STUDENT RESPONSES TO THE PROJECT

Through observations in classes and during work in the villages, as well as through interviews with students each semester, field workers found strong evidence of students' benefit from, and support of, the project. Moreover, these data confirmed the positive images that teachers and village leaders were giving about the project. Observational data showed that in classrooms where teachers made serious efforts to implement the project, students were developing skill and confidence in:

1. framing questions that would elicit significant data related to the issues being studied;
2. talking with adults and asking questions effectively during interviews;
3. organizing, analyzing, and interpreting data from investigations and interviews;
4. presenting the information to community members in a public forum;
5. working with adults, including community leaders, to identify a forest-related-problem and formulate strategies to resolve it.

In addition, evidence from teachers about test results on school cluster and district examinations showed that students improved in their knowledge pertaining to relevant concepts about trees, forests, forest product usage, and related issues such as flooding and drought in the region's monsoon-type climate. Thus, students' knowledge of relevant concepts from science and social studies advanced as a result of this content. Evidence from teachers about school-cluster testing, common in Thailand, showed that students scored at least as well as students in the traditional curriculum on externally generated tests for their grade-level.

Perhaps of greater importance was the boost in motivation and self-confidence exhibited by students of the more active teachers. All of these students interviewed each semester showed a very high level of enthusiasm for the project activities. All reported that they felt their actions were having an important impact on their communities. All felt proud of their own personal accomplishments in reading, data analysis, public speaking, and talking with important adults in the community. All reported that they found the project very interesting, enjoyable, and important to their learning. All reported that they were pleased with the improvements in their self-confidence. Only one student, of more than one-hundred interviewed, expressed concern that the program would not adequately prepare him for later academic study. However, in spite of the positive findings from most students, a different picture was found with students of the teachers who did not engage effectively in the project. These students tended to be uncertain about the intentions of the project and their role in it. While they enjoyed going to the village and forest for activities, they were not reaping the same benefits as students of teachers who were following project guidelines more diligently.

## 5.4. COMMUNITY ACTIONS

By the end of the fifth semester of pilot study, the project had an impact in all communities where teachers, villagers and students developed an activity to address a forest related problem. The most common activities involved tree planting and forest management. These were active projects involving large numbers of community members with school students, teachers, adult leaders, and community members working side by side.

Another type of project involved prevention of forest burning. In the region, some people gain a portion their livelihood by controlled burning of forest floor vegetation to chase out small animals, which are then captured and eaten or sold to others. Moreover, many people believed that controlled burning of the forest floor resulted in increased growth of mushrooms during and after the rainy season. These sources of food and income benefit some low-income people, but the controlled burning is detrimental to forest regeneration as it tends to reduce new growth of trees. As a result of this age-old process, many forests in the area are in a degraded state with aging trees and little or no re-growth.

Forest burning was identified as a problem, and became a focus of activity in four communities during the dry-season each year. Because this was only a supplemental food and income source for part of the year, elimination of forest burning was not perceived as a major difficulty for the persons who were engaged in it. Therefore, an adult education campaign was mounted to alert community members of the detrimental effect of forest burning. Teams were organized to monitor any burning activities and to put out fires when they were noted. It was found that people from neighboring villages continued the practice, essentially "poaching" in the local community's forests. Thus, adjacent communities were forced to work together to bring a stop to this detrimental practice.

During the pilot study, significant changes resulted in forest practices in nearly all of the participating communities. The projects were small, and they suffered from lack of budgetary support. Virtually all of the work associated with the projects was done on a volunteer basis. Village leaders, community members, teachers, school administrators, and students participated. School and community collaboration and attitudes were at a very positive level in all of the venues of this project. Interviews and focus group activities provided evidence of the high levels of involvement and positive attitudes in the communities and schools.

Variation did exist in communities' responses, however. Some were torn by factions which limited ability to work cooperatively and effectively on projects. Some communities were much more proactive than others. Where administrative support and teacher leadership has been strong, and where effective communication had occurred between school staff and community members, projects were more visible and effective. Where administrators and teachers had not been as effective in initiating activities with the community, projects were not as successful.

The following quote summarizes the perceived benefits of this program by community members: One Village Headman said at a public meeting with community members and students, "I used to worry about who would take care of the forests and trees after I died because no one else seemed concerned abut them. Now I do not worry any more as I know these students and community members will care for the forests."

## 6. Lessons for Other Nations

Several lessons were learned as we developed, implemented and evaluated this project that can be of benefit to people in Thailand and in other nations who intend to embark on a school-community project like this one. These are listed below. Each is described briefly to highlight the lesson and its importance:

1. Project development should address a recognized need: This lesson is so obvious that it may be perceived as unnecessary to describe it. However, because local community members are often suspicious of governmental agendas, the importance of the project in addressing a clearly recognized need in local communities is essential. If the value of the project is not clear to community members, gaining support for it will be difficult, resulting in extensive expenditures of time and energy.
2. The project leadership team should have recognized status and expertise: Any person who is involved as a participant in the project should have confidence that the project leadership team is knowledgeable in relevant subject areas, well informed in pedagogical knowledge and working in the best interest of those served. This is important at all levels, from community and school to governmental agencies. At the higher levels, such as the Ministry of Education, it is important that project leadership team members are recognized as having the needed knowledge and skills to carry out the project. Moreover, it is essential that they have access to relevant information and sufficient influence in the agency to engender the support needed in terms of human and fiscal resources to carry out the project.
3. Teachers need opportunities to learn new content knowledge and teaching methods: Projects like this one require many substantial changes on the part of teachers. As indicated earlier in this paper, teachers needed to develop new knowledge about trees and forests at both practical and theoretical levels. In addition, they needed to learn new instructional techniques and new skills in working outside of school in their communities and in forests. They also needed to develop new skills of interacting with community members and in guiding students in new levels of activity and interaction. These new demands that the project placed on teachers required extensive learning on their part.
4. Develop and maintain a clear curriculum focus: This is a correlate of the previous item (3). In this project, studies of local forestry problems were connected directly to the curriculum, providing students a more effective way to learn and understand key concepts. This approach differs fundamentally from efforts to teach agricultural subjects in primary or secondary schools, or to involve students in harvesting crops, digging fishponds, or participating in

community infrastructure supports on an *ad hoc* basis. The success of enlisting parental support rests in part on the fact that they could see the connection between what students were learning and the expectations of the school's academic program. This link often is missing in the kind of school-community projects just described. As a result, parents become concerned about how such activities will affect their children's chances to continue on to further studies. They resent what they see as "free labor." Moreover, teachers fully recognize that such activities limit the time available to teach academic content. The close connection between the curriculum, local case-studies, and subsequent school-community activities characterizing the Social Forestry, Education, and Participation Project, thus, represents a significant advance over previous efforts to link schools and communities around the theme of "project learning."

5. Ongoing staff development and support is essential: As important as workshops are in staff development, they are not sufficient. Teachers need ongoing staff development and support to address the practical problems of implementation. Moreover, teachers need continuing help in turning ideas and techniques learned in workshops and elsewhere into practical actions in classrooms and communities. This support can take many forms. Supervisors can work with teachers one-on-one in their classroom or in the community. Teachers can work together in small teams to provide peer coaching and practical problem solving. Supervisors can facilitate the effective action of small teacher-teams to enhance their effectiveness. However in all of these cases, time and resources are involved. It is essential that opportunities be provided for this to occur so teachers have time to do this. This may involve additional compensation, released time, or changing the school schedule so that time is made available for teachers to work collaboratively in advancing their own professional development. Evidence is accumulating that shows the benefits of continuous staff development on students' learning and motivation. However, without this continuing support, most efforts at reform will fail.
6. Support of school administrators is fundamental to project success: Principals and other school administrators typically enjoy respect and influence in local communities. Moreover, they have access to resources that are not always available to teachers. Therefore, it is important these leaders are engaged in, and enthusiastically support, the planning, development and implementation of projects like the Social Forestry, Education and Participation Project at the local level. The school principal can facilitate interactions about the project with community leaders, parents and other members of the community. Endorsement of the project by the principal can aid in gaining support for it from teachers, students, and parents.
7. Supplemental budget/school level budget must be provided to implement school-community projects: A collaborative project between schools and communities usually entails some additional expenses not included in the

regular school budget. In the case of this project, the amount of money involved was very small – money for lunches when community members are invited to school, money to support travel to other sites to see demonstrations of effective school-community collaboration. However, lack of funds committed to extra expenditures was an inconvenience for teachers, community members, and administrators. Sometimes it was perceived as a point of contention between teachers and administrators. More critically, it became a point of contention between school personnel and Ministry officials that diminished teachers' morale and their enthusiasm for the project.

A small amount of money for extra expenses such as materials and transportation costs would be a worthwhile investment that demonstrates goodwill toward, and appreciation of, teachers who work hard on such projects. It would be a tangible demonstration of support for the project from leadership in the Ministry.

8. A means of providing continuous, effective feedback to project leaders about problems in implementation is essential: Complex projects, like the Social Forestry, Education, and Participation Project, require continuous monitoring to assure that implementation is going smoothly and impediments are addressed and circumvented before they become serious obstacles. As already mentioned, project coordinators need to be sure that teachers have needed supplies and support to work effectively. Moreover, continuous monitoring will allow coordinators to identify deficiencies in planning and staff development, as well as unforeseen impediments to project success in time to plan ways of addressing problems.

   Because of long distances between Ministry headquarters and the schools, differences between teachers and Ministry officials in communication styles, and the over-extension of Regional staff of the Ministry, the feedback loop from schools to project leadership in this project had some flaws that needed attention. This is an important area that can have positive consequences when it is working effectively, or negative ones when it is flawed. Thus, it is an important dimension to address in planning for complex projects.

9. Coordination within and among agencies is both essential and difficult: For the Social Forestry, Education, and Participation Project, interagency collaboration was necessary due to the interdisciplinary nature of the work being done. This project was initiated by the Ministry of Education; and some collaborative planning was carried out in conjunction with the Royal Forestry Department and the Department of Community Development. In retrospect, more interagency planning should have occurred. The project planning team made assumptions about teachers' facility and willingness in enlisting support from local officials of these agencies that were not valid.

   First, some difficulties occurred due to logistical problems in making contact with agencies. More importantly, primary teachers did not want to "bother" agency personnel with what they perceived as simple requests. Teachers' perceptions of status differences between themselves and personnel in these agencies interfered with effective communication. In some cases where interactions did occur between teachers and agency personnel, the technical language of the latter group tended to limit effective communication between them and teachers. The lesson learned here is that interaction and communication between teachers, students, and community member *and* agency personnel, is an issue that needs far greater attention than we gave it in initial planning and during implementation. Moreover, agency personnel need to give some attention to their style of communication, including their technical jargon, when working with teachers, students, and community members.

10. Staff development activities should include community leaders: This project required that teachers learn new knowledge and skills pertaining to teaching, applied science regarding forests and trees, and working with community members. It also required community leaders to develop new knowledge about applied science and how to work effectively with administrators, teachers, and students. Staff development was provided for teachers, but no similar efforts were made to provide staff development for community leaders. A promising approach to staff development could include bringing teachers and community leaders together on areas of common interest – technical content about forests and trees and effective strategies for school and community interaction – during part of the staff development workshops.
11. The project focus needs to be maintained: Dedicated teachers are continuously trying to improve their teaching and to make their classes more interesting to students. This positive tendency became an obstacle during the second and third year of the pilot study. In spite of strong evidence of students' continuing interest in the project, a few teachers wanted to shift the focus of the project from forests to other topics because they were afraid students' interest would decline as a result of continued focus on the same topic. Maintaining continued focus on forest-related problems in the community was perceived as *essential* for at least two important reasons: forest related problems are inherently long term and they require technical knowledge to resolve effectively. Therefore it is counter-productive to only engage in them for a short duration.
12. The project produced many intrinsic rewards for all participants: Throughout the pilot study for this project, the leaders were continuously pleased with the dedication and high level of activity of the teachers, students, and community members. In trying to understand the source of this "energy," we came to an important conclusion: The project provided many rewards for all involved. Teachers saw that students were engaged in work that was perceived as important. Community members were pleased to see students working with adults to understand and address serious problems with forests. Students saw themselves as doing important work for their community and they also recognized that they were learning important knowledge and skills that appeared more beneficial than some components of the standard curriculum.

Principals saw the benefits of students' applied work and positive school and community relations. In short, the project was appealing to all concerned. As a result, motivation was partly self-generating because of the multiple rewards that resulted from its implementation.

13. Extrinsic rewards needed for teachers who demonstrate real change: Such rewards compensate, in part, the many additional hours required by a project like this. In spite of the immediate and long-term positive feedback from students, principals, and community members, teachers came to see that few rewards were likely to emanate from the Ministry of Education. While some teachers did receive double in-grade promotions, others who worked equally hard at change did not. Some teachers wanted to use their work on this project to apply for "master teacher" status, only one achieved this level. In projects like this, it is essential to plan from the beginning about how to provide multiple rewards for teachers who implement change effectively.
14. Implementing a complex project represents a significant challenge: This lesson seems obvious, but it is worth stating nevertheless. Complex projects that require major changes in the way teachers teach demand ongoing support over an extended duration to initiate, nurture and maintain the changes. Further, in changing the teaching and learning process, other issuer about curriculum, assessment, organization, and rewards also emerge. In this project, the addition of community participation in teaching and learning, as well as the collaborative involvement of schools in community development, brought new issues to the forefront, many of which accounted for failures of previous attempts that are recorded in the international literature on school change.

This project suggests that implementing the lessons described above can promote important change in all these areas. However, it must be remembered that:

- change comes incrementally and unevenly over time,
- that not all teachers change their practice at the same rate, and
- communities vary in their ability to respond to collaborative projects.

For those who embark similar initiatives, it is important to keep this in mind.

The lessons described in the foregoing pages were drawn from the pilot development for the Social Forestry, Education, and Participation Project. This project was very successful in achievement of its goal of bringing school children and adults together into a collaborative effort to reverse deforestation in local communities. The successes of the project were manifold in achieving very significant changes in teaching, learning, and productive interaction for the teachers, students, and community members. Given the complexity of the project, it is a testimony of the dedication of teachers and the responsiveness of all participants to an opportunity for community improvement. The fact that so much was accomplished with so few added resources is an important lesson of great significance. The fact that the project appeared to provide self-energizing motivation for diverse participants is of great significance as well.

## 7. Going to Scale

Following conclusion of the pilot study in 1996, the Ministry of Education expanded the project to over 40 schools in the same region of the country. According to Ministry officials, expansion of this project at the national level is intended.

Going to scale with a complex project such as this, which involves new content, interactions, and teaching approaches, is not automatic. Just because the project has seen success in the pilot study does not mean that others can easily accommodate its new content and approach. A substantial body of knowledge and experience with dissemination of innovations and supporting changes in teaching supports this claim. The more divergent the new approach is from current practice, the more support that is needed to help teachers gain the knowledge, skills and confidence to utilize it effectively and willingly. Therefore, in going to scale with this project, the amount of staff development, supervisory support, and peer collaboration will be very close to the amount needed in the pilot study. This is a substantial investment. However, the consequences in improved teaching, learning, and school-community relations are only part of the multiple benefits to be derived. Improved environmental quality and improved interactions between adults and youth in communities are two other important outcomes that are worthy of noting. Thus, the expenditures will yield many benefits that outweigh the costs.

## References


Campbell, D. and Olson, J.: 1991, *Framework for environment and development,* CASID Occasional Paper Series Number 10, Center for Advanced Study of International Development, East Lansing.

Gallagher, J.: 1993, "Secondary science teaching and constructivist practice", in K. Tobin (ed), *The practice of constructivism in science education,* Earlbaum. Hillsdale NJ.

Namfa, B.: 1999, *Learning a new approach to teaching in a traditional context: A case of Thai primary school teachers making fundamental changes in their practice,* Unpublished doctoral dissertation, Michigan State University, East Lansing.

Sykes, G., Wheeler, C., Scott, J., and Wilcox, S.:1996, *Professional community among teachers: How does it form? An inquiry and case study*, Unpublished manuscript, College of Education, Michigan State University, East Lansing, MI.

Wheeler, C., Gallagher, J., McDonough, M. and Soopokakit-Namfa, B.:1997a, "Improving school community relations in Thailand", in W. Cummings and P. Altbach (eds), *The challenge of Eastern Asian education: Implications for America,* State University of New York Press, Albany.

Wheeler, C., McDonough, M., Gallagher, J., Soopokakit, B. and Duongsa, D.:1997b, "Linking school change to community participation in community forestry: A guided innovation in Thailand", in D. Chapman, L. Mahlck, and A. Smulders (eds), *From planning to action: Government initiatives for improving school-level practice,* Pergamon, Oxford.

Wheeler, C. Raudenbusch, S., and Pasigna, A.: 1989, *Policy initiatives to improve school quality in Thailand: An essay on implementation, constraints, and opportunities for school improvement,* Harvard Institute for International Development, Bridges Report Series. No. 5. Cambridge, MA.

# POLLUTION PREVENTION: A NEW PARADIGM FOR ENGINEERING EDUCATION

P. L. BISHOP

*Department of Civil and Environmental Engineering, University of Cincinnati, P.O. Box 210071, Cincinnati, OH 45221-0071, USA*

**Abstract**. It is estimated that our materials-dominated society consumes about 10 metric tons of raw materials per person per year in the production of consumer goods. Within 6 months of extraction or production of these materials, 94% of them become waste. More efficient manufacturing practices are needed to lessen the demands for raw materials and to reduce the amounts and toxicity of waste materials. It is estimated that 70% of this waste material could be eliminated through better design decisions and reuse of materials. Engineering education has evolved into fairly segregated disciplines which focus on narrowly defined design and manufacturing functions, often without consideration of the environmental consequences of these functions. This is no longer the case in industry, however, where pollution prevention and waste minimization have become very important. Unfortunately, most of our engineering graduates are not prepared to step into a role where "green engineering" principles are espoused. We must quickly incorporate the "green engineering" principles into the engineering curriculum in all disciplines to ensure that all engineering graduates understand the environmental and economic consequences of engineering decisions. This paper describes methods that can be used to introduce the principles of pollution prevention, environmentally conscious products, processes and manufacturing systems. Students will learn the impacts of wastes from manufacturing and post-use product disposal, environmental cycles of materials, sustainability, and principles of environmental economics. Materials selection, process and product design, and packaging are also addressed.

**Keywords:** curriculum, course development, pollution prevention, waste minimization, green engineering, design for the environment

## 1. Introduction

It is estimated that our materials-dominated society consumes about 10 metric ton of raw materials per person per year in the production of consumer goods. Within 6 months of extraction or production of these materials, 94% of them become residual material that is disposed of as waste. More efficient practices for using materials in manufacturing are needed to lessen the demands for raw materials and to reduce the amounts and toxicity of waste materials. It is estimated that 70% of this waste material could be eliminated through better design decisions and reuse of materials.

As currently structured, engineering education has evolved into fairly segregated disciplines which focus on narrowly defined design and manufacturing functions without consideration of the environmental consequences of these functions. This is no longer the case in industry, however, where pollution prevention and waste minimization have become very important. This rapidly changing industrial emphasis came about initially because of regulatory pressure, but now is driven primarily by economics. Industries are striving to minimize waste generation at the source, to reuse more of the waste materials that are generated, and to design products for easier disassembly and reuse after their useful life is completed. The overall objective is to minimize "end-of-pipe" treatment, although some waste treatment will always be needed. This new environmental ethic in manufacturing is labeled "Pollution Prevention," "Green Engineering" or "Environmentally Conscious Engineering." It is a collection of attitudes, values and principles that result in an attempt to reduce the rate at which we adversely impact the environment in the practice of the engineering profession.

Pollution prevention involves a holistic approach to waste management. Rather than waiting until after a waste is produced and then attempting to make it innocuous, the pollution prevention approach is to look at the entire life of a product, from extraction of the raw materials from the earth through manufacturing to product use and finally to product disposal and possible recycling or reclamation, in order to find ways to minimize all environmental impacts. This may mean changing to use of materials that are less toxic or that are less scarce in the earth, to more efficient manufacturing processes or ones that demand less energy, to new product designs that make recycling after use easier, or to new packaging materials that reduce the amount of packaging going to landfills or incinerators.

## 1.1. POLLUTION PREVENTION HIERARCHY

Pollution prevention (P2) is a term used to describe production technologies and strategies that result in eliminating or reducing waste streams. The U.S. EPA defines pollution prevention as:

> "the use of materials, processes, or practices that reduce or eliminate the creation of pollutants or wastes at the source. It includes practices that reduce the use of hazardous materials, energy, water or other resources and practices that protect natural resources through conservation or more efficient use."

Thus, pollution prevention includes both the modification of industrial processes to minimize the production of wastes, and the implementation of sustainability concepts to conserve valuable resources.

Integrated industrial production and waste management provides the flexibility to use an almost limitless variety of waste minimization, waste treatment and waste disposal techniques. Until recently, only the last two were seriously considered. This changed with passage of the Pollution Prevention Act (PPA) of 1990. The preamble to the PPA says:

> "The Congress hereby declares it to be the national policy of the United States that pollution should be prevented or reduced at the source whenever feasible; pollution that cannot be prevented should be recycled in an environmentally safe manner, whenever feasible; pollution that cannot be prevented or

recycled should be treated in an environmentally safe manner whenever feasible; and disposal or release into the environment should be employed only as a last resort and should be conducted in an environmentally safe manner."

Thus, Congress made pollution prevention a *national policy*, rather than just a desired goal, and established a hierarchy for determining how pollution should be managed. It established source reduction as the preferred method to be used for waste management, if it is feasible and cost effective. Congress realized that all pollution could not be eliminated through source reduction alone, and set recycling (and presumably other methods of waste reuse such as reclamation) as the preferred alternative for management of residuals that remain after all viable source reduction measures are taken. Anything that remains after these steps should be treated to render it less hazardous and more environmentally friendly. Disposal into secure chemical landfills or direct release into the environment is allowed only as a last resort. The waste management hierarchy can be depicted as shown in Figure 1.

## 1.2. EDUCATIONAL NEEDS

Industry has bought the concept of pollution prevention, because they have seen the economic benefits resulting from it. However, most of our engineering graduates are not prepared to step into a role where "green engineering" principles are espoused. It is essential that we quickly incorporate the "green engineering" principles into the engineering curriculum in all disciplines to ensure that all engineering graduates are aware of environmental issues and understand the environmental and economic consequences of engineering decisions. The goal of this educational change should be to reduce the necessity for "end-of-pipe" treatment by incorporating, at all stages of engineering, measures which minimize wastes and permit recycling and reuse. A knowledge of pollution prevention principles should allow the engineer to include environmental consequences into decision processes in the same way that economic and safety factors are now considered. Eventually, we must extend this way of thinking to others in the decision making process, including management, but this probably will not be successful until engineers buy into it willingly.

A new interdisciplinary educational program incorporating Environmentally Conscious Engineering concepts in all undergraduate engineering curricula is being developed at the University of Cincinnati in order to meet industry's needs for engineers knowledgeable of pollution prevention. The purpose of this program is to reorient our undergraduate programs in all engineering disciplines to incorporate society's interest in environmental quality and sustainability into engineering education starting at the freshmen level and continuing throughout the curriculum. This will culminate in incorporation of concern for the environment in design projects in the senior design courses in all disciplines. The objective is to encourage a stronger environmental ethic among engineering students, especially those with a manufacturing or production focus.

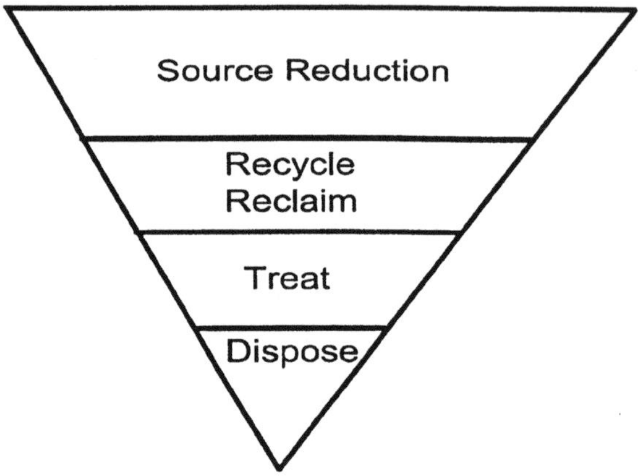


Fig. 1. The pollution prevention hierarchy

As a part of this new educational paradigm, a course on pollution prevention has been developed for senior and graduate engineering students. To support this course, a textbook was written to introduce the principles of pollution prevention, environmentally conscious products, processes and manufacturing systems. In this course, students learn the impacts of wastes from manufacturing and post-use product disposal, environmental cycles of materials, sustainability, and principles of environmental economics (Bishop, 2000). Materials selection, process and product design, and packaging are also addressed. This book, entitled *Pollution Prevention: Fundamentals and Practice*," has been published by McGraw-Hill Co.

This paper describes the cross-disciplinary curriculum development that is under way at the University of Cincinnati, with particular emphasis on the senior/graduate level course that has been offered to our engineering students for the past four years. It is deliberately designed for a mixed engineering discipline audience, rather than specifically for environmental engineering students, because the included information is essential for all engineers, be they process or product designers, industrial engineers or environmental engineers. The classes have all had a mix of academic backgrounds, which has led to an expanded input of ideas on how to best minimize industrial waste production.

## 2. Curriculum Structure

A new interdisciplinary educational program incorporating Environmentally Conscious Engineering concepts in all undergraduate engineering curricula is being developed at the University of Cincinnati in order to meet industry's needs for engineers knowledgeable of pollution prevention. Because of the difficulty in adding new material to already overly crowded course contents, these curriculum changes consist primarily of assimilating P2 concepts into existing courses beginning in the freshman year and continuing in courses taken in later years, and development of a new elective course on pollution prevention for senior engineering students to take along with their senior design course. Because incorporation of these materials into existing courses will only be successful if the faculty become knowledgeable of and comfortable with pollution prevention concepts, workshops are being designed to teach faculty how to incorporate environmental concerns into their classes and to teach environmentally conscious engineering.

The goal of this initiative is to produce graduates from our College with an environmental ethic which is incorporated into their basic engineering approach to design and production. The students will understand environmental processes and the impacts of their engineering decisions on these processes. They will be able to answer questions directed at the ultimate fate of products after their useful life is over, and will be able to investigate ways to make the product or byproducts more recyclable. In particular, they will understand the environmental impacts of their decisions, and they will be able to make better decisions which cause minimal damage to the environment.

A significant focus of this program is the interdisciplinary approach taken. The program is designed by an environmental engineer, but is primarily directed at engineers in the product design and manufacturing sectors. These students are not turned into environmental engineers. Rather, they are provided with enough knowledge of environmental fates and effects to make educated decisions about the environmental implications of their designs or manufacturing processes, and then learn how they can design or manufacture products for minimal environmental impact.

## 3. Pollution Prevention Course Structure

A senior/graduate level course entitled "Environmentally Conscious Engineering" has been offered annually for the past four years to a mix of students from all engineering disciplines. This course was originally developed with funding from the Ohio Environmental Protection Agency. The objective of this course is to introduce the principles of environmentally conscious products, processes and manufacturing systems. Students learn the impacts of wastes from manufacturing and post-use product disposal, environmental cycles of materials, sustainability, and principles of environmental economics. Materials selection, process and product design, and packaging are also addressed. There are a number of major topics that can be covered in this course (Table I).

TABLE I

Topics potentially covered by an "Environmentally Conscious Engineering" course.

1. Introduction to Pollution Prevention
2. Properties and Fates of Environmental Contaminants
3. Industrial Activity and the Environment
4. Environmental Regulations
5. Improved Manufacturing Operations
6. Life-Cycle Assessment
7. Pollution Prevention Economics
8. Waste Audits
9. Recycling Options
10. Design for the Environment
11. Residuals Management
12. Incentives for Waste Minimization
13. Municipal Pollution Prevention Programs
14. Sustainability

There is usually not enough time to cover all of these topics in one course, so the actual course syllabus is dictated by the make-up of the class participants. For example, a class dominated by environmental engineering students might already have sufficient knowledge of topics 2, 3, 4 and 11 so that these could be covered in a cursory fashion, allowing most of the course to focus on areas the students may not be familiar with, such as the organizational and technical aspects of pollution prevention, manufacturing operations, life-cycle assessment and waste audit procedures, and design for the environment. A class dominated by mechanical engineers, on the other hand, might focus more on the environmental impacts of their design decisions and life-cycle assessment procedures than on manufacturing processes. In a mixed class, a selected number of topics could be covered in detail, with background readings being used to familiarize the student with necessary information from the other topics. This mixed-audience course would introduce the concepts of pollution prevention, describe the consequences of pollution emissions, and present methods for setting up a pollution prevention program and assessing its effectiveness.

Any course of this type should include an introduction to what pollution prevention is and why it is needed, a discussion of the industrial impacts on the environment that led to the need for a pollution prevention initiative, an overview of environmental regulations pertaining to industrial waste management, and an introduction to environmental economics, which is often the basis for environmental management decision-making. Of particular importance is a discussion of "sustainability," as maintaining a sustainable environment through intelligent business decision-making is a critical aspect of the rationale for pollution prevention.

Following are brief descriptions of a few of the pollution prevention course topics described in Table I. A complete slide presentation package for a course of this type, comprising over 800 PowerPoint slides, can be found on the Internet at http://www.cee.uc.edu/~pbishop/.

## 3.1. LIFE-CYCLE ASSESSMENT

Probably the most important concept that must be conveyed in a course on waste minimization is that of life-cycle assessment. The objective of life-cycle assessment (LCA) is to evaluate the environmental effects associated with any given activity, from the initial gathering of raw material from the earth until the point at which all residuals are returned to the earth (Vigon *et al.*, 1993). This evaluation includes all side-stream releases to the air, water and soil from the production of the raw materials (including water and energy), the use of the product and its final disposal, as well as from the processing of the product itself. LCAs are used to identify and measure both "direct" (e.g., emissions and energy usage during manufacturing processes) and "indirect" (e.g., energy use and impacts caused by raw material extraction, product distribution, consumer use and disposal) impacts (Nash and Stoughton, 1994). LCA is a systematic approach that provides a true measure of the impact of a particular product or process. Unlike an environmental audit of an industrial process, which focuses on one particular facility and usually only on the activities that occur at that site, LCA looks at the linked interactions of the firm with the actions of its suppliers and customers. The result is a total cradle-to-grave analysis of the environmental impact of a product.

The principal components of life-cycle assessment have been established, and the procedures refined, by the Society of Environmental Toxicology and Chemistry (SETAC). Life-cycle assessments usually use three interdependent stages, or subsystems, to express the cradle-to-grave inputs and outputs (Fava *et al.*, 1993). These are: Life-Cycle Inventory Assessment, Impact Assessment, and Improvement Assessment. These can be used independently or combined to achieve proper pollution prevention decisions.

Any course on pollution prevention should include a thorough discussion of how life-cycle assessments are conducted, probably using case studies as examples of how they are performed and how their output can be used.

## 3.2. IMPROVED MANUFACTURING OPERATIONS

The basis for a thorough understanding of how waste minimization can be employed in an industrial setting is an appreciation of how industrial production works. Except possibly for some mechanical or industrial engineers, the industrial process is often a mystery to most engineering students. The objective of the manufacturing process is to transform an idea into a saleable product. There are many ways by which this can be achieved, though. The traditional way is to use what is known as "sequential engineering," in which product design and engineering departments work independently, developing the technical requirements (e.g., materials and size) and final design details; the manufacturing, testing, quality control and service groups generally only see the design in its almost complete state. This can lead to many inefficiencies if it is later decided that the product or manufacturing process needs modification. A more efficient manufacturing philosophy is known as "concurrent engineering." Also referred to as simultaneous engineering, life-cycle engineering, integrated product development, or team design, this provides a systematic and integrated approach to the design of products. The purpose is to ensure that the decisions made in the design stage result in a minimum overall cost during its entire life-cycle. It is essential that students learn modern industrial management approaches that can lead to minimization of waste production during the manufacturing stage and during the useful life of the product, and that result in a product that can be more easily remanufactured or recycled after its useful life is over.

It is also important for students to obtain an understanding of the various manufacturing processes and equipment used in the production of a product. In order to be able to recommend an improved manufacturing system, the student must understand the uses, capabilities, advantages and disadvantages of a variety of industrial processes such as chemical reactors, heat exchangers, evaporators/dryers, crystallizers, distillation columns, etc. Design of these units is typically the arena of the chemical engineer, but all other engineers dealing with pollution prevention must also have at least a rudimentary understanding of them if they are to make informed decisions.

## 3.3. POLLUTION PREVENTION PLANNING

Successful reduction of waste generation by industry requires careful planning in order to ensure that the best pollution prevention activities are carried out. However, pollution prevention planning is not enough; the P2 projects must actually be implemented. Too often, P2 planning results in plans that languish on a shelf because other higher priorities for the capital appeared or because the company did not have an overall environmental management system in place that was coordinated with their more common business management system. Therefore, it is essential to set up and implement effective pollution prevention programs. The techniques for doing this should be stressed in all pollution prevention courses.

Pollution prevention planning requires a detailed understanding of how a company does business and how it makes its products. The resulting plan should provide a mechanism for a comprehensive and continuous review of the company's activities as they pertain to environmental issues. This task can appear daunting at the beginning, especially to management which may not be familiar with many of the concepts involved. The amount of information and data required for a good pollution prevention plan may also give many pause. However, the task should not be as onerous as it may at first appear. Many of the procedures used are actually analogous to those commonly used to run a business; only the nomenclature may be different. Because pollution prevention activities can have a major impact on the day-to-day operation of a business, pollution prevention planning should be done in parallel with other business planning.

The pollution prevention course should describe how the planning process is conducted and how the results are compiled to develop an Environmental Management System for the company. It should also describe the environmental audit and environmental inventory procedures.

### 3.4. DESIGN FOR THE ENVIRONMENT

There are a number of ways by which a company can reduce the pollution resulting from the manufacture or use of a product. For example, it may evaluate the process used to manufacture the product, with the goal of reducing process wastes; it may examine the materials used in manufacture and investigate the efficacy of using less harmful materials or ones that are more sustainable; it may redesign the product so that it is more readily recyclable or more easily disassembled for reuse/recycle; it may choose to redesign the product or the manufacturing process so that its manufacture or use requires less energy; it may redesign the product to extend its useful life; or it may attempt to minimize potential pollution from the final disposal of the product after its use. Each of these options has been given a name, such as Design for the Environment (DfE), Design for Recycle (DfR), or Design for Disassembly (DfD). The collective term for all of these options has been labeled Design for X (DfX), where X refers to whatever objective has been selected.

Design for the Environment has been defined as "a practice by which environmental considerations are integrated into product and process engineering design procedures" (Allenby, 1991). On analysis of this definition, it readily becomes obvious that DfE and life-cycle assessments are intimately intertwined. DfE involves the refinement of a manufacturing process or of the product itself in order to reduce pollution.

There are a large number of DfE topics that can be addressed in a pollution prevention course, depending on the audience. Among these are Green Chemistry, which evaluates the use of alternative synthetic pathways for producing a material in a more benign way, use of alternative and safer feedstocks, alternative reaction conditions, and design of safer chemicals; Pinch Analysis, which can be used to minimize energy, process chemical or water needs in a manufacturing process; Design for Disassembly, which considers ways to design a product to make it easier to disassemble the product after use, refurbish and recycle the parts, or recover the materials for reuse; and Packaging, which examines ways to minimize the need for packaging and to recycle that packaging that is required.

An example of how procedures developed for a specific engineering application can be used in a different way for pollution prevention is the use of pinch analysis to minimize pollution from an industrial process. Pinch analysis was originally developed as a way to minimize heating or cooling requirements in an industrial process by establishing a heat exchange network (HEN), in which waste heat or cooling water in one part of the plant could be used in another where it is needed. Its use has now expanded to include minimization of water and process chemical usage and waste production. Pinch analysis is a holistic design approach based on process integration, in which the process is conceived of as an integrated system of interconnected units and streams, rather than as a series of individual and independent steps. In this way, materials or energy that are needed or available in one process step can be supplied by or made available to another process step. Pinch analysis is based on rigorous thermodynamic principles that are used to predict heat and materials flows through the processes making up the manufacturing system. In many instances, water usage or waste production rates have been cut by as much as 50 percent, with only minor plumbing changes required. The use of pinch analysis in pollution prevention can easily be incorporated into a P2 course.

### 3.5. FUGITIVE EMISSIONS

By applying the principles of pollution prevention, a company can usually greatly decrease the pollution emanating from its facilities. However, there is another major source of pollution at many facilities that may be overlooked. An average sized manufacturing plant may have 3,000 to 30,000 components such as pumps, valves, compressor seals and pipe flanges that may leak. Larger facilities may have as many as 100,000 of these connection points. Large refineries may have over 250,000 pieces of potentially leaking equipment. Well maintained, non-leaking equipment emits very little process fluids. However, it is common in a large facility, even a well maintained one, for some equipment seals, packing materials or gaskets to leak, resulting in unintentional releases of process liquids and gases. These unintentional releases due to leaking equipment are referred to as fugitive emissions. Leakage from an individual component may be small, but the cumulative effect from thousands of components can be major. These leaks may be continual leakage of small amounts of process fluids due to faulty process equipment, or sudden, major leaks due to equipment failure. This leakage presents an enormous task for the company to continually monitor for possible leaks or potential equipment failure and to rectify any leaks that are found.

To be complete, a pollution prevention course should include a description of the sources of fugitive emissions, procedures for estimating the magnitude of the losses from the components, and approaches that can be used to remediate serious fugitive emission sources.

### 3.6. COMPUTER SIMULATIONS

Actual pollution prevention activities by industry can be very time consuming and data intensive. A P2 plan may encompass many small changes to an industrial process that may be physically separated within the production process. Evaluating all of these individually and then attempting to integrate them into a coherent answer is usually beyond the scope of an educational exercise. Consequently, it is often difficult for a student to develop a clear understanding of the benefits of the pollution prevention approach. This can be overcome, however, if an industrial simulation package is used. An example of such software is SuperPro Designer®, a process simulation software package developed by Intelligen, Inc. (Scotch Plains, NJ). It can be used by any chemical, biochemical or environmental engineer to evaluate a manufacturing process with the intent of optimizing the process and minimizing pollution. It can also be used to evaluate or design wastewater treatment facilities. Its operation is very straight forward, and students quickly learn how to use it to evaluate easy to intricate pollution prevention options.

## 4. Summary

Many industries have already begun to use the concepts of pollution prevention in their operations, because they have seen the economic benefits resulting from it. However, most of our engineering graduates are not prepared to step into a role where "green engineering" principles are espoused. It is essential that we quickly incorporate the "green engineering" principles into the engineering curriculum in all disciplines to ensure that all engineering graduates are aware of environmental issues and understand the environmental and economic consequences of engineering decisions. Coverage of pollution prevention topics in environmental engineering curricula in the United States is now mandatory to meet the Accreditation Board for Engineering and Technology (ABET) minimum curricular requirements. Other areas throughout the world are also encouraging this new educational paradigm. This paper summarizes the need for a pollution prevention course for all engineering students and suggests topics that can be included to prepare engineers from all disciplines to work in an environmentally conscious engineering setting.

## References

Bishop, P. L.: 2000, *Pollution Prevention: Fundamentals and Practice*, McGraw-Hill, Inc., New York.

Fava, J. A., Consoli, F., Denison, R., Dickson, K., Mohin, T., and Vigon, B.: 1993, *Guidelines for Life-Cycle Assessment: A "Code of Practice"*, The Society of Environmental Toxicology and Chemistry Workshop Proceedings, Pensacola, FL.

Nash, J., and Stoughton, M. D.: 1994, "Learning to live with life cycle assessment," *Environmental Science & Technology* **28**, 236-237.

Vigon, B. W., Tolle, D. A., Cornaby, B. W., Latham, H. C., Harrison, C. L., Boguski, T. L., Hunt, R. G., and Sellers, J. D.: 1993, *Life-Cycle Assessment: Inventory Guidelines and Principles*, U.S. EPA, EPA/600/R-92/245, Cincinnati, OH.

# TOWARDS EXCELLENCE IN ENVIRONMENTAL EDUCATION A VIEW FROM THE UNITED STATES

B. SIMMONS

*Department of Teacher Education, Northern Illinois University, DeKalb, IL 60115, U.S.A*

**Abstract.** Within the United States, a nationwide debate has been raging over how to best provide quality education for all learners. Much of this debate, spurred by poor test scores and other measures of achievement, has centered on the development of national, state and local standards and assessments for the core disciplines (e.g., mathematics, science, geography). For the most part, environmental education has been left out of this debate and out of the various standards development initiatives. Whether one agrees philosophically with academic standards or not, these standards are determining what is being taught in the classroom. By 1993, environmental education in the United States found itself in a conundrum. It has always been argued that environmental education should be interdisciplinary, infused throughout the curriculum. However, with the new standards, environmental education was in real danger of becoming marginalized. To address this situation, the North American Association for Environmental Education (NAAEE) initiated the National Project for Excellence in Environmental Education.

**Keywords**: environmental education, environmental standards

## 1. Introduction

In any discussion of environmental education in the United States it must be understood that it is a grassroots movement, characterized by literally thousands of educators working in schools, colleges, nature centers, zoos, museums, government agencies, and non-governmental organizations. It must also be remembered that education in the United States is decentralized. There is no national curriculum; there are no national exams. Each state determines how schools will function. Some states have a state-mandated curriculum and statewide adoption of textbooks. Others allow each school district to determine its own curriculum and select its own teaching materials. Consequently, when the education system in the United States is discussed, let alone environmental education in the United States, that discussion must be framed in terms of general trends.

Although the field has struggled with defining environmental education in a meaningful way, the most commonly accepted working definitions continue to draw heavily from the Belgrade Charter (UNESCO, 1976) and the Tbilisi Declaration (UNESCO, 1978). As the field has evolved, the principles promoted in these two documents have been researched, critiqued, revisited, and expanded. They still stand as a strong foundation for a shared view of the core concepts and skills that environmentally literate citizens need.

*Water, Air, and Soil Pollution* **123:** 517–524, 2000.

Although at times it may seem difficult for everyone to agree upon the exact wording of a definition, the practice of environmental education in the United States is characterized by some essential elements (Disinger and Monroe, 1994):

- Environmental education is based in knowledge about ecological and social systems. It draws on and integrates knowledge from disciplines that span the natural sciences, social sciences, and humanities.
- Environmental education considers humans and their creations to be a part of the environment. Along with biological and physical phenomena, EE considers social, economic, political, technological, cultural, historical, moral, and aesthetic aspects of environmental issues.
- Environmental education emphasizes the critical thinking and problem-solving skills needed for informed personal decisions and public action.
- Environmental education emphasizes the role of attitudes, values, and commitments in shaping environmental issues. It acknowledges that environmental issues are not strictly scientific in nature. Recognizing the feelings, values, attitudes, and perceptions at the heart of environmental issues is an essential step in understanding them, and a precursor to accepting responsibility for exploring, analyzing, and resolving them.

The purpose of this paper is to discuss the relationship between environmental education and education reform in the United States and to describe the efforts of the National Project for Excellence in Environmental Education to develop *Excellence in Environmental Education – Guidelines for Learning (K-12)* (North American Association for Environmental Education,1999).

## 2. Environmental Education and Education Reform

Standards, accountability, assessment, transdisciplinary learning, and systemic change each describe one of the many themes of the current education reform movement in the United States. Although the development of academic standards at both the national and state levels is only one of the pieces of education reform, it has garnered much public attention and scrutiny.

The calls for standards setting were first heralded with the publication, in 1983, of *A Nation at Risk.* It became common to call into question the very structure of American education. In at least partial response to the concerns raised in *A Nation at Risk*, each of the core curriculum areas (i.e., science, geography, mathematics, English-language Arts, history, civics) developed a set of voluntary national standards. These standards, many of which have been adapted or adopted at the state level, delineate the knowledge and skill bases of their respective fields. They are designed to define what students should know and be able to do in order to be considered geographically literate, scientifically literate, mathematically literate, etc. by the time they graduate from secondary school.

Although obviously written to address the needs of specific discipline based areas, these standards do, to one degree or another, address environmental education interests. Taken singly, the standards of any one discipline allow for environmental learnings. For example, ecological knowledge such as the components of Earth's physical systems: the atmosphere, lithosphere, hydrosphere, and biosphere; how Earth-sun relations affect conditions on Earth; and the physical characteristics of places (e.g., landforms, bodies of water, soil, vegetation, and weather and climate) are included within *Geography for Life: National Geography Standards* (1994). Similarly, understandings of measurement, patterns and relationships, and statistics and probability, all elements of *Curriculum Evaluation Standards for School Mathematics* (1989), are also important to environmental literacy.

Conversely, environmental education programs can be used to meet discipline-based standards. Because environmental education is by its very nature interdisciplinary, it can help students meet the high standards set by the traditional school disciplines (e.g., science, civics, geography, history). Additionally, integrated throughout the curriculum, environmental education has the potential of furthering the general education reform agenda. Conley (1993) argues that education reform must address the past failure to teach process skills which "led to the inevitable fragmentation of knowledge into 'infobits,' and to graduates who appeared unable to apply much of what they had learned to real-world situations." Environmental education with its focus on developing a well-informed, responsible citizenry "has the potential as an exemplary vehicle for what many believe all of education should consider its primary function: furthering the development of higher-order skills – critical thinking, creative thinking, integrative thinking, problem-solving" (Disinger, 1993).

Although environmental education can effectively and efficiently facilitate the learning of specific concepts and process skills, it also provides an often missed opportunity for synthesis of materials that crosses disciplinary boundaries, connecting learnings to create a whole. The explicit focus of environmental education on the integration of knowledge and skills is one of the primary distinguishing factors between it and a traditional view of curricular disciplines. Because environmental education is, by its very nature, interdisciplinary, the synthesis of learnings across subject material is a deliberate and essential outcome.

Environmental education has the potential of linking the K-12 curriculum, providing the opportunity to meet the requirements of the core disciplines by creating a comprehensive and cohesive program of study. With this said, it must be emphasized that environmental education is more than a useful theme that can tie units of learning together or an effective pedagogy that makes learning more meaningful. Environmental education is essential education. Environmental literacy must be a goal of our society, and environmental education must play an integral role throughout our educational system – at the national level, at the state level, and in each and every classroom.

## 3. National Project for Excellence in Environmental Education

The National Project for Excellence in Environmental Education, sponsored by the North American Association for Environmental Education (NAAEE), was initiated in 1993 to provide an opportunity for environmental education to become a voice in the national education reform agenda. The National Project for Excellence in Environmental Education is a multi-year program designed to establish guidelines for the development of balanced, scientifically accurate, and comprehensive environmental education programs and to identify and provide examples of high quality environmental education practice. The Project has initiated four interrelated efforts: 1) publication of *Environmental Education Materials: Guidelines for Excellence* (1996); 2) creation of a three volume series of educators' resource guides to quality environmental education materials (*The Environmental Education Collection – A Review of Resources for Educators*); 3) development of *Excellence in Environmental Education – Guidelines for Learning (K-12)* (1999); and 4) development of a set of recommendations for the preparation of teachers and other environmental educators.

### 3.1. EXCELLENCE IN ENVIRONMENTAL EDUCATION – GUIDELINES FOR LEARNING (K-12)

*Excellence in Environmental Education – Guidelines for Learning (K-12)* was developed to provide students, parents, educators, administrators, policy makers, and the public a set of common voluntary guidelines for environmental education. The guidelines support state and local environmental education efforts by:

- Setting expectations for performance and achievement in fourth, eighth, and twelfth grades;
- Suggesting a framework for effective and comprehensive environmental education programs and curricula;
- Demonstrating how environmental education can be used to meet standards set by the traditional disciplines and to give students opportunities to synthesize knowledge and experience across disciplines; and
- Defining the aims of environmental education.

*Guidelines for Learning* has been developed over the last four years with the input of literally thousands of teachers, school administrators, environmental educators, scientists, and parents, as well as from a variety of professional organizations and government agencies. Developed through an extensive process of review and comment, they set a standard for high-quality environmental education in schools and other educational settings across the country. They draw on some of the best thinking in the field and its rich history to outline the core ingredients for environmental education.

## 3.2. ESSENTIAL UNDERPINNINGS OF ENVIRONMENTAL EDUCATION

Environmental education builds from a core of key principles that inform its approach to education. Some of these important underpinnings are:

- *Systems* – Systems help make sense of a large and complex world. A system is made up of parts that can be understood separately. The whole, however, is understood only by understanding the relationships among the parts. The human body can be understood as a system; so can galaxies. Organizations, individual cells, communities of animals and plants, and families can all be understood as systems. And systems can be nested within other systems.
- *Interdependence* – Human well being is inextricably bound with environmental quality. Humans are a part of the natural order. We and the systems we create—our societies, political systems, economies, religions, cultures, technologies—impact the total environment. Since we are a part of nature rather than outside it, we are challenged to recognize the ramifications of our interdependence.
- *The importance of where one lives* – Beginning close to home, learners forge connections with, explore, and understand their immediate surroundings. The sensitivity, knowledge, and skills needed for this local connection provide a base for moving out into larger systems, broader issues, and an expanding understanding of causes, connections, and consequences.
- *Integration and infusion* – Disciplines from the natural sciences to the social sciences to the humanities are connected through the medium of the environment and environmental issues. Environmental education offers opportunities for integration and works best when infused across the curriculum, rather than being treated as a separate discipline or subject area.
- *Roots in the real world* – Learners develop knowledge and skills through direct experience with the environment, environmental issues, and society. Investigation, analysis, and problem solving are essential activities and are most effective when relevant to the real world.
- *Lifelong learning* - Critical and creative thinking, decision-making, and communication, as well as collaborative learning are emphasized. These skills are essential for active and meaningful learning, both in school and over a lifetime.

## 3.3. HOW THE *GUIDELINES FOR LEARNING* ARE ORGANIZED.

*Excellence in Environmental Education – Guidelines for Learning (K-12)* offers a vision of environmental education that makes sense within the formal education system and promotes progress toward sustaining a healthy environment and quality of life. Guidelines are suggested for each of three grade levels – fourth, eighth, and twelfth. Each guideline focuses on one element of environmental literacy, describing a level of skill or knowledge appropriate to the grade level under which it appears. Sample performance measures illustrate how mastery of each guideline might be demonstrated. The guidelines are organized into four strands, each of which represents a broad aspect of environmental education's goal of environmental literacy.

*3.3.1. Strand 1: Questioning and Analysis Skills*

Environmental literacy depends on learners' ability to ask questions, speculate, and hypothesize about the world around them, seek information, and develop answers to their questions. Learners must be familiar with inquiry, master fundamental skills for gathering and organizing information, and interpret and synthesize information to develop and communicate explanations.

I. Questioning
II. Designing investigations
III. Collecting information
IV. Evaluating accuracy and reliability
V. Organizing information
VI. Working with models and simulations
VII. Developing explanations

*3.3.2. Strand 2: Knowledge of Environmental Processes and Systems*

An important component of environmental literacy is understanding the processes and systems that comprise the environment, including human systems and influences. That understanding is based on knowledge synthesized from across traditional disciplines. The guidelines in this strand are grouped in four sub-categories.

Strand 2.1 – The Earth as a physical system

I. Processes that shape the Earth
II. Changes in matter
III. Energy

Strand 2.2 – The living environment

I. Organisms, populations, and communities
II. Heredity and evolution
III. Systems and connections
IV. Flow of matter and energy

Strand 2.3 – Humans and their societies

I. Individuals and groups
II. Culture
III. Political and economic systems
IV. Global connections
V. Change and conflict

Strand 2.4 – Environment and society

I. Human/environment interactions
II. Places
III. Resources
IV. Technology
V. Environmental Issues

*3.3.3. Strand 3: Skills for Understanding and Addressing Environmental Issues*

Skills and knowledge are refined and applied in the context of environmental issues. These environmental issues are real-life dramas where differing viewpoints about environmental problems and their potential solutions are played out. Environmental literacy includes the abilities to define, learn about, evaluate, and act on environmental issues. This strand is subdivided in two.

Strand 3.1 - Skills for analyzing and investigating environmental issues

I. Identifying and investigating issues
II. Sorting out the consequences of issues
III. Identifying and evaluating alternative solutions and courses of action
IV. Working with flexibility, creativity, and openness

Strand 3.2 – Decision-making and citizenship

I. Forming and evaluating personal views
II. Evaluating the need for citizen action
III. Planning and taking action
IV. Evaluating the results of actions

*3.3.4. Strand 4: Personal and Civic Responsibility*

Environmentally literate citizens are willing and able to act on their own conclusions about what should be done to ensure environmental quality. As learners develop and apply concept-based learning and skills for inquiry, analysis, and action, they also understand that what they do individually and in groups can make a difference.

I. Understanding societal values and principles
II. Recognizing citizens' rights and responsibilities
III. Recognizing efficacy
IV. Accepting personal responsibility

## 4. A Final Thought

Taken together, these *Guidelines for Learning* create a vision of environmental literacy. A knowledgeable, skilled, and active citizenry is a key to resolving the environmental issues that promise to face us in the years to come. For each environmental issue there is not just one right answer or solution – there are many perspectives and much uncertainty. A quality environmental education program cultivates the ability to recognize uncertainty, envision alternative scenarios and adapt to changing conditions and information. This translates into a citizenry that is better able to address its common problems and take advantage of opportunities, whether environmental concerns are involved or not. *Excellence in Environmental Education – Guidelines for Learning (K-12)*, in particular, and the National Project for Excellence in Environmental Education, in general, cannot "create" an environmentally literate citizenry. The project is aimed at producing a series of tools that might help educators develop effective, locally relevant environmental education programs. Environmental literacy does not just happen. It requires a concerted effort of all those who care about quality education and support the notion that students need to be prepared to make informed decisions as individuals, as consumers, as workers and as members of society.

## References


Conley D.T. :1993, *Roadmap to Restructuring: Policies, Practices and the Emerging Visions of School*, ERIC.

*Curriculum and Evaluation Standards for School Mathematics*.:1989, National Council of Teachers of Mathematics, Reston,VA.

Disinger J.:1993)Environmental Education in the K-12 Curriculum: An Overview, in R.Wilke (ed), *Environmental Education Teacher Resource Handbook*, Kraus International Pub, Milwood, NY.

Disinger J. and Monroe M.:1994, *Defining Environmental Education*, An EE Toolbox Workshop Resource Manual, University of Michigan, National Consortium for Environmental Education, Ann Arbor, MI.

*Geography for Life*: *National Geography Standards*.:1994, National Geographic Research and Exploration, Washington, D.C.

National Commission on Excellence in Education.:1983, *A Nation at Risk: The Imperative for Education Reform*, U.S. Government Printing Office, Washington, D.C.

North American Association for Environmental Education.:1999, *Excellence in Environmental Education – Guidelines for Learning (K-12)*, NAAEE, Rock Spring, GA.

UNESCO .:1976, Belgrade Charter. *Connect*, **1** (1), 1-2.

UNESCO.:1978 Final Report Intergovernmental Conference on Environmental Education, *Connect*, **3** (1), 1-8.

# ENVIRONMENTAL AND AGRICULTURAL LITERACY EDUCATION

D. HUBERT[1], A. FRANK[1] and C. IGO[2]

[1] *Southwest Center for Agricultural Health, Injury Prevention and Education, University of Texas Health Center at Tyler, 11937 US Highway 271, Tyler, TX, USA 75708-3154*

[2]*Department of Agricultural Education, Communications, and 4-H Youth Development, Oklahoma State University, 448 Agriculture Hall, Stillwater, OK 74078-6031*

**Abstract.** Educational offerings that utilize environmental and agricultural themes can reinforce basic education for students in kindergarten through twelfth grade (K-12) while also teaching about the environment and agricultural methods and products. A curriculum guide about the environment and food and fiber production was created for K-12 teachers. These materials were evaluated for their effectiveness in increasing student knowledge using elementary classes in several states. It was noted through pre/post tests that younger students, in general, made greater gains. Within five thematic areas, the greatest overall improvement was shown in themes related to the environment. Environmental topics covered, all in the context of agricultural themes, included the need to preserve shared natural resources including land, water, and air as well as the managing of the ecosystem and the use of non-renewable energy resources. Clearly such classroom guides have utility in teaching young students about environmental issues and their relationship to other important topics. This guide and its corresponding Website enhance both opportunities to transmit new knowledge as well as assess performance and impact on behavior.

**Keywords**: agricultural education, agricultural literacy, curriculum, environmental education, food and fiber systems literacy

## 1. Introduction

Educational topics inclusive of agriculture and environmental topics are neither new nor unique. Agricultural education professor and historian John Hillison recently published findings on the history of integrating agricultural themes or concepts into academic areas of study, particularly science. He noted that the primary school agricultural curriculum was preceded in the latter part of the 1800s by a nature-study movement that aimed at bringing reality to science lessons in elementary school classrooms. This movement was in reaction to the use of less interesting methods of teaching science, and noted nature-study usage in the states of New York and Massachusetts (Hillison, 1998). Although there was no indication that student learning of science principles increased, it may be inferred that inclusion of nature-study topics made the subject matter more relevant to students' lives in the still very rural United States.

As countries develop and move to more urbanized societies, basic knowledge and understanding of the natural environment and its interrelated systems appears to have declined. Whatever the reason, from video games to other less outdoors-oriented activities, urban youth populations in general appear to have

*Water, Air, and Soil Pollution* **123:** 525–532, 2000.

lost connections with the natural environment and the respect and admiration associated with the understanding of its systems through participation or interaction. This loss of ecological and biological knowledge gained through hands on involvement has raised the level of concern from both agricultural and environmental organizations, among others, especially within the United States. The purpose of this paper is to heighten awareness to a recently developed framework for improving food and fiber literacy in K-12 students.

## 2. Agricultural and Environmental Literacy Development

As America's urban sectors expanded and engulfed more farmland during the 1970s, 1980s, and early 1990s, agricultural organizations realized a lack of knowledge and understanding of agriculture and agricultural processes by the general public could have been part of the problem of this urban sprawl and loss of productive acreage. The continued encroachment by cities into productive, arable lands seemed to emphasize the lack of respect for farming as small farmers were either forced or bought out of their profession. There appeared to be hope that creating a greater public understanding of agriculture would increase agriculture's importance in America. This, in turn, might offset the agricultural near-sightedness acquired by land developers and the large support of a generally agriculturally illiterate public. An organized and increased effort to re-educate Americans, beginning with elementary students, was beginning to take shape.

In 1988 the National Research Council's Committee on Agricultural Education in Secondary Schools proposed that an agriculturally literate person would understand the food and fiber system in relation to its history, economic, social, and environmental significance (National Research Council, [NRC] 1988). Additionally the Committee recommended that "all students should receive at least some systematic instruction about agriculture beginning in kindergarten or first grade and continuing through twelfth grade" and "the subject matter ...about agriculture be broadened...to include the utilization of environmental and resource management."

According to Terry *et al.* (1996), the need for societal knowledge about agriculture is based on two primary factors. First, as consumers of agricultural goods, people need to understand basic principles of food and fiber sources, marketing, distribution, and nutrition. Secondly, they hypothesized that because of the roles citizens play in policy decisions, people need to understand the impact of agriculture upon society, the economy, and the environment. There is a similar need for the understanding of environmental positions associated with agriculture as well. In this context, controversial agricultural and environmental issues are often the results of competing factions and there is an equal significance for environmentally and agriculturally literate populations throughout the world.

Some assistance in creating agriculturally literate populations can be found in the use of mass media to provide correct information about agricultural and environmental issues with science knowledge providing the best facilitating means to understanding. Rogers (1983) and Terry (1994) indicated the most important factor contributing to consumer awareness and understanding about science and technology is mass media (as cited by Vestal and Briers, 1999). Recently, increased media coverage of environmental controversies appears to have magnified the natural resource connections between the environment and agriculture. This media awareness has not been entirely positive for the natural resource utilizing industries of agriculture, forestry, and fishing, and their corresponding efforts to create sustainable societies. Often public perception, or misperception, is affected by television and periodical coverage overdramatizing these industries' activities and impacts on non-renewable resources.

However, the apparent public disapproval of some natural resource utilizing industries portrayed in the press may be found not only in mass media's focus on the negative influence of agricultural production on the environment, but also in the lack of understanding of these matters by journalists. Analysis by Terry *et al.* (1996) of agriculture/environment and public policy news articles revealed a high percentage of unfavorable, judgmental sentences and a high degree of journalistic bias in them. Of the three most popular news periodicals in U.S. circulation in 1995, *Time*, *Newsweek*, and *U.S. News and World Report*, only 13 articles were printed with agricultural issues as the main topic during the entire year. Eight of the agricultural articles were classified by the researchers as environmentally related. Situations such as this (the largest means of information dissemination contributing to public misinterpretation) provide the motivation to engender societies that can synthesize, analyze, and communicate basic information about agriculture and the environment. However, for these societal changes to evolve, we must focus on learners when they are most susceptible to new thoughts and ideas.

## 3. Curricula/Teaching Resources About Agriculture and the Environment

### 3.1. A GUIDE TO FOOD AND FIBER SYSTEMS LITERACY

The challenge for educators in infusing food and fiber systems literacy into core academic subjects is recognizing existing connections. Connecting biology and life science, as well as environmental science to agriculture is easy. Other core academic subjects can be more challenging to infuse with agricultural topics. Without detailed guidance for attaining knowledge about agriculture and the environment, students, as they mature into adults, will be asked to make decisions about matters they know little about. It was determined that students of all ages, if presented information in a systematic manner, would become better decision-making adults in matters relating to agriculture and the environment.

*A Guide to Food and Fiber Systems Literacy* (referred to as the *Guide*) was planned to facilitate these types of challenges. Developed at Oklahoma State University, the *Guide* was the culmination of four years of work in developing and testing a curriculum framework of agricultural themes, standards and benchmarks, and supporting material needed to produce agriculturally literate students. The *Guide* also includes explanatory narrative needed for implementing Food and Fiber Systems Literacy in schools. Themes are sectioned into *Understanding Food and Fiber Systems*; *History, Culture, and Geography*; *Science, Technology, Environment*; *Business and Economics*; and *Food, Nutrition, and Health.* Grade specific benchmarks guide teachers to make appropriate academic connections to agriculture and environmental topics, with sample lessons included to facilitate this process. It is hoped that implementation of the *Guide* by schools and districts across all grade levels will produce better-educated students so their agricultural and environmental issue decision-making capabilities will be enhanced. To facilitate distribution of the Guide, an Internet Website was developed for teachers and other interested educators and administrators (http://food_fiber.okstate.edu).

The use of the *Guide* to provide direction for classroom instruction about agricultural and environmental topics and ultimately increase student learning in these areas was studied. A pilot testing of the *Guide* was conducted in three K-8 schools during the 1997-98 academic year in the states of California, Montana, and Oklahoma (one school per state). These case studies included 366 students, 177 students, and 257 students, respectively. The researchers used a pretest and posttest methodology to determine gains in students' food and fiber knowledge for the grade groupings K-1, 2-3, 4-5, and 6-8. These grade groupings correspond with the grade-grouped benchmarks in the *Guide*. Using SAS for analysis procedures, statistically significant differences in pretest and posttest group results were found. Table I provides a summary of these results.

Pearson's Product Moment Correlation Coefficients were computed to determine whether a relationship existed between these knowledge score differences and the number of instructional connections teachers made to Food and Fiber Systems, i.e., the number of times teachers referred to agricultural topics in lessons. Table II summarizes the result of the analysis.

Both the Montana site and the Oklahoma site showed a strong correlation, 0.621 and 0.586 respectively, between the test score differences and the number of instructional connections made by teachers, with the Oklahoma site returning a significant statistical difference. Pooling the Montana and Oklahoma data to create a composite yielded a 0.603 correlation coefficient and the computed difference was statistically significant as well. Knowledge score increases of 10 percent or better were seen when the number of reported connections rose to 20 or above. California data were not included in the correlation due to structural differences between the California school and the other two schools.

TABLE I

Students' food and fiber knowledge levels as measured by pretest and posttest scores

| | | Pretest | | Posttest | | | | |
|---|---|---|---|---|---|---|---|---|
| State: | Grade | n | Mean | n | Mean | Difference | F-value | p |
| CA: | K-1 | 15 | 54.8 | 12 | 57.3 | +2.5 | 0.86 | 0.3555 |
| MT: | | 54 | 72.1 | 50 | 88.8 | +16.7 | 74.75 | 0.0001* |
| OK: | | 53 | 77.3 | 53 | 86.1 | +8.8 | 21.33 | 0.0001* |
| CA: | 2-3 | 42 | 71.0 | 39 | 76.8 | +5.8 | 8.83 | 0.0032* |
| MT: | | 38 | 75.6 | 35 | 89.4 | +13.8 | 46.28 | 0.0001* |
| OK: | | 73 | 79.3 | 72 | 88.4 | +9.1 | 41.24 | 0.0001* |
| CA: | 4-5[a] | --- | --- | --- | --- | --- | ----- | ------ |
| MT: | | 49 | 67.2 | 47 | 71.2 | +4.0 | 5.13 | 0.0239* |
| OK: | | 75 | 66.1 | 74 | 72.7 | +6.6 | 15.11 | 0.0001* |
| CA: | 6-8[b] | 502 | 31.8 | 315 | 29.3 | -2.5 | 21.45 | 0.0001* |
| MT: | | 50 | 63.7 | 45 | 62.4 | -1.3 | 0.23 | 0.6315 |
| OK: | | 67 | 57.9 | 58 | 55.0 | -2.9 | 1.53 | 0.2157 |

Note. df for all calculations was 1. *p<0.05

[a] There was no 4-5 component in CA

[b] There were no 6th grade participants in CA

TABLE II

Correlation of differences in pretest and posttest scores to instructional connections by site

| Site | n | reported connections | Pearson r | p |
|---|---|---|---|---|
| Montana | 8 | 14-28 | 0.621 | 0.1003 |
| Oklahoma | 13 | 5-27 | 0.586 | 0.0353* |
| Composite | 21 | 5-28 | 0.603 | 0.0038* |

*p < 0.05

The California school did not include all grade levels and was based on a village concept. [The village concept refers to the assignment of groups of students to specific teachers whereby the groups have continuous educational contact with each other over a two-year period. Villages may contain from five to seven groups. Teachers are representative of each core subject area (e.g., Science, Language Arts, Physical Education, Mathematics, etc.) and hold weekly, common classroom planning sessions. This provides opportunity to use content themes within lessons from class to class in the village.] The California teachers did submit reports, however the village concept prevented applicable correlation tests between a particular group or class of students and the number of instructional connections those students received.

The FFSL Framework was organized around five thematic areas: *Food and Fiber Systems—Understanding Agriculture; History, Culture, and Geography; Science—Agricultural and Environmental Interdependence; Business and Economics; and Food, Nutrition, and Health.*

With each site using an infusion approach to implementing FFSL, the data were combined to provide a composite view of the thematic area analysis. That composite information was presented in Table III.

TABLE III

F-Value Comparison Of Composite Pretest And Posttest Differences By Grade Groups Within Theme Areas For California, Montana, and Oklahoma Sites

| Theme and grade grouping | F-value | p |
|---|---|---|
| *Understanding Agriculture* | | |
| K-1 | 15.5 | 0.0001* |
| 2-3 | 11.01 | 0.0001* |
| 4-5[a] | 42.71 | 0.0001* |
| 6-8 | 19.54 | 0.0001* |
| *History, Culture, and Geography* | | |
| K-1 | 1108.58 | 0.0001* |
| 2-3 | 33.33 | 0.0001* |
| 4-5[a] | 52.83 | 0.0001* |
| 6-8 | 290.48 | 0.0001* |
| *Science and Environment* | | |
| K-1 | 202.96 | 0.0001* |
| 2-3 | 0.00 | 0.9820 |
| 4-5[a] | 79.96 | 0.0001* |
| 6-8 | 14.09 | 0.0002* |
| *Business and Economics* | | |
| K-1 | 4.80 | 0.0295* |
| 2-3 | 22.56 | 0.0001* |
| 4-5[a] | 18.76 | 0.0001* |
| 6-8 | 0.40 | 0.5254 |
| *Food, Nutrition, and Health* | | |
| K-1 | 59.88 | 0.0001* |
| 2-3 | 145.27 | 0.0001* |
| 4-5[a] | 24.21 | 0.0001* |
| 6-8 | 5.92 | 0.0151* |

Note. df for all calculations was 1. *$p < 0.05$
[a] there was no 4-5 component in CA – data represent only MT and OK

With only two exceptions, all grade groups within each theme area showed statistically significant differences between pre- and post test results. Within the Science and Environment theme, the 2-3 grade-group produced a zero F-value, which yielded a 0.98 significance score. The 6-8 group, within the Business and Economics theme, also showed no statistical significance, producing an F-value of less than one.

Though not generalizable outside of this study, these results may suggest that when implemented consciously, the *Guide* may influence student learning about agriculture and the environment.

### 3.2. AGRICULTURE AS ENVIRONMENTAL SCIENCE

Prior to development of the *Guide*, Israel also had recognized the importance of tangible, realistic instruction that uses agriculturally based instruction, albeit 20 years earlier and specifically in science. In the 1970s, according to Blum (1985), Israeli schools introduced a new Agriculture as Environmental Science (AES) curriculum. The AES was similar to Rural Studies in other countries but was adapted to the urbanization of Israel and the environmental crisis. The AES curriculum emphasized the application of science to relevant and interesting situations. The developers intended to enhance students' attitudes toward the subject of agriculture and the environment by putting it into a context that was relevant and useful to their lives. Blum also indicated that the AES, inquiry-type curriculum was successfully used to change student perception of the usefulness of a subject matter with increased levels of cognitive understanding and of concept formation.

## 4. Conclusions and Implications

An understanding of environmental and agricultural issues have a potentially significant role on the well being of society, and ultimately within the political process. Food and fiber are essential for life and survival of human species. Unfortunately, there is a less than adequate appreciation in many industrialized societies of the importance of sound environmental and agricultural policy.

Since education is a major component of processes to bring about changes in attitudes, behavior, and adoption of innovations (Rogers, 1995), the development of an easy to use, field-tested, broad-based curriculum guide should be a valuable addition to the teaching armamentarium of K-12 teachers in their classrooms. The *Guide* presented here, and its related Website, have the potential to educate students and teachers thus bringing about the desired changes. Test results to date confirm an increase in knowledge is possible, the first necessary step to bring about change in attitudes and behaviors.

Further adaptations of this guide and its further implementation hold great promise for the future. Since all societies are dependent on agriculture for their survival, agricultural and environmental policies will become increasingly more important over time. Educational materials such as these will assist with general Food and Fiber literacy efforts, and assist with continuing to keep the world a safe, and habitable planet.

## References

Blum, A.: 1985, "Assessment of subjective usefulness of an Environmental Science curriculum," *Science Education* **66 (1)**, 25-34.

Hillison, J.: 1998, "Agriculture in the classroom: Early 1900s style," *Journal of Agricultural Education* **39 (2)**,11-18.

Igo, C. and Leising, J.: 1999, "Assessing agricultural literacy: A case study approach," in: *Proceedings of the 1999 Southern Agricultural Education Research Meeting: Breaking New Ground in Agricultural Education,* Memphis, Tennessee, 164-176.

Leising, J., Igo, C., Hubert, D., Heald, A., and Yamamoto, J.: 1998, *A Guide to Food and Fiber Systems Literacy: A compendium of standards, benchmarks, and instructional materials for grades K-12,* Oklahoma State University, Stillwater, Oklahoma.

National Research Council: 1988, *Understanding Agriculture: New Directions for Education*, National Academy Press, Washington D.C.

Rogers, E.: 1995, *Diffusion of Innovations*, The Free Press. New York City, New York.

Terry, R., Dunsford, D., Lacewell, B., and Gray, B.: 1996, "Evaluation of information sources about agriculture," in: *Proceedings of the 1996 National Agricultural Education Research Meeting: Partnerships for Success Through Research in Agricultural Education*, G. Wardlow and D. Johnson (eds.), vol XXIII, Cincinnnati, OH, 215-226.

Vestal, A. and Briers, G.: 1999, "Knowledge, attitudes, and perceptions of journalists for newspapers in metropolitan markets in the United States regarding food biotechnology," in: *Proceedings of the 1999 Southern Agricultural Education Research Meeting: Breaking New Ground in Agricultural Education*, Memphis, Tennessee, 153-163.

# THE UPCOMING CHALLENGE: TRANSBOUNDARY MANAGEMENT OF THE HYDRAULIC CYCLE

E. FEITELSON

*Department of Geography, The Hebrew University of Jerusalem, Mount Scopus, Jerusalem 91905, Israel*

*Email: msfeitel@mscc.huji.ac.il*

**Abstract**. The increase in population and subsequent demand for food will lead to rising demand for water. These, in turn, will lead to increasing utilization of transboundary water resources. In the past treaties have focused primarily on the utilization of freshwater surface resources, in particular rivers. Most of the treaties dealt only with water abstractions and, in some cases, with in-stream uses, mainly navigation and hydro-electricity. However, a hydraulic cycle view suggests that *transboundary* water resources include not only freshwater flows, but also return flows (direct or as effluents), lakes and reservoirs, aquifers, and precipitation. Moreover, water quality changes along the cycle, and effects the potential and cost of utilization. As water resources would have to accommodate increasing and diversifying demand, better management of all parts of the hydraulic cycle would be needed. This paper argues that as a result of these observations, and the increasing tendency toward decentralization of authority and in some areas separatist trends, this century would be marked by a need to establish increasingly intricate transboundary management structures, that would address all facets of the hydraulic cycle. This argument is demonstrated for the Israeli-Arab case.

**Keywords**: groundwater, international water, Israeli-Arab water, wastewater, water management institutions

## 1. Introduction

The appropriation of water resources for human use is growing rapidly. Between 1950 and 1990 alone water use worldwide more than doubled, from approximately 1400 $km^3$ to some 3000 $km^3$ (Raskin *et al.*, 1996). This amount accounts for only 8% of the average annual runoff. However, due to the vast spatial and temporal discrepancies between availability of runoff and human demand patterns, many regions already utilize most of the readily available runoff. As a result, in many of these areas there is increasing competition over water. Such competition is made ever keener by the apparent deterioration in water quality and the growing realization of the environmental importance of in-stream use. As projections for this century suggest that these trends will continue, we can expect in this century (and millennium - if the trends are not reversed) that the extent of stressed freshwater resources would continue to rise. To address such stress, an increasing array of tools would have to be used (Postel, 1992).

*Water, Air, and Soil Pollution* **123:** vii, 2000.

Water resources do not conform to administrative and political boundaries. In 1978 the Centre for Natural Resources Energy and Transport identified 214 international river and lake basins, 48 more than it identified 20 years previously. Biswas (1993a) argues that this often-quoted number is probably an undercount, as it is based in part on maps at a small scale, does not account for groundwater flows, is derived by an approach that suffers from several technical limitations, its definitions of what constitutes an international basin do not necessarily reflect the area that may be most important from a management perspective, and the data on which it is based is dated. In particular, it does not account for the many new countries that have been established since, particularly in eastern Europe. Boundaries complicate the management of resources, as they create discrepancies between spheres of control and natural systems. The likelihood that stressed international water bodies would be degraded may be greater, therefore, than intra-national resources.

As the demand for water increases, additional resources are appropriated. The result of such increasing utilization is in many cases reduced availability of freshwater to downstream and in-stream uses and users, and detrimental effects on the quality of the water remaining for other users. Moreover, as the demand for water increases and the local, exclusively national, sources are fully developed, it is likely that the main resources that would be further developed would be international in character (Biswas, 1993b). If this is indeed the case such development may cause international tensions and conflict to arise among the different users (be they riparian or not). It may also lead to the degradation of such resources, if cross-boundary management regimes are not agreed upon and implemented.

The potential conflicts regarding transboundary water resources have led to the signing of a large number of treaties. Wolf (1998a) notes that approximately 300 treaties have been signed since 1814 that deal with non-navigational issues of water management, flood control, hydropower projects, and allocations for consumptive and or nonconsumptive uses in international basins. Almost all of the 145 treaties he analyzed dealt with river systems. Only few international treaties address water quality issues (Shmueli, 1999). Yet, the history of human development is replete with the story of deterioration of water resources, that are aggravated as numbers rise (Ponting, 1992). Past treaties are unlikely therefore to address future issues comprehensively.

Terrestrial water, which is the focus of most studies, discussions, disagreements and agreements, is but one part of the great hydrological cycle, from the oceans to the atmosphere to the land and back again. Actually, in any given time freshwater resources are but a few percent of the total water resources. Moreover, of the total freshwater resources, almost two thirds are locked in glaciers and permanent snow cover, and are thus not included in the conventional water discourse. As a result, the options for utilizing additional parts of the hydraulic cycle are receiving increasing attention.

This paper suggests that if current trends continue (and there is no indication at this point that they would not) this century (and perhaps millennium) would be marked by the increasing need to address the international character of water flows, and the fact that all water flows are part of the great hydrological cycle. The management of water resources within this context would require more complex management structures. This argument is advanced by focusing on the Israeli-Arab case, as this region is one of the most water-stressed (using the water per capita based indices), and hence faces issues related to water scarcity somewhat earlier than more water abundant regions.

## 2. Human Interventions in the Hydraulic Cycle

The hydraulic cycle is described usually as a natural sequence through which water passes, including all the physical states - gaseous, liquid and solid - as well as the transformations among these states (for example, evaporation, precipitation, freezing and melting). In this cycle, described by the solid arrows in Figure 1, water passes from the atmosphere to oceans and terrestrial systems. In the terrestrial systems water flows between the surface, soil and aquifers.

Humans have intervened in the hydraulic cycle since the transition to agriculture in the Neolithic era, when farmers began to divert some of the flowing water for irrigation. During the last millennium these interventions have grown increasingly sophisticated, widespread, and larger in scale (the broken arrows in Figure 1).

The most widespread human intervention is in surface flows, often through the construction of dams. The reservoirs behind such dams often serve a myriad of uses - such as irrigation, power generation, flood control, municipal supply, groundwater recharge and recreation. By changing water flow patterns, dams affect ecosystems, alter groundwater and surface water regimes, change land use in wide areas and affect the local micro-climate. Canals and pipelines divert surface water from their channels and, in some cases, out of their natural drainage basins to distant urban or agricultural use. The extraction of surface water, particularly for irrigation, affects the extent of evaporation and hence the quantity and quality of the water remaining in rivers and lakes.

Human intervention is not limited any longer to surface flows. In many parts of the world groundwater is increasingly the main source of water. Abstractions

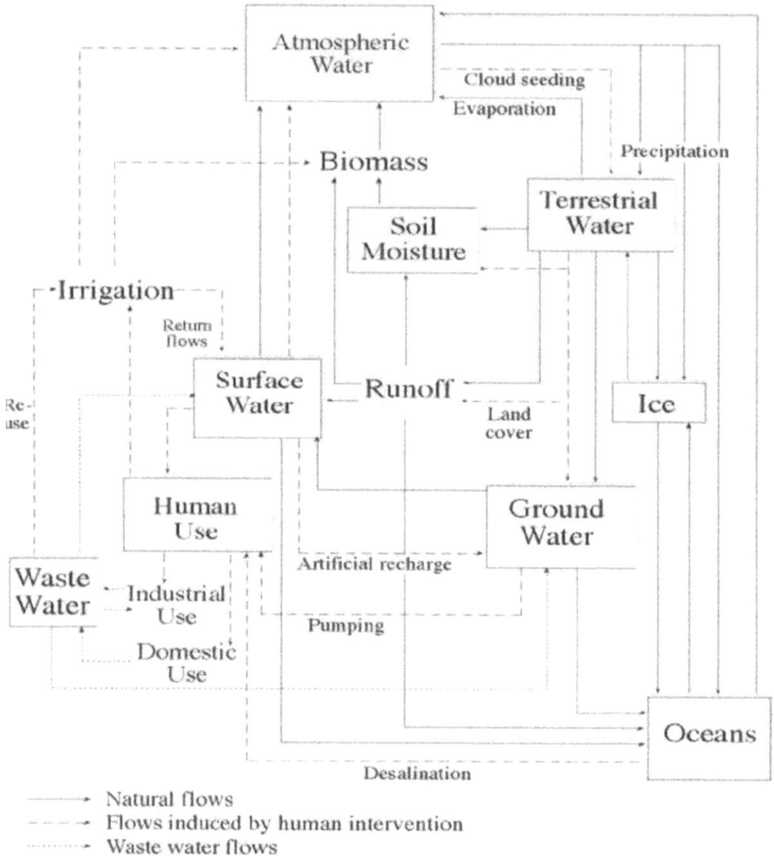


**Figure 1: Human Interventions in the Hydraulic Cycle**

from groundwater affect water levels in the aquifers, and subsequently spring discharge and the probability of salinization. Water levels and quality in the aquifers are also affected by land use and development patterns. Rising urbanization combined with increasing decentralization within metropolitan areas, rising levels of motorization and subsequent increase in roads, parking facilities and other ancillary transport services, increase the extent of impervious areas - affecting the amount of runoff and recharge. Increasing amounts of wastewater, solid waste and hazardous materials may also affect the quality of water percolating down to the aquifer. In recent decades humans also affect atmospheric water. Clouds are seeded to produce rain or divert storms. Long-range air pollution acidifies precipitation, and subsequently lakes and rivers.

Today, the scale of human intervention is such that in many parts of the world a significant percentage of terrestrial water flows in and out of human use (Raskin *et al.*, 1996). These amounts can be enhanced by desalinization of sea water, and are reduced by increased evaporation. Human use, and particularly domestic and industrial usage, affects water quality. As a result the water emanating from such uses is termed wastewater, water that is not suitable for other uses without treatment. Therefore, in Figure 1, a third set of flows, wastewater flows, is added to the flows induced by human intervention. Wastewater often flows into rivers and lakes or percolates down to the aquifers, thereby polluting surface and groundwater resources. Wastewater can also be re-used beneficially, mostly for irrigation, after appropriate treatment. In other words, in an increasing number of cases, wastewater can be seen as a potential resource and not merely as a hazard.

Humans have devised many ways to manage common pool resources, such as water (Ostrom, 1990). However, most of the structures created for this purpose have been at the local level, and dealt with situations that are less stressful than those that arise today or are likely to arise in the future. As the scale of human intervention in the hydraulic cycle increases so does the likelihood that they cross some boundary. This has been recognized by a few international lawyers (Teclaff, 1991a). However, most of these interventions have not been addressed in international law, or in the treaties regarding water issues signed between different countries. Moreover, several important points of interaction, such as estuaries, have not received adequate attention in the international law literature (Hayton, 1991). Overall, international law has focused on the allocation of water, and has not provided an appropriate base for cooperation, or for a holistic managerial view of the hydraulic cycle, or even of basins (Benvenisti, 1996; Teclaff, 1991b). Yet, if the current trends continue these will have to be addressed in the future. To understand the need, complexity and options we turn now to the Israeli-Arab case.

## 3. Cross-Boundary Water Issues Between Israel and her Neighbors

The Middle East is considered the most water-stressed region of the world, with probably the lowest water availability per capita, and highest use per resource ratio (Raskin *et al.*, 1996)[1]. Moreover, Raskin *et al* (1996) show that under conventional development scenarios conditions the use to resource ratio is likely to rise significantly within the next generation (from 58% in 1990 to 75% in 2025 and 94% in 2050). Within the Middle East, Israel, Jordan and the Palestinians seemingly face the most difficult situation, as they are supplied by

[1] Use to resource ratio is the ratio between average annual use and average natural long-term (past) recharge or flows.

already depleted fully utilized sources. Moreover, most of these sources, in particular the Jordan River and its tributaries and the Mountain Aquifers are shared by more than one entity. As a result the allocation and management of these resources has received significant attention in the literature (for example, Kliot, 1994; Wolf, 1993).

The actual interactions between the parties in this region are, however, more intricate than suggested in most studies. They include most of the human interventions in the water cycle identified in Figure 1. In this section these interactions are briefly reviewed, in order to identify some of the management issues they raise.

### 3.1. THE JORDAN RIVER AND ITS TRIBUTARIES

The Jordan River basin is currently shared by Israel, Jordan, Lebanon, Syria as well as the Palestinians. The allocation of water in the Jordan basin has been the focus of contention, discussions and negotiation since the early fifties (Lowi, 1993; Shuval, 1998). The upper Jordan River supplies approximately a third of Israel's annual water use, and is the source for Israel's National Water Carrier conveying water to the south (see Figure 2). The Yarmouk River is the principal source of supply to the densely populated parts of Jordan and through the East Ghore Canal to the Jordan Valley. The Syrians have dammed some of the Yarmouk's tributaries as part of a large-scale irrigation development effort in southern Syria. The Palestinians lay claim to 200 MCM of the 700 MCM that Israel abstracts from the Jordan's basin (Soffer, 1994).

Water allocations were the topic of Article 6 in the 1994 Israeli-Jordanian Peace Treaty. In this treaty the two sides did not opt for the seemingly straightforward separation option (as was done in the Indus case), whereby each receives the water of several tributaries and manages it as it sees fit. Rather, in addition to augmenting Jordan's water supply, Israel agreed to provide Jordan with *de facto* storage services for 20MCM. The parties also agreed to desalinate the saline water diverted from Lake Kinneret (Sea of Galilee), of which Jordan is to receive 50% (10 MCM). In addition, the agreement allows Israel to continue utilization of an aquifer in the Arava Valley, and establishes a management framework.

This recent agreement addresses, therefore, the joint development of future resources, as well as a series of issues beyond the straightforward division of surface flows. It does not specify, however, any measures for coping with drought situations, something that proved a source of contention during the last year (1998-9), which had an extremely dry winter, and does not show real concern for the ecological and environmental implications of water abstractions for the lower Jordan River and the Dead Sea. This agreement also does not address the Syrian, Lebanese and Palestinian claims. These will need to be addressed in future negotiations (Shuval, 1998).

## 3.2. THE MOUNTAIN AQUIFERS

Israel shares with her neighbors several aquifers. These are the Mountain Aquifers underlying the West Bank and Israel, the Arava Aquifer, the Hermon Aquifer feeding the Jordan River and the Gaza aquifer (Gross & Soffer, 1996). So far agreements have been reached with regard to the Arava and Gaza aquifers. The most important of the shared aquifers, however, are the Mountain Aquifers. These aquifers, shown in Figure 2, are the highest quality storage within Israel's water system, and the sole source of water for Palestinians on the West Bank. They include three major sub-basins. The western and northeastern aquifers are fully utilized, mostly by Israel. In the Oslo B accords, signed in

Figure 2: The Hydraulic Interactions between Israel and her Neighbors

September 1995, Israel agreed to a small increase in Palestinian use of the northeastern aquifer, and to substantial development of new sources in the eastern aquifer.

These karstic, limestone aquifers are susceptible to pollution, and to salinization from subterranean saline water bodies, if they are over pumped. Therefore, it has been long recognized that they should be managed judiciously, and that pumping from them should be controlled. Such restrictions were imposed by Israel during its period of absolute control over the whole three sub-aquifers (1967-1995). The introduction of boundaries between Israel and the Palestinian entity above these aquifers presents new challenges, as neither party can manage the aquifer alone and yet maintain its quality (and hence storage capacity). Therefore some level of joint management is necessary (Feitelson, 1996). This was realized in the Oslo B accords, where a coordinated management structure, including a Joint Water Committee and Joint Supervision and Enforcement Teams, were established.

The joint (or coordinated) management of shared aquifers requires that multiple issues be addressed (Feitelson *et al.*, forthcoming). These include protection of the aquifer by control of land use, wastewater recycling, and waste disposal sites and methods, control of pumpage, monitoring of water levels and quality, data collection, enforcement of restrictions, settlement of disputes regarding extraction or land use implications, adjustments of pumpage to variations in natural recharge and the establishment of appropriate institutional and legal structures to carry out the various tasks. As a result of the complexity of these tasks, and the usually false perception that water in the aquifers does not flow (and hence does not cross borders), there has been only scant experience with transboundary management of shared aquifers.

Since 1993 an Israeli-Palestinian team has been working to identify options for joint management of shared aquifers, and tailoring these options for the Israeli-Palestinian case. This team has advanced five basic options, each of which can be developed over time (Feitelson & Haddad, 1998). These options vary according to their goals and basic rationale. Consequently, they differ in terms of the actions they would undertake and the sequencing of the actions. However, the implementation of any such structure is fraught with pitfalls, all of which have to be addressed if any successful structure is to be established (Haddad *et al.*, 1999). The range of issues that has to be addressed as part of any effort to manage shared aquifers in a sustainable manner ranges, therefore, far beyond the questions of sustainable yield, as manifested in abstraction and recharge rates.

### 3.3. ENHANCEMENT PROJECTS

The supply of water in any given region can be enhanced in a wide variety of ways. Allan (1994) has shown, for example, that the Middle East has been able to support its burgeoning population through the import of 'virtual water' in the form of food and agricultural products. However, there are additional options, many of which have been tried or discussed within the Israeli-Arab context.

*3.3.1. Water Transfers*

Israel has been conducting large scale inter-basin water transfers since the mid-fifties. As can be seen in Figure 2, Israel has built over the years a national water distribution system as an integrated grid, that also augments supply to parts of the West Bank and the Gaza Strip. Following the Israel-Jordan peace treaty, initial connections between the Israeli and Jordanian system have also been executed. One problem that emerged in the discussion of such transfers between Israel and her neighbors is the allocation of capital and operating cost for ongoing transfers as well as compensation for past investments.

In addition to these connections, enabling water transfers among different parts of Israel, and between it and her immediate neighbors, there have been several proposals to bring in additional water from further afield. In particular, there have been suggestions for transporting water from Turkey, Lebanon and the Nile (Shuval, 1992). While none of these has advanced beyond the conceptual stage, they are indicative of possibilities for transboundary water transfers that can help redress the spatial discrepancy between demand and supply patterns, and the acute scarcity in Israel, Jordan and Palestine.

*3.3.2. Cloud Seeding*

Cloud seeding to enhance rainfall has been attempted in Israel since the early seventies. Studies of the effect of this effort have indicated an increase of 5-15% in rainfall. However, these results have been disputed. Unsubstantiated evidence suggests that the seeding also had a beneficial effect in Jordan, and perhaps Syria. The lack of sufficiently clear evidence of beneficial effects within Israel, and budget cuts in the last few years have resulted in practical termination of this effort. Regardless, this effort indicates some of the difficulties inherent in positive sum efforts that have a stochastic cross-border effect. Specifically, there is a disincentive to invest in such efforts, as the party that invests in the effort cannot be assured of reaping all the benefits. Consequently, efforts that may have a positive regional effect may be under-funded, or terminated, because of lack of support in the intra-national budget allocation.

*3.3.3. Flood Control and Artificial Recharge*

Other efforts that may have positive stochastic transboundary effects are flood control and artificial recharge schemes. These schemes usually involve damming of tributaries. A small number of such schemes were built within Israel. Recently, a proposal for such a scheme on the West Bank has been studied. This scheme suggested that winter flows affecting the Tel Aviv metropolitan area be captured upstream, within the West Bank, and recharged into the Mountain Aquifer. In a cross-boundary setting, such schemes face particular problems, as the loss of land, cost and maintenance would be borne in the West Bank, while the flood mitigation benefits would be felt downstream within Israel, and the distribution of recharge benefits would be a function of the allocation of the water of the specific wells positively affected by the additional recharge. Moreover, pollution issues during recharge could lead to cross-border pollution problems.

### *3.3.4. Desalination*

Large scale desalination was first proposed in Israel by the U.S. in the mid-sixties. However, it was found to be too costly at the time. Following the massive immigration wave of the early nineties (when over 800,000 new immigrants poured into the country within eight years) and rapid economic growth in the early nineties, sea water desalination was proposed again in the nineties as a major element in Israel's water future. This proposal has been incorporated into the Israeli long term plans (Schwartz, 1996).

As water stress in the land-locked West Bank and in Jordan is expected to worsen, it is possible to augment water supply to those areas from desalination plants sited along Israel's Mediterranean coast[2]. Clearly, any such scheme would have to be part of an agreement between the partners. The major issue in any desalination project are its costs. As most of Israel's major population centers lie along the Mediterranean, desalination plants can therefore supply water at relatively lower cost. In contrast, supply to the West Bank or Jordan would require significant additional pumping costs. Such cost may undermine this option. Therefore, it is possible that desalination would be a feasible regional input only within a more comprehensive management plan, whereby Israel would provide the Palestinians and/or Jordanians with cheaper water, possibly supported by donors, and substitute them with desalinated water, thus reducing the overall cost of provision for all parties. Naturally, implementation of such a comprehensive scheme would require a much more complex financial, institutional and accounting framework. One possibility for addressing these issues is the establishment of a regional water market. However, water trading will lead to higher, real water prices, hence encouraging conservation and substitution away from irrigation, and thus is likely to delay the need for desalination (Arlosoroff, 1997).

---

[2] Alternatively, inland sites drawing sea water from the Mediterranean can be conceived, or plants can be sited in the Palestinian Gaza Strip, and the water produced be transported across Israel. For this paper the important point that in all the options Israel would have to be a party, as at the very least water to or from the desalination plants would have to traverse its territory.

*3.3.5. Water Conservation and Demand Management*

The least obvious, but perhaps the most readily available way to enhance water availability is water conservation and demand management. Israel has been famous for its success in increasing the product of water in irrigation through water conservation, primarily drip irrigation. Still, Israel can achieve much greater saving if demand management and water conservation in the urban sector would be implemented. Water conservation also holds much promise in Jordan, where urban water losses are significant.

Clearly, these methods should be implemented within each country, and thus do not have an obvious cross-boundary aspect. However, as the emphasis in many of the water negotiations is on need (see Wolf, 1998b), and the ability to address needs can be affected by the extent to which these measures are implemented, they have an indirect effect on international water negotiations. In the Israeli-Arab context these measures were raised as part of the multi-lateral negotiation track.

## 3.4. WASTEWATER MANAGEMENT AND RE-USE

Wastewater is both a potential source of pollution, threatening water quality in the major reservoirs and streams, and the cheapest significant source for additional water for irrigation. In Israel most of the coastal steams are currently polluted by wastewater. Concurrently, wastewater has been extensively used for irrigation, mainly of industrial crops. Since 1990 a major effort was undertaken in Israel to upgrade existing treatment plants and build new ones. As a result, it is expected that the quality of wastewater available for re-use will improve within the next few years. One of the possible uses for part of the recycled water according to current plans is the rehabilitation of coastal streams.

No secondary wastewater treatment plants have been built on the West Bank until recently. As most of the population centers in the West Bank are situated above the Mountain Aquifers' recharge areas, the sewage flows threaten the shared resource. Moreover, in several locations, sewage flows will cross the boundaries between Israel and the nascent Palestinian entity, regardless of the final delineation of these boundaries. The main points where such flows cross boundaries are near the towns of Jenin, Tul Karem and Qalkilia, and in the Jerusalem metropolitan region.

The sewage of Tul Karem pollutes the Alexander stream that runs through the Emek Hefer regional council to the sea. A plan has been prepared for rehabilitating this stream by using recycled water. Subsequently, the regional council has reached an agreement with the city of Tul Karem (situated in the West Bank) whereby it would assist in the building and maintenance of an internationally funded treatment plant for Tul Karem's wastewater, and would re-use the wastewater that cannot be re-used by the Palestinians. While this agreement has not been implemented yet, it is indicative of the potential for local level agreements regarding both the treatment of cross-boundary effluents, and the re-use of the recycled water.

The Jerusalem metropolitan region is comprised of an intricate web of Israeli and Palestinian settlements and neighborhoods. It is the only metropolitan area in Israel not served by a secondary level treatment plant, until 1999. As it sits astride the main water divide between the Mediterranean and the Dead Sea, and several sub-basins, all current plans suggest that the metropolitan area's wastewater should be treated in several plants. Today three plants are planned to treat the wastewater of Jerusalem and the Bethlehem area. In a study of the management possibilities in such a cross-order situation total separation, coordination, comprehensive metropolitan management and privatization options were compared. The analysis of these options suggested that privatization of the treatment systems may have a special advantage in a cross-boundary situation, as it may help de-politicize the wastewater issue. Moreover, the two sides have to work together in order to obtain the best deal from a third party, often a multi-national firm (Feitelson, and Abdul-Jaber, 1997).

The study of Jerusalem's wastewater treatment options identified enforcement, financing, flexibility and accountability issues as key variables for addressing transboundary sewage flows. This study did not analyze, however, the re-use options of the treated wastewater. One option for integrating the wastewater and transborder aquifer management issues was raised as part of the previously noted joint management of shared aquifers study. Essentially, a freshwater for recycled water trading option was proposed, whereby Palestinians would receive additional freshwater for domestic use, contingent upon the return of a pre-specified percentage as recycled water at an agreed upon quality level (Feitelson, 1998). This option, as well as the Emek Hefer-Tul Karem case, highlights the importance of integrating return flow quality considerations, financing and distribution issues in agreements regarding transboundary water resources.

## 4. Discussion: A New Transboundary Management Agenda

The Israeli-Arab case reflects worldwide trends in water issues, and provides insights regarding the unfolding agenda for the future. As in other parts of the world, the agenda initially focused on the use of surface water, primarily its allocation among the riparian states. This agenda has only recently been widened to include groundwater abstraction and quality. Still, the issues addressed under the current agenda are unlikely to reflect the full scope of issues that would have to be addressed in the future. This can be seen in all facets of the Israeli-Arab case.

In the case of the Jordan River, inadequate attention was given to the environmental and ecological importance of in-stream use, or to the possibilities for rehabilitating the river. These issues are already receiving some attention in the international water scene (Teckalf & Teckalf, 1994). An important feature in the Israeli-Jordanian treaty is the provision of storage services by Israel to Jordan. Such services are of particular importance in a semi - arid climate. Management of such services may well be part of the future agenda, especially if the reservoir itself is shared. In this case innovative approaches to the management of water storage may be called for, such as the capacity sharing approach advanced by Dudley and Musgrave (1988). Other elements in the water cycle that did not receive sufficient attention in past or current agreements are return flows and their implications for water quality. Yet, return flows are central to the management of rivers in many parts of the world. They should be, therefore, an integral part of future treaties.

The description of the water related interactions between the parties in the Israeli-Arab case shows that the extent of interactions among the parties is much wider than the question of the allocation of surface water. Actually, the allocation of the Jordan River's water is the only issue already addressed in a peace treaty between two major riparians. Therefore, it can be expected that the focus of discussion in future negotiations would gradually shift to other elements of the water cycle, most notably groundwater, water quality and wastewater issues. Yet, as the discussion of the Israeli-Arab context suggests, these issues cannot be addressed alone. Any coordinated or joint management structure for shared aquifers requires that a wide array of issues would be addressed as part of the joint management agreement (Haddad *et al.*, 1999). Moreover, the more advanced the level of management the more elements that need to be addressed within such a structure. The same is true also for wastewater issues. In the Israeli-Palestinian case, for example, wastewater treatment issues are related to groundwater protection, water quality, water allocations, agricultural development, public health and stream rehabilitation. It is likely, therefore, that the future agenda would have to address a much wider set of issues, pertaining to more parts of the hydrological cycle than the current agenda does.

Surface water, groundwater and wastewater quantity and quality issues differ in the level at which they should be addressed. Transboundary surface water management often requires a basin-wide approach. Only at this scale can the best use of water be made to all concerned, the need for which would only increase. Aquifer basins often differ markedly from river basins. Moreover, their edges are not as easily discerned. Thus the determination of an appropriate spatial scale of management, and hence for transboundary agreements, is more difficult than in the case of surface water. Wastewater issues, in contrast, are determined by human actions. Yet, the scope is mostly local or regional. The methods used to analyze such situations have to be tailored accordingly (Haruvy, 1996). The implication of these observations is that management structures that address the multiplicity of factors, such as agreements pertaining to rivers fed by several aquifers along which there are many wastewater treatment plants, would have to be multi-layered. They would have to include different elements at different spatial scales, and in some cases imbed within them an array of local or regional agreements.

The brief discussions in the previous sections, and the studies on which they are based, suggest that not only should future agreements pertain to additional elements of the water cycle, but also that there are many issues that should be addressed but have not received adequate attention in previous Israeli-Arab negotiations, or in most international treaties signed to date. Consequently, the agenda of future international water treaties would be different than that faced by past generations. Table I compares the agenda of past treaties, as seen in the impressive data base compiled by Aaron Wolf with the issues of the new agenda.

TABLE I

The Current and Future International Water Agenda

| Facet | Current Treaties | Future Agenda |
|---|---|---|
| Principal Focus | water supply; hydropower | water quality; return flows; pollution |
| Parts of Water Cycle Included | mainly rivers and lakes | surface water; groundwater; recycled water |
| Signatories | mostly bilateral | increasingly multilateral |
| Non Water Linkages | few, mostly related to money | many, including land use, waste management & money |
| Enforcement | rare (20% of treaties) | increasingly incorporated |
| Monitoring and Data Sharing | appears in over half of treaties | will appear in most agreements |
| Allocation Methods | simple 'equitable' one-time allocation of freshwater | multi-dimensional time sensitive allocations differentiated by quality levels |

Source for current treaties: Hamner J. and Wolf A., Patterns in international water treaties: the transboundary freshwater dispute database, *Colorado Journal of International Environmental Law and Policy* 1997 Yearbook, cited in: Wolf (1998a).

The main differences between past treaties and the new agenda is likely to be in their principal focus. While past treaties focused primarily on water supply and hydropower the new agenda is likely to focus much more on water quality, resource management and wastewater issues. The new agenda will also address many parts of the water cycle hardly addressed before, in particular groundwater. The differences, however, are not likely to be limited to the principal issues to which the treaties pertain. The treaties of the next century are also likely to differ in terms of the measures and specifications they propose and the ancillary issues they address. For example, as more costly sources are tapped, issues of affordability are likely to become increasingly prominent, something that may result in greater emphasis being placed on financial issues.

The treaties may also be increasingly multilateral as the multiplicity of interactions and the full spatial ramifications are recognized.

One of the most important differences may be in the allocation principles of water. Wolf (1998b) has already noted a shift in international treaties from rights to need based allocations. When combined with the likely increase in prominence of affordability issues, this trend suggests that socio-economic considerations will become a more important consideration in water allocations than is the case today. Moreover, water allocations should not be limited to water quantities. Rather, an allocation should be specified by several parameters including the quality of water received and the quantity and quality of return flows to which the recipient is obliged, the timing of abstraction and discharge, the allowed use, priorities of abstractions and redlines for abstraction by use (Feitelson, 1998). These specifications would pertain to the wastewater part of the cycle as well. As competition over all parts of the cycle becomes keener and environmental awareness rises, it can be hoped that within the next century treaties would address these parameters.

One of the ancillary issues that has not been addressed in most treaties, but is likely to become increasingly important is the relations between land use development patterns, runoff and water quality. This relationship is well known. However, it was mostly discussed at the local, intra-national levels. Yet, the rising scale of development, growing concern over water quality and groundwater recharge may well force this issue into the international agenda as well.

## 5. Conclusions

Within this millennium, and probably even within this century, progressively larger parts of the world will face the water issues currently faced by Israel and her immediate neighbors. As a result it can be expected that the lessons of the Israeli-Arab case will serve as an important precedent to other regions. In particular, it seems that additional parts of the water cycle will be utilized in an ever wider set of circumstances. These will increasingly include transboundary situations.

The challenge that will be faced, therefore, is to establish management structures that will be able to address this progressively more complex situation. These management structures would by necessity be more complex than those established under current treaties. They would have to include new and innovative combinations of local, regional, national and supra-national bodies, incorporating both the public and private sectors as part of the management regimes. They would be manifest in an increasingly varied and complex set of institutional structures governing transboundary flows of different parts of the hydraulic cycle. These structures would have to address a wider set of issues than addressed under current transboundary regimes, as manifest in current treaties in force. In this paper a very tentative attempt was made to identify some elements of the new agenda that would have to be addressed by these institutions. However, it is more than likely that reality would be even more challenging than the set of issues raised herein.

## Acknowledgements

This paper benefited from Shaul Arlosoroff's helpful comments, as well as the ongoing joint collaboration with him over the years. The figures were prepared by Michal Kidron. However, only the author bears any responsibility for the ideas expressed herein, as well as all remaining errors.

## References

Allan, J. A.: 1994, Economic and political adjustments to scarce water in the Middle East, in J. Isaac and H. Shuval, (eds), *Water and Peace in the Middle East*, Elsevier, Amsterdam, 375-387.

Arlosoroff, S.: 1997, Water trading and pricing issues in the Middle East: Israel as a case study, in: M. Haddad and E. Feitelson, (eds) *Joint Management of Shared Aquifers: The Third Workshop*, The Palestine Consultancy Group and the Harry S Truman Institute for the Advancement of Peace, The Hebrew University of Jerusalem, Jerusalem, 32-47.

Benvenisti, E.: 1996, Collective Action in the Utilization of Shared Freshwater: The Challenges of International Water Resources Law, *The American Journal of International Law* **90**, 384-415.

Biswas, A. K.: 1993a, Management of international waters: problems and perspectives, *Water Resources Development* **9**, 167-181.

Biswas, A. K.: 1993b, Water for sustainable development in the twenty first century: a global perspective, in A. K. Biswas, M. Jellali and G. Stout, (eds), *Water for Sustainable Development in the 21st Century*, Oxford University Press, Delhi.

Dudley, N. and Musgrave, W.: 1988, Capacity sharing of water reservoirs, *Water Resources Research* **24**, 649-658.

Feitelson, E.: 1996, Joint Management of Groundwater Resources: Its Need and Implications , in J. Cotran. and C. Mallet, (eds), *The Arab-Israeli Agreements: Legal Perspectives,* Kluwer Law International, London

Feitelson, E.: 1998, Water Rights within a Water Cycle Framework, in E. Feitelson, and M. Haddad, (eds), *Joint Management of Shared Aquifers: The Fourth Workshop*, The Harry S Truman Institute for the Advancement of Peace and the Palestine Consultancy Group, Jerusalem

Feitelson, E. and Abdul-Jaber, Q.: 1997, *Prospects for Israeli-Palestinian Cooperation in Wastewater Treatment and re-use in the Jerusalem Region*, The Jerusalem Institute for Israel Studies and the Palestine Hydrology Group, Jerusalem.

Feitelson, E. and Haddad, M.: 1998, A Stepwise Open-ended Approach to the Identification of Joint Management Structures for Shared Aquifers, *Water International* **23**, 227-237.

Gross, D. and Soffer, A.: 1996, *International Groundwater and International Rules: The Case of the Middle East and the Agreement Between Israel and Jordan and Between Israel and the Palestinians about Gaza Strip*, Occasional Papers on the Middle East (New Series) No. 12. University of Haifa (in Hebrew).

Haddad, M., Feitelson, E., Arlosoroff, S. and Nasseredin, T.: 1999, *Joint Management of Shared Aquifers: An Implementation-Oriented Agenda*, Final report Phase II, The Palestine Consultancy Group and the Harry S Truman Institute for the Advancement of Peace, The Hebrew University of Jerusalem, Jerusalem.

Haruvy, N.: 1996, Wastewater reuse - regional considerations, paper presented at the 36th congress of the European Regional Science Association, 26-30 August, Zurich.

Hayton, R.: 1991, Reflections on the estuarine zone, *Natural Resources Journal* **31**, 123-138.

Kliot, N.: 1994, *Water Resources and Conflict in the Middle East*, Routledge, London.

Lowi, M.: 1993, *Water and Power: The Politics of Scarce Resource in the Jordan River Basin*, Cambridge University Press, Cambridge.

Ostrom, E.: 1990, *Managing the Commons*, Cambridge University Press.

Ponting, C.: 1992, *A Green History of the World*, Penguin, New York.

Postel, S.: 1992, *The Last Oasis: Facing Water Scarcity*, Earthscan, London.

Raskin, P., Hansen E. and Margolis R. M.: 1996, Water and sustainability: global patterns and long-range problems, *Natural resources Forum* **20**, 1-15.

Shmueli, D.: 1999, Water quality in international river basins, *Political Geography* **18**, 437-476.

Shuval, H.:1992, Approaches for resolving the water conflicts between Israel and her neighbors: A regional water for peace plan, *Water International* **17**, 133-143.

Shuval, H.: 1998, Water and security in the Middle East: The Israeli-syrian water confrontations as a case study, in L.G. Martin (ed), *New Frontiers in Middle East Security*, St. Martin's Press, New York, 183-213.

Soffer, A.: 1994, The relevance of the Johnston Plan to the reality of 1993, in J. Isaac. and H. Shuval.(eds), *Water and Peace in the Middle East*, Elsevier, Amsterdam, 107-121.

Tecklaff, L.: 1991a, Introduction, *Natural Resources Journal* **31**, 7-9.

Tecklaff, L.: 1991b, Fiat or custom: the checkered development of international law, *Natural Resources Journal* **31**, 45-73.

Tecklaff, L. and Tecklaff, E.: 1994, restoring river and lake basin eco-systems, *Natural Resources Journal* **34**, 905-932.

Wolf, A.: 1993, The Jordan watershed: past attempts at cooperation and lessons for the future, *Water International* **18**, 5-17.

Wolf, A. T.: 1998a, Conflict and cooperation along international waterways, *Water Policy* **1**, 251-265.

Wolf, A. T.: 1998b, From rights to needs: water allocations in international treaties, in E. Feitelson, and M. Haddad, (eds), *Joint Management of Shared Aquifers: The Fourth Workshop* , The Harry S Truman Institute for the Advancement of Peace and the Palestine Consultancy Group, Jerusalem.

# THE QUEST FOR WATER EFFICIENCY - RESTRUCTURING OF WATER USE IN THE MIDDLE EAST

P. BEAUMONT
*Department of Geography, University of Wales, Lampeter, SA48 7ED, UK*

**Abstract**. Despite widespread aridity in the Middle East, human societies have had access to adequate water resources until the second half of the twentieth century. With the rapid growth of population from around 150 million in 1950 to almost 400 million in 2000 great pressure has been exerted on existing water resources so that in certain countries these are now totally committed. Examination of water use indicates that each cubic metre of water utilised in the industrial and service sectors generates at least 200 times more wealth than in the agricultural sector. This suggests that as water shortages increase many countries will be best served by the reallocation of irrigation water to meet the growing water needs of the urban regions.

**Keywords**: irrigation, reallocation of water resources, urban water demand, water, wealth creation

## 1. Introduction - A Region out of Water

For the first fifty years of this century all of the countries of the Middle East were able to sustain their economies through access to indigenous water resources. Although the region has always experienced seasonal and often annual water shortages, the inhabitants had been able to cope except in the most severe droughts (Beaumont *et al.*, 1988). The main reason for this was that prior to the twentieth century, local communities had only been using a fraction of the renewable water resources which were available.

During the twentieth century and particularly after the Second World War, the region has seen a massive growth in population numbers. In the 1950s the total for the Middle East region was around 150 million, while by 1995 it had reached 390 million. It has been this growth in numbers which has eventually led to the utilisation of much of the available water in the Middle East. Since the Second World War governments in many countries have harvested the unused surface waters of major rivers through the construction of dam and reservoir complexes (Biswas, 1994; Kliot, 1994; Kolars and Mitchell, 1991). This has permitted the expansion of irrigated areas and also the development of urban systems. Turkey, Morocco, Egypt, Israel and Iran have been in the vanguard of such activities, but almost all states have undertaken some form of water development project. Groundwater reserves have also been exploited through pumped wells. These wells have tapped groundwater resources which might previously have discharged unused into the sea, basins of inland drainage or into rivers. The result has been that by the end of the twentieth century most of the renewable water resources of the region have been committed to human use.

*Water, Air, and Soil Pollution* **123:** 551–564, 2000.

The date when a country fully utilises its renewable water resources marks an interesting landmark, but does not represent the end of economic development. Kuwait ran out of water in the 1950s, but was able to obtain a new and reliable water supply from desalination which for more than forty years has permitted the economy to grow and develop. Israel ran out of renewable water in the late 1960s and has subsequently maintained its growth largely through the mining of ground water, the use of reclaimed sewage waters and the reallocation of already committed water (Lonergan and Brooks, 1994). For Jordan the crucial time looks set to occur in the next five years or so (Salameh and Bannayan, 1993). Other countries such as Turkey, Morocco and Iran still have unused water resources available, but even here regional shortages will grow as population numbers continue to increase (Bagis, 1994). Estimates suggest that Middle Eastern population numbers will increase from the 1995 figure of 390 million to 727 million in 2025 (Population Reference Bureau, 1995). Certainly within the first few decades of the twenty-first century many of the countries of the Middle East will be using all of their renewable water resources. Beyond this time new water management strategies will be required.

## 2. Water Use in Society

Throughout the early growth of Middle Eastern societies and indeed up to the late nineteenth century almost 90 per cent of all water usage would have been for irrigation. It is only in the twentieth century, with the advent of Western urban systems being imposed on traditional Middle Eastern towns, that domestic/industrial use of water has become much more important (Beaumont, 1980). This has also been accompanied by the growing urbanisation of the region. At the beginning of the twentieth century only about 10 per cent of the population of the Middle East lived in towns and cities (Blake and Lawless, 1980). By the end of the twentieth century the average figure is in excess of 50 per cent and in some countries it is more than 80 per cent.

In the twenty-first century it will be the urban communities which will decide the future development of water usage. It is here that most of the wealth creating parts of the economy will be located, and it is here also where the political elite reside. Urban water usage will definitely increase as a result of increasing population numbers, rising standards of living, industrialisation and a growing service sector. However, the actual amounts of water needed in modern urban systems are small compared with those required for irrigation. In the majority of countries these urban water needs are only a fraction of the renewable water resources which are available. The problem is that in many countries most of the water resources are already committed for irrigation use.

## 3. Water Management Strategies

In a modern society it is essential that water is utilised to achieve maximum returns in terms of overall value to a country's economy. In dryland countries where water shortage occurs, this imperative is even greater. Water management consists of two separate, but related aspects. These are supply management and demand management. In most countries, since modern water management techniques were introduced, the emphasis has been on supply management. What has almost always occurred is that as demand for water has risen within a country the water management agencies have sought to meet this need by the provision of new water sources and water delivery systems.

This position has been further complicated by the fact that certain countries have followed, for strategic reasons, policies of food security (Falkenmark *et al.*, 1998). This has meant that these countries have often attempted to become self-sufficient in basic foodstufffs through an increasing commitment to irrigated agriculture. This has led to ever greater water demands and increasing environmental problems as cultivation is extended on to ever more marginal lands. Saudi Arabia's attempt to increase wheat production through government subsidies provides the best example of this type of policy (Beaumont, 1999). The fact that it had to be curtailed illustrates that even the richest countries of the region have difficulties in implementing such strategies.

The situation in the late 1990s is that many countries are now experiencing increasing difficulties in supplying current water needs and it is obvious that future needs in all sectors of their economies will not be able to be met within a decade or two. As a result, increasing attention is being given to demand management and water reallocation as a means of ensuring the maximum productivity of the available water. With this approach, priority is given to those uses which will provide the greatest economic benefit. Policy decisions like these have to be implemented at governmental level as they can have crucial impacts on the overall performance of the country's economy.

## 4. Current Water Use in the Middle East

Accurate data on the water used by the countries of the Middle East are difficult to obtain and vary in quality from one country to another. FAO has compiled tables showing water use which are the most comprehensive data set available (FAO Aquastat, 1998). The data show that in all countries agricultural water use dominates and accounts for between 56 per cent and 99 per cent of all consumption. For the region as a whole, it represents 88 per cent of total water usage. Average irrigation water use is around 9,500 cubic metres per hectare. Domestic water use averages 7 per cent for the region and industrial use 5 per cent.

Work by Shuval (1992) has shown that an advanced country like Israel can survive and prosper with a per capita domestic water usage value of around 100 cubic metres each year. In contrast, Beaumont (1997) quotes figures for the USA revealing that urban water demand there is around 255 cubic metres/capita/annum. These two figures will be referred to as "minimum water requirements" and "maximum water requirements" respectively as they are thought likely to bracket any future demand levels for water in the urban areas of the Middle East.

By combining population data with FAO water statistics it is possible to calculate current water use per capita for domestic uses in the countries of the Middle East (Table I). However, these figures have to be treated with caution as many uses of domestic water are probably under-recorded. The data show that the small oil rich states of the Gulf already have domestic water use figures in excess of 100 cubic metres per annum. If the figure for the UAE is correct this value is even above the average for the USA. In contrast, there are many countries, often with substantial populations, where water use is considerably less than 50 cubic metres per annum.

With growing urbanisation and rapid population growth, it is likely that there will be a substantial increase in water demand in urban regions over the next few decades. Estimates have been made of the likely water demand of the countries of the Middle East in the year 2025 based on population predictions from the Population Reference Bureau (Beaumont, 1997) (Table I). These water demand figures were based on all the population of the country enjoying a standard of water use comparable with values from advanced urban societies. The "minimum water requirement" in Table I is based on a figure of 100 cubic metres per person per annum, which is roughly equivalent with water consumption in Israel in the late 1990s. The "maximum water requirement" is based on data of 255 cubic metres/capita/annum from the USA in the early 1990s.

The differences between current domestic usage and the estimated "minimum water requirements" for urban usage in 2025 reveal massive variations from one country to another (Table I). For example, in Afghanistan and Yemen domestic demand might rise by more than fifteenfold while in Israel it might increase by only 1.3 times. In UAE it might actually fall. In general the poorer the country in 1995 the greater will be the water demand which will have to be supplied to meet consumption rates of around 100 cubic metres per capita per annum (minimum water requirements) in 2025. What is interesting is that if a figure of 100 cubic metres per person per year is to be met for the population of every country in the region, domestic water consumption will rise from a figure of 21,805 million cubic metres in the early 1990s to 72,730 million cubic metres in 2025. This represents a 3.3 fold increase or in absolute terms the need to provide an extra annual volume of water equivalent to the annual discharge of the River Tigris.

TABLE I

Domestic water use in the Middle East in 1995 and estimates of "minimum" and "maximum" water demand for the year 2025

| Country | Domestic use-million $m^3$-1990s | Population million 1995 | Domestic water-$m^3$/cap/yr-1995 | Population million 2025 | Minimum water needs-million $m^3$/yr 2025 | Maximum water needs-million $m^3$/yr 2025 |
|---|---|---|---|---|---|---|
| Afghanistan | 261 | 18.4 | 14.2 | 41.4 | 4140 | 10578 |
| Algeria | 1120 | 28.4 | 39.4 | 47.2 | 4720 | 12060 |
| Bahrain | 94.3 | 0.6 | 157.2 | 1.1 | 110 | 281 |
| Egypt | 3100 | 61.9 | 50.1 | 97.9 | 9790 | 25013 |
| Iran | 4395 | 61.3 | 71.7 | 106.1 | 10610 | 27109 |
| Iraq | 1280 | 20.6 | 62.1 | 52.6 | 5260 | 13439 |
| Israel | 597 | 5.5 | 108.5 | 8.0 | 800 | 2044 |
| Jordan | 214 | 4.1 | 52.2 | 8.3 | 830 | 2121 |
| Kuwait | 201 | 1.5 | 134.0 | 3.6 | 360 | 920 |
| Lebanon | 368 | 3.7 | 99.5 | 6.1 | 610 | 1559 |
| Libya | 500 | 5.2 | 96.2 | 14.4 | 1440 | 3679 |
| Morocco | 543 | 29.2 | 18.6 | 47.4 | 4740 | 12111 |
| Oman | 56 | 2.2 | 25.5 | 6.0 | 600 | 1533 |
| Qatar | 65.9 | 0.5 | 131.8 | 0.7 | 70 | 179 |
| Saudi Arabia | 1517 | 18.5 | 82.0 | 48.2 | 4820 | 12315 |
| Sudan | 800 | 28.1 | 28.5 | 58.4 | 5840 | 14921 |
| Syria | 530 | 14.7 | 36.1 | 33.5 | 3350 | 8559 |
| Tunisia | 261.4 | 8.9 | 29.4 | 13.3 | 1330 | 3398 |
| Turkey | 5200 | 61.4 | 84.7 | 95.6 | 9560 | 24426 |
| UAE | 500 | 1.9 | 263.2 | 3.0 | 300 | 767 |
| Yemen | 201 | 13.2 | 15.2 | 34.5 | 3450 | 8815 |
| Total | 21804.6 | | | | 72730 | 185827 |

Source - FAO Aquastat, 1998; Population Reference Bureau, 1995; Beaumont, 1997.

## 5. The Wealth Created by Water Use

In any discussion of water efficiency a crucial factor is just how much wealth does a cubic metre of water generate in different parts of a country's economy. No detailed statistics dealing with this type of information are currently available. However, it is possible to calculate figures of water value if the FAO data on water use by sector (agriculture; domestic; industry) are combined with information on GDP generated by the major sectors of the economy (agriculture; industry; services). For agriculture and industry the calculations are straightforward as it is merely a case of dividing the wealth created in monetary terms by the volume of water utilised in that sector.

The service sector of the economy does, however, provide difficulties as it is extremely difficult to assess just how much water is used in this sector of the economy. For the most part water use in the service sector is confined to what might be defined as domestic uses. These include personal hygiene, food preparation, cleaning and toilet flushing. The amounts of water used tend to be small even when compared with industrial usage. Very little water is actually used as part of an individual service with the obvious exception of the cleaning of clothes and similar items. Any estimate of water used in the service sector is bound to be arbitrary, but it seems reasonable to conclude that per capita use will not exceed 60 per cent of the national per capita domestic consumption. It has to be remembered that the people who work in the service sector will still consume considerable quantities of water at home when they are not working. If such a figure of 60 per cent of average domestic use is accepted as reasonable it is possible to estimate the amount of water used by the service industries provided that the number of workers in the sector is also known.

For these calculations FAO water data were used with economic information on GDP by sector and employment in the service sector derived from the World Factbook 1998 which is published by the CIA (USA). For certain countries, other data sets had to be utilised. It is accepted that the quality of the data will vary from one country to another, but the results are considered to provide an overview of the wealth which each cubic metre of water creates within the Middle East region (Table II).

The results show that the wealth created by a cubic metre of water in the three sectors of the economy varies considerably. It is particularly low in agriculture, where for the region as a whole it averages US$1.86/cubic metre. The highest value of US$9.89 is found in the West Bank and the lowest value of US$0.40 in Afghanistan. The vast majority of the figures are below US$2.00/cubic metre. With the industrial sector the regional average for wealth creation is US$532.77/cubic metre of water. However, there is a very large range from a low of US$18.58 in Egypt to a high of US$1887.69 in Kuwait. It is difficult to distinguish any overall pattern though the oil rich countries tend to record high figures. Data for the service sector have to be the most suspect as a major assumption is made in the calculations. More data sources are used, for example, in employment statistics for the service sector, and each of these data sets may introduce new errors. Nevertheless, the figures do provide some interesting insights. Throughout the Middle East the average wealth created by the service sector is US$649.37/cubic metre of water. The maximum value is US$2,036.3 in Morocco and the minimum is US$154.88 in Lebanon.

Overall the data indicate that the wealth creation/cubic metre of water in the industrial and service sectors is basically similar, with the service sector tending to record the higher value in most individual countries. What is striking, however, is the way in which both the industrial and service sector contributions/cubic metre of water are often hundreds of times higher than those produced by agriculture. For example, the regional value for industry is more

TABLE II

Wealth generated by the use of water in each sector of the economy – US dollars/cubic metre

| Country | Agriculture | Industry | Services |
|---|---|---|---|
| Afghanistan | 0.40 | | 268.62 |
| Algeria | 5.35 | 88.53 | 385.30 |
| Bahrain | 0.61 | 315.15 | |
| Egypt | 0.96 | 18.58 | 685.53 |
| Iran | 1.22 | 92.55 | 511.57 |
| Iraq | | | |
| Israel | 1.49 | 120 | 687.10 |
| Jordan | 1.68-1.91 | 138-207 | 360.10 |
| Kuwait | | 1887.69 | 237.37 |
| Lebanon | 0.69 | 69.92 | 154.88 |
| Libya | 0.48 | 209.00 | 516.30 |
| Morocco | 1.47 | 109.66 | 2036.30 |
| Oman | 0.45 | 389.47 | |
| Qatar | 0.53 | 653.57 | |
| Saudi Arabia | 0.81 | 492.18 | 411.07 |
| Syria | 2.18 | 53.04 | 1512.00 |
| Sudan | 0.52 | 22.60 | 707.07 |
| Tunisia | 2.90 | 183.74 | 1158.36 |
| Turkey | 2.54 | 92.55 | 511.57 |
| UAE | 1.16 | 149.05 | 211.01 |
| West Bank | 9.89 | | |
| Yemen | 1.77 | 400.00 | |
| Average | 1.86 | 532.77 | 649.37 |
| Maximum | 9.89 | 1887.69 | 2036.30 |
| Minimum | 0.40 | 18.58 | 154.88 |

Source - The calculations shown in the table were derived from data sets from two major sources. The water data were obtained from FAO Aquastat, 1998 and the economic and employment information from the CIA publication World Factbook, 1998. Water data for Israel were obtained from the Israel Statistical Yearbook, 1997.

than 280 times that of the agricultural sector, while that of the service sector is almost 350 times greater. What the data clearly show is that if water availability is a limiting factor then it makes economic sense to transfer water from agriculture to the industrial or service sectors.

However, the position is not usually that straightforward. In a number of poorer countries, a significant number of people still earn their livelihood through agriculture which has often been the traditional basis of a country's economy. For many countries the agricultural sector still has a major role to play in wealth creation for the economy as a whole, and yet at the same time it has to be admitted that irrigated agriculture is a wasteful use of water in the long-term. In such cases what is needed is a strategy which recognises the crucial role which agriculture continues to play as a creator of wealth which can then be invested to promote self-sustaining growth in the industrial and service sectors. In this instance irrigated agriculture must be regarded as a temporary necessity, which though wasteful in water terms, will generate the necessary wealth to help a country make the transition from a traditional agricultural economy to an urban economy based largely on services and industry.

## 6. Coping with Urban Water Demands in 2025

The crucial issue is how easy will it be for the countries of the Middle East to cope with the estimated increase in domestic water demands for the year 2025. This demand can be met from utilising new sources of surface water or ground water; from the reallocation of existing resources such as irrigation water, and from the desalination of seawater or saline groundwater. When current water use is expressed as a percentage of the internal and external water resources of a country it provides an assessment of the volume of water resources which might still be developed in the future (Table III). This table shows that of all the countries in the region Turkey is the best endowed with water resources and is currently using only about 17 per cent of its total amount. Even with increased domestic supply up to the year 2025 this figure only increases to 19.3 per cent.

From this table it is possible to identify a number of what might be regarded as "water-rich" countries, which even with increased domestic demands could still be using less than two-thirds of their total water resource base in 2025. These include Afghanistan, Algeria, Iran, Iraq, Lebanon, Morocco, Sudan, Syria and Turkey. However, it has to be remembered that the most suitable dam sites have already been developed and that subsequent projects will be both expensive and difficult to construct. Although a country might theoretically have ample water resources available, developing the schemes necessary to harvest this water might take between one and two decades. As a result. countries might be forced to look to other solutions to their water problems.

A number of countries are already consuming more than 200 per cent of their natural water resources. This can only be achieved by the mining of ground water or by the provision of new water by desalination. These include Bahrain, Kuwait, Libya, Qatar, Saudi Arabia and the UAE. In the future most of these countries will continue to meet their needs through new desalination capacity as they all have access to cheap energy supplies.

The countries likely to experience the greatest difficulties in meeting their water needs in the future are those which are currently using between 70 per cent and 150 per cent of their water resources. These are Egypt, Israel, Jordan, Oman, Tunisia and Yemen (Table IV). For such countries there is little opportunity for developing further surface or groundwater resources and so they are left with the choice of desalination or the reallocation of already existing water supplies. For a number desalination will be too expensive, leaving reallocation of water use as the only sensible alternative.

TABLE III

Current and predicted water use as a percentage of renewable water resources

| Country | Total water use in 1990s-million $m^3$ | Total as percent of internal + external water resources in 1990s | Extra domestic water needed for 2025 - million $m^3$ | Predicted total water use for 2025 - million $m^3$ | Predicted total as percent of internal + external water resources in 2025 |
|---|---|---|---|---|---|
| Afghanistan | 26110.0 | 40 | 3879 | 29989 | 45.9 |
| Algeria | 4500.0 | 31 | 3600 | 8100 | 55.8 |
| Bahrain | 239.2 | 206 | 15.7 | 254.9 | 219.6 |
| Egypt | 55100.0 | 95 | 6690 | 61790 | 106.5 |
| Iran | 70034.0 | 51 | 6221 | 76255 | 55.5 |
| Iraq | 42800.0 | 57 | 3980 | 46780 | 62.3 |
| Israel | 2031.0 | 123 | 203 | 2234 | 135.3 |
| Jordan | 984.0 | 112 | 616 | 1600 | 182.0 |
| Kuwait | 538.0 | 2690 | 159 | 697 | 3485.4 |
| Lebanon | 1293.0 | 29 | 242 | 1535 | 34.4 |
| Libya | 4600.0 | 767 | 940 | 5540 | 923.0 |
| Morocco | 11045.0 | 37 | 4197 | 15242 | 51.1 |
| Oman | 1223.0 | 124 | 544 | 1767 | 179.0 |
| Qatar | 284.9 | 538 | 4.1 | 289 | 545.2 |
| Saudi Arabia | 17018.0 | 709 | 3303 | 20321 | 846.7 |
| Sudan | 17800.0 | 20 | 4680 | 22480 | 25.3 |
| Syria | 14410.0 | 55 | 2820 | 17230 | 65.8 |
| Tunisia | 3075.0 | 75 | 1069 | 4144 | 101.1 |
| Turkey | 31600.0 | 17 | 4360 | 35960 | 19.3 |
| UAE | 2108.0 | 1405 | -200 | 1908 | 1272.0 |
| Yemen | 2932.0 | 72 | 3249 | 6181 | 151.8 |

Source - FAO Aquastat, 1998; Population Reference Bureau, 1995; Beaumont, 1997.

It has to be accepted, though, that any of the countries of the region might choose reallocation of water use as a solution to its future domestic water needs. It has already been shown that this future urban demand seems likely to be up to 51,000 million cubic metres by 2025. This represents about 18.6 per cent of total irrigation water use in the Middle East. In itself, this does not seem too large a figure for the region to cope with. Reallocation of water resources, though, might not be as simple as water which may currently be used in parts of the country where there are few urban regions. If suitable rivers and aquifers do not exist, transporting the water to urban demand areas may prove difficult.

Assuming that water can be reallocated from irrigation to other uses within a country it is interesting to calculate just how much water would be needed to meet the estimated domestic demands of countries for the year 2025 and also what effects these transfers would have on irrigated agriculture. For this exercise all countries within the Middle East are considered, including those which might be able to develop new water sources. Even for these countries reallocation of water from one use to another might be the most practical solution to their water problems in the short term (Table IV).

In two countries, Algeria and Yemen, the extra volume of water required to meet the expected demand is greater than the current irrigation water usage (Table IV). If it was a case of simple substitution then irrigated agriculture would cease in both countries. In reality, though, Algeria does have major mountain regions where higher precipitation totals and so extra sources of water may be able to be utilised. With Yemen the situation could become critical within a short time as new resources are limited. Here the decline of irrigated agriculture seems likely.

TABLE IV

Extra water needed to meet estimated "minimum water needs" for the year 2025 and expressed as a percentage of current irrigation use

| Country | Irrigation water usage in 1990s- million $m^3$ | Extra water needed to meet predicted domestic water demand (minimum) in 2025- million $m^3$ | Extra water needed as percent of current irrigation usage |
|---|---|---|---|
| Afghanistan | 25849.0 | 3879 | 15.0 |
| Algeria | 2700.0 | 3600 | 133.3 |
| Bahrain | 135.1 | 15.7 | 11.6 |
| Egypt | 47400.9 | 6690 | 14.1 |
| Iran | 64155.0 | 6221 | 9.7 |
| Iraq | 39380.0 | 3980 | 10.1 |
| Israel | 1297.0 | 203 | 15.7 |
| Jordan | 737.0 | 616 | 83.6 |
| Kuwait | 324.0 | 159 | 49.1 |
| Lebanon | 875.0 | 242 | 27.7 |
| Libya | 4000.0 | 940 | 23.5 |
| Morocco | 10180.0 | 4197 | 41.2 |
| Oman | 1148.0 | 544 | 47.4 |
| Qatar | 210.6 | 4.1 | 2.0 |
| Saudi Arabia | 15308.0 | 3303 | 21.6 |
| Sudan | 16800.0 | 4680 | 27.9 |
| Syria | 13600.0 | 2820 | 20.7 |
| Tunisia | 2727.5 | 1069 | 39.2 |
| Turkey | 22900.0 | 4360 | 19.0 |
| UAE | 1408.0 | -200 | |
| Yemen | 2700.0 | 3249 | 120.0 |

Source - FAO Aquastat, 1998; Beaumont, 1997.

Jordan appears to be the next most vulnerable country as 83.6 per cent of current irrigation demand would be needed to supply the potential 2025 domestic demand. Jordan is already using more than its internal renewable water supply through the mining of ground water, so no other obvious water sources are available. The inevitable conclusion is that there will be a decline of irrigated agriculture within the country which might pose serious problems for the economy as a whole. One possible solution might be the desalination of sea water for urban use. However, production costs are high and the long distribution pipelines from the Gulf of Aqaba would be very expensive to build and maintain.

A number of countries have estimated demands which suggest that between 35 and 50 per cent of the current irrigation usage would be needed to meet future domestic demand. These include Kuwait, Morocco, Oman and Tunisia (Table IV). Kuwait is a special case and could solve its problems easily with extra desalination capacity. Morocco contains substantial highland areas and so in the longer term might be able to develop extra water sources. Oman and Tunisia would have to settle for at least some reallocation of irrigation water to meet future urban needs.

Lebanon, Libya, Saudi Arabia and Syria register figures between 20 and 30 per cent, which would suggest that in the period up to 2025 their water needs could be met. In the longer term Lebanon has at least 100 million cubic metres of extra water which could be obtained from the Litani, while for Syria further water supplies in the north could probably be developed. Libya and Saudia Arabia are extensively dependent on ground water and are already withdrawing water at a faster rate than that at which it is renewed. Inevitably groundwater production will have to be cut back. However, both countries are oil-rich and could meet all their new domestic demands with desalinated water if they so wished.

The final category includes countries where the extra water demand is less than 20 per cent of current irrigation demand. It is a large group including Bahrain, Egypt, Iran, Iraq, Israel, Qatar and Turkey. Bahrain and Qatar seem likely to choose the desalinated solution. Egypt has no other water than the Nile available but could probably begin to use its irrigation water more efficiently. This might be achieved by applying lower amounts of water per unit area, together with the selection of crops and cropping patterns which use less water. If it did this, the extra domestic demand up to 2025 could be met.

Iran and Turkey have large areas of highlands where not all water resources are yet developed and so coping with extra demands should not be a problem. Iraq is in an interesting situation as the water available to it in the Euphrates river is set to decline substantially as a result of the development of irrigation projects in Turkey and Syria. Iraq will be able to meet its growing domestic demand as it still controls a large portion of the flow of the Tigris River. However, the shrinkage of the irrigated area may have to be much higher than the figure of 10 per cent suggested in the table. Finally, with Israel the extra water demand of 15.7 per cent of current irrigated usage will pose severe strains as no extra water sources are available. The irrigated area in Israel seems bound to contract, but the effect on the economy does not look likely to be very significant as a result of the very small contribution which the agricultural sector makes to the economy as a whole.

In summary, the domestic water supply situation for the Middle East in the year 2025 does not appear to be as serious a problem as many people have suggested. Many countries look as if they will be able to meet their enlarged domestic water needs without too much of a loss of irrigated land in the short-term. In the long-term, though, beyond the year 2025 greater sacrifices of irrigated land do seem inevitable.

## 7. The Jordan Basin

Specific problem areas remain with perhaps the most important one being the Jordan basin. Here it is obvious that both Israel and Jordan will face ever growing pressures on their already overutilised water resources. These difficulties are compounded by the fact that the Palestinians of the West Bank and Gaza are also suffering greatly from water shortages as the result of Israeli occupation. One of the main difficulties of equitable water allocation in the Jordan basin at the present time is that the various political units are at different stages of economic development. Israel is moving towards a mature Western economy and its agricultural sector has declined from 6 per cent in 1969 to c.2 per cent in 1997 of total GNP. On the other hand both Jordan (6%) and Syria (28%) still possess significant agricultural sectors and lack the balanced industrial activity which is needed for sustained and stable development. However, all these three units are much better off economically than the West Bank and the Gaza regions.

These economic differences mean that it is difficult at the present time to envisage a single water allocation system for the Jordan basin which can cope with the wide divergences in economic activity. For example, the introduction of water markets, which is undoubtedly an economically efficient way of distributing water, is likely to favour the advanced economy of Israel at the expense of other political units (Becker and Zeitouni, 1998). This is because Israeli industry and services would be able to pay much higher rates for water than West Bank farmers. The only way an equitable system could be developed would be if elements of a water market system were introduced in parallel with a water allocation system based on economic indicators such as the GNP per inhabitant of the individual political units. With this approach the basic allocation of water would be much higher for those units with low GNP per capita. This would ensure that adequate water is available to farmers for irrigation in the West Bank and Gaza Strip as this is the best way in which wealth can be created to stimulate growth in other sectors of the economy. The remaining water resources would be allocated through free water markets. As GNP per capita rises in the poorer economic units the protected water allocations for these regions would be phased back. Eventually, when GNP per capita values were similar a free water market system could be allowed to prevail.

## 8. Conclusion

In the long-term the growing urban water demands of the Middle East can only be met from three sources. These are the utilisation of unused renewable water sources, the desalination of sea water and the reallocation of irrigation water to more productive uses. For many countries the first option is no longer possible, and for many others it will provide water for only a decade or two. Desalination of sea water is a solution, but an expensive one. However, in the very long-term it seems likely that it will become ever more important as other water sources are fully utilised. It does have the very great advantage that the amounts of water which can be produced are limitless. Finally, the reallocation of irrigation water seems to be the most likely immediate solution to water demand problems over the next two decades. In almost all countries ample water is available in this sector of the economy and for many the transfer to urban usage will not present serious physical problems. On the other hand such actions will intensify the decline of rural areas and put even more pressure on the urban regions.

A crucial test for all the governments of the region over the next two decades will be the provision of employment opportunities for the growing populations of the urban regions. It seems unlikely that industrial employment will be able to be provided for a large proportion, and so it will be left to the service sector, in both its formal and informal aspects, to sustain the majority of the jobs which will be needed. An encouraging fact from a water efficiency point of view is that such jobs require relatively little water.

## References

Bagis, A. I. (ed): 1994, *Water as an element of cooperation and development in the Middle East,* Hacettepe Universtiy and Friedrich-Naumann Foundation, Ankara.

Beaumont, P.: 1980, *Urban water problems*, in G.H. Blake and R.I. Lawless (eds), *The changing Middle Eastern city,* London, Croom Helm, 230-250.

Beaumont, P.: 1997, *Water and armed conflict in the Middle East - Fantasy or Reality?* in N.P. Gleditsch (ed), *Conflict and the environment,* Dordrecht, Kluwer Academic Publishers, 355-374.

Beaumont, P.: 1999, *Irrigation,* in M. Pacione (ed), *Applied Geography: Principles and Practice,* London, Routledge, 172-187.

Beaumont, P., Blake, G. H. and Wagstaff, J. M.: 1988, *The Middle East - a geographical study,* David Fulton Publishers, London.

Becker, N. and Zeitouni, N.: 1998, *Water International* **23**, 238-243.

Biswas, A. K. (ed): 1994, *International waters of the Middle East - from Euphrates-Tigris to Nile,* Oxford University Press, Bombay:

Blake, G. H. and Lawless, R. I. (eds): 1980, *The changing Middle Eastern city,* Croom Helm, London.

CIA: 1998 *World Factbook,* Washington, D.C. Internet - http://www.cia.gov

Falkenmark, M. *et al.*: 1998, *Ambio* **27**, 148-154.

FAO Aquastat: 1998, Internet - http://www.fao.org

Kliot, N.: 1994, *Water resources and conflict in the Middle East,* Routledge, London.

Kolars, J. and Mitchell, W. A.: 1991, *The Euphrates River and the Southeast Anatolia Development Project,* Southern Illinois University Press, Carbondale.

Lonergan, S. C. and Brooks, D. B.: 1994, *Watershed - the role of fresh water in the Israeli-Palestinian conflict,* International Development Research Centre, Ottawa.

Population Reference Bureau.: 1995, *World Population Data Sheet 1995,* Population Reference Bureau, Washington, D.C.

Salameh, E. and Bannayan, H.: 1993, *Water resources of Jordan - present status and future potentials,* Friedrich Ebert Stiftung, Amman.

Shuval, H. I.: 1992, *Water International* **17**, 133-143.

# TURKISH WATERS: SOURCE OF REGIONAL CONFLICT OR CATALYST FOR PEACE?

G. E. GRUEN
*Middle East Institute, Columbia University, 420 West 118 Street, New York, NY 10027*

**Abstract**. Unlike most Middle East countries which are highly dependent on water from sources originating in other countries or on desalination, Turkey is naturally endowed with relatively abundant water resources. The Turkish government has assigned the highest priority to completing its massive $32 billion Southeastern Anatolia Project (GAP), consisting of 22 dams and 19 hydroelectric power plants on the Euphrates and Tigris rivers. Scheduled for completion in 2005, GAP will generate 27 billion kilowatt hours of hydroelectric power and will divert water from the Atatürk Dam reservoir through the two giant Şanliurfa Tunnels into a canal system to irrigate 1.7 million hectares in south-eastern Anatolia just north of the Syrian border. For Turkey, GAP will not only provide food and energy for a growing population, but is the crux of a comprehensive and sustainable economic development plan designed to end instability and reduce out-migration by radically transforming the feudal economic and social structure of this poor and largely Kurdish inhabited region of the country. Syria and Iraq, the downstream riparians in the Tigris-Euphrates Basin, also have rapidly growing populations and ambitious development plans. They contend that GAP will greatly diminish and degrade their water supply in future years. The severe current drought conditions in Syria and Iraq have added urgency to their demands for a greater share of the rivers' flow. This article examines the legal, political, military and technological strategies employed by the parties to advance their interests. After reviewing efforts to achieve a negotiated solution, we examine various Turkish proposals to foster regional peace by exporting water from other Turkish rivers to Cyprus, Israel, the Gaza Strip, Jordan, and other Arab countries.

**Keywords**: international rivers, regional cooperation, Tigris-Euphrates Basin, water exports, water law

## 1. Introduction

Although Turkey, Syria and Iraq have been engaged in intermittent diplomatic negotiations and technical discussions over the waters of the Euphrates River for decades, they have been unable to agree on a permanent tripartite treaty that would set the terms for "sharing" (Syrian and Iraqi terminology) or "allocating" (Turkish terminology) the river's flow among the three riparians. Because of the drought conditions affecting much of the region, declared to be the most severe in 60 years in Syria, Israel and Jordan, and the most "catastrophic" in nearly a century in Iraq (Khalil, 1999), this chronic controversy has now become acute. The increase in demand due to rapidly growing populations, ambitious development plans, and rising economic expectations, all add urgency to the need to develop reasonable and efficient means of utilizing this precious and limited resource. Since the Tigris and Euphrates flow through the territory of three states, both logic and equity argue for achieving a cooperative approach by the three riparians. However, issues of national sovereignty, historical grievances and conflicting interests have impeded collaboration and have led to the perception that this is a zero sum game. While the total average annual flow of the Euphrates is 35.58 billion cubic meters (BCM), to which Turkey contributes more than 88 percent and Syria less than 12 percent, the combined planned future consumption of Turkey (18.42 BCM), Syria (11.30 BCM), and Iraq (23 BCM) totals 52.92 BCM. As more water is drawn off for irrigation and industrial uses by the upper riparians, less remains available for downstream users.

It is natural that Turkish President Süleyman Demirel, an engineer who early in his career headed the country's Dam Department (1954-55) and the State Hydraulic Works (1955-60), proudly calls the $32 billion Southeast Anatolia Project (GAP) "a symbol of national achievement" that will enable Turkey "to assume its historical place in the world scene as a leading country fully able to complete the most advanced projects using the most modern technology" (Demirel, 1992). This integrated system of 22 dams and 19 hydroelectric power plants on the Euphrates and Tigris will generate 27 billion kilowatt hours of hydroelectric power – saving energy-poor Turkey some 28 million tons of oil imports annually, and divert sufficient water to irrigate an additional 1.7 million hectares (Bagis, 1989).

Since the planned development projects of Turkey, Syria and Iraq together far exceed the projected annual flow in the Tigris-Euphrates Basin, this has resulted in friction among them. Water disputes have at times even brought the parties to the brink of conflict. In the early 1970s, Iraq threatened to go to war with Syria because construction of the Tabqa dam and the filling of Lake Assad reservoir temporarily deprived Iraq of some of the Euphrates' flow. Soviet and Saudi mediation in 1975 helped avert hostilities (Kienle, 1990,). In response to the imminent threat of diminished flow to both countries by Turkey's impounding of Euphrates water to fill the Atatürk Dam, the rival Ba'athist regimes in Damascus and Baghdad finally signed an agreement on April 16, 1990, under which Syria would receive 42% and Iraq 58% of the river's annual flow from Turkey, regardless of quantity (Beschorner, 1992).

Earlier, in December 1980, Turkey and Iraq had established a Joint Technical Committee (JTC) for information exchange and to "decide the methods and procedures which lead to a definition of the reasonable and appropriate amount of water that each country needs" from the two rivers. Although Syria had been invited by Ankara to participate, it did not join the JTC meetings until 1983. The JTC met fairly regularly for seven years, until talks were suspended after the Iraqi invasion of Kuwait in 1990 (Bilen, 1996). Basic political disagreements among the parties have colored their approach even to technical issues.

This article will examine the legal, political, military and technological strategies employed by the parties to advance their interests and will review different Turkish proposals to export water from to various areas in the region.

## 2. History of Negotiations

Under the Treaty of Lausanne of 1923, which established the relations of the new Republic of Turkey with its neighbors, Ankara agreed to consult Iraq before undertaking any hydraulic works. In 1946, Turkey and Iraq had signed a Protocol for the Control of the Waters of the Tigris and Euphrates and Their Tributaries. They agreed then that flood control dams and storage facilities would most effectively be built upstream on Turkish territory. They promised to exchange hydrological and meteorological data daily during flood periods. Cooperation worked well until 1964, when Turkey completed plans to construct the Keban Dam and Power Plant on the Euphrates. Ankara submitted the plans to Syria and Iraq, and in meetings with their technical experts, pledged to supply 350 $m^3$/sec downstream of the dam, assuming that there was sufficient natural flow. In a tripartite meeting in Baghdad in September 1965, Turkey first proposed creation of a Joint Technical Committee to study the entire Tigris-Euphrates basin. At that time, Syria endorsed a Turkish proposal that the Euphrates and Tigris flow should be considered together so that if Euphrates water was insufficient to meet all the irrigation needs of the three riparians, some Tigris flow would be diverted and channelled into the Euphrates. Iraq strongly opposed this and insisted on discussing only the Euphrates. However, after 1980, Syria changed its position and has since then joined with Iraq in insisting that each river be considered separately (Bilen, 1996).

In 1975 Turkey approached the World Bank for funding for the Karakaya Dam on the Euphrates. Bank experts conducted a technical study and concluded that if Turkey maintained an average monthly discharge of 500 cubic meters per second ($m^3$/sec) at the point it flowed into Syria, the existing downstream requirements for power generation and irrigation and future growth could be met. Turkey and the Bank agreed on this formula, which was termed the "Rule of 500." Ankara informed Syria and Iraq and offered to conclude a tripartite agreement to monitor its compliance. But when both Syria and Iraq raised objections, the Bank deferred funding. Turkey then financed both this dam and the massive Atatürk Dam largely on its own. The Turkish and Syrian prime ministers met in July 1987 and concluded a Protocol of Economic Cooperation, Article 6 of which gave approval, but only on a provisional basis, to the World Bank formula: "During the filling up period of the Atatürk Dam reservoir and until the final allocation of the waters of the Euphrates among the three riparian countries, the Turkish side undertakes to release a yearly average of more than 500 $m^3$/sec at the Turkish-Syrian border," adding that whenever the monthly flow fell below this level, Turkey would make up the difference during the following month. (Turkey, 1987).

The Atatürk Dam began to produce electricity in July 1992 and the first of the two giant irrigation tunnels was completed in November 1994. At first only some 30 $m^3$/sec were being diverted, or less than 10 percent of the system's ultimate capacity, but this is gradually being increased as the necessary irrigation canals are completed. It is the cumulative effect of the entire GAP project that worries Turkey's southern neighbors, especially when they read forecasts in the Turkish press that the country's growing domestic water needs will eventually require Ankara to reduce the quantity below 500m$^3$/sec. Turkey contends that through the JTC it meets its obligation to keep Syria and Iraq informed and denies their claim to veto or restrict projects in Turkey.

In December 1992, Syria launched a diplomatic offensive in the Arab League to put pressure on Turkey and urged League members not to finance Turkish water projects. The Syrian démarche charged that while Ankara had concluded agreements on common waters with Russia, Bulgaria and Greece, Turkey refused to sign a "just and reasonable agreement" with Syria and Iraq (*Cumhuriyet*, Dec. 25 and *Turkish Daily News*, Dec. 26, 1992). The timing of the Syrian complaint was attributed to the signing by the Turkish Government of an agreement a few days earlier for construction of the Birecik Dam and hydropower plant. This dam, the fourth largest on the Euphrates, was to be built by a consortium of Turkish, German, Belgian, Austrian and French firms.

Dismissing the Arab reaction as politically motivated rather than based on facts, Turkish officials have pointed out that the Birecik dam and the projected Karkamiş dam closer to the Syrian border were intended primarily for hydroelectric power generation. Irrigation and evaporation losses in the reservoirs would be more than made up by the benefits accruing to Syria and Iraq from the increased storage capacity provided by this and the other Turkish dams. Not only would they provide a reserve during years of drought, but they would also regularize the flow throughout the year by smoothing out the sharp seasonal fluctuations in the river's natural flow. Ankara noted that during the drought of 1991, when the flow of the Euphrates had dropped to 190 m$^3$/sec, Turkey continued to provide Syria with 500 m$^3$/sec by releasing water from the Keban and Karakaya dams, causing a one year delay in the filling of the Atatürk Dam (Yetkin, 1993). More recently, Baki Ilkin, the Turkish Ambassador to the U.S., confirmed in February 1999 that despite the current drought in the region, Ankara was fulfilling its commitment to supply Syria with 500 m$^3$/sec. Asserting that "Syria has ample water running from the Euphrates River," he added: "We are prepared to allocate water to Syria, we have no intention of using the waters of the Euphrates as a bargaining chip or as a threat towards Syria." He also claimed that, "at the moment, Syrians get more water per capita than do Turks" (Ilkin, 1999).

Seeking to improve Turkish-Arab relations in the wake of the controversy in December 1992 over the planned Birecik Dam, Premier Demirel travelled to Damascus and met with President Hafez al-Assad the following month. The Syrians expressed the view that since the Joint Technical Committee, after 16 meetings, had failed to meet the parties' expectations, the issue be taken up at the political level. At the end of the discussions on January 20, a joint communiqué was issued declaring that "pursuant to the 1987 Protocol, the foreign ministers of the two countries would assign top officials to achieve, before the end of 1993, a final solution determining the allocation to the parties from the waters of the Euphrates river." (*Newspot*, January 28, 1993.) To reassure the Syrians, Demirel declared: "There is no need for Syria to be anxious about the water issue. The waters of the Euphrates will flow to that country whether there is an agreement or not." Officials in Ankara noted that following Iraq's invasion of Kuwait, Turkey rejected the advice of some of its NATO allies to pressure Iraq by reducing the water flow. However, in compliance with United Nations sanctions, Turkey did close the oil pipeline from Iraq to Turkey's Mediterranean coast, at a cumulative cost to Turkey's economy that has been estimated by Turkish officials as exceeding $30 billion.

### 3. Security and Water Issues Linked

The January 1993 communiqué also stated that Syria and Turkey had reiterated their determination not to permit any activity on their respective territories "detrimental to each other's security." The 1987 water protocol had been implicitly linked to another protocol, signed at the same time, under which Syria promised to cooperate with Turkey on issues of security along their 877-kilometer border and pledged not to "permit" anti-Turkish activities within its borders. Ankara had demanded this in exchange for its water supply commitment because the Syrian regime had over the years permitted various militant groups opposed to the Turkish regime to operate from bases in Syria itself and in the Syrian-controlled Beqaa valley in Lebanon. These included Dev-Sol (a revolutionary leftist Turkish group), ASALA (an Armenian extremist group) and most significantly, the Kurdistan Workers Party (PKK), originally a Marxist-Leninist Kurdish group outlawed in Turkey, whose original aim was to carve out an independent Kurdish state from Turkey's eastern Anatolian provinces and unite it with neighboring Kurdish areas. Since 1984, the PKK, listed as a terrorist group by the U.S. and other Western countries, has engaged in a guerrilla war against Turkish forces in southeast Turkey and has also employed tactics of intimidation and terrorist attack against Turkish school teachers and other civilians. By the beginning of 1999, this protracted conflict had resulted in more than 30,000 killed, and a cost of over $50 billion, according to official Turkish estimates (Barkey and Fuller, 1998; *Turkish Times*, 1999).

Syria failed to live up to its 1987 commitment. Although the PKK center of operations was moved from Syria itself to Lebanon's Beqaa valley, which remains under Syrian military control, Abdullah (Apo) Ocalan, the PKK political leader continued to reside and move freely within Syria, despite repeated Turkish demands for his extradition. Upon his return from Damascus in January 1993, Demirel recounted to his True Path Party colleagues that it was Syrian President Assad himself who had assured him that Syria was not behind any terrorist activity against Turkey. "Instead of asking Assad why Abdullah (Apo) Ocalan, the leader of the outlawed PKK, was still in Syria," Demirel said, "I presented Assad with some addresses, telephone numbers and post office boxes in Damascus, Aleppo and Latakia which belong to Apo." (Quoted by Yetkin, 1993). Ilnur Çevik, editor-in-chief of the *Turkish Daily News*, who accompanied Demirel, confirmed that Assad took the note, feigned surprise, and promised to investigate (Çevik, 1993).

It was not until October 1998, after increasingly blunt Turkish official threats to take military action against Syria unless it ended all support for the PKK, that the "serious crisis" in Turkish-Syrian relations was defused as a result of intensive mediation by President Hosni Mubarak of Egypt and high level interventions by the Iranians and Americans, who sent "the right message" to Damascus. Ocalan was finally expelled from Syria, and after wandering for months from country to country, was eventually captured by Turkish agents in Nairobi, Kenya, brought to Istanbul and placed on trial for treason in May 1999. On October 20, 1998, high level representatives of Turkey and Syria met in Adana, Turkey and concluded a special security agreement, in which Syria explicitly pledged not to support the PKK in any way. Both sides promised to cooperate in improving security along their common border and established a hotline between their military commanders to prevent unintended incidents. Typical of the cautiously positive Turkish response was the comment of Ambassador Ilkin: "I am delighted to say that Syria has been able to respond to our requests – very belatedly – yet better late than never." He added that it was "absolutely imperative" that Syria "implement the understanding in full," if the two countries were "to move forward together, hand in hand" to start a new chapter of positive relations, which he envisaged as encompassing "economic cooperation, technical cooperation, more trade volume, more cultural contacts." While there have been some recent discussions on bilateral trade and tourism, according to Syrian and Turkish officials at the United Nations with whom the author spoke in early June 1999, there have not yet been any joint discussions on water issues. The capture of Ocalan and the latest Syrian-Turkish agreement, if faithfully implemented, means that Damascus will no longer have the PKK card to play in its attempts to disrupt the implementation of the GAP projects. Indeed, if the PKK insurrection is finally ended, and if Ankara adopts wise policies to address the legitimate grievances of its citizens in the Kurdish regions, this will free up some $7 billion annually in the Turkish budget, which can be allocated to developing the necessary infrastructure in the underdeveloped GAP area. If instability in the region ends, Turkish and foreign businesses will be less reluctant to invest in industrial and agricultural ventures in the southeast.

To return to our review of the efforts to resolve the water disputes by diplomatic means, a Syrian delegation came to Ankara for negotiations in May 1993, but after three days no progress had been achieved. Ankara proposed that in addition to the Euphrates, the flow of the Orontes should be discussed. Ankara complained that the Syrians were utilizing virtually all of the flow of the Orontes

(Asi in Turkish), an international river that flows northward from Lebanon through Syria, before passing through the Turkish province of Hatay (Alexandretta) and flowing into the Mediterranean at Antakya (Antioch). The Syrians categorically refused to discuss the "Arab waters" of the Orontes with Turkey, since Damascus still does not recognize Turkish sovereignty over Alexandretta, which was detached from Syria in 1939 after a disputed plebiscite conducted by the French mandatory authorities. Iraq was invited to join the next negotiating session, scheduled for June 1993. The Iraqi delegation came, but the Syrians, without explanation, stayed away. The Iraqi delegation supported the basic Syrian position, reiterated the notion of a mathematical division of the Euphrates' flow and demanded that the quantity released by Turkey be raised to 700 $m^3$/sec. No agreement was reached in 1993.

As implementation of the GAP project has progressed, Syria and Iraq have reiterated their complaints. Thus, for example, Syria, in December 1995, and Iraq, in January 1996, sent notes of protest to Ankara claiming that the Birecik dam would reduce the flow of the Euphrates and that Turkish irrigation activities had already polluted the river. Ankara responded with notes denying and refuting the allegations. Most recently, in May 1999, Syria issued a formal protest to the British Foreign Office over the government's plan to guarantee a £1 billion credit to the British firm Balfour Beattie to construct the Ilisu dam on the Tigris river. Jordan also protested, on behalf of Iraq, claiming that the dam project was "a breach of international law" since Baghdad had not been consulted, and that the use of water for irrigation in Turkey which then passes to Syria and Iraq "will pollute the flow with agro-chemicals and pesticides." As a result, "millions of Iraqis will be denied their right to clear water." UK Defence Forum, a British think-tank advising the government on regional risks, warned that support for the project might involve Britain "in armed conflict between Syria, Iraq and Turkey over the right to water from the Tigris." In response to the various protests, joined by British Liberal Democrats and the environmental group Friends of the Earth, Trade Minister Brian Wilson responded: "A great deal of care is being taken to ensure that proper relocation and compensation arrangements are drawn up and implemented for the local population, and that water quality and water flow issues are fully addressed" (Brown, 1999).

## 4. Opposing Legal Views and UN Efforts To Codify International Water Law

Underlying Arab charges that Turkey is violating international water law is their view of the legal status of the Euphrates River. Iraq and Syria consider it to be an international watercourse which is to be treated as an integrated whole. Furthermore, Iraq contends that the basic injunction against causing "appreciable" harm bars upper riparians from reducing the natural flow to established downstream users, who have "historically acquired rights," without their consent. The official Turkish position is that international rivers are only those which form the border between two or more riparians. Ankara regards the Euphrates as a transboundary river, under Turkey's exclusive sovereignty until it flows across the border into Syria. It is only after the Euphrates joins the Tigris in lower Iraq to form the Shatt al-Arab, which serves as the border between Iraq and Iran that it becomes an international river. In August 1991, when Demirel was prime minister, he summed up the Turkish position as follows: "Water is an upstream resource and downstream users cannot tell us how to use our resource. By the same token oil is an upstream resource in many Arab countries and we do not tell them how to use it." (*The Middle East*, August 1991). The Arab states reject the analogy, arguing that oil is like other mineral resources that stay in one place until drilled and pumped out by human effort, while water flows naturally to downstream riparians unless interrupted by human intervention.

Turkish officials used to cite the Harmon Doctrine, named for the U.S. Attorney-General who in a dispute with Mexico in the late 19th century asserted a similar absolute American sovereign right to utilize the Rio Grande (Rio Bravo). However, by 1942 the Legal Adviser of the State Department acknowledged that no recent treaty still supported the Harmon Doctrine (Hackworth, 1964). Dante Caponera, the international legal authority who in 1966 drafted the International Law Association's Helsinki Rules on the Uses of the Waters of International Rivers, similarly told the author, in December 1992, that no international arbitral decision supports the Harmon Doctrine. In an effort to reconcile the conflicting claims of upper riparians for more water for a growing population and economic development against the historical rights of lower riparians, the International Law Association asserted the principle of "equitable" and "optimal" utilization subject to the requirement that no "appreciable" (later changed by the ILC to "significant") harm be done to other riparians. The International Law Commission (ILC) of the United Nations, after many years of work adopted a set of Draft Articles on the Law of the Non-Navigational Uses of International Watercourses, and defined an international watercourse as "a watercourse, parts of which are situated in different States." The ILC did not distinguish between "international and transboundary rivers," thereby tending to support the Syrian and Iraq view. While the ILA and ILC drafts set out a series of criteria to be used in determining what was an equitable division in a given case, the problem in practice has been that different states emphasize certain principles over others (Sela,1999).

The UN General Assembly on May 21, 1997, passed a resolution adopting the ILC-prepared draft Convention on the Law of the Non-Navigational Uses of International Watercourses. The vote was 103 in favor to 3 against (Turkey, China and Burundi) with 27 abstentions. The Convention was to be open for signature for three years and was to go into effect upon ratification by 35 countries, or only 18% of the UN's current 185 members (United Nations, 1999). [By December 1998 only 11 states had done so.] Syria has been a strong supporter of the

Convention and had unsuccessfully proposed mandatory submission of unresolved water disputes to the International Court of Justice. Syria's UN Ambassador, Dr. Mikhail Wehbe, claimed that since the Convention now embodied the "norms of international law" it had become part of the customary law binding on all states (Wehbe, 1999). This point was elaborated upon in a lengthy essay by Dr. Badre Kasme, Syria's legal expert on water in Geneva (Kasme, 1998). Syria did succeed in having included in Article 33 of the draft, dealing with the settlement of disputes, a provision that if the parties were unable to resolve their dispute within six months of a call for negotiations, a commission of inquiry could be created on the demand of a single party. Although the findings of the commission were not necessarily binding, and the submission of the dispute to arbitration or judicial decision still required approval of both, the presumption was that the findings of an impartial body would carry considerable moral weight.

In explaining Turkey's vote against the resolution, Ambassador Huseyin Celem said that as a Framework Convention, it should have only set out general principles and not establish "a mechanism for planned measures." Such a practice had "no basis in international law," created an obvious inequality between states by in effect giving "a veto right" to lower riparians over the development plans of upper riparians, and that instead of setting out compulsory rules for dispute settlement, the Convention should have left this to the discretion of the states concerned. Turkey also objected to the failure of the Convention to make any reference to "the undisputable principle of the sovereignty of the watercourse States over the parts of international watercourses situated in their territory." [Similar objections were raised by the representative of China and several other members (UNGA, Press Release GA/9248, 21 May 1997.)] Moreover, Turkey believed that the Convention should have "established the primacy of the principle of equitable and reasonable utilization over the obligation not to cause significant harm." Therefore Turkey would not sign the Convention, which would not have "any legal effect for Turkey in terms of general and customary international law" (Celem,1997).

What is equitable? Gruen (1993) notes that Kolars and Mitchell estimate that the total average natural flow of the Euphrates measured at Hit, Iraq is around 33,000 Million Cubic Meters ($Mm^3$/yr), which translates to an average of 902 $m^{3/}$sec for the period 1924-1973, i.e., before the construction of recent major dams. Beaumont estimates that the 500 $m^3$/sec that Turkey is releasing to Syria amounts to 15,768 $Mm^3$/yr, or roughly half of the river's natural flow. He concludes that since some 90% of the river's water originates in Turkey, in terms of both "international precedent" and "natural justice, it does not seem unfair that Turkey should be able to utilize up to one half of the water which is generated within its borders" (Beaumont, 1992). The Arab view appears based on a simple calculation: There are three riparian states, each entitled to an equal "share," thus limiting Turkey to only one-third (or about 300 $m^3$/sec) and leaving the remaining two-thirds to be divided by Syria and Iraq.

Basing themselves on the ILC endorsed principles of "optimum, reasonable and equitable utilization of the water," Turkish officials contend that climactic and other factors make it more economical for Turkey to concentrate on food production and to exchange the surplus for Iraqi oil and Syrian gas. Turkey has offered to pay the cost of a scientific survey of the optimum uses of the region's water resources and has offered to work jointly with Syria on developing water saving technologies and cleaning up of irrigation return flow before it reaches Syria. In a meeting with the author in April, 1999, Dr. Kasme and Ambassador Wehbe acknowledged that the Syrian land was not as productive as Turkey's, but that was precisely why Syria required more water and therefore the natural flow of the river should be maintained. Syria favored the introduction of water saving technology, such as drip irrigation, but this required capital investment and it would take time to introduce new technology and change cultural patterns. Dr. Wehbe contended that the ILC had ruled that comparative economic output was not to be a criterion for allocating international waters, and Ambassador Wehbe stressed that food security was an important principle to be maintained, adding that if Syria was deprived of sufficient water to irrigate its land, there would be migration out of the rural areas, social dislocation and unemployment (Wehbe and Kasme, 1999). Sela recently noted that the "Syrian economy is largely dependent upon agriculture," accounting for 25 percent of the labor force and nearly 30 percent of the GDP (Sela,1999).

GAP officials are aware of the importance of providing on-site training to Turkish farmers in the GAP area in modern farming methods, including the use of water-saving techniques, in providing the necessary technology and equipment for drip irrigation, "to promote farm mechanization in proper combination with the application of fertilizer, agro-chemicals and irrigation water," and in promoting marketable crops "effective in overcoming adverse agro-ecological conditions (Gruen, 1993). Turkey recently concluded an agreement with Israel under which the International Cooperation Department (MASHAV) of Israel's Foreign Ministry brings GAP decision-makers and regional directors to participate in multi-disciplinary rural development training courses in Israel, and Israeli teams of experts in rural development and immigrant absorption are working with Turkish officials to introduce cooperative management concepts, in addition to modern agricultural and animal husbandry techniques to assist the modernization process in the GAP region (Arbel, 1998). They have begun to apply in the GAP region the drip irrigation techniques and other environmentally-friendly technologies that had been introduced by American and Israeli experts in the Moslem Turkic Central Asian Republics of the Commonwealth of Independent States. Israeli irrigation experts from Kibbutz Gvat working with Uzbeki cotton farmers have already demonstrated that drip irrigation techniques not only reduce environmental damage from over-intensive cotton farming, but also improve crop yields. Already in 1992, Beaumont was one of the international experts who urged Turkey to reconsider the furrow and flood irrigation methods originally planned for the GAP area and substitute sprinkler, trickle and drip systems (Beaumont, 1992). Turkish diplomats and agricultural and hydrological experts have been urging Iraq and Syria to do likewise, adding that the available supply would be adequate if they ended their wasteful traditional open-channel and flood irrigation practices (Bağiş *et al.*, 1993).

## 5. Turkish Water Exports to Promote Regional Peace

Shortly after the signing of the Israel-PLO agreement on September 13, 1993, Turkish Foreign Minister Hikmet Çetin reiterated Ankara's position that "as part of the peace process," Turkey was prepared to supply water to Israel and its Arab neighbors from sources on the southern coast of Anatolia, "such as the Manavgat, Ceyhan and Seyhan Rivers." In fact, he told the author in New York on September 28, 1993, he doubted that lasting peace between Israel and Syria, Jordan and the Palestinians could be achieved without the addition of Turkish water supplies. When Çetin visited Israel in November 1993, he received a positive response in principle from Foreign Minister Shimon Peres. This was not a new idea. In fact it represented a scaled-down version of the late President Turgut Özal's grandiose proposal in 1986 of a "Peace Water Pipeline," a $21 billion project to bring water from the Seyhan and Ceyhan Rivers via two pipelines to supply nearly six million cubic meters of water daily to the major cities in Syria, Jordan and the Arab Gulf states. In the face of Arab objections, Turkish officials explained that export of water to Israel, which was originally included in the plan, would have to await Arab-Israeli peace (Gruen 1994). Nevertheless, Syria in its December 1992 démarche to the Arab League, continued to allege that the Turkish Peace Pipeline proposal was "a plot to give Israel large quantities of water." The project never got off the ground, because the oil-rich Arab states in the Gulf, who had been expected to fund the pipeline, claimed that gas-fuelled desalination was economically cheaper and politically preferable, since one of their upstream neighbors might turn off the pipeline tap. (Yet large desalination plants are also potentially vulnerable, as the Gulf Arabs learned in the conflict after Saddam Hussein's invasion of Kuwait.)

Shortly after the start of the Arab-Israel peace process in Madrid in October 1991, Turkey joined the multilateral working group on water resources. Senior Turkish officials, including President Demirel and Prime Minister Tansu Çiller at first expressed support for a shorter pipeline, estimated to cost $5 billion, to convey Turkish water to Syria, Jordan, Israel and the Palestinians. However, Syria – through whose territory any pipeline would have to go – has refused to attend any of the multilateral committees, demanding that Israel make a commitment to total withdrawal from the Golan Heights before Damascus would even consider discussing any regional cooperation with Israel. Noting that such a pipeline "could supply both Israel and Syria," Ciller stressed during her visit to Israel in November

1994, "But first we need peace" (Silver). Although the election of Ehud Barak as Israel's prime minister in May 1999 may lead to early resumption of the Israeli-Syrian talks broken off in 1996, as of this writing Damascus has still refused to join the multilateral groups, including the one on water.

The emphasis in Ankara has therefore shifted to giving priority to plans to convey water from uncontestedly purely Turkish rivers to Cyprus. The island is suffering from a prolonged three-year drought, which has exacerbated the long term problems of a falling water table and increasing seawater intrusion due to over-pumping of the aquifers to meet the growing demands of the local population in both the Greek and Turkish parts of the island and to sustain the lucrative and expanding tourist industry. Christos Marcoulis, a senior official in the Cyprus Republic's water development department, declared in June 1999: "We can't rely on rainfall. Over the past three years the dams have become exhausted and aquifers haven't been replenished. Development of new water resources is a priority" (Hope, 1999). The initial response of the Greek Cypriote authorities has been to offer concessions for the construction of desalination plants, which have thus far contributed 40,000 $mc^3$ of drinking water daily. The Turkish Republic of Northern Cyprus (TRNC) has turned to water imports as the main answer. In the Summer of 1998, a Norwegian-Turkish joint venture began to deliver water from a spring near Alanya on Turkey's southern Mediterranean coast to Güzelyurt in Turkish-controlled northern Cyprus. The water is being brought by tugboats, towing "giant cucumbers," plastic bags 390-ft long and 84-ft wide when inflated. Initially the bags will carry 10,000 $m^3$ (about 10,000 tons) of water each, but plans are to increase the capacity to 20,000. If current technical problems can be overcome, the project will transport 3 MCM the first year, and if "cucumber convoys" can be put together, the capacity will rise to 7 MCM annually. But this quantity, while easing the shortage of drinking water, with many homes currently getting fresh water only once or twice a week, will not be sufficient to meet the farmers' needs for irrigation. The Norwegians anticipate a price of Nkr4.2 a cubic meter (around 50 US cents).

A more significant project that will provide a long term solution and may also help promote Turkish-Greek cooperation on the long-divided island is a 78 km. suspended underwater pipeline (of 1600 mm diameter HDPE) that will transport drinking and irrigation water from the Dragon River on Turkey's Mediterranean coast to Güzelyurt in northern Cyprus. The project which has received official approval in Ankara, is being undertaken by a consortium of Turkish and European firms headed by Alarko Holding of Istanbul, one of Turkey's largest conglomerates and experienced construction companies. According to information provided to the author by Dr. Uzeyir Garih, President of Alarko, the pipeline will be able to provide 75 MCM of water annually at an estimated cost of between 25 and 34 cents. This is considerably less than the estimated cost of the plastic balloon process and around one-third the cost of desalination plants, with per cubic meter costs still generally between 75 cents and over $1 (Garih, 1999). The total cost of the pipeline project is estimated at $225 million, for which international financing is being sought.

The most exciting part of the Dragon-Güzelyurt pipeline project is its potential contribution to Turkish-Greek confidence-building and practical cooperation. According to Ali Budak, project manager of Alarko's Cyprus project, "In the feasibility study, the quantity of 75 million $m^3$ per year is foreseen for the year 2025." However, since the current water needs in the Turkish part of Cyprus is only 30 million $m^3$, "the extra amount of water that will be transported every year after the realization of the Peace Water Project can be used for the whole island." Dr. Üzeyir recalls that many years ago the late President Özal had called him late one night to urge him to develop such a water-sharing proposal as a means of using Turkish water not only to meet the practical needs of the Cypriote population but to serve as a catalyst to building confidence among the long hostile Greek and Turkish local communities. Özal died in 1993 before he could act on this idea. But it has been vigorously pursued by President Demirel. In mid-December 1998 an informal group of businessmen from Turkey, Greece and both the Greek and Turkish communities in Cyprus met in Istanbul and issued a statement welcoming the Alarko pipeline project. Unfortunately, following the revelation that PKK leader Ocalan had been aided by Greece and even been hosted by the Greek ambassador in Kenya, the Turkish members walked out in protest. The talks have not yet resumed.

Most recently, Alarko has proposed a similar underwater pipeline project to convey water from the Manavgat River near Antalya to a site along Israel's eastern Mediterranean coast, from whence it can be conveyed in Israel's existing network to anywhere in the country and potentially also to Palestinian territory or the West Bank. It could also replace some of the water Israel is committed to supply to Jordan under their 1994 Treaty. While the distance is greater than to Cyprus, a pipeline similar to the one being constructed to Cyprus will be employed and anchored around 100 meters below the sea surface. Dr. Garih estimates that the total cost of building a project that will bring 50 MCM annually will be between $200 and $300 million. Assuming that the Turkish Water Authority agrees to charge the Israeli buyer 4 to 5 cents per $m^3$, and the project can obtain international financing for a 20 -25 year loan at 4%, "the water will cost less than 30 US cents" (Garih letter to Gruen, April 26,1999). In order to obtain international funding for either the Cyprus or the Israel pipeline projects, agreements have to be concluded by the parties, including reliable long term supply contracts and formal legal mechanisms to insulate the agreement from political interference in case there are changes in government.

If Greek and Turkish Cypriotes can reconcile their differences and if the new Barak Government in Israel succeeds in its efforts to conclude lasting agreements with Lebanon, Syria and the Palestinians, then former Prime Minister Shimon Peres' optimistic vision of a new Middle East may finally have a chance. Long a supporter of Turkish-Israeli cooperation as means to transform the Middle East,

Peres would frequently point out that the European Union had begun with limited cooperation by two longtime enemies, France and Germany, in the fields of coal and steel. Similarly, he was confident that a new peaceful Middle East could grow out of cooperation in the fields of tourism and water resources.

## References


Arbel, M.: 1998, Interview by the author with the director of MASHAV's training department, Jerusalem.

Bağiş, A.: 1989, *G.A.P. Southeastern Anatolia Project: The Cradle of Civilization Regenerated*, Interbank, Istanbul; personal communications from professors Bağiş and Ilter Turan and Turkish officials who asked not to be identified, Istanbul and Ankara, July 1993.

Barkey, H. and Fuller, G.: 1998, *Turkey's Kurdish Question*, Rowman & Littlefield, Lanham, MD.

Beaumont, P.: February 1992, "Water - A Resource Under Pressure," in G. Nonneman (ed), *The Middle East and Europe: An Integrated Communities Approach*, for the EC Commission, Federal Trust for Education and Research, London..

Beschorner, N.: 1992, *Water and Instability in the Middle East*, Adelphi Paper No. 273, International Institute for Strategic Studies, London, 41-42

Bilen,Ö: 1997, *Turkey and Water Issues in the Middle East*, Southeastern Anatolia Project (GAP) Regional Development Administration, Ankara.

Brown, P.: May 8, 1999, "Britain is warned not to fund Turkish dam," *The Guardian*, London.

Celem, H.: May 21, 1999. Ambassador's Statement in the UN General Assembly, Unpublished text.

Çevik, I.: July 1993, personal interview with the author, Ankara.

Demirel, S.: 1992, Introduction to *GAP: Southeastern Anatolia Project*, General Directorate of Press & Information, Turkish Republic, Ankara.

Garih, U.: May and June 1999, letters and technical documents provided G. Gruen.

Gruen, G.:1992, *The Water Crisis: The Next Middle East Conflict?* The Simon Wiesenthal Center, Los Angeles, CA.

Gruen, G.: 1993, "Recent Negotiations Over the Waters of the Euphrates and Tigris," International Water Resources Association, *Proceedings of the International Symposium on Water Resources in the Middle East: Policy and Institutional Aspects*, University of Illinois at Urbana-Champaign, *Pre-Symposium Proceedings*, 100-108.

Gruen, G.: 1994, "Contribution of Water Imports to Israeli-Palestinian-Jordanian Peace," in J. Isaac and H. Shuval (eds), conference proceedings, *Water and Peace in the Middle East*, Elsevier Science Publishers, B.V.,Amsterdam, 273-88.

Hackworth, G.: 1964, Memorandum cited in M. Whiteman (ed), *Digest of International Law* **3**, 950.

Hope, K.: June 7, 1999, "Expanding tourism is putting severe strains on an essential commodity," *Financial Times*, Cyprus Survey.

Ilkin, B.: 1999, "Amb. Ilkin: 'Your Enemy Need Not Be My Enemy,'" *The Turkish Times* February 15, 1999 (reprinting his interview with *Middle East Insight*).

Kasme, B.: 1998, "L'Obligation de Reglement des Differends Relatifs aux Cours d'Eux Internationaux," in E. Yakpo and Boumedra (eds), *Liber Amircorum--Mohammed Bedjaoui*, Kluwer Law International, 1-22.

Khalil, A.: June 2, 1999, Iraq's representative to the FAO, quoted in AP dispatch from Baghdad.

Kienle, E.: 1990, *Ba'th versus Ba'th, Syrian-Iraqi Relations since 1968*, 96-100.

Sela, A.: 1999 (ed), "Water Politics," in *Political Encyclopedia of the Middle East*, Continuum, New York.

Silver, E.: Dec. 21, 1994, "Middle East Interview: The Ottoman Empress," *Jerusalem Post*.

Turkey, Republic of: 1987, *Protocol on Matters Pertaining to Economic Cooperation between the Republic of Turkey and the Syrian Arab Republic*, Official Gazette.

*Turkish Times:* June 1, 1999, "Cost of PKK Violence: 30,000 Lives, $50 Billion," United Nations, General Assembly, Press Release GA/9248, 21 May 1997 includes summary of statements and votes; text of draft resolution: document A/51/L.72; text of Convention, UNGA, Sixth Committee (Legal) (doc.A/51/869).

Wehbe, M. and Kasme, B.: April 16, 1999, Interview by the author at the Syrian UN mission, New York.

Yetkin, M.: January 26, 1993,"New Inroads in Relations with Syria," *Turkish Probe*, 4-7.

# INTERACTIONS AMONG SCIENTISTS, MANAGERS AND THE PUBLIC IN DEFINING RESEARCH PRIORITIES AND MANAGEMENT STRATEGIES FOR MARINE AND COASTAL RESOURCES: IS THE RED SEA MARINE PEACE PARK A NEW PARADIGM?

M. P. CROSBY[1], A. ABU-HILAL[2], A. AL-HOMOUD[3], J. EREZ[4], and R. ORTAL[5]

[1]*National Oceanic and Atmospheric Administration, Washington, DC, USA;* [2]*Marine Science Station, Aqaba, JORDAN;* [3]*Aqaba Regional Authority, Aqaba, JORDAN;* [4]*Inter-University Institute, Eilat, ISRAEL;* [5]*Nature and National Parks Protection Authority, Jerusalem, ISRAEL*

**Abstract.** It has been assumed that marine habitats and resources, especially, are almost unlimited, and that if one habitat became degraded or a particular fisheries resource depleted, there always would be another to replace it. Therefore, natural resource management principals are beginning to include human motivation and responses as part of the marine and coastal systems that are being studied and managed. Managers of marine resources face the challenge of balancing conservation and development objectives in the context of the inherent uncertainty of natural systems and the political and social pressures of human systems. Natural resource managers, scientists and the general public seem to share a vision for the future as a world in which societal and economic decisions will be strongly coupled with an increasingly comprehensive understanding of the environment. This in turn will lead to both socio-economic health and ecosystem health. A paradigm shift is being seen in the evolution of the role of scientists in society from simply observers of the natural world with tenuous linkages to resource managers and the public, to partners in modern society's quest for answers to pressing questions related to sustainable use and conservation of natural resources. A US Agency for International Development supported, joint effort between the US National Oceanic and Atmospheric Administration, the Government of Israel and the Hashemite Kingdom of Jordan to conduct a comprehensive research and monitoring program directed at the new Binational Red Sea Marine Peace Park will be a pioneering effort to employ and test this new paradigm.

**Keywords**: coral reef, Gulf of Aqaba, management, marine protected area, Red Sea, research, sustainable use

## 1. Introduction

The importance of coral reef ecosystems may be seen in their numerous ecological, aesthetic, economic and cultural functions. Not only is the coral reef structure itself composed of and built by a diversity of organisms, but the reef structure serves as the basis for one of the highest diversity ecosystems in the world (Talbot, 1994; Maragos *et al.*, 1996). When speaking of biological diversity, it is indeed appropriate to refer to coral reef ecosystems as the rain forests of the marine realm. Of the 33 phyla that exist in marine habitats, 30 are found in coral reefs. Diversity of reef building corals and other reef organisms tends to diminish in an east to west direction across the tropical latitudes of the Indian Ocean; however, the Red Sea supports higher levels of species diversity than any Eastern Indian Ocean region. There are approximately 50 genera of coral reef building corals existing in the Middle East Seas region (see Figure 1 for the location of major coral reef sites in the Middle East).

One manifestation of the poor management of natural resources is the loss of biological diversity (see Norse, 1993; Eichbaum *et al.*, 1996; Maragos *et al.*, 1996). The problem of biodiversity loss and ineffectual management of natural resources has many aspects: social, economic, cultural, managerial and scientific (Solbrig, 1991). The term "biodiversity" can also have several meanings. UNEP, in the Convention on Biological Diversity (June 5, 1992), define "Biological Diversity" as the variability among living organisms from all sources including, *inter alia*, terrestrial, marine and other aquatic ecosystems and the ecological complexes of which they are part; this includes diversity within species, between species and of ecosystems. The widespread use of the term "biological diversity" began around 1980 (see Norse, 1993), with the most widely used definition of biological diversity (Norse *et al.*, 1986) including three levels of organization: genetic, species, and ecosystem.

Coral reefs ecosystems have formed in over 100 countries around the world (Birkeland, 1997). Coral reefs have far greater productivity than other marine systems, surpassing 7,000 g C $m^2$ $yr^{-1}$ (Odum *et al.*, 1959; Helfrich and Townsley, 1965). Reef fishes, sea urchins, coralline algae and many additional species of plants and animals contribute to healthy reef ecosystems and play a significant role in helping maintain the resilience, stability, and accelerated coral reef recovery following natural and anthropogenic disturbances.

Although difficult to quantify, encroachment into marine and coastal areas, and destruction of coastal and marine habitats, is evident all over the world (NRC, 1990; Goldberg, 1994; NRC, 1995), along with an associated increasing demand on the natural resources of these areas. Habitats are being lost irretrievably to the construction of harbors and industrial installations, to the development of tourist facilities and mariculture, and to the growth of settlements and cities. Still other marine habitats, such as coral reefs, are being significantly altered and damaged on a routine basis through exploitation of natural resources (i.e., commercial and recreational fishing, recreational diving and boating, oil and mining activities). If unchecked, this trend will lead to global deterioration in the quality and productivity of the marine environment (GESAMP, 1990).

The ability of coral reef ecosystems to exist in balanced harmony with other naturally occurring competing/limiting physico-chemical and biological agents has been severely challenged in the last several decades by the dramatically increased negative and synergistic impacts from poorly managed anthropogenic

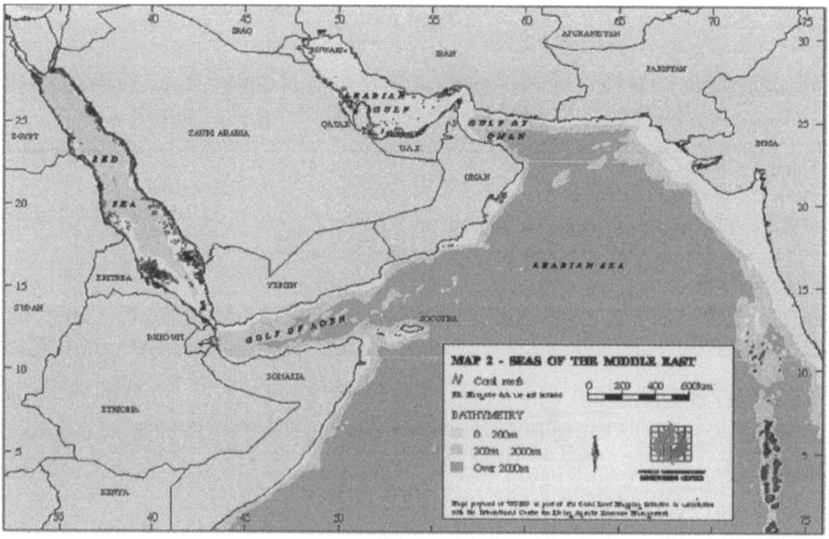


Fig. 1. Location of major coral reef sites in the Middle East

activities (Hughes, 1994; Maragos *et al.*, 1996). Placing a monetary figure on reefs is difficult, however a recent settlement in Egypt assessed a value of US$1,765 on a $m^2$ of coral reef (Spurgeon, 1992). In a similar settlement in the US, a court assigned the value of a $m^2$ of coral reef at US$2,833 (Mattson and DeFoor, 1985). In the Middle-Eastern Seas, the major threats to reefs stem from (1) urbanization and tourism, (2) oil exploitation and transport, and (3) industrial pollution and sewage discharges. Population growth and tourism are increasing throughout the region. Degradation of coral reefs is also occurring from heavy souvenir collection (Fouda, 1995). Table I lists examples of both natural and anthropogenic causes of stress in coral reef ecosystems in the Gulf of Aqaba.

The Gulf of Aqaba (180 km long, 20 km wide) is a semi-enclosed sea with many unique natural and physical features, and geographically isolated by the narrow Strait of Tiran (Figure 2). While much of the Gulf is deep (more than 1,800 meters deep), the northern sector has a relatively shallow shelf adjacent to the major population centers. Ancient coral reef formations characterize most of the beach areas along the coasts of the Gulf while patches of living coral reefs (as many as 140 stoney coral species) fringe parts of the steep-sided edge of the Gulf in the north. Coral reefs in the Gulf of Aqaba are thought to represent the northern latitudinal limit ($29^0$ 32') for reef corals in the Western Indo-Pacific

TABLE I

Some examples of natural and anthropogenic causes of stress in Gulf of Aqaba coral reef habitats (Maragos *et al.*, 1996; Crosby, 1997; and Loya *et al.*, in press)

| |
|---|
| Natural Stresses |
| • Periodic extreme low tides |
| • Gastropod (*Drupella cornus*) predator outbreaks |
| • Crown-of-Thorns starfish (*Acanthaster spp.*) predator outbreaks |
| • Earthquakes |
| • Wave action |
| • Periodic freshwater runoff |
| Anthropogenic Stresses |
| • Pressure from population increases, including migration; and intensified uses |
| • Depletion of fish stocks and destructive fishing methods (traps, nets, spearfishing, dynamite fishing, etc.) |
| • Collection of corals and ornamental reef species |
| • Untreated domestic sewage and industrial effluent |
| • Excessive non-point source pollution, e.g., from agricultural runoff and contamination of aquifers |
| • Ship-based pollution; including oil, plastics and bilge water |
| • Land based and urban construction activities including dredging, filling, and increased siltation |
| • Breakage of corals by divers, watercraft and ships |

region. However, Schumacher *et al.* (1995) reported that coral growth rates at Aqaba are the highest observed at these latitudes, and suggested that "the reefs at Aqaba do not mark the northernmost ecophysiological outpost of the Indian Ocean and that reefs would occur at even higher latitudes, if the Gulf extended further north." These coral reefs and adjacent coastal zone are readily accessible by the public. The Gulf enjoys the presence of nearly 1,000 species of fish, and the diversity of fish and coral species combined with calm seas and clear water, make the Gulf an increasingly popular tourist destination.

Diurnal tides are normally minimal with a range in the order of one meter or less. Currents, specifically, and circulation, generally, appear to be largely wind-driven, with additional influence from tides, density gradients, and evaporation (USAID, 1992). The constriction of the Gulf from the flushing dynamics of open ocean circulation systems is a major pollution concern, especially in the north where the major population centers of Aqaba and Eilat are located. However, the physical oceanography of the Gulf is still not well understood.

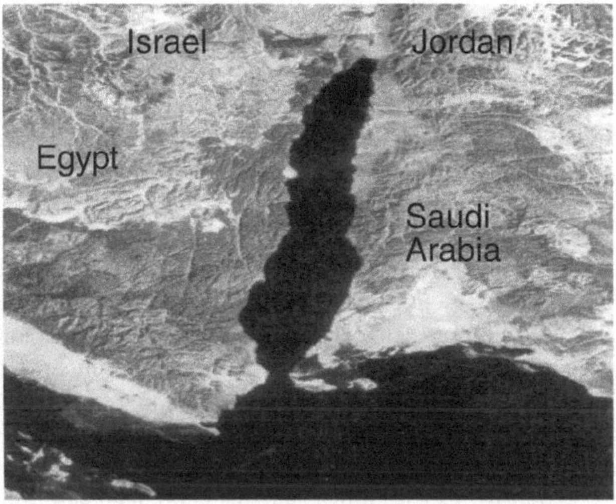


Fig. 2. The Gulf of Aqaba

Achieving sustainable human use of natural marine resources is a complex process of balancing competing human needs with various natural resource management goals at all spatial and temporal scales (Crosby, 1997). Among the most important of these goals are:

- Promoting sustainable economic production.
- Conserving natural marine resources and biodiversity.
- Sustaining cultural values that are intimately linked with the marine environment.
- Providing basic information to understand the condition of the marine environment.

Humankind is perhaps only now grasping the concepts needed to manage relations between people and the oceans (Kelleher and Kenchington, 1991; Crosby, 1994; Kelleher *et al.*, 1995; Eichbaum *et al.*, 1996). An increasingly important mechanism to advance the protection of marine biodiversity is the new generation of marine and coastal protected areas (MACPAs) (Eichbaum *et al.*, 1996).

The need for better management of MACPAs in particular, and the broader seascape in general, is inescapable (Kelleher *et al.*, 1995; Eichbaum *et al.*, 1996). Marine habitats and resources, especially, have been assumed to be almost unlimited, and that if one habitat became degraded or a particular fisheries resource depleted, there always would be another to replace it.

One of the key recommendations that was put forth by Ludwig *et al.* (1993) for future natural resource management principals is to include human motivation and responses as part of the system to be studied and managed. We should seek to avoid the "tragedy of the commons" (Hardin, 1966) in our utilization of marine and coastal resources. In doing so, however, natural resource decisions have traditionally emphasized achieving the management goals of a particular agency, organization or interest group in addressing particular human needs (such as the need for timber, water, fisheries or recreation) in a particular management unit (such as public rangelands, water management districts, or public marine and coastal waters). The shortsightedness of this approach has become apparent in recent years as the demand for marine and coastal resources has increased.

Today's MACPAs have evolved considerably from the traditional model of the terrestrial park, and can achieve far more than the protection of critical habitats and endangered species. Additional important niches that modern MACPAs fill are public education and outreach, as well as serving as a physical reminder of the social, economic and ecological benefits of marine and coastal resource protection. By employing a framework for the application of "adaptive management," MACPAs can establish and maintain feedback loops between science and policy (Crosby, 1997). Finally, multiple-use MACPAs address the differing sets of objectives of a myriad of stakeholders, thereby providing a framework for resolving conflict between various users of marine and coastal ecosystem services.

This paper sets out to present these new approaches within the context of a cooperative research, monitoring and management program which has been initiated by the US National Oceanic and Atmospheric Administration, the Government of Israel and the Hashemite Kingdom of Jordan.

## 2. The Red Sea Marine Peace Park: Moving Towards a New Paradigm?

If one were to review the basics of classic scientific method in the context of MACPAs (see Crosby, 1997), it is clear that the classic approach lacks mechanisms for data and information exchange between MACPA managers, policy-makers, scientists, NGOs and other interest groups. Contrasting perspectives between scientists, managers and policy makers in how they view and deal with similar parameters is a likely reason for this barrier to meaningful information exchange being difficult to overcome (Table II).

However, even under the more recent paradigm for the role of science in our society, which has created somewhat of a partnership between funding agencies and the research community, results of science studies are often not transferred to MACPA managers and policy makers in a user-friendly format such as described by Gault (1997; for a more detailed discussion, also see Crosby, 1997). MACPA managers are rarely included in designing scientific studies to ensure products will be of use in their decision making processes, and specialists capable of translating technical information to MACPA managers and expressing management and policy needs/gaps in a scientific forum are lacking.

Few would question that high quality, objective science that applies the classic scientific method is a key to understanding how marine and coastal ecosystems function and interact. However, a scarcity of truly MACPA management-oriented science programs exist (e.g., Crosby and Golde, 1993; Crosby, 1994; Crosby and Beck, 1995; GESAMP, 1996). Ideally, MACPA management-oriented research should (Crosby, 1997):

- Address specific and fundamental needs of MACPA managers by focusing on the ecological and socio-economic impacts of management strategies in MACPAs.
- Produce information that will serve to build more cohesive partnerships between managers, scientists, special interests, and the public at large for the development, implementation and operation of MACPAs.
- Focus on critical management strategies that influence ecological, economic and sociological sustainability in marine and coastal environments.
- Foster analyses and recommendations for dealing with other current and emerging management issues on sustainability of marine and coastal resources.
- Illustrate how sociological, cultural and economic factors (i.e., the human dimension) can be integrated into natural science analyses of MACPAs.

In order to protect and manage our natural resources in MACPAs, we must learn how natural factors interacting with human activities in a particular MACPA determine changes in biota and the ecosystems on which they depend. Though studies detailing the ecological impacts of MACPAs are limited, there are examples. For instance, fish populations have increased and other marine resources have improved considerably in Looe Key and Key Largo, FL, USA since the designation of these areas as National Marine Sanctuaries (Bohnsack, 1991; Causey, 1991). Protective management strategies in a short-lived MACPA in Sumilan Island, Philippines maintained high abundance of fishes in the MACPA and significantly higher yields to fisherman from areas adjacent to the MACPA (Alcala and Russ, 1990). Within 18 months of the opening of this MACPA to fishing, significant reductions of fishes occurred (54% decline in total yield of numerous species of reef fishes) and catch per unit effort was significantly reduced. Traps and gill nets damaged the habitat and reduced populations of non-targeted species. Species densities, biomass and diversity were found to be significantly greater for rocky substrates within the Scandola MACPA in Corsica when compared with similar habitats immediately adjacent to the MACPA (Francour, 1991).

TABLE II

Contrasting perspectives between scientists and policy makers in how they view and deal with similar parameters. (see Crosby, 1997)

| Ends and Means | Scientists | Policy makers |
|---|---|---|
| *Goal, purpose:* | seek truth | public welfare, represent constituents |
| *Basic orientation:* | understand, explain | act, decide |
| *Mechanisms:* | unbiased methods, impersonal | adversarial, opposing interests, highly personal |
| *Real-world Affairs:* | problems to be solved | problems to be resolved |
| *Currency:* | knowledge, expertise | power, influence |
| *View of Personal Judgment:* | mistrustful, stick to data | essential - need to act before being certain |
| **Time and Attention Spans** | **Scientists** | **Policy makers** |
| *For system:* | long, incremental | short - must act now |
| *For individual:* | next grant, tenure | next election |
| *Cognitive demands:* | depth, detail, little range | huge range, little depth |
| *Attention span:* | long | short - situational press |
| **Accountability & Rewards** | **Scientists** | **Policy makers** |
| *Responsible to:* | standards, peer | constituency |
| *Real-world accountability:* | low | high |
| *Rewarded for:* | experimenting | being right |
| *Career Incentives:* | successful research, publishing | power, orchestrating positive outcomes |
| *Idealized mode of action:* | autonomy | being "team player" |
| **Communicate & Interact** | **Scientists** | **Policy makers** |
| *Primary means:* | written | personal, face-to-face |
| *Role of interpersonal skills:* | low relevance to work quality | heart of work and effectiveness |
| *Value Characteristics in colleagues:* | knowledge, analytical power, creativity | loyalty, judgment, knowledge |
| *Language, imagery:* | precise, impersonal, technical | understandable, inclusive, personal |

To facilitate the incorporation of new and existing knowledge into MACPA management and policy decisions, a new paradigm for the interaction and role of integrated, multi-disciplinary science, management and education/outreach efforts must be developed (Figure 3). Implementing this new paradigm requires that applied research and technical development become a priority within the research community. Scientists and managers must work together as a team to identify and understand the ecological, economic and social driving forces behind the loss of marine biodiversity and the destruction of marine and coastal ecosystems.

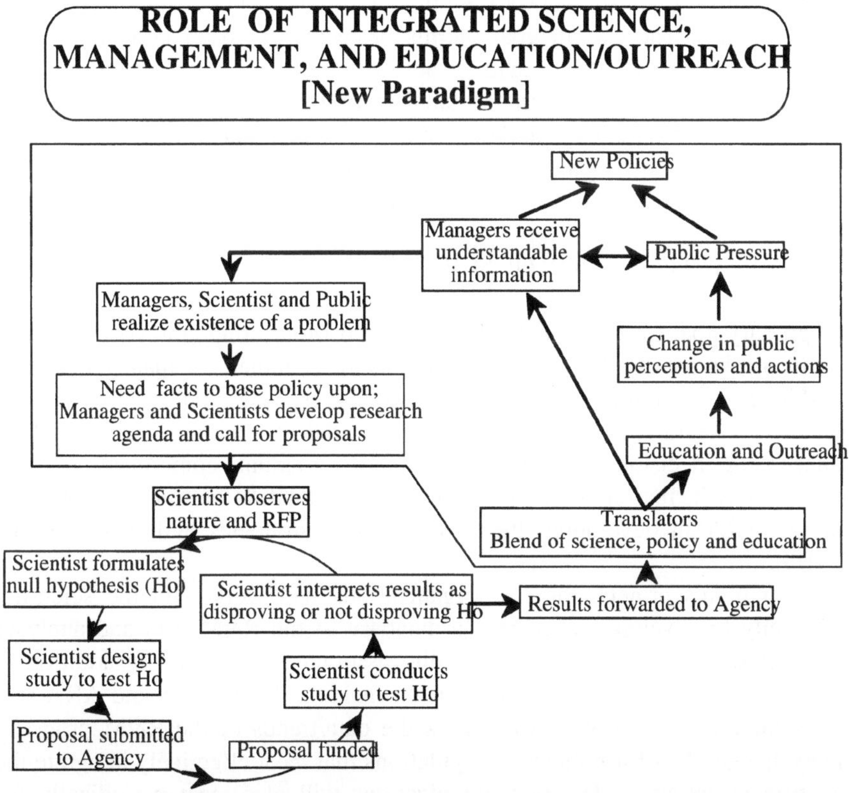


Fig. 3. New paradigm for the interaction and role of integrated, multi-disciplinary science, management and education/outreach efforts (Crosby, 1997)

A scientific approach must be coupled with the development of improved MACPA management strategies in order to fully sustain vital marine ecological systems. There is a need to make the full spectrum of data and information required to better understand and conserve MACPA biodiversity and ecosystem processes available and accessible to scientists, decision makers, and the public.

However, it is essential that user-friendly information and validated analytical models that can be used to explore the possible consequences of alternative MACPA management and policy decisions be available.

A pioneering effort to employ and test this new paradigm in a MACPA is being implemented in the Red Sea Marine Peace Park (RSMPP). Israel's Nature and National Parks Protection Authority (NNPPA) and Inter-University Institute, and Jordan's Aqaba Region Authority (ARA) and Marine Science Station have joined together in collaboration with the U.S. National Oceanic and Atmospheric Administration (NOAA) and the Agency for International Development (USAID) to initiate a cooperative research, monitoring and management program to address pressing environment and development issues in the RSMPP. The overall goal of this project is to foster cooperation and collaboration among Jordan and Israel in studying, managing, promoting awareness of, and protecting their shared marine resources. The two countries plan to address this goal by implementing a long-term monitoring and research program, and by participating in cooperative management and education/outreach activities related to the RSMPP.

## 2.1. RSMPP INTEGRATION OF SCIENCE AND MANAGEMENT

The primary goal for the RSMPP Research and Monitoring (R&M) program is to deliver relevant data to the management team. Together, the management and R&M team will jointly discuss trends and issues. The primary goal for the management team is to seek ways to integrate the data and analyses into their management decision making process. An essential component of that integration is increased communication and coordination between the NNPPA and the ARA in developing their management strategies and plans for the RSMPP.

Through this project, the two organizations will work with the scientific community to develop and manage a database of the R&M data and analyses that meet standard quality control/quality assurance guidelines. In addition to using the data for their individual decision making, the NNPPA and ARA will hold semi-annual workshops to discuss the data/trends of the R&M program, and exchange ideas for policies or regulations that most effectively integrate the scientific information. The two organizations will also meet periodically (as needed) on specific management, science, education/outreach questions or issues that arise, such as specific development projects or emergency situations. Finally, the two organizations (with the Inter-University Institute in Eilat, Israel and the Marine Science Station in Aqaba, Jordan) will strengthen their regular communication by increasing their use of communication technology (more Internet access, conferencing equipment, etc.).

### 2.2. RSMPP RESEARCH AND MONITORING

As the primary goal of the R&M portions of this project is to deliver relevant data to the managers of the RSMPP that will enable them to better manage the associated habitats, logic dictates that first order principals related to these natural resources must be addressed prior to initiating other ecosystem "process" studies. The three highest priorities of the R&M portions of this study, therefore, will initially be focused on:

- determining the basic water circulation patterns that impact the RSMPP,
- completing a comprehensive mapping of all significant coral reefs within the RSMPP with an initial base-line assessment of condition, and
- developing a framework for, and initiating a coordinated long-term monitoring for basic coral reef ecosystem parameters such as percent cover and diversity of corals, sponges, algae and numbers and diversity of associated fish and macro-invertebrates, as well as basic physico-chemical parameters.

While the project will initially focus principal efforts on completion of the above priorities, they are only a part of a more comprehensive research program dealing with physiological ecology, more sophisticated modeling of physical dynamics, etc. that has been developed and will be initiated during the three year project period. It is critical, however, that the basic questions of "what and where are the resources", "what condition are these resources in", and "in what direction, where and when does the major driver of flushing, food availability and larval dispersal occur" be answered first in order to provide the foundation for future and more in-depth research studies, as well as the vital information required to develop and implement a coordinated RSMPP Management Plan.

A crucial element in establishing the long-term sustainable use and conservation of coral reef ecosystems is increasing scientific understanding of their processes through enhanced mapping and assessment, management-oriented research and the establishment of comprehensive and coordinated coral reef monitoring programs (Crosby *et al.*, 1996; Crosby and Maragos, 1995), such as that being implemented in the RSMPP. It is hoped that the RSMPP R&M program will lead to a training program for NNPPA and ARA park personnel to become proficient with the techniques and eventually take charge of the monitoring effort on a long term basis.

## 3. Vision for the Future

In order to move towards a vision for the future of a world in which societal and economic decisions will be strongly coupled with an increasingly comprehensive understanding of the environment, the paradigm for managing MACPA resources and ecosystem health must shift from a fragmented to an integrated approach, from a site-specific to an ecosystem-wide context, and from a reactive to pro-active mode (Crosby, 1997). It is realized that ecosystem health problems are complex and that solutions will require a comprehensive, multi-disciplinary approach.

Successful implementation of the new paradigm is built on a foundation of improved communication and partnership and cannot achieve its goals without integrating its activities with key players in the management of ecosystems. Coordination is also critical to information sharing and joint scientific research.

A stronger public education program (as per the new paradigm) will help residents understand the impact of land use change on associated marine habitats. In the example presented here, it will also elucidate the role of the RSMPP in examining the health of marine and coastal ecosystems, and the connection between the terrestrial coastal zone and the marine environment. Ultimately, however, humans decide how resources will be used, attribute value or importance to various parts of the environment, and determine what will be conserved or destroyed.

The key to controlling the loss of long-term sustainability of healthy marine and coastal ecosystems is understanding how these decisions are made in the realms of social, economic and political systems. Therefore we must also identify and understand the economic and social driving forces behind the loss of biodiversity and the destruction of these ecosystems. In addition, there is a need to develop incentives to maintain and enhance ecological goods and services. In order to facilitate the incorporation of new and existing knowledge into management and policy decisions, user-friendly information and validated analytical models that can be used to explore the possible consequences of alternative management and policy decisions are required. Lack of information or uncertainty freezes political action (Wurman, 1990).

Despite the historic lack of a global understanding of the intimate linkages between sustaining natural marine ecosystems and sustainable human use of these ecosystems, a growing number of trustees for the natural resources and habitats of MACPAs, foresee both economic health and ecosystem health in these ecosystems. To attain this vision, we must develop a strategy for environmental stewardship that involves all stakeholders. A key component of such a strategy would be promotion of healthy ecosystems by ensuring that economic development is managed in ways that maintain marine biodiversity and long-term productivity for sustained use of our marine and coastal natural resources and ecosystems. Working in partnership at global, national and local scales, MACPA institutions and agencies should integrate their operational management activities with an increased level of monitoring and assessments (to provide better information for decision-makers), education and outreach (for better public understanding), and pro-active research efforts (to enhance our predictive capabilities for changes in MACPA ecosystems).

## References


Alcala, A. C. and Russ, G. R.: 1990, *J. Cons. int. Explor. Mer.* **46**:40-47.

Birkeland, C.: 1997, in C. Birkeland (ed), *Life and death of coral reefs.* New York, Chapman and Hall, Chapter 1: pp. 1-22.

Bohnsack, J. A.: 1991, *Proc. Amer. Fish. Soc.*, San Antonio, Texas. p 128.

Causey, B. D.: 1991, *Proc. Amer. Fish. Soc.,* San Antonio, Texas. p 129.

Crosby M. P. and Beck, A. D.: 1995, *Natural Areas Journal* **15**:12-20.

Crosby, M. P.: 1994, *Marine Protected Areas and Biosphere Reserves: 'Towards a New Paradigm'.* in D. J. Brunkhorst (ed), Australian Nature Conservation Agency, Canberra, Australia, pp. 45-65.

Crosby, M. P.: 1997a, in *Report of the Middle East Seas Regional Strategy Workshop for the International Coral Reef Initiative.* NOAA, Silver Spring, MD, USA, pp. 6-7 and A23-30.

Crosby, M. P.: 1997b, in M. P. Crosby, D. Laffoley, C. Mondor, G. O'Sullivan and K. Geenen, *Proceeding of the Second International Symposium and Workshop on Marine and Coastal Protected Areas, July, 1995.* Office of Ocean and Coastal Resource Management, National Oceanic and Atmospheric Administration, Silver Spring, MD, USA, pp. 10-24.

Crosby, M. P. and Golde, H. M.: 1993, *A review and synthesis of the first decade of research in the National Estuarine Research Reserve System.* Technical Memorandum #26. Office of Ocean and Coastal Resource Management, National Ocean Service, National Oceanic and Atmospheric Administration. Washington, D.C.

Crosby, M. P. and Maragos, J. E.: 1995, in J. E. Maragos, M. N. A. Peterson, L. G. Eldredge, J. E. Bardach and H. F. Takeuchi (eds), *Marine and coastal biodiversity in the tropical island Pacific region. Vol I: Species systematics and information management priorities.* East West Center, Honolulu, HI, pp. 303-316.

Crosby, M. P., Gibson, G. R. and Potts, K. W. (eds.): 1996, *A Coral Reef Symposium on Practical, Reliable, Low Cost Monitoring Methods for Assessing the Biota and Habitat Conditions of Coral Reefs, January 26-27, 1995.* Office of Ocean and Coastal Resource Management, National Oceanic and Atmospheric Administration, Silver Spring, MD, USA. 80 pages.

Eichbaum, W. M., Crosby, M. P., Agardy, M. T. and Laskin, S. A.: 1996, *Oceanography* **9**: 60-70.

Fouda, M. M.: 1995, *Regional report, Middle East Seas: Issues and activities associated with coral reefs and related ecosystems.* Prepared for the 1995 International Coral Reef Initiative Workshop, Dumuguete City, Philippines, May 1995. 50 pages.

Francour, P.: 1991, *Rev. Ecol.* **46**: 65-81.

Gault, C.: 1997, in M. P. Crosby, D. Laffoley, G. O'Sullivan and K. Geenen, *Proceeding of the Second International Symposium and Workshop on Marine and Coastal Protected Areas, July, 1995.* Office of Ocean and Coastal Resource Management, National Oceanic and Atmospheric Administration, Silver Spring, MD, USA. 247 pages, pp. 35-41.

GESAMP (IMO/FAO/UNESCO-IOC/WHO/IAEA/UN/UNEP Joint Group of Experts on the Scientific Aspects of Marine Environmental Protection): 1996, *The contributions of science to coastal zone management.* Rep. Stud. GESAMP. 66 pages.

GESAMP: 1990, *The State of the Marine Environment;* UNEP Regional Seas Reports and Studies No. 115.

Goldberg, E. D.: 1994, *Coastal zone space - Prelude to conflict?* UNESCO, Paris, 138 pages.

Hardin, G.: 1966, *Science* **162**:1243-1248.

Helfrich, P. and Townsley, S. J.: 1965, in F. R. Fosberg (ed), *Man's place in the island ecosystem.* 10th Pac. Sci. Cong. (1961). Bishop Mus. Press, Honolulu. 264 pages, pp. 39-56.

Hughes, T. P.: 1994, *Science* **265**: 1547-1551.

Kelleher, G. and Kenchington, R.: 1991, *Guidelines for establishing marine protected areas.* IUCN, Switzerland, 79 pages.

Kelleher, G., Bleakley, C., and Wells, S.: 1995, *A global representative system of marine protected areas, Vol. I.* IUCN, Switzerland. 219 pages.
Loya, Y., Al-Moghrabi, S. M. Ilan, M. and Crosby, M. P.: (in press), *Proceedings of the Hawaii Coral Reef Monitoring Workshop-A tool for management.* East West Center, Honolulu, HI.
Ludwig, D., Hilborn, R. and Walters, C., 1993, *Science* **260**: 17-36.
Maragos, J. E., Crosby, M. P. and McManus, J.: 1996, *Oceanography* **9**: 83-99.
Mattson, J. S. and DeFoor, J. A. II.: 1985, *J. Land Use and Env. Law* **1**: 295-319.
National Research Council.: 1990, *Managing troubled waters - The role of marine environmental monitoring.* National Academy Press, Washington, D.C. 125 pages.
National Research Council.: 1995, *Science, Policy, and the coast.*National Academy Press, Washington, D.C. 85 pages.
Norse, E. A.: 1993, *Global marine biological diversity: A strategy for building conservation into decision making.* Island Press, Washington, D.C. 383 pages.
Norse, E. A., Rosenbaum, K. L., Wilcove, D. S., Wilcox, B. A., Romme, W. H., Johnson, D. W. and Stout, M.L.: 1986, *Conserving Biological Diversity in our National Forests.*The Wilderness Society, Washington, D.C.
Schumacher, H., Kroll, D. K. and Reinicke, G. B.: 1995, *Beitraege zur Paleontologie* **20**: 89-97.
Solbrig, O. T.: 1991, *From genes to ecosystems: A research agenda for biodiversity.* International Union of Biological Sciences. Paris, France. 124 pages.
Spurgeon, J. P. G.: 1992, *Mar. Pollut. Bull.* **24**: 529:536.
USAID: 1992, *Gulf of Aqaba Environmental Data Survey.* Irrigation Support Project for Asia and the Near East, U.S. Agency FOR International Development. 35 pages.
Wurman, R. S.: 1990, *Information Anxiety.* Bantam Books, NY. 355 pages.

# THE PRODUCTION OF MICROBIOLOGICALLY SAFE EFFLUENTS FOR WASTEWATER REUSE IN THE MIDDLE EAST AND NORTH AFRICA

D. D. MARA
*School of Civil Engineering, University of Leeds, Leeds LS2 9JT England*

**Abstract.** Wastewater treatment in waste stabilization ponds (WSP) is a very efficient, low cost, low maintenance process which is well able to produce an effluent quality that meets the World Health Organization's recommendations for wastewater reuse for crop irrigation. Treatment in anaerobic and facultative ponds is required for restricted irrigation, with further treatment in maturation ponds for unrestricted irrigation. However, it is shown that the land requirements for the latter are at least twice that for the former. Unrestricted irrigation should, therefore, only be selected if it is economically viable. The use of wastewater storage and treatment reservoirs (WSTR), after pretreatment in anaerobic ponds, is advantageous in that it permits the whole year's wastewater to be used for irrigation, so allowing a much greater quantity of crops to be produced. A hybrid WSP-WSTR system can be used which produces effluents safe for both restricted and unrestricted irrigation.

**Keywords:** Middle East, North Africa, reuse, storage reservoirs, waste stabilization ponds, wastewater treatment

## 1. Introduction

The reuse of treated wastewater for crop irrigation is becoming more common in arid and semi-arid areas, including those in the Middle East and North Africa. Treatment of the wastewater prior to such use is especially important: firstly, from the point of view of public health, in order to minimise, even eliminate, the risk of human disease transmission; and secondly in order not to militate against local social and religious sensibilities.

### 1.1. PUBLIC HEALTH ASPECTS

The World Health Organization and the World Bank have reviewed the public health aspects of crop irrigation with domestic wastewaters, and have made the following recommendations for the microbiological quality of treated wastewaters used for this purpose (Shuval *et al.*, 1986; WHO, 1989):

- Restricted irrigation: no more than one human intestinal nematode egg per litre;
- Unrestricted irrigation: as above, plus no more than one thousand faecal coliform bacteria per 100 millilitres.

Restricted irrigation refers to the irrigation of crops not for direct human consumption, and so covers the irrigation of industrial crops (such as cotton), fodder crops and those processed prior to consumption (such as wheat); whereas

*Water, Air, and Soil Pollution* **123:** 595–603, 2000.

unrestricted irrigation covers all crops for direct human consumption, including those eaten raw (such as salad crops). The human intestinal nematodes are *Ascaris lumbricoides* (the human roundworm), *Trichuris trichiura* (the human whipworm) and *Anclyostoma duodenale* and *Necator americanus* (the human hookworms). Reliable techniques now exist for counting such low numbers of eggs (Ayres and Mara, 1966).

Shuval *et al.* (1997) have quantified the annual health risks to an individual which result from the consumption of raw salad crops irrigated with wastewaters treated to various faecal coliform (FC) levels, and these are compared to the EPA's (1989) acceptable annual risk of waterborne disease, as follows:

a) EPA's (1989) acceptable annual risk of waterborne disease: $10^{-4}$
b) Consumption of raw salad crops irrigated with raw wastewater (FC ~ $10^7$ per 100 ml):
   - annual risk of hepatitis A: $10^{-2}$
c) Consumption of raw salad crops irrigated with wastewater treated to the WHO (1989) guideline level of 1,000 FC per 100 ml:
   - annual risk of hepatitis A $10^{-6} – 10^{-7}$
   - annual risk of rotavirus infection $10^{-5} – 10^{-6}$

Thus, as noted by WHO (1989), irrigation with untreated wastewaters is dangerous; but irrigation with wastewaters treated to 1000 FC per 100 ml is safer than drinking potable water by 1-3 orders of magnitude.

WHO (1989) noted that in most situations in warm climates the most appropriate treatment option for meeting these effluent quality requirements was waste stabilization ponds. To this we would add wastewater storage and treatment reservoirs. Both these treatment technologies are discussed in detail below.

## 1.2. SOCIAL AND RELIGIOUS ASPECTS

Ali (1987) and Farooq and Ansari (1983) discuss, from an Islamic perspective, whether treated wastewater might be used for irrigation. They both quote the *fatwa* (legal ruling) given by the religious scholars of the Islamic Council of Research and Consultation in 1979 which declared that treated wastewater could be used for all purposes provided it met the required health standards. The religious scholars argued that "unclean water" might regain its original characteristics of "cleanliness" by, for example, the following:

- the water by itself causes its unclean characteristics to disappear;
- clean water is added in sufficient quantity to dilute the unclean characteristics; and/or
- the unclean characteristics are removed by the passage of time or by the effect of the sun or winds, etc.

There is a striking similarity between the first and especially the last of these and wastewater treatment in waste stabilization ponds.

## 2. Waste Stabilization Ponds

Waste stabilization ponds (WSP) are shallow man-made basins into which wastewater flows and from which, after a retention time of several days (rather than several hours as in conventional treatment processes), a well treated effluent is discharged. WSP systems comprise a series of anaerobic, facultative and maturation ponds, or two or more such series in parallel. In essence, anaerobic and facultative ponds are designed for BOD removal and maturation ponds for pathogen removal, although some BOD removal occurs in maturation ponds and some pathogen removal in anaerobic and facultative ponds. Brief descriptions of the advantages and perceived disadvantages of WSP, and of the function and performance of these three different types of ponds, are given below. Design procedures are given in Mara and Pearson (1998).

### 2.1. ADVANTAGES OF WSP

The main advantages of WSP are:

- low (and usually the least) costs, both capital and operation and maintenance (O&M);
- simple maintenance requirements; and
- no energy input (other than that which arrives at the pond surface from the sun).

In most parts of the world, including the Arab world, these are normally sufficient to justify the adoption of WSP in most cases. Processes such as activated sludge are often too complex for satisfactory local O&M, and they are generally too expensive.

### 2.2. PERCEIVED DISADVANTAGES OF WSP

WSP are often regarded as having two clear disadvantages: high land requirements and, in arid and semi-arid areas especially, a high water loss through evaporation.

The large land requirement of WSP is often viewed as a disadvantage. Yet this is seldom a real disadvantage since (1) often sufficient land is available at low cost; (2) even if the cost of the available land is high, WSP may still be the least cost option: a World Bank study (Arthur, 1983), based on the city of Sana'a, found WSP to be the least cost option up to land prices of US$50,000 – 150,000 per ha, depending on the value of the discount factor used (5-15 percent); and (3) the land purchased for WSP can be an excellent real estate investment in itself: for example, the city of Concorde, in California, purchased land for WSP in 1955 at US$50,000 per ha; in 1975 the land was worth US$187,500 per ha in 1955 dollars, equivalent to a real rate of return of 6.8 percent per annum (Oswald, 1976).

Arthur's (1983) methodology for comparing wastewater treatment options is strongly recommended as it explicitly takes into account land costs, and it is the only rational way of selecting the least cost option.

Up to 20 percent of the influent wastewater flow can be lost by evaporation. However, even in arid regions where the value of water is high, it is doubtful that the value of the wastewater lost in this way from WSP exceeds the cost of, for example, the electricity which would be required to treat the wastewater in an alternative way (activated sludge, for example). Thus evaporative losses from WSP should be viewed as part of the price paid for treating the wastewater in the least cost way – a disadvantage certainly, but not one that should militate against the use of WSP.

## 2.3. EFFLUENT QUALITY REQUIREMENTS

WSP easily meet most quality requirements for discharge into rivers – for example, they readily satisfy the European Union directive on urban wastewater treatment ($\not>$25 mg filtered $BOD_5$ per litre and $\not>$150 mg suspended solids per litre) (CEC, 1991). However, the principal advantage of WSP is that they remove excreted pathogens to a very high degree (much higher than that achieved by other processes, unless these are supplemented by effluent chlorination – a process which in itself can be highly damaging to the environment). WSP can easily achieve the following removals of excreted pathogens:

- bacteria 6 $\log_{10}$ units
- viruses 3 $\log_{10}$ units
- parasites 100 percent

This high microbiological efficiency of WSP means that WSP effluents can be safely used for crop irrigation, as WSP can be easily designed to meet the WHO (1989) recommendations for the microbiological quality for treated wastewaters used to irrigate crops.

## 2.4. ANAEROBIC PONDS

Anaerobic ponds are 2-5 m deep and receive such a high organic load (usually > 100 g BOD/$m^3$ d, equivalent to > 3000 kg/ha d for a depth of 3 m) that they contain no dissolved oxygen and no algae (although occasionally a thin film of mainly *Chlamydomonas* may be seen at the surface). They function much like open septic tanks, and their primary function is BOD removal. They work extremely well in warm climates: a properly designed and not significantly underloaded anaerobic pond will achieve around 60 percent BOD removal at

20°C and as much as 75 percent at 25°C. Retention times are short: for wastewater with BOD of up to 300 mg/l, one day is sufficient at temperatures > 20°C.

Designers have in the past been afraid to incorporate anaerobic ponds in case they cause odour. Hydrogen sulphide, formed mainly by the anaerobic reduction of sulphate by sulphate-reducing bacteria such as *Desulfovibrio* spp., is the principal potential source of odour. However, in aqueous solution, hydrogen sulphide is present as either dissolved hydrogen sulphide gas ($H_2S$) or the bisulphide ion ($HS^-$), with the sulphide ion ($S^{2-}$) only really being formed in significant quantities at high pH. At pH values normally found in well designed anaerobic ponds (around 7.5), most of the sulphide is present as the odourless bisulphide ion. Odour is only caused by escaping hydrogen sulphide molecules as they seek to achieve a partial pressure in the air above the pond which is in equilibrium with their concentration in it (Henry's law). Thus, for any given total sulphide concentration, the greater the proportion of sulphide present as $HS^-$, the lower the release of $H_2S$. Odour is *not* a problem if the recommended design loadings (for example, $\not>$300 g $BOD_5/m^3$ day at 20°C and above) are not exceeded and if the sulphate concentration in the raw wastewater is less than 500 mg $SO_4^{2-}$/l (Gloyna and Espino, 1969).

## 2.5. FACULTATIVE PONDS

Facultative ponds are designed for BOD removal on the basis of a relatively low surface loading (100-400 kg BOD/ha d) to permit the development of a healthy algal population as the oxygen required for BOD removal by the pond bacteria is mostly generated by algal photosynthesis. Due to the algae, facultative ponds are coloured dark green, although they may occasionally appear red or pink (especially when slightly overloaded) due to the presence of anaerobic purple sulphide-oxidising photosynthetic bacteria. The algae that predominate in the turbid waters of facultative ponds are the motile genera, such as *Chlamydomonas, Pyrobotrys* and *Euglena*, as these can optimise their position in the pond water column in relation to incident light intensity and temperature more easily than non-motile algae (such as *Chlorella*, although this is fairly common in facultative ponds). The concentration of algae in a healthy facultative pond depends on loading and temperature, but it is usually in the range 500-2000 µg chlorophyll *a* per litre.

As a result of the photosynthetic activities of the pond algae, there is a diurnal variation not only in the concentration of dissolved oxygen, but also pH since at peak algal activity carbonate and bicarbonate ions react to provide more carbon dioxide for the algae, so leaving an excess of hydroxyl ions with the result that the pH can rise above 9, which rapidly kills faecal bacteria.

Helminth eggs, which can number up to 2000 per litre of wastewater depending on the endemicity of intestinal nematode infections, are removed by sedimentation and thus most egg removal occurs in the anaerobic and facultative ponds. If the final effluent is to be used for restricted irrigation, then it is sensible to check whether the facultative pond effluent contains ≤ 1 egg per litre. Often this is the case but, depending on the number of helminth eggs present in the raw wastewater and the retention times in the anaerobic and facultative ponds, it may be necessary to incorporate a single, short-retention-time maturation pond to ensure that the final effluent contains at most only one egg per litre.

## 2.6. MATURATION PONDS

A series of maturation ponds receives the effluent from the facultative pond, and the size and number of these is governed mainly by the required bacteriological quality of the final effluent; for unrestricted irrigation this is ≤ 1000 faecal coliforms per 100 ml (WHO, 1989). Maturation ponds usually show less vertical biological and physicochemical stratification than facultative ponds, and are well oxygenated throughout the day. Their algal population is thus much more diverse than that of facultative ponds, with non-motile genera tending to be more common; algal species diversity increases, and total algal numbers decrease, from pond to pond along the series.

Research in northeast Brazil at 25°C (Oragui *et al.*, 1987) has shown that faecal bacterial pathogens such as *Salmonella* spp. and *Campylobacter* spp. were absent when the FC count was 7000 per 100 ml, and *Vibrio cholerae* was totally removed in a series of WSP when the FC count was 60,000 per 100 ml (Oragui *et al.*, 1993). These results lend much credence to the safety of the WHO guideline value of ≤ 1000 FC per 100 ml for unrestricted irrigation (see also Mara, 1995).

## 2.7. WSP DESIGN FOR CROP IRRIGATION

Using the WSP process design procedures given in Mara and Pearson (1998), and assuming a typical domestic wastewater with 300 mg $BOD_5$ per litre, 500 nematode eggs per litre and $1 \times 10^8$ faecal coliforms per 100 ml, and a design temperature of 20°C, the following WSP combinations and retention times are required:

(a) Restricted irrigation

| | |
|---|---|
| Anaerobic pond | 1 day |
| Facultative pond | 7 days |

(b) Unrestricted irrigation

| | |
|---|---|
| Anaerobic pond | 1 day |
| Facultative pond | 7 days |
| Maturation ponds | 12 days |

i.e., 8 days for restricted irrigation *vs.* 20 days for unrestricted irrigation.

Thus designers need to be very careful in deciding to opt for unrestricted irrigation: the additional land area requirement is very significant – in this case, which is not atypical, 2.5 times that required for restricted irrigation.

In financial terms, designers should only opt for unrestricted irrigation if the additional value of the crops grown (i.e., the value of the "unrestricted" crops less that of the "restricted" crops) exceeds, on a net present value basis, the cost of providing the additional maturation ponds needed to reduce faecal coliform numbers to no more than 1000 per 100 ml.

## 3. Wastewater Storage and Treatment Reservoirs

Wastewater storage and treatment reservoirs (WSTR), also called effluent storage reservoirs, are especially useful in arid and semi-arid areas. They were developed in Israel to store the effluent from a WSP system during the period when it is not required for irrigation (Juanico and Shelef, 1994). It is thus a method of conserving effluent so that, during the irrigation season, a greater area of land can be irrigated. For restricted irrigation the wastewater is pretreated in an anaerobic pond and then discharged into a single 5-15 m deep WSTR with a retention time equal to the number of months during which the effluent is not required for irrigation. This strategy complies with the WHO guideline for restricted irrigation since any helminth eggs settle out in the anaerobic pond and the WSTR.

If the WSTR effluent is to be used for unrestricted irrigation, then it should contain ≤ 1000 faecal coliforms per 100 ml, which the above single WSTR cannot achieve, at least not during the irrigation season (Liran *et al.*, 1994). Instead three (occasionally four) sequential batch-fed WSTR in parallel are required (Mara and Pearson, 1992; see also Shelef and Azov, 1999). These receive anaerobic pond effluent and are each operated on a sequential cycle of fill, rest and use, with faecal coliform die-off to < 1000 per 100 ml occurring during the fill and rest periods.

Research in northeast Brazil (Mara *et al.*, 1996) has shown that sequential batch-fed WSTR are very efficient at removing faecal coliforms: at temperatures of 25°C die-off to < 1000 per 100 ml throughout the whole reservoir depth of 6 m occurred 3 weeks into the rest phase. WSTR were found to behave much like deep facultative ponds with an algal biomass of around 500 μg chlorophyll *a* per litre.

WSTR are a very flexible system of wastewater treatment and storage. Juanico (1995) details several arrangements, including two WSTR in series, with effluent from the first being used for restricted irrigation and that from the second for unrestricted irrigation. An alternative "hybrid" WSP-WSTR system is to treat the wastewater in anaerobic and facultative ponds, the effluent from the latter being discharged into a WSTR during the non-irrigation season, but used for restricted irrigation during the irrigation season when the WSTR contents are used for unrestricted irrigation (Mara and Pearson, 1999).

## 4. Conclusions

Waste stabilization ponds are a highly efficient, low-cost method of wastewater treatment well suited to environmental conditions in the Middle East and North Africa.

Anaerobic and facultative ponds together can treat domestic wastewaters to meet EU discharge standards, and also the WHO recommendation for wastewater reuse for restricted irrigation. Maturation ponds (2-4 in series) are needed to comply with the WHO recommendation for unrestricted irrigation.

In arid and semi-arid areas the use of wastewater storage and treatment reservoirs permits all the wastewater produced throughout the year to be safely used during the irrigation season. This allows at least twice the amount of land to be safely irrigated and thus at least double the quantity of crops to be produced.

## References


Ali, I.: 1987, *Journal of Irrigation and Drainage Engineering, American Society of Civil Engineers* **113** (2), 173-183.

Arthur, J. P.: 1983, *Notes on the Design and Operation of Waste Stabilization Ponds in Warm Climates of Developing Countries,*The World Bank, Washington, DC.

Ayres, R. M. and Mara, D. D.: 1966, *Analysis of Wastewater for Use in Agriculture: A Laboratory Manual of Parasitological and Bacteriological Techniques,* World Health Organization, Geneva.

Council of the European Communities: 1991, *Official Journal of the European Communities* No. L 135/40-52 (30 May).

EPA: 1989, *Federal Register* **54** (124), 27486-27491.

Farooq, S. and Ansari, Z. I.: 1983, *Environmental Management* **7** (2), 119-123.

Goyna, E. F. and Espino, E.: 1969, *Journal of the Sanitary Engineering Division, American Society of Civil Engineers* **95** (SA3), 607-628.

Juanico, M.: 1995, The effect of the operational regime on the performance of wastewater storage reservoirs, Paper presented at the 3rd IAWQ International Specialist Conference on Waste Stabilization Ponds, João Pessoa, Brazil, 27-31 March.

Juanico, M. and Shelef, G.: 1994, *Water Research* **28**, 175-186.

Liran, A., Juanico, M. and Shelef, G.: 1994, *Water Research* **28**, 1305-1314.

Mara, D. D.: 1995, *Water Quality International* (3), 29-30.

Mara, D. D. and Pearson, H. W.: 1998, *A Design Manual for Waste Stabilization Ponds in Mediterranean Countries*, Lagoon Technology International Ltd, Leeds.

Mara, D. D. and Pearson, H. W.: 1992, *Water Science and Technology* **26** (718), 1459-1464.

Mara, D. D. and Pearson, H. W.: 1999, *Water Research.* **33** (2), 591-594.

Mara, D. D., Pearson, H. W., Oragui, J. I., Cawley, L. R., de Oliveira, R. and Silva, S. A.: 1996, *Wastewater Storage and Treatment Reservoirs in Northeast Brazil,*TPHE Research Monograph No. 12, Department of Civil Engineering, University of Leeds, Leeds.

Oragui, J. I., Curtis, T. P., Silva, S. A. and Mara, D. D.: 1987, *Water Science and Technology* **19** (Rio), 569-573.

Oragui, J. I., Arridge, H., Mara, D. D., Pearson, H. W. and Silva, S. A.: 1993, *Water Research.* **27**, 727-728.

Oswald, W. J.: 1976, in E. F. Gloyna, J. F. Malina and E. M. Davis (eds), *Ponds as a Wastewater Treatment Alternative,* University of Texas at Austin, Austin, Texas, 257-272

Shelef, G. and Azov, Y.: 1999, Meeting stringent environmental and reuse standards by an integrated pond system aimed at the 21st century, Paper presented at the 4th IAWQ International Conference on Waste Stabilization Ponds, Marrakech, Morocco, 20-23 April.

Shuval, H. I., Lampert, Y. and Fattal, B.: 1997, *Water Science and Technology* **35** (11/12), 15-20.

Shuval, H. I., Adin, A., Fattal, B., Rawitz, E. and Yekutiel, P.: 1986, *Wastewater Irrigation in Developing Countries: Health Effects and Technical Solutions,* Technical Paper No. 51, The World Bank, Washington, DC.

WHO: 1989, *Health Guidelines for the Use of Wastewater in Agriculture and Aquaculture,* World Health Organization, Geneva.

# ARE THE CONFLICTS BETWEEN ISRAEL AND HER NEIGHBORS OVER THE WATERS OF THE JORDAN RIVER BASIN AN OBSTACLE TO PEACE? ISRAEL-SYRIA AS A CASE STUDY

H. I. SHUVAL
*Kunin-Lunenfeld Professor of Environmental Science*
*Fredy and Nadine Herrmann School of Applied Sciences*
*The Hebrew University of Jerusalem, Israel, email: hshuval@vms.huji.ac.il*

**Abstract.** Are the conflicts over water resources between Syria, Lebanon and Israel who share the transboundary waters of the Jordan River Basin a major obstacle to the peace process? The Syrians and Lebanese have in the past claimed as their own all of the sources of the Jordan River which arise in their territory. International water law provides a strong legal basis to assure the water rights and continued use of water by a downstream riparian, such as Syria's use of the Euphrates which arises in Turkey and similarly, Israel's use of the Jordan River, based on prior use in an international river basin. This paper will evaluate the water security implications for Israel of a possible peace agreement with Syria and Lebanon which would involve Israel forgoing the continued use of those amounts of water from the Jordan River that were approved by the Israel Government under the so-called Johnston Plan of 1956 – 35 million $m^3$/yr. for Lebanon from the Hasbani Springs and 42 $mm^3$/yr. from the Banias Springs and Jordan River for Syria. The maximum replacement cost for that amount of water by desalination of seawater may be about \$0.70/$m^3$ or, some \$56,000,000 per year. This is not a great amount of money as part of the price for peace. The paper also shows that Israel does not have to hold on to the entire area of the Golan Heights to assure its water security and that a 1-3 km water security zone along the Syrian side of the international border under joint and international inspection can be an effective water security measure to assure inspection, monitoring and control of all the sources of the Jordan River and Lake Kinneret vital to Israel's water security. The scare stories published by some groups in Israel, that Syria could, after a peace agreement is signed, covertly divert essentially all of Israel's water resources derived from the sources of the Jordan River or 30% of the country's water supply are highly unrealistic. Such major works could not go undetected and Syria and the world recognize that such an act would be viewed as a *causus belli* and as an act of war by Israel. In an era of peace, development of the shared water resources of the Jordan and continuous water systems in a program of regional cooperation can bring benefits to all of the partners on the Jordan River Basin.

**Keywords**: international river basins, international water law, Israel, Johnston Plan, Jordan River, Lebanon, Middle East, riparian rights, Syria, transboundary water, water conflicts

## 1. Introduction

There is a popular perception shared by some journalists and political leaders that the issue of water security is so existential that the conflicts over water between Israel and her neighbors concerning the ultimate fate of the shared transboundary water resources of the Jordan River Basin and the Mountain

*Water, Air, and Soil Pollution* **123:** 605–630, 2000.

Aquifer are so deep and intractable that they alone will be one of the major obstacles to peace between Israel, Syria, Lebanon, Jordan and the Palestinians and might even lead ultimately to exacerbation of the conflict between the countries of the region. Dr. Butrus Butrus Ghali, former Foreign Minister of Egypt and former Secretary-General of the United Nations, has said that the "next war in the Middle East will be over water." Other Middle Eastern leaders, including the late King Hussein of Jordan have in the past made public statements containing similar dire predictions about future wars over water in the Middle East. Journalists, political scientists and veteran water experts, have quoted, re-quoted and reformulated the "water wars" hypothesis so often, that it has become accepted by many layman and politicians as one of the conventional wisdoms of the Middle East geopolitics. Joyce Star's *Water Wars* (1991), Gleick's *Water, War and Peace in the Middle East* (1994) and Bulloch and Darwish's *Water Wars* (1993) are but a few examples of this apocalyptic view.

Since the reopening of peace talks between Syria and Israel in December 1999, the issues revolving about the fate of the water resources of the upper Jordan River are particularly relevant and timely.

This paper will examine the issues of the shared Jordan River Basin water resources at stake between Israel, Syria and Lebanon as a case study in an attempt to evaluate whether or not the conflicts of interests are indeed so great that they are intractable or that a basis for an accommodation is nevertheless feasible. The paper will be devoted mainly to an analysis of the past and present water conflicts between Syria and Israel who share the transboundary waters of the upper Jordan River Basin (Figure 1) based in part on an earlier study prepared for the Harvard Center for Middle Eastern Studies (Shuval, 1998).

These nations will hopefully be attempting to reach an accommodation over their conflicts in the peace process initiated by the United States and Russia at the Madrid Conference in 1992 which has assumed new momentum with the peace initiatives of Prime Minister of Israel, Ehud Barak elected in 1999 and the declarations of Syria's President Hafas El Assad that it is Syria's strategic goal to achieve peace with Israel based on the principle that "territories taken by force in war should be returned as a condition for peace". These developments have received new impetus by the initiatives of US President Bill Clinton in December 1999.

However, to provide a framework for a better understanding of the Syria-Israel case study a general evaluation of the water resources available to the five riparian nations on the Jordan River Basin – Syria, Lebanon, Jordan, Israel and the Palestinians – will be presented since they are closely interrelated.

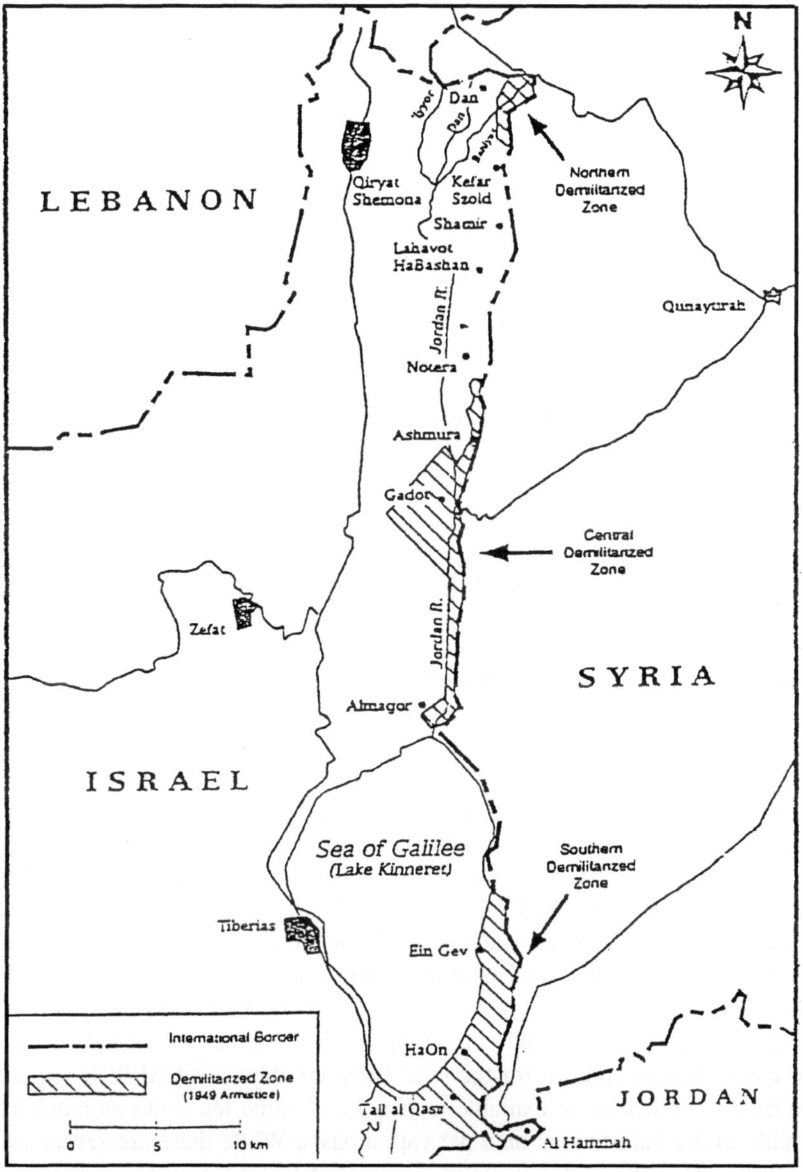


Fig. 1. Upper Jordan River Basin showing the sources of the Jordan with the 1923 International Border Between Israel and Syria and the 1949 Truce Lines and Demilitarized Zones. Source: From Peace with Security: Israel's Minimum Security Requirements in Negotiations with Syria, by Ze'ev Shiff, Policy Paper No. 34 (Washington D.C.-The Washington Institute for Near East Policy, 1993) p.8.

## 2. Water Scarcity Exacerbates the Problems

According to the study of the Population Action International (Engelman and LeRoy, 1993), many countries in the Middle East now face or will be facing severe water shortages as their populations grow and their water resources remain fixed. The intensity of the differences over water resources in the Jordan River Basin appear to be particularly grave since three of the five partners to the disputed waters – Jordan, the Palestinians and Israel – face serious, long term, water problems, particularly when considering the expected doubling of populations within the next thirty years or so.

It has been suggested by various researchers (Falkenmark, 1992; Gleiek, 1991) as well as by the World Bank that for a country to be considered as having sufficient water for all purposes it would be desirable to have at its disposal at least 1000 cubic meters/person/year(CM/P/Yr.). This estimate apparently assumes that this amount of water is required to assure enough water for agriculture to provide self-sufficiency in the production of most food for local consumption. On the other hand, I have estimated, that the absolute minimum water requirement (MWR) for essential domestic/urban/commercial and industrial needs for a truly arid country, with little or no allocation of fresh potable water for agriculture or food production, is a little more than 10% of that figure or some 100 to 125 CM/P/Yr.(Shuval, 1992).

Israel's estimated potential renewable fresh water resources for the year 2000, assuming a return to normal mean rainfall for the region after the extremely severe drought of 1998-99, are about 270 CM/P/Yr. (Israel Hydrological Service, 1998) with somewhat less for the Hashemite Kingdom of Jordan at 200 CM/P/Yr. It has also been estimated that the current figure for the year 2000 for the Palestinians in the West Bank and Gaza is about 90 CM/P/Yr.

The two upstream Jordan River riparians, Syria and Lebanon, currently have considerably more abundant water supplies at their disposal and will not face the same sort of conditions of extreme water scarcity in the future even with a doubling of the population, that will be faced by their three less fortunate downstream neighbors. The potential total water resources available to Syria on a per capita basis estimated for the year 2000 are about 900CM/P/Yr., or some three times as much as is available to Israel and some ten times as much as is available to the Palestinians on a per capita basis. While there are severe water shortages in Damascus, this is mainly a result of lack of proper water transport infrastructure within Syria which could pump available water to the capital. The estimated water resources available to Lebanon in the year 2000 are estimated at some 1200 CM/P/Yr., or four time greater than Israel and some twelve times greater than available to the Palestinians on a per capita basis.

For purposes of a graphic comparison of what is often referred to as the water stress index, Figure 2 shows the estimated annual fresh water availability in the year 2000 per capita in CM/P/Yr. for Israel, Jordan, Syria, Lebanon, and the Palestinians as well as for Turkey – the country with the most bountiful water resources of the region.

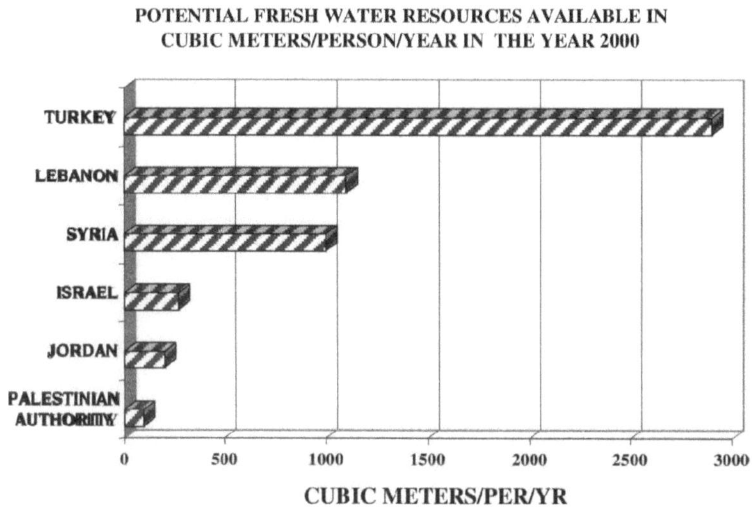


Fig. 2. Water Stress Index on the Jordan River Basin

From the above analysis, with all of its tentative and possibly inaccurate estimates of the availability of water resources some 25-30 years in the future, one thing is clear however, Jordan and Israel will have serious shortages of water and will have available to them just about the minimum of 125 CM/P/Yr., considered by many as the Minimum Water Requirement (Shuval, 1992) for human survival to meet all the needs of domestic/urban/commercial and industrial uses at a reasonable hygienic level and standard of living. The Palestinians will face a still more extreme degree of water shortage with less than half of the amount of water per person considered essential for a minimum hygienic standard of living –that is unless they achieve greater water allocations from Israel and possibly their northern neighbors in the framework of the peace agreement and programs for regional cooperation. It is noted that Lebanon and Syria are estimated still to have available to them significantly more water than the absolute minimum for survival based on the MWR concept.

## 3. The 1956 "Johnston Plan" Proposals for Allocation of the Jordan Waters Between Syria, Lebanon, Jordan and Israel

In order to understand one of the important background issues in the current Israel-Syrian water conflict it is essential to go back to the year 1955 when Israel started to construct its National Water Carrier (NWC). The plan was to transport water in a 108 inch pipeline, for a total of some 300 km from the Jordan River in the north to the arid south and Negev. This huge project was the flagship endeavor of the Israel water master plan to increase irrigated agriculture and food production and establish homes and farms for the hundreds of thousand of refugees who had arrived in Israel from the concentration camps in Europe after World War II and other areas where there was economic and political unrest. The Syrians objected to Israel constructing the initial diversion canal at Gesher B'not Ya'akov, a point on the Jordan River which was then in the Demilitarized Zone established in the 1949 truce between Israel and Syria. Syrian tanks fired on the Israel construction workers and equipment. In order to prevent an armed conflagration, US President Dwight Eisenhower called for a cease fire and appointed Eric Johnston as his personal envoy and roving ambassador to seek a comprehensive program to develop the Jordan River's water resources 'on a regional basis'. Johnston skillfully avoided discussions of water rights and succeeded in achieving consensus among Israeli, Jordanian, Syrian and Lebanese water experts at the technical level as to the amounts of water each of the riparians could rationally use for agricultural development schemes, with particular emphasis on promoting refugee resettlement projects for both sides.

Stevens (1965) reports that the following was the proposed apportionment of the Jordan River Waters under Ambassador Johnston's final proposal which later became known as the "Johnston Plan" (Table I). It must be stated at this point that there are a number of different interpretations as to the final water allocations included in the Johnston proposals. It is worthwhile noting , that the amount of water allocated by the "Johnston Plan" to Syria of up to 132 million cubic meters/year (MCM/Yr.) and Lebanon of up to 35 MCM/Yr., were exactly the amounts requested by the Technical Committee of the Arab League which under Egyptian leadership formulated the "Arab Plan" for allocation of the Jordan Basin waters.

Of the 132 MCM/Yr allocated to Syria, up to 20 MCM/Yr. was to come from the Banias and up to 22 MCM/Yr. from the main stream of the Jordan River. The Jordan water was "to be delivered by Israel" for irrigation of nearby farms along the eastern banks of the Jordan, since Israel had, according to the international border, full and sole access to the Jordan River. Up to an additional 90 MCM/Yr. was to come from the Yarmuk River which arises and flows through Syria. Those are the amounts of water that Lebanon and Syria themselves demanded and claimed that they needed, and could rationally use in the limited agricultural areas in the vicinity of the Jordan sources. These figures proposed by the Arab League and approved fully by the Johnston Plan, are worth remembering when it comes to the discussions of the Lebanese and Syrian claims and demands in the water negotiations which are part of the peace talks that will hopefully take place. It is also worth noting that of the 480 MCM/Yr allocated by the Johnston Plan to the Hashemite Kingdom of Jordan, it was clearly understood that some 150-200 MCM/Yr. would be transferred to the West Bank of the Jordan, through a siphon under the river to the proposed West Ghor Canal for the resettlement of the Palestinian refugees in agricultural communities. This is the basis of the current Palestinian claim for an allocation of that amount of Jordan River water.

TABLE I

Volume of the Jordan River Basin's Flow Apportioned between the States in the Final Form of the "Johnston Plan" ( Stevens, 1965)
(in million cubic meters/ year - MCM/Yr.)

| | |
|---|---|
| Jordan | 480 MCM/Yr |
| Syria | 132 MCM/Yr (42MCM/Yr from Banias & Jordan) |
| Lebanon | 35 MCM/Yr |
| Israel | 466* MCM/Yr |
| Total Annual Flow | 1113 MCM/Yr |

* The residual flow is Israel's share of the total flow, given that the above listed amounts were claimed as necessary by the other states. Israel's share would vary according to the flow conditions of the river system.

Israel was ambivalent, at first, about the Johnston proposals since they were allocated a much smaller share of the Jordan waters than they felt they should rightfully have. Wishart (1990) quotes recently declassified internal US State Department documents which indicate that in June 1955 Israel agreed to the basic terms of the plan that Johnston had drawn up. From Brecher's (1974) study of the documents and minutes of cabinet meetings, it is revealed that in the final internal Israeli debate which approved the Johnston Plan, Foreign Minister Moshe Sharett, Levi Eshkol, head of the Israel water negotiating team and former Prime Minister, David Ben-Gurion who had just returned to the Government as Defense Minister, after his self imposed "retirement" in Sde Boker, all supported the comprehensive view that acceptance of the Johnston Plan would, in the long run bring Israel major geopolitical, economic and strategic advantages including a potential opening of cooperation, *de facto* recognition and ultimately peace agreements with her Arab neighbors –Syria, Lebanon and Jordan.

On the other hand, the technocrats, water experts, water engineers and the settlement and agricultural lobby led by Engineer Simcha Blass, Israel's veteran visionary water planner, took the narrow view and bitterly opposed the plan on the grounds that Israel was being deprived of vital water resources under the deal (Blass, 1960). The broader view of economic, strategic and security advantages of cooperating with the United States and hopefully the Arab neighbors, outweighed the disadvantages concerning the exact amount of water that might be available to Israel under the Johnston Plan and the potential limitations of security and sovereignty over the Lake Kinneret (Sea of Galilee). Without the massive financial support from the United States and the world community, Israel would not have been able to develop its water resources and it opted for potential strategic advantages and political realism rather than the narrow water technocrat/agricultural lobby view which would have led to confrontation and a stalemate.

Despite the acceptance of the Johnston Plan by Israel and the official Arab representatives at the technical level, the plan failed to win the official approval of the Arab Governments and the Arab League (Wishart, 1990). On 11 October 1955, the Technical Committee of the Arab League forwarded the Unified Johnston Plan, which it had approved, to the Political Committee of the Arab League where it failed to win approval. Brecher (1974) and Lowi (1990) cite Arab concern "that their agreement would imply indirect recognition of the Zionist state".

Wishart (1990) concludes that at the political level "the Arabs were reluctant to accept a plan that involved *de facto* recognition of Israel, acquiescence to Israel's development goals and the possibility of a United States security pact with Israel." Wishart reasons that the Arabs apparently opted to break off the water negotiations with the United States, since they did not feel a serious need to develop or utilize the waters of the Jordan River Basin or any pressure about water development in general and felt that they had more to lose politically, than they would gained economically, particularly since each one of the Arab States felt that they could eventually develop their water resources on their own and without need of any regional cooperation projects.

On the practical level, however, informal agreement to comply with the Johnston formula both by Israel and Jordan did provide the basis for the major American financial assistance to Israel in the construction of its National Water Carrier (NWC), which enabled Israel to develop important irrigation projects in the south and in the Negev, and to Jordan in the construction of the Eastern Ghor Canal (now known as the Abdullah Canal) providing for major irrigation development projects along the previously barren eastern banks of the Jordan River. Both countries have cooperated informally ever since in allocations of Yarmuk water along the lines of the Johnston proposals. It was understood that under this arrangement Jordan would eventually build the Western Ghor Canal and supply the Palestinians on the West Bank with 150-200MCM/Yr.

Israel meanwhile continued to plan and work on construction of the National Water Carrier, but shifted the point of water diversion from the controversial site in the Demilitarized Zone at Gesher B'Not Yaakov to one at Eshed-Kinerot (near Tabcha) on the shores of Lake Kinneret, thus avoiding the issue of construction in the Demilitarized Zone and a direct military confrontation with Syria. However, the issues of water allocation with Syria and Lebanon remained unresolved. The Israeli NWC was completed and put into operation in 1965.

## 4. The Water Issues Involved in the Current Syrian-Israel Peace Negotiations

At the time of this writing, at the early stages of the current Israel-Syrian peace negotiations, there is no clear indication that water issues per se, have been discussed directly or indirectly. However, one critical water related issue has emerged indirectly in connection with the opening Syrian and Israeli positions on the question of the final borders. The official Syrian position, as stated by President Assad and in the press, is Syria expects the peace agreement with Israel to be based on an Israeli return to the borders that existed between Israel and Syria on June 4, 1967 prior to the outbreak of the "Six Day War". This is presumably based on the cease fire lines established in the Armistice Agreement of 1949 between Israel and Syria at the end of the 1948 war, Israel's War of Independence, and additional areas occupied through military actions afterwards by the Syrian armed forces. For example, the cease fire line of the Armistice Agreement of 1949 included within Israeli control the entire 10 meter strip along the eastern fringe of Lake Kinneret as delineated by the 1923 International Border between Mandatory Palestine and Syria and the Hamat Gader ( El Hama) Springs area contiguous to the Yarmuk River. The Syrian army occupied these areas by force after the Armistice Agreement. These lines were never clearly established and there are several different interpretations of them. However, the June 4$^{th}$ lines clearly included within Syrian control critical water resource areas on the western side of the 1923 International Border between Syria and Mandatory Palestine, never previously controlled by Syria, which were captured and occupied by the Syrian Army during its attack against the newly founded State of Israel in 1948 and in the period afterward up to 1967 (Hof, 1999).

The stated policy of the Israel Government under Prime Ministers Rabin and Peres, and repeated by Prime Minister Ehud Barak after his election in 1999, concerning the Golan Heights is that "the depth of the peace will determine the depth of withdrawal". President Assad and the Syrian press have stated on many occasions that they base their demand to restart the negotiation at the point they were left off by the late Prime Minister Yitchak Rabin, whom they claim informed them unofficially through the American Secretary of State at the time,

Warren Christopher, that Israel would be prepared to withdraw from the *entire* Golan in exchange for all of the security arrangements they demanded. Israel officially denies that such a commitment was ever made and in November 1999 the US State Department spokesman Mr. Rubin, officially denied that the Americans ever passed on such as commitment to President Assad (Ha'Aretz, November 7, 1999). However, press reports, have suggested unofficially, that based on the precedent of its peace agreements with Egypt and Jordan, where a return to the recognized international borders provided the basis for the agreement, Prime Minister Ehud Barak of Israel might be prepared to consider that in the case of Syria as well, the 1923 international borders that existed between Syria and Mandatory Palestine in 1948 before the war, might provide the point of departure for the negotiations with some minor adjustments (Ha'aretz, November 10,1999).

The Israelis point out that the cease fire lines of 1949 and latter Syrian armed occupation of critical water sensitive areas along the Jordan and Lake Kinneret where the result of aggression and military conquest and that Syria cannot logically demand from Israel to give up the Golan Heights captured during the 1967 war while allowing Syria to hold strategically important areas on the western side of the international border, that it captured in the 1948 war and afterwards. Israelis point out that the very same international border of Mandatory Palestine provided the basis for the peace agreements with Jordan and Egypt and that only by a return to the international border with Syria will there be symmetry and the principle that land taken in war by either side should be returned as part of the peace agreement. While there are other important strategic considerations in the debate about the return to the international border, the water issue is one of the most critical.

## 5. The Issue of the Borders with Syria Based on Strategic "Water Security" Considerations

There are a number of groups in Israel who oppose giving up part or all of the Golan Heights to Syria and the withdrawal from most of the areas of the West Bank and Gaza to the Palestinians in return for peace. They differ greatly in motivation and ideology. With some, security considerations are of uppermost importance, while with others religious/nationalist/ideological and personal considerations as home owners, farmers and settlers dominate.

Another approach has been inspired by concerns over "water security" and protecting Israel's water sources which arise in the Golan. This approach motivated a study of the possible alternative borders with Syria which would provide Israel with complete control over its current water resources, thus assuring Israel of "water security" on its border with Syria and Lebanon. These resources, which total some 330 MCM/Yr., include the flow of the Banias of about 120 MCM/Yr. and surface flow to the Jordan from the side wadis of the Golan Heights which can contribute some 30-40 MCM/Yr. in rainy years as well as the flow of the Hasbani of 150MCM/Yr. which arises in Lebanon.

In 1991 a study was made on this question by the Jaffe Center for Strategic Studies of Tel Aviv University, under the direction of the late General (Reserves) Aaron Yariv, former head of Israel Defense Army's Intelligence in cooperation with Tahal-Water Planning for Israel, Israel's leading quasi-governmental water resources planning agency.

The study was never released to the public, for what was originally claimed as security considerations, but which according to Shiff (1993) apparently were mainly politically motivated. A fairly detailed report, leaked to the press by Shiff, has revealed that the Jaffe Center/Tahal study evaluated the possibility of drawing up *water security borders* with Syria which would assure that all the strategic water elements, both ground water and surface water, that Israel is currently utilizing would remain in Israel control. Figure 3. shows these proposed water security borders which are based on the newspaper reported version of the Jaffe Center/Tahal map (Shiff, 1993). This water security map would include within the borders of Israel the key water security areas of importance to Israel in both Syria and Lebanon.

This would include the water tributaries of the Jordan River – the Banias and Hasbani, the El Hama (Hamat Gader) springs, the side wadis of the Golan Heights which can be dammed up for water storage and could divert water from the Jordan water basin, as well as the entire area contiguous to the Jordan River and Lake Kinneret. According to this map, Israel would not have to hold on to most of the Golan from a water security point of view.

Another, possibly more feasible alternative arrangement that has been suggested, which I personally view as promising and that might achieve the same goals, would be for Israel to agree that the areas to be formally returned to official Syrian sovereignty would be up to the international border of 1923. However, as a condition for a peace agreement, a special status water security zone under joint and/or international management, inspection and control should be established. This water security zone would include a strip of some 1-3 kilometers in width all along the entire eastern side of the border within Syria which would include all the main water sources and assure that there would be no direct Syrian access to the Jordan River, the Banias Springs, El Hama (Hamat Gader) or to the shores of Lake Kinneret. Except for the amount of water which would be allocated under agreement to Syria, the remainder of the water from these sources would continue to flow freely into the Jordan River for Israeli use as in the past (Figure 4).

The justification for either of the above arrangements is Israel's legitimate concern that Syria and Lebanon might once again attempt to divert the sources of the Jordan River in order to prevent Israel's use of the Jordan water as they did in 1965. International Law gives special weight to the fact that once one of the partners to a dispute has made a serious violation of international water law which gravely threatens a country's main source of water, there is a strong case in support for requiring special arrangements and assurances in a peace treaty to guarantee that such a violation will not be repeated.

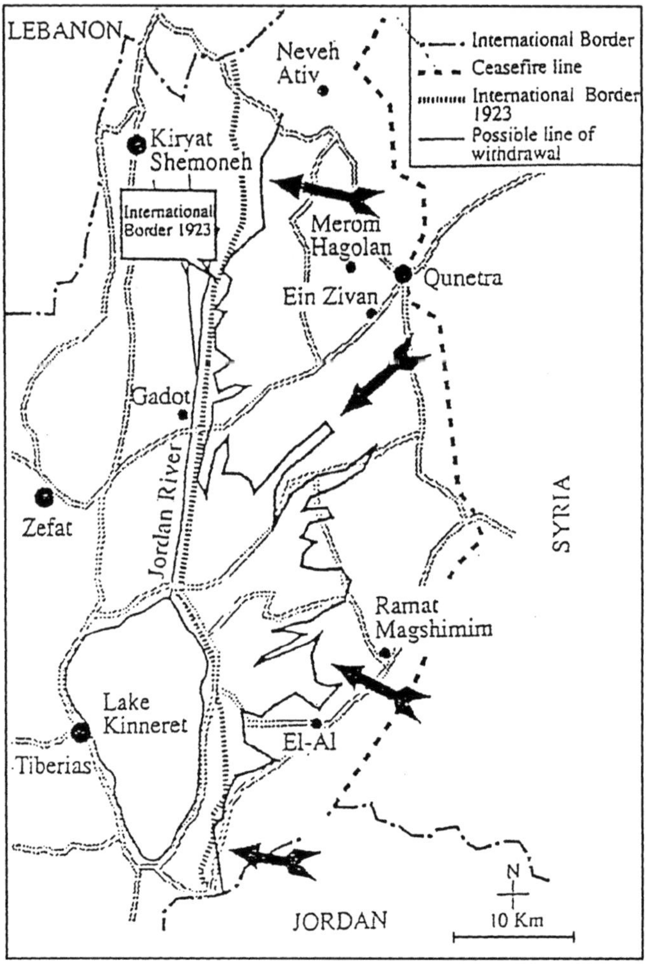


Fig. 3. Suggested Possible Lines of Israeli Withdrawal from the Golan Heights Based on Water Security Considerations. After the Report by the Jaffe Institute of Strategic Studies of Tel Aviv University and Tahal-Water Planning for Israel. Source : Ha'aretz Daily Newspaper Ltd. October 8, 1993 , part 2.

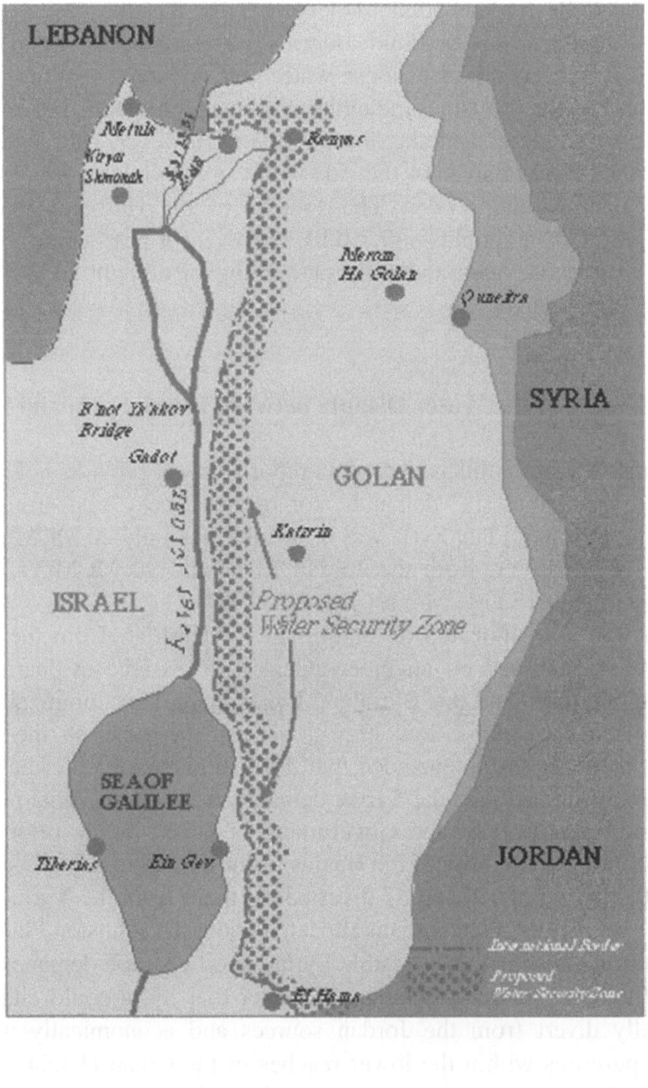


Fig. 4. The Proposed Water Security Zone: The Golan Heights, the Upper Jordan River and Lake Kinneret showing the 1923 International Border between Israel and Syria. A Schematic Presentation of the Proposed Water Security Zone (hatched area) along the Eastern Side of the International Border with Syria which could be under Syrian sovereignty but with joint and/or international inspection, control and patrols which would assure Israel's water security without the need to hold on to the Golan.

In spite of the above, it should be pointed out that the official international Syrian/Palestine (Israel) borders of 1923, which according to press reports are apparently the basis for the point of departure for the discussion of borders proposed by Prime Minister Ehud Barak's government as a basis for an agreement, covers a good part of these water security zones that are of special importance and interest to Israel including no direct contiguous border between Syria and Lake Kinneret and the Jordan River. However, the actual Banias sources and the drainage areas of the side wadis in the Golan are on the Syrian side of the international border. The sources of the Hasbani River are in Lebanon. These Golan Heights side wadis, however, at best would yield some 30 MCM/Yr. in good years and little or nothing in draught years. They are hardly worth arguing over.

## 6. Size and Scope of the Water Dispute between Israel, Syria and Lebanon

As of the time of writing this paper, press reports do not provide a clear picture on Syrian expectations or demands for a settlement concerning water. According the Johnston Plan, Syria was to be allocated only 20 MCM/year from the Banias, 22 MCM/Yr. from the Jordan River and 90 MCM/Yr. from the Yarmuk for a total of 132 MCM/Yr. ( Brecher, 1974).

Lebanon was also allocated an additional 35MCM/Yr. from the Hasbani. These figures were based on an independent Arab League evaluation of the amount of agricultural land that Syria and Lebanon could economically irrigate with waters taken at those low elevations. At the time of the Johnston negotiations in 1955, Syria demanded that those amounts of water be allocated for its use. Johnston included the Syrian demands in full in his final plan which was apparently approved by the Government of Israel. While prior to 1967, Syria had not utilized the flow of the Banias or the Jordan to any great extent, it has during the period of 1974-1987 diverted far more from the Yarmuk and its groundwater sources than the amount allocated under the Johnston Plan.

As a first rough estimate of possible Syrian and Lebanese demands, it could be assumed that the minimum amount of water that Syria could claim that it could usefully divert from the Jordan sources and economically utilize for agricultural purposes within the lower reaches of the Golan Heights would be the 42 MCM/Yr. requested by Syria in 1955 and included in the Johnston Unified Plan. That would be enough water to irrigate some 5,000-10,000 ha. There is, however, a serious question whether that much flat irrigable agricultural land is actually available in the lower Golan or area contiguous to the Jordan River on the Syrian side. Lebanon might also be expected to demand the 35 MCM/Yr. allocated to it under the Johnston Plan although here too, it is questionable if it could effectively use the water without major pumping to other more distant areas. This is a total of close to 80MCM/Yr.

One scenario assumes that under a peace agreement, Israel will return all or most of the area of the Golan Heights to Syria, in exchange for adequate security arrangements including demilitarization of the Golan, full diplomatic relations, an agreement on water and peaceful cooperation. While there was only very limited Syrian irrigated agriculture on the Golan prior to its occupation in 1967, Israeli settlements with the help of the "Mei Golan" Water Cooperative and the Mekorot Water Company have meanwhile developed irrigated agriculture on the Golan, mainly through the construction of some 15 small dams which can supply some 30 MCM/Yr. of water in good years. Practically no water is collected by these dams in drought years. Prior to the Israeli occupation of the Golan in 1967, most of this water would normally have drained into Lake Kinneret and become available for use by the Israel National Water Carrier. In addition a well field pumping some 6.5 MCM/Yr. has been developed at Alonie Bashan in the central Golan at an elevation of 600 m above sea level (Amon, 1994). It can be assumed that under a peace agreement which returns all or most of the Golan to Syria, these waterworks in the Golan Heights would revert to Syria and that Israel would view the 40-45 MCM/Yr. of Golan water as part of the Jordan River Basin waters allocated to Syria under the original Johnston Plan. This could be seen as meeting the full allocation assured Syria under the Johnston Plan.

Pumping water up from the Lake Kinneret, which is at minus 210 meters below sea level or from the Banias Springs at plus 350 meters to the agricultural areas on the central Golan Heights at levels ranging from up to 800-1000 meters above sea level is expensive. It has been estimated that the energy cost alone is about $\$0.20/m^3$ on average (Amon, 1994). Including capital cost the total cost of the water might come to about $0.25-0.30/CM. While water at this price is too expensive and totally unfeasible for Syrian agriculture, it might eventually be feasible for urban use. The Syrian press has reported on plans to resettle the Golan with some 400,000 refugees after its return to Syria under a peace agreement with Israel. It not possible at this time to validate how realistic such plans may be, but it is questionable that there is an economic basis for settling that many people based on the limited agricultural resources of the Golan.

There are at this time little if any other uses for water in southern Syria other than simple gravity flow irrigation and labor intensive agriculture. Most of the cultivable land in southern Syria that might use the disputed waters of the Jordan sources is at an elevation some 500-1000 meters above the water sources of the Jordan and Lake Kinneret. Pumping water up some 500- 1000 m from the lake or Jordan sources, to the upper Golan Heights and to southern Syria would be out of the question from an economic point of view for normal agricultural uses.

However, the Syrians may have in mind a project supplying Jordan water for domestic purposes to Damascus which suffers from serious water shortages. Considering the distance and height to which the water would have to be pumped, the cost would be considerable and most likely more expensive than closer alternative sources that do not require lifting the water to such heights, that could supply water for Damascus.

At one stage in the early water conflicts between Israel and Syria in the 1950s Syria claimed that all of the water of the Jordan River is derived from rainfall in Syria and Lebanon and thus is "Arab Water" that belongs to those two upstream countries. Under such a rationale the Syrian maximum claim might include the entire flow of the Banias of 120 MCM/Yr. and the flow of 30-40 MCM/Yr. from the Golan side wadis for themselves and the entire flow of the Hasbani, of 150 MCM/Yr. for the Lebanese, for a total of some 300 MCM/Yr. They might justify these claims based on the argument that the sources of these tributaries to the Jordan River arise in Syria and Lebanon and thus are fully their property. However, prior to the occupation of the Golan in 1967 and prior to the establishment of Israel in 1948, neither Syria nor Lebanon actually used much water from those sources, which have always flowed naturally downstream in the Jordan River and for the past 40 years or so have been fully exploited by Israel.

The Syrians would have difficulty in justifying such a claim under modern precepts of international water law which does not recognize upstream, source countries, as the sole and absolute owners of all water flows on an international river basin. Just the opposite, international law gives considerable weight to the rights of downstream users to continue their use of that portion of an international water basin which they have previously used for human uses and economically productive purposes (Caponera,1992). The UN approved version of the International Law Commission report gives priority to the rights of historical or prior use and considers depriving a downstream riparian that currently uses the water for economic and social uses as unacceptable since it will result in "appreciable damage" to the current user.

An outstanding example of such a situation, with which Syria is fully familiar, are the undeniable historic rights of Syria itself, which derives much of its waters from the Euphrates river that emanates in the territory of its upstream neighbor – Turkey. Another well know example is that of Egypt. International water law fully recognizes the historic rights of Egypt to the use of the waters of the Nile River, which it currently uses and has used for thousands of years, despite the fact that essentially 100% of its flow emanates from upstream countries.

Another factor in international water law that would weigh heavily against such a Syrian claim would be the fact that the Syrian overall water resources potential per capita is estimated for the year 2000 to be some 900 CM/P/Yr. or more than three times that of Israel's 270 CM/P/Yr. Syria would have a difficult time proving that it has an overriding objective need for the additional allocation of water resources it never, in fact, used, as compared to its water poor downstream neighbors. As pointed out previously, even the 1955 Arab League water plan accepted by Syria, Lebanon and Ambassador Johnston only claimed 20 MCM/Yr. for Syria from the Banias and 22MCM/Yr. from the upper Jordan with only 35 MCM/Yr. for Lebanon from the Hasbani.

## 7. What Will Be the Basis for an Agreement on the Water Issues?

It is beyond the scope of this paper to anticipate the outcome of the direct negotiations between Israel, Syria and Lebanon on the peace agreement in general, and the issue of borders and water agreements, in particular. However, it would not be unreasonable to assume, as suggested above, that Syria and Lebanon will, in the first instance stake a maximum claim to the full estimated 300 MCM/Yr. from the upper Jordan headwaters which are derived from sources in Syria and Lebanon (150MCM/Yr. from the Banias and side wadis of the Golan and 150 MCM/Yr. from the Hasbani), even though they never used significant amounts of those waters themselves.

While such a maximalist water claim may be used to gain leverage on other points in the negotiations, the economic motivation and objective needs for Lebanon and Syria to gain major water allocations at the sources of the Jordan would apparently not be great. A pragmatic "compromise" proposal on the Syrian/Lebanese side would not be so unlikely if this analysis is correct. Such a Syrian/Lebanese compromise might be the demand that they be allocated their share of Jordan water as defined under the Johnston Plan, which was approved by the Israel Government in 1955 - that is 35MCM/Yr. for Lebanon and 42 MCM/Yr. for Syria. The Syrians have already taken more than their share of the Yarmuk. as defined by the Johnston Plan. It should be pointed out that the Jordan - Israel agreement used the allocations of the Johnston Plan as the point of departure for their negotiations.

## 8. Agricultural, Social and Economic Implications of a Compromise on Water Based on the "Johnston Plan"

Although I take no position on the feasibility or justification of such a proposal, it is presented here as one possible example and illustration in order to examine the water resources, agricultural, social and economic implications for Israel of a possible Syrian-Lebanese proposal for a compromise settlement. Let us assume that Syria and Lebanon propose to reduce Israel's use of the headwaters of the Jordan River sources by about 80 MCM/Yr., which are the volumes of water that theoretically would have been allocated to them under the Johnston

Plan. This represents some 5% of Israel's current annual renewable fresh water resources of some 1600 MCM/Yr. This would, in the first instant, result in a direct reduction of the highly subsidized water allocation to agriculture in Israel, leading to about a 10% reduction in the supply of fresh water of good potable quality to the agricultural sector. With Israel's deep ideological commitment to support its agricultural base such a reduction of the water allocation to agriculture would be painful indeed and would most likely be opposed by the water and agricultural lobbies as well as raise questions in the minds of part of the Israeli public as to the potential threat to Israel's water and food security.

Let us examine if indeed, a 10% a cut in the water allocation to agriculture would be, in reality, a threat to Israel's food security? The Food and Agriculture Organization of the UN (1989) has estimated that it takes some 1000 tons or $m^3$ of water to produce 1 ton of wheat or grain and some 16,000 tons of water to produce one ton of meat. Thus, the import of these products can be considered the import of "virtual water" – the huge amounts of water which are imbedded within such food staples (Allan, 1995). The FAO has reported that the amount of water required to grow all of the basic food required for an individual is somewhere between 1000-2000 CM/P/Yr. In Israel, after allocating some 125 CM/P/Yr. to the urban/commercial/ industrial sector, the remaining 145 CM/P/Yr. goes to Israel's agriculture sector. This amount of water is only about 10% of what the FAO estimates is needed to grow all of the food needs of an individual. Thus, it can be seen that today Israeli agriculture can produce only a very small percent of the basic food needs of the country, even if it devoted all of the current available agricultural water for that purpose. However, since Israeli agriculture naturally attempts to optimize profits by growing crops with the highest economic return, it in fact exports a high portion of its high value agricultural production including flowers and exotic high quality vegetables and fruit. Israel's food security for the past 20-30 years or so has not been based on its limited water resources and local agricultural production, but on the ability of its commerce, tourism and industry to earn enough money for the national economy to allow for the unrestricted import of inexpensive "virtual water" in the form of food staples which provide a high percent-about 85-90% of the calories intake and food needs consumed by the country's population. Since today Israel's agriculture plays only a minimal role in the country's food security, a 10% reduction of water to agriculture is primarily a financial issue for farmers, not a food security issue for the country.

At this point we must ask what will be the real economic and social impact of such a reallocation of water resources to Syria and Lebanon as part of the price of peace? The Harvard Middle East Water Project (HMEWP) (Fisher,1996; Shuval,1995) has made an evaluation of the value of water to the Israel economy and has developed the concept of water markets for the Israeli, Palestinian and Jordanian economies. In our work on this project we have, among other things, evaluated the economic value of the water involved in the water disputes between Israel and the Palestinians. In other word, we have monitized the size of the water in dispute and found that it was not great. As Professor Fisher puts it, when the water dispute is viewed in monitory terms it is easier to see that, "Such a sum of money is small enough for countries to negotiate over rather than to go to war over" (Fisher, 1996).

However, the HMEWP has not made a study of the economics of water disputes between Israel, Lebanon and Syria, so that my own very tentative analysis presented here can only be considered as a most preliminary attempt to illustrate the type of thinking involved. The earlier stages of the studies by the HMEWP in which the author participated have shown that the maximum current value of the Jordan River water for Israeli agriculture has been estimated roughly at about $\$0.20/M^3$ (Fisher, 1996). Thus, it can be estimated that the net current loss to Israel of forgoing 80 MCM/Yr. of Jordan water for agricultural use in the central or southern regions of Israel would be about $16 million/year. This is a relatively small amount of money or about 0.0016% of the Israeli Gross Domestic Product (GDP) of about $100 billion/year in 1999. This would have little or no economic or social impact on the overall strength of the Israel economy.

Let us, however, assume that at some time in the future, say in the year 2010, the 80 MCM/Yr. of water that hypothetically Syria and Lebanon demand and divert, would be needed for domestic/urban/industrial use along the coast of Israel at Tel Aviv and would have to be replaced by seawater desalination in the south of Israel. It is not unreasonable to assume that with technological improvements already on the horizon, there will be a significant reduction in the cost of desalination. Thus the predicted replacement cost for that amount of water for Israel by desalination along the Mediterranean coast by that time would be about $\$.70/M^3$ or some $56,000,000 per year. Some experts predict that desalination costs will be even lower than that. However, these optimistic estimates neglect to include additional costs such as transport of the desalinated water from the plant to the distribution system, operational storage, and mixing and treatment facilities. In any event, even this amount of money, that might be forfeited at some distant date in the future, is not great.

However, by increasing the total available fresh water resources by seawater desalination for domestic/urban/industrial use, there will be a proportional increase in wastewater flow from urban areas and the increased potential for recycling and reuse of some 65% of that additional flow. This might provide an additional 50 MCM/Yr. for agricultural use. Thus, the total reduction of water to Israeli agriculture will be only about 30 MCM/Yr. This is indeed an insignificant cost as part of the price of peace which would make little or no economic impact on the Israeli economy or agriculture.

In this way, using the Harvard approach, the approximate size of the dispute in monetary terms can be estimated. Of course there are many necessary refinements in the actual economic simulation model used by the HMEWP in order to obtain a more accurate estimate, but as a rough first look, Professor Fisher agrees that the above figure is not far off.

While Israel's water and agricultural lobby may well see such a compromise and the agricultural and economic implications involved as unacceptable, it can be pointed out that such a price for an accommodation on the water issue would be only a small part of the total expense of reaching a comprehensive peace agreement with Syria involving withdrawal from most or all of the Golan Heights. These costs will include many billions of dollars in compensation for the settlers and investors who built homes, farms and factories on the Golan and who must be moved and resettled in other areas, the relocation of army bases and the construction of special security and early warning arrangements included in the peace agreement. It is beyond the scope of this paper to discuss the geopolitical, economic and social advantages for Israel and the region at large of reaching a peace agreement with Syria and Lebanon which are obviously very great , but one thing is clear – reaching a reasonable compromise on the water issue will have very little strategic or economic impact on Israel as compared to the other issues associated with the peace process.

In conclusion, this approach would suggest that in rational social and economic terms, when the water dispute is converted to financial terms, such a dispute between the three countries over such a small annual amount of money is hardly enough to justify an end to the peace negotiation or starting a "Water War". It might be argued that that would be a relatively small price to pay to achieve an overall peace and security settlement. It must be emphasized that water is, after all, an economic good which can be replaced and purchased in unlimited quantities by seawater desalination at a price which is, more or less, the price that Israelis now pay for urban water supply.

## 9. Regional Cooperation in Developing Additional Water Resources

Various studies have shown that on complex multinational watersheds regional cooperation is essential in optimizing the development of the water resources for the benefit of all of the riparians. Rogers (1993) has shown, based on the experience in the India-Pakistan water dispute and others, that in most situations cooperation by the riparians in the development of the water resources can benefit all the partners to the river basin dispute. He proposes approaches based on game theory concepts which provide solutions for economic cooperation on international river basins "based on maximizing the total net benefits that could be derived from the utilization of the basin, given the total resources available for the development".

It is beyond the scope of this paper to discuss in detail possible advantages that can be gained by all partners, through the development of regional water projects under conditions of peace in the region. Some of these have been described elsewhere (Shuval, 1992; Kally in Assaf *et al.*, 1993 and Kally, 1990). They might include the following:

## 9.1. PURCHASE OF WATER FROM LEBANON

Purchase of water from Lebanon and the construction of pipelines from the Litani and Awali rivers could supply water directly or indirectly to Jordan, The Palestine Authority (PA) and Israel. A simple 10 km tunnel from the Litani could deliver the flow of the Litani River in Lebanon directly to the Jordan River which could transport it to downstream users. The water of the Awali River might be diverted in a similar manner. Lebanon might be able to supply as much as 500 MCM/Yr. from these sources for a period up to 25-30 years, which would cover the economic life of the projects which would bring a reasonable profit to Lebanon and relatively cheap water to the downstream partners. At the moment, most of this water flows to the sea, since there is little suitable land for agricultural utilization in the proximity of those two rivers. Estimates indicate that the cost of water from such a project would be about one third the cost of desalination of seawater. An alternative alignment for a relatively inexpensive Awali/Litani pipeline would be along the Lebanese coast directly to Israel where it could be pumped up to the West Bank for Palestinian use or used directly by Israel which would in exchange transfer water from the mountain aquifer for Palestinian use. Water projects such as these from Lebanon may well be the least expensive and most feasible both from and engineering and geopolitical point of view in an era of peace.

## 9.2. DAMS ON THE YARMUK

The construction of a dam or dams on the Yarmuk could supply electrical power to Syria and Jordan and water to Jordan and the PA. This project might make possible the construction of the Western Ghor Canal which was originally planned by Jordan to transport Yarmuk water across the Jordan River for the benefit of the water-short Palestinians on the West Bank. While not inexpensive it might be feasible from a geopolitical point of view since it would not require direct Israeli participation, but would require Israeli agreement as the downstream riparian on the Yarmuk.

## 9.3. TRANSPORT OF WATER FROM TURKEY

Such a project would involve purchase of water from Turkey, which could be supplied to Syria, Jordan, the PA and Israel through a regional overland or undersea pipeline system or alternatively by sea transport, in refurbished oil tankers or with special large plastic bag/tankers called "Medusa Bags". About 1000-2000 MCM/Yr. or more might be supplied to the area by Turkey through such projects. Of course, a project involving Syrian cooperation would require that Turkey, Syria and Iraq reach an accommodation over their long standing disputes over the Tigris and Euphrates Rivers. The concept of an under-sea pipeline from Turkey is not as far fetched technically as might appear at first hand, since there are numerous successful examples of such sea bottom pipelines for gas, oil and water. Such a pipeline which would pass through international waters would overcome most of the geopolitical objections to such a project. The undersea pipeline would be attractive to Turkey since, with it, it could supply water to the Turkish Zone in Northern Cyprus, which suffers from a severe water shortage.

Another attractive low cost alternative, first proposed by us in 1993 (Assaf *et al.,* 1993) would be based on an agreement between Turkey and Syria for the purchase of an increased supply to Syria through the natural river systems. Syria could then increase the flow of the Yarmuk to Jordan which would transport a portion to the Palestinians through a pipe under the Jordan River as envisioned in the original Western Ghor Canal. Another option would be for Syria to transport the Euphrates water in a new pipeline which could also supply water-short Damascus, to the headwaters of the Hasbani in Lebanon for release into the Jordan River system with the understanding that this would also lead to an increased allocation for the Jordanians and Palestinians. This option would not involve any major pipeline construction work and might be reasonably economical. Of course all of the beneficiaries would have to share in the costs and the payments to Turkey for the water sold to the project. All of these projects could only be considered in an era of stable peace and mutual trust.

## 9.4. ECONOMIC COOPERATION THROUGH WATER MARKETS

The approach developed in the early stage of the HMEWP has clearly shown that in the case of Israel, Jordan and the Palestinians, economic cooperation within the context of a limited water market approach, among the riparians sharing the water resources, is more beneficial to all of partners than going it alone (Fisher, 1996). Other studies on the economic approach have reached similar conclusions.

While the engineering feasibility and price of water from the Litani and Awali has been estimated to be promising, the geopolitical aspects and economics of the other two water import projects are less clear and require further study (Kally in Assaf *et al.,* 1993). While they undoubtedly will be more expensive they still may possibly prove to be less expensive than desalination. Another factor that might weigh in favor of water transport or pipeline projects is that once the capital investment has been made, the pumping and other operational costs and energy requirements are relatively low. Desalination, on the other hand, has very high energy demands and would result in a very heavy long term commitment to import fuel which is not naturally available to Israel, the Palestinians or Jordan. This is in itself a serious security factor that must be considered. It must also take into consideration that, in the long run, the costs of imported fuel may well increase significantly which could, with time, change the optimistic estimates of the cost of seawater desalination.

## 10. Conclusions

In this paper, we have shown that it may be possible to reach a reasonable accommodation with Syria and Lebanon over their water conflicts with Israel which would take into consideration both realistic Syrian needs and interests and Israel's water needs and deep water security concerns, without resulting in a serious negative agricultural or economic impact on Israel. We have also shown that it is not necessary for Israel to hold onto the entire area of the Golan Heights since a 1-3 km water security zone under joint and/or international inspection and control on the Syrian side of the border would ensure Israel's water security needs.

It is not clear that Israel's agriculturally-oriented water establishment and water negotiators are ready at this stage to accept the pragmatic economic and geopolitical approach as presented in this article. Some may oppose an accommodation with Syria that involves any reduction of Israel's absolute control over its water resources on the Golan or any commitment to share the water sources of the Jordan River system with Syria and Lebanon. There are groups in Israel that claim that a country must physically maintain an absolute hold over all of its water sources to assure its water security. The term "control over the water sources" is essentially a code word for some of these groups, who want to maintain full political control over all or most of the Golan.

However, experts in international water law would hold that the claims that only by physical occupation of the territories, which serve as a source of its water resources, can a country assure its water rights, is not generally supported by the normal practice of peaceful nations or international water law. If this were so, Iraq and Syria would be justified in taking over the vast water sources in Turkey to assure the continued flow of the Tigris and Euphrates Rivers. Egypt would have to occupy most of Sudan, Ethiopia and some eight other African countries which are the source of the Nile. Holland would have to take over much of Germany and France to assure its control over the Rhine River, its major source of water. There are numerous examples of agreements between nations living in peace on international rivers and transboundary water sources who have achieved agreed upon modes for joint inspection, monitoring and environmental control which assure the water rights and the protection of the environment for each partner through joint management and cooperation. Even long term enemies such as India and Pakistan have reached peaceful accommodations over their bitter water conflicts.

However, whatever final agreement is reached on the division of the waters of the Jordan River Basin, and the fate of the Golan Heights, it will be essential to establish a joint Jordan River Management Board (JRMB) which will have the task of assuring mutual inspection, monitoring and control on both sides of the final borders to assure that all parties to the agreement take no more than their agreed upon share and that agreed upon pollution control measures to protect the quality of the water sources be strictly enforced. The JRMB would hopefully also be given the task of developing joint regional cooperation projects and their management. The peace agreement must also include methods of resolving disagreements and disputes at various levels and stages mainly by direct negotiations, but including additional procedures for conflict resolution such as facilitation, mediation, arbitration and if all of those fail to resolve the dispute, there should be a final obligation to resort to binding arbitration or adjudication before an agreed upon court.

Those in Israel who are prepared to accept a major territorial compromise with Syria on the Golan Heights, in return for full peace with adequate security arrangements and diplomatic, economic and social normalization, have accepted the reality that it will most likely be necessary to reach an agreement which could formally return to Syrian sovereignty most or even all of the Golan up to the recognized Syrian-Israel international borders of 1923. However, a reasonably high degree of water security for Israel could be assured if this peace agreement would contain a special proviso that would assure that there be special areas –water security zones – of joint and/or international management and control. These areas – a 1-3 km wide strip along the Syrian side of the international border, for special joint and/or international inspection and control – would include critically sensitive water source areas such as the Hasbani, El Hama and Banias Springs which are the main tributaries of the Jordan River. These special water security zones should include as well a sufficiently wide strip of land contiguous to the eastern banks of Lake Kinneret and Jordan River to be monitored and patrolled by Israel and/or an international force, in which there would in effect be no Syrian activity. The area involved is a relatively minor portion of the total area of the Golan Heights. In this way Israel will be able to assure monitoring and control of most of the important water sources and assure that there will be no direct Syrian access to the Jordan River and Lake Kinneret.

Hopefully, the sides to the water conflicts on the Jordan can achieve an agreement over water which assures each of them their appropriate water rights, national interests and needs based on a reasonable degree of water sharing coupled with strict joint inspection, monitoring, control and management of critically sensitive water sources.

## Acknowledgement

Special thanks must be given to Professor Frank Fisher of the Massachusetts Institute of Technology and head of HMEWP, Professor Lenore Martin, of the Center for Middle Eastern Studies, of Harvard University, Mr. Moshe Yizraeli, of the Office of the Israel Water Commissioner, and Professor Nurit Kliot, of Haifa University for their detailed review of the manuscript in its early stages and many insights and helpful suggestions. The author was awarded the *1999 International Water Resources Association Distinguished Lecturer Award* for an earlier version of this paper, presented at the 7th International Conference of the Israel Society for Ecology and Environmental Quality Sciences, Jerusalem, June 1999.


## References


Allen, J. A.: 1995, "The Political Economy of Water: Reasons for Optimism, but Long-Term Caution," in *Water in the Jordan Catchment Countries: A Critical Evaluation of the Role of Water and Environment in Evolving Relations in the Region,* London: SOAS, 35-59.

Amon, Y.: 1994, A Profile of "Mei Golan"- 1994, *Mayim V'Hashkaya (Water and Irrigation),* June 1994 (in Hebrew)

Assaf, K., Al Khatib, N., Kally, E. and Shuval, H.: 1993, *A Proposal for the Development of a Regional Water Master Plan: Prepared by a Joint Israeli-Palestinian Team,* Israel Palestine Center for Research and Information, pp. 192.

Ben, A.: 1996, "The water problems between Israel and Syria can be solved pragmatically," *Ha'aretz,* Jan. 19, 1996 (in Hebrew).

Blass, S.: 1960, *Mai - Miriva v Maas* (Waters of Conflict and Water Development), Massada, Tel Aviv (in Hebrew).

Braverman A.: 1994, *Israel Water Study for the World Bank,* Ben-Gurion University of the Negev and Tahal Consulting Engineers Ltd., August 1994 (R-94-36-1), Beersheba.

Brecher, M.: 1974, *Decisions in Israel's Foreign Policy,* Oxford University Press, Chapter 3, 173-224.

Bulloch, J. and Darwish, A.: 1993, *Water Wars: Coming Conflicts in the Middle East,* Victor Gollancz, London, pp. 224.

Caponera, D.: 1992, *Principles of Water Law and Administration*, Balkema, Rotterdam.

Engelman, R. and LeRoy, P.: 1993, "Sustaining Water: Population and the Future Renewable Water Supplies," *Population Action International,*Washington D.C., pp. 56.

Fisher, F.:1996, The Economics of Water Dispute Resolution, Project Evaluation and Management: An application to the Middle East, *Int. Jour. of Water Resources Development* **11,** 377-390.

Falkenmark, M.: 1992, "Fresh Water: Time for a Modified Approach," *Ambio* **15**, 192-200.

Food and Agriculture Organization: 1989, *Food and Agriculture Organization Production Yearbook,* Volume **43**, Rome.

Gleick Peter, H.: 1991, "Water and Conflict," *International Security* **18**, 79-112.

Gleick, P. H.: 1994, Water, War, and Peace in the Middle East. *Environment* **36,** 6-42.

Israel Hydrological Service:1998, *Status of Water Resources -Fall 1998*, Israel Hydrological Services Annual Report, Water Commission, Jerusalem,

Hof, F. C.: 1999, Blurry Lines -The Origins of the Israel Syria Border Controversy, *Jerusalem Post,* December 17, 1999.

Kally, E.: 1990, *Water in Peace*, Sifriat Hapoelim Publishing House and Tel Aviv University (in Hebrew).

Kliot, N.: 1994, *Water Resources and Conflict in the Middle East,* Routledge, London, pp. 450.

Lowdermilk, W.C.: 1944, *Palestine Land of Promise*, Harper Bros., New York, pp. 236.

Lowi, M.: 1989, *The Politics of Water: The Jordan River and the Riparian States,* McGill Studies in International Development **35**, McGill University, Montreal.

Naff, T. and Matson, R.: 1984, *Water in the Middle East - Conflict and Cooperation*, Westview Press, Boulder Co. and London.

Naff, T.: 1995, Conflict and Water in the Middle East, in P. Roger (ed), *Water in the Arab World,* MIT University Press.

Rogers, P.: 1993, The value of Cooperation in Resolving International River Basin Disputes, *Natural Resources Forum,* May 1993.

Shiff, Z.: 1993, The Censored Report Revealed, *Ha'aretz*, 8 October 1993 (in Hebrew).

Shuval, H. I.: 1992, Approaches to resolving the water conflicts between Israel and her Neighbors – a Regional Water for Peace Plan, *Water International* **17,** 133-143.

Shuval, H.: 1995, An Economic Approach to the resolution of Water Conflicts: *Proceedings of Workshop on Joint Management of Aquifers Shared by Israeli and Palestinians,* Truman Institute for Peace - The Hebrew University of Jerusalem, Israel.

Shuval, H. I.: 1998, Water and security in the Middle East: The Israeli-Syrian water confrontation as a case study, in L. G. Martin (ed), *New Frontiers in Middle East Security,* St. Martin Press, New York, 183-213.

Soffer, A.: 1992, *Rivers of Fire*, Haifa University Press, pp. 258 (in Hebrew).

Soffer A.: 1994, The Relevance of the Johnston Plan to the Reality of 1993 and Beyond, in J. Isaac and H. Shuval (eds), *Water and Peace in the Middle East,* Elsevier Press, Amsterdam, 107-122.

Star, J.:1991, Water Wars, *Foreign Policy* **82,**17-36.

Stevens, G. G.: 1965, *Jordan Water Partition,* Stanford University Press, Stanford CA

Wishart, D. M.: 1990, The breakdown of the Johnston negotiations over the Jordan waters, *Middle Eastern Studies* **26,** 536-546.

Wolf, A.: 1994, A Hydropolitical History of the Nile, Jordan and Euphrates River Basins, in A. K. Biswas (ed), *International Water of the Middle East,* Oxford University Press, Bombay, 5-43.

MIX
Papier aus verantwortungsvollen Quellen
Paper from responsible sources
FSC® C105338

If you have any concerns about our products,
you can contact us on
**ProductSafety@springernature.com**

In case Publisher is established outside the EU,
the EU authorized representative is:
**Springer Nature Customer Service Center GmbH**
**Europaplatz 3, 69115 Heidelberg, Germany**

Printed by Libri Plureos GmbH
in Hamburg, Germany